Art, Aesthetics, and the Brain

Joseph P. Huston is Professor of Physiological Psychology at the University of Düsseldorf, Germany. Joseph has published more than 400 books and scientific articles in areas of behavioral neurosciences with a focus on brain mechanisms of addiction, memory, and reward. He is chief editor of the journals *Behavioural Brain Research* (Elsevier) and *Reviews in the Neurosciences* (Degruyter) and the book series *Handbook of Behavioral Neuroscience* (Elsevier).

Marcos Nadal is Assistant Professor at the Department of Psychology of the University of the Balearic Islands, Spain. His main research focuses on the evolution, neural correlates, and function of cognitive and affective processes involved in aesthetic preference and art appreciation, as well as moral judgment and metaphoric language.

Francisco Mora is Professor of Human Physiology at the Faculty of Medicine, University Complutense of Madrid, Spain. He has written numerous articles for scientific journals including *Science* and *Nature* and several books, among them *The Hot Brain* (MIT Press, 2000) and *Neurocultura* (Alianza Editorial, 2007).

Luigi F. Agnati is Professor of Human Physiology at the University of Modena, Italy, and Adjunct Professor at the Karolinska Institutet, Sweden, where he received the Honoris Causa Degree in Medicine. He has published more than 620 papers in refereed international journals and seven textbooks.

Camilo J. Cela-Conde is Emeritus Professor of the University of the Balearic Islands, Spain. His publications include *On Genes, Gods and Tyrants* (Springer, 1989), *Human Evolution: Trails from the Past* (with Francisco J. Ayala, Oxford University Press, 2007) and *Processes in Human Evolution: The journey from early hominins to Neandertals and Modern Humans* (with Francisco J. Ayala, Oxford University Press, 2017).

Art, Aesthetics, and the Brain

Edited by

Joseph P. Huston

Marcos Nadal

Francisco Mora

Luigi F. Agnati

Camilo J. Cela-Conde

OXFORD
UNIVERSITY PRESS

OXFORD
UNIVERSITY PRESS

Great Clarendon Street, Oxford, OX2 6DP,
United Kingdom

Oxford University Press is a department of the University of Oxford.
It furthers the University's objective of excellence in research, scholarship, and education by publishing worldwide. Oxford is a registered trade mark of Oxford University Press in the UK and in certain other countries

First published 2015
First published in paperback 2019

Impression: 1

Published in the United States of America by Oxford University Press
198 Madison Avenue, New York, NY 10016, United States of America

British Library Cataloguing in Publication Data
Data available

Library of Congress Cataloging in Publication Data
Data available

ISBN 978–0–19–967000–0 (Hbk.)
ISBN 978–0–19–882523–4 (Pbk.)

Printed in Great Britain by
Bell & Bain Ltd., Glasgow

Praise for *Art, Aesthetics, and the Brain*

"Humans are the only species to create art objects and manipulate their physical surroundings in order to elicit aesthetics experiences. Why is this the case? How do we do it? What kind of psychological and neurobiological mechanisms are responsible for this unique behavior? A growing number of psychologists, neurobiologists, and evolutionary biologists have, in recent years, begun unearthing answers to these tantalizing questions. This superbly curated book, which collects state of the art chapters on this research, will give the reader the best currently available introduction to our understanding of why Homo sapiens is the aesthetic species."

Martin Skov, Copenhagen Business School and Copenhagen University Hospital Hvidovre, Denmark

"Art, Aesthetics, and the Brain takes readers to the provocative quest where art and science interlace to unravel the human mind. This book is a fascinating convergence of recent research findings in art, humanities, psychology, and neurobiology that is also a pleasure to read."

Tomohiro Ishizu, Research Fellow, University College London, UK

"This book is unmatched in its range of coverage, with multichapter sections on visual art, dance, and music. Readers looking for comprehensive contemporary coverage of topics in individual chapters will not be disappointed."

Aaron Kozbelt, *Evolutionary Studies in Imaginative Culture*

"Written by the leading researchers in the fields of psychology, neurophysiology, neuroimaging, aesthetics, and philosophy, this book would serve the needs of anyone from graduate students to well-informed and practicing researchers in various disciplines. It is indeed a real tribute to editors and contributors in bringing neuroscience terminology closer to a wider audience."

Jasmina Stevanov, *Perception*

Preface

John Ruskin, the nineteenth-century English critic and advocate of Romantic thought and aesthetics, believed that "Greatness in art . . . is not a teachable nor gainable thing, but *the expression of the mind of a God-made great man*" (Ruskin 1888, p. 141). Although today we no longer regard art as a divine gift, art and art-related concepts still retain the transcendent connotations they acquired during the eighteenth and nineteenth centuries. It was at that time, for instance, that philosophers and other writers forged the image of the artist as a misunderstood, bohemian, and eccentric individual, possessed with extraordinary creative powers. Individual freedom and creativity defined the essence of an artist who was expected to seek inspiration in the solitude of nature. Art—the fine art—became associated with a refined and elevated sort of contemplative pleasure, an aesthetic pleasure. Artworks turned into autonomous objects, free from all functional purpose, and intended solely for aesthetic contemplation. This was, specifically, a kind of rational, calm, quiet, undemonstrative contemplation, expected from any educated person. Before the eighteenth century, however, the artist was not viewed as alienated or detached from his social and cultural context (Barasch 1998). There was nothing inherently incomprehensible about artists, no mystery that could not be grasped by human understanding. And art was as comprehensible as craft. It was an integral part of people's lives; a way to engage with their social, moral, religious, and recreational interests, as it still is in many traditions and non-Western societies.

Much of the twentieth-century artistic avant-garde eschewed and contested the elevation of art and artists to an almost inscrutable mystical status, but these connotations have proven difficult to shake off completely. Moreover, they have significantly hampered scientific attempts to understand artistic and aesthetic behavior. Almost a century ago, Munro lamented that

> any proposal to apply the methods of natural science to aesthetics is apt to be met with skepticism and indifference. The reply is sure to be forthcoming that its problems are beyond the reach of scientific investigation: that value in art is largely a subjective affair, and hence not susceptible to objective generalization; that aesthetic feelings are too subtle and indescribable to be analyzed in scientific terminology, too diverse and unpredictable to be formulated in universal laws. If they are to be rationally grasped at all, it can be only through a philosophic imagination which is itself a kind of poetry. (Munro 1928, p. 14)

Fortunately, almost 150 years worth of scientific research since Fechner's (1876) inauguration of empirical aesthetics has clearly shown how ill-founded those prejudices were. Art is understandable in scientific terms. In fact, given that art and aesthetics rely upon perceptual, cognitive, and affective processes—just as any other product of the human mind—they are accessible to comprehension, treatable from a scientific perspective, and cannot be ignored in a comprehensive study of the human mind and its biological foundations (Arnheim 1966).

Furthermore, after brushing aside the transcendent connotations that accumulated around the notions of art and aesthetics, it is easy to see that the object of study of scientific aesthetics is not a trivial, superficial, frivolous, or elitist form of behavior. On the contrary, all human societies have developed aesthetic systems that use diverse forms of visual representation, body art, music, literature, or performance to convey culturally important meaning (Anderson 1989), and have done so probably since the origin of our species (e.g. Henshilwood et al. 2011). Thus, "All human life is filled with works of art of every kind—from cradlesong, jest, mimicry, the ornamentation of houses, dress and utensils, up to church services, buildings, monuments, and triumphal processions. It is all artistic activity" (Tolstoy 1904, p. 51). Art and aesthetics, thus, are inherent constituents of human cognition.

Scientific aesthetics is today a thriving and respected research field in its own right. It has made substantial contributions to our basic understanding of some of the unique features of the human mind (Kandel 2012; Nadal and Skov 2013), and to practical issues related to consumers' decisions (e.g. Ramsøy and Skov 2014), social judgment (e.g. Leder et al. 2010), and mate choice (e.g. Fink et al. 2006), to cite just a few examples. In fact, if the hallmark of cognitive neuroscience is to "study humans with brain science techniques in concert with cognitive science methodologies" (Gazzaniga 1984, p. vii), bringing together the cognitive and neural levels of explanation of behavior into a single approach (Churchland and Sejnowski 1988; Posner and DiGirolamo 2000), then scientific aesthetics can properly be viewed as a domain of cognitive neuroscience, and cognitive science in general.

The present volume attempts to bring together some of the recent approaches that have sought to understand art and aesthetics in scientific terms. In doing so, it follows the path laid down by earlier books, such as Skov and Vartanian's *Neuroaesthetics* (2009), Shimamura and Palmer's *Aesthetic Science* (2012), and Chatterjee's *The Aesthetic Brain* (2014). It differs from them, however, in several ways. First, this volume highlights the broad range of possible ways in which cognitive neuroscientific methods and concepts can be used to understand the biological foundations of art and aesthetics. Second, it brings together general and specific approaches. The general chapters wrap around the specific contributions, providing foundations and integrating conclusions; the specific contributions deal with issues related to concrete art forms, or factors influencing the relation between art, aesthetics, and the brain. Third, it attempts to move away from the traditional bias that emphasizes visual art, and acknowledges the stimulating study of music and dance. Fourth, it includes work by contributors with substantially different backgrounds, from philosophy and art history to neuroscience and engineering.

If "the central challenge of science in the twenty-first century is to understand the human mind in biological terms" (Kandel 2012, p. xiv), the aim of this book is to bring us a little closer to understanding some of humanity's greatest achievements in such terms. It does not aim to provide the last word on this topic, or to settle great debates. On the contrary, we hope it will add fuel to the debates and interdisciplinary collaboration that makes the cognitive neuroscience of art and aesthetics the challenging and exciting field it is today.

Finally, the editors of this volume would like to express their deepest gratitude to the authors who have contributed their valuable work, without whom there would be no book with which to pursue our aims. We would also like to thank Charlotte Green, Victoria Mortimer, and Martin Baum at Oxford University Press, whose commitment, guidance, and support from beginning to end made it possible to turn this project into a reality.

References

Anderson, R.L. (1989). *Art in small-scale societies* (2nd edn). Englewood Cliffs, NJ: Prentice Hall.

Arnheim, R. (1966). *Toward a psychology of art*. Berkeley, CA: University of California Press.

Barasch, M. (1998). *Modern theories of art, 3. From Impressionism to Kandinsky*. New York, NY: New York University Press.

Chatterjee, A. (2014). *The aesthetic brain. How we evolved to desire beauty and enjoy art*. New York, NY: Oxford University Press.

Churchland, P.S. and Sejnowski, T.J. (1988). Perspectives on Cognitive Neuroscience. *Science* **242**, 741–5.

Fechner, G.T. (1876). *Vorschule der Ästhetik* Leipzig: Breitkopf und Härtel.

Fink, B., Grammer, K., and Matts, P.J. (2006). Visible skin color distribution plays a role in the perception of age, attractiveness, and health in female faces. *Evolution and Human Behavior* **27**, 433–42.

Gazzaniga, M.S. (ed.) (1984). *Handbook of Cognitive Neuroscience*. New York, NY: Plenum Press.

Henshilwood, C.S., d'Errico, F., van Niekerk, K., et al. (2011). A 100,000-year-old ochre-processing workshop at Bolmbos Cave, South Africa. *Science* **334**, 219–22.

Kandel, E.R. (2012). *The age of insight. The quest to understand the unconscious in art, mind, and brain, from Vienna 1900 to the present*. New York, NY: Random House.

Leder, H., Tinio, P.P.L., Fuchs, I.M., et al. (2010). When attractiveness demands longer looks: The effects of situation and gender. *Quarterly Journal of Experimental Psychology* **63**, 1858–71.

Munro, T. (1928). *Scientific method in aesthetics*. New York, NY: W.W. Norton.

Nadal, M. and Skov, M. (2013). Introduction to the Special Issue: Toward an interdisciplinary neuroaesthetics. *Psychology of Aesthetics, Creativity, and the Arts* **7**, 1–12.

Posner, M.I. and DiGirolamo, G.J. (2000). Cognitive neuroscience: Origins and promise. *Psychological Bulletin* **126**, 873–89.

Ramsoy, T.Z. and Skov, M. (2014). Brand preference affects the threshold for perceptual awareness. *Journal of Consumer Behaviour* **13**, 1–8.

Ruskin, J. (1888). *Modern painters* (Vol. III). New York, NY: John Wiley and Sons.

Shimamura, A.P. and Palmer, S. (eds) (2012). *Aesthetic Science. Connecting minds, brains, and experience*. New York, NY: Oxford University Press.

Skov, M. and Vartanian, O. (eds) (2009). *Neuroaesthetics*. Amityville, NY: Baywood.

Tolstoy, L. (1904): *What is art?* New York, NY: Funk and Wagnalls.

Contents

Section 3 **Cognitive neuroscience of dance**

Section 4 **Cognitive neuroscience of music**

Section 7 **Integrative approaches**

Contributors

Luigi F. Agnati
Department of BioMedical Sciences
University of Modena
Italy

Francisco J. Ayala
Department of Ecology and Evolutionary Biology
University of California
Irvine, California, USA

Elvira Brattico
Brain and Mind Laboratory
Department of Biomedical Engineering and Computational Science
Aalto University School of Science
Cognitive Brain Research Unit
Institute of Behavioural Sciences
University of Helsinki
Finnish Center of Excellence in Interdisciplinary Music Research
Department of Music
University of Jyväskylä
Finland

David Brieber
Cognitive Science Research Platform
University of Vienna
Austria

Beatriz Calvo-Merino
Department of Psychology
City University
London, UK
Universidad Complutense Madrid
Spain

Camilo J. Cela-Conde
University of the Balearic Islands
Spain

Anjan Chatterjee
Department of Neurology and Center for Cognitive Neuroscience
University of Pennsylvania
Philadelphia, Pennsylvania, USA

Julia F. Christensen
Human Evolution and Cognition (IFISC-CSIC) and Department of Psychology University of the Balearic Islands
Spain

Kathleen A. Corrigall
Department of Psychology
MacEwan University
Edmonton, Canada

Emily S. Cross
Wales Institute for Cognitive Neuroscience
School of Psychology
Bangor University
Gwynedd, Wales
Behavioural Science Institute
Donders Institute for Brain, Cognition and Behaviour
Radboud University
Nijmegen, The Netherlands

Kirill Fayn
School of Psychology
University of Sydney
Australia

W. Tecumseh Fitch
Department of Cognitive Biology
University of Vienna
Austria

Kjell Fuxe
Department of Neuroscience
Karolinska Institutet
Stockholm, Sweden

David Freedberg
Pierre Matisse Professor of the History of Art
Columbia University
New York, NY, USA

Gernot Gerger
Department of Basic Psychological Research and Research Methods
Faculty of Psychology
University of Vienna
Austria

Spas Getov
Wellcome Trust Centre for Neuroimaging and Institute of Cognitive Neuroscience
University College London
UK

Diego Guidolin
Department of Molecular Medicine
University of Padova
Padova, Italy

Barbara G. Goodrich
Department of Philosophy
University of Colorado at Denver
USA

Alexander J. Huston
Freelance Architect
Berlin, Germany

Joseph P. Huston
Center for Behavioral Neuroscience
Institute of Experimental Psychology
University of Düsseldorf
Germany

Alumit Ishai
Professor of Cognitive Neuroscience
Department of Neuroradiology
University of Zurich
Switzerland

Corinne Jola
INSERM-CEA Cognitive Neuroimaging Unit, NeuroSpin Center
Gif-sur-Yvette, France

Ulrich Kirk
Virginia Tech Carilion Research Institute
Roanoke, Virginia, USA

Christoph Klein
Department of Child and Adolescent Psychiatry and Psychotherapy
University of Freiburg
Germany
School of Psychology
Bangor University
Wales

Stefan Koelsch
Department of Educational Sciences and Psychology
Cluster of Excellence "Languages of Emotion"
Freie Universität Berlin
Germany

Helmut Leder
Department of Basic Psychological Research and Research Methods
Faculty of Psychology
University of Vienna
Austria

Suzee E. Lee
Memory and Aging Center
Department of Neurology
University of California, San Francisco
USA

Moritz Lehne
Department of Educational Sciences and Psychology
Cluster of Excellence, "Languages of Emotion"
Freie Universität Berlin
Germany

Francisco Mora
Universidad Complutense

Marcos Nadal
University of the Balearic Islands
Spain

Marcus T. Pearce
School of Electronic Engineering and Computer Science
Queen Mary College, University of London
London, UK

Robert C. Pepperell
Professor of Fine Arts
Cardiff Metropolitan University
Wales

Edmund T. Rolls
Oxford Centre for Computational Neuroscience
Oxford, UK
www.oxcns.org

Raphael Rosenberg
Department of Art History
University of Vienna
Austria

E. Glenn Schellenberg
Department of Psychology
University of Toronto
Canada

William P. Seeley
Department of Philosophy
Bates College

Paul J. Silvia
Department of Psychology
University of North Carolina at Greensboro
USA

Oshin A. Vartanian
University of Toronto
Scarborough, Canada

Indre V. Viskontas
Department of Psychology
University of San Francisco
Department of Humanities and Sciences
San Francisco Conservatory of Music
USA

Gesche Westphal-Fitch
Department of Cognitive Biology
University of Vienna
Austria

Joel S. Winston
Wellcome Trust Centre for Neuroimaging and Institute of Cognitive Neuroscience
University College London
UK

Dahlia W. Zaidel
Department of Psychology
University of California, Los Angeles
USA

Abbreviations

AAA	Assessment of Art Attributes
ACC	anterior cingulate cortex
AD	Alzheimer's disease
AIM	affect infusion model
AON	action observation network
BH	basal hypothalamus
BOLD	blood oxygen-level-dependent
bvFTD	behavioral variant fronto-temporal dementia
CNS	central nervous system
CRDI	continuous response digital interface
CSF	cerebrospinal fluid
DLPFC	dorsolateral prefrontal cortex
DMN	default mode network
dPM	dorsal premotor
EBA	extrastriate body area
EBL	emotional body language
ECF	extracellular fluid
EDA	electrodermal activity
EEG	electroencephalograph
EMG	electromyography
ERAN	early right anterior negativity
ERP	event-related potential
ESM	experience sampling methods
FBA	fusiform body area
FFA	fusiform face area
FI	frontal insula
FM	functional module
fMRI	functional magnetic resonance imaging
FTD	fronto-temporal dementia
HR	heart rate
IAPS	International Affective Picture System
IFG	inferior frontal gyrus
IPL	inferior parietal lobe
ITG	inferior temporal gyrus
LPP	late positive potential
LRP	lateralized readiness potential
MEG	magnetoencephalography
MFC	medial frontal cortex
MIDI	Musical Instrument Digital Interface
MHC	major histocompatibility complex
MMN	mismatch negativity
MOBO	music of black origin
MTG	middle temporal gyrus
MTL	medial temporal lobes
MUSIC	Mellow, Unpretentious, Sophisticated, Intense, Contemporary
NAcc	nucleus accumbens
nfvPPA	nonfluent variant primary progressive aphasia
OFA	occipital face area
OFC	orbitofrontal cortex
PCC	posterior cingulate cortex
PCUN	precuneus
PET	positron emission tomography
PLV	phase locking value
PMd	dorsal premotor
PPA	primary progressive aphasia
pSTS	posterior superior temporal sulcus
PVN	paraventricular nucleus
rCBF	regional cerebral blood flow
rTMS	repetitive TMS
SCL	skin conductance level
SCR	skin conductance responses
SIT	stimulus-independent thoughts
SLEA	sublenticular extended amygdala
SOA	stimuli onset asynchronies
SON	supraoptic nucleus
SPL	superior parietal lobe
STN	solitary tract nucleus
STOMP	Short Test of Music Preferences
STS	superior temporal sulcus
svPPA	semantic variant primary progressive aphasia
TCI	Temperament and Character Inventory

TMS	transcranial magnetic stimulation
TPJ	temporo-parietal junction
TTCT	Torrance Test of Creative Thinking
VBM	voxel-based morphometry
VEN	von Economo's neuron
VLSM	voxel-based lesion-symptom mapping
VMPFC	ventromedial prefrontal cortex
vPMc	ventral premotor cortex
VTA	ventral tegmental area
VT	volume transmission
WT	wiring transmission

Section 1

Foundational issues

Chapter 1

Neuroculture: A new cultural revolution?

Francisco Mora

1.1 Introduction

What is culture? The definition of this concept has certainly proved as elusive as one for happiness, and little consensus amongst researchers can be found (Mora 2012a). Over half a century ago, Kroeber and Kluckhohn (1952), making a brief summary of the more than 160 definitions of culture, concluded that "[c]ulture consists of patterns, explicit and implicit, of and for behaviour acquired and transmitted by symbols, constituting the distinctive achievements of human groups, including their embodiments in artifacts; the essential core of culture consists of traditional (i.e. historically derived and selected) ideas and especially their attached values; culture systems may, on the one hand, be considered as products of action, and on the other as conditioning elements, of further action." I would like to highlight the historical feature within this definition, the fact that cultures appear over time, renewing, in these sequences, the readings of values and norms that governed social interactions.

From this derives the question, are we, now, on the verge of a new culture? Both humanistic and scientific thinkers suggest as much. For instance, in a public lecture at the V Philosophy and Epistemology international conferences celebrating the 30th anniversary of the Piaget Institute in Portugal (1999), in which he was honored, George Steiner said almost prophetically, "All cultures are mortal, all religions too. All are mortal cultural events as mortal are the men who produce them. And it is now that there is a transition period. We are now entering a post religion era. Christianity will die as Marxism did. What will fill the void? What awaits us? What will be born?" As I recall, Steiner did not elaborate very much beyond justifying these questions historically and philosophically but he certainly emphasized the possible emergence of a new way of thinking in which critical and analytical thinking would play major roles.

Edward Wilson stated in his book *The Future of Life* (2003) that it was time to change to a culture of life-sustaining permanence of both ourselves and the biosphere. He advocated looking within the mind, gesturing towards the inner workings of the brain to discover its functioning via thoughts, emotions, and feelings out of which the creation and the succession of different cultures have been moulded.

Science—that is, physics and astronomy, biology and biological evolution—is producing a revolutionary change in our vision of the human world. Many people see in science the protagonist of this new period announced by Steiner. It is in this new period or culture that new answers to old and universal questions such as those related to the nature of mind and consciousness, the social and moral nature of man, and, most particularly, what drives us to make and appreciate art (Levallois et al. 2012; Mora 2007), are anticipated to be found. These changes will be based on the knowledge of how the human brain operates because it is the brain integrated and interacting with the rest of the body (Mora et al. 2012b) and with its physical and social environment that determines everything man does and expresses in his behavior, feelings, thoughts, or beliefs (Mora 2007).

1.2 Neuroscience and the human brain

These days, neuroscience is starting to be viewed as a field whose scope extends beyond the understanding of neural mechanisms to trying to achieve comprehension of the psyche and human nature itself; a combination of brain and mind, biological entity and psyche. Neuroscience and, in particular, cognitive neuroscience, are offering tremendous insight into both these aspects of humankind. We are today facing rapid and continuous expansion in our knowledge about the human brain. It is fairly well established that the brain is a permanently changing organ in both structure and function, continuously absorbing and transforming information from the environment (Jenkins et al. 1990; Mora et al. 2007). This flow of information is not just a passive processing in which an immutable area of the brain or a distributed system in the brain process it using an unalterable code. The brain is a neuronal plastic organ in which neuronal processes are constantly feeding back to alter the processor—the brain and the person attached. We know today that these plastic changes are generated in neurons and pathways that release neurotransmitters and activate receptors which, in turn, will activate genes, new synthesis of proteins, new receptors, and other membrane components which, finally, result in morphological and physiological changes of neurons and of neuronal circuit activity (Mora 2007). It is in this dynamic way that the chemistry and physics of the brain is transformed into biochemistry, which in turn transforms the anatomy and physiology of the brain, on into behavior, in a continuum in time and space (Mora 2008).

Considerable neurobiological evidence now exists about the activity of different sensory systems, in particular the visual system, and the perception and consciousness of what is visually perceived (Zeki 1993). In this respect, vision has contributed to improving the understanding of the brain–mind problem (Zeki 1993; Crick 1994). In addition, a considerable amount of neurobiological research and studies in computational neural connectivity using powerful computers (Proteus) and graph theoretical analysis have given rise to hypotheses surrounding the organization and dynamics of complex neural circuits (networks) in the cerebral cortex which form the neural basis for mental functions (Bullmore and Sporns 2009). Research on large-scale models has also provided new anatomical and functional concepts such as those of the criticality hypothesis (Beggs 2008; Chialvo 2010; Halmovichi et al. 2013; Moretti and Muñoz 2013). Important new research projects have begun which will increase our understanding of the human brain. Looking back, the first Human Brain Project started in the 1980s and tried to provide a comparison of data obtained through different methodologies at different levels of analysis, hoping to throw some light on complex cognitive processes from the most basic pillars, molecules, cells, and circuits (Shepherd et al. 1998). Today, new projects are underway. The Big Brain Project seeks to forge new pathways in neuroanatomy (they have already obtained more than 7400 cuts of a brain and the images obtained have been scanned) (Amunts et al. 2013). The Human Connectome Project aims to build a map of the anatomical and functional connectivity of a healthy human brain (National Institutes of Health, University of Washington at Saint Louis, University of Minnesota, Harvard University, and UCLA 2013), and the Blue Brain Project, whose final objective is to study the structure of the neocortex creating a molecular simulation (2013), are both major works in progress.

Hopefully the data obtained from these new research projects will provide an important advance in our understanding about how the brain works and, in parallel with progress in imaging techniques, will advance our knowledge about the brain–mind problem, referring specifically to perceptions, emotions, feelings, abstractions and ideas, thoughts, learning, and memory, and also consciousness and self-consciousness (Mora 2000). Particularly in reference to "neuroculture", all this progress will have an important effect on our knowledge of the neural basis for neuroeconomics and sociology (Levallois et al. 2012), ethical behaviors (Moll et al. 2005, 2008), politics (Fowler and Screiber 2008; Cortina 2011), art (Cela-Conde et al. 2011), and religion (Mora 2011, 2012b).

1.3 Biological evolution and the brain and the mind

As Theodosius Dobzhansky pointed out, nothing can be understood in biology unless it is analysed from the perspective of evolution (Dobzhansky 1973). There is not doubt that evolution is one of the basic elements of that continuum which constitutes man. The human brain is the result first, of several millions of years of evolution, and then a product of natural selection over thousands of generations. Reconstructing how the brain has evolved would be one of the most valuable pieces of work for an integrative advance in neuroscience. Without such a reconstruction of such a history we will hardly understand the meaning and functioning of the human brain and therefore its products, in this case, neuroculture (the relation between the sciences that study the functioning of the brain and culture) (Edelman and Mountcastle 1978; Mora 2007).

In this context it is worth mentioning that among the different biological species, man has the greatest brain volume in relation to body weight (Jerison 1973) with the highest encephalization coefficient (EQ 7) compared to his predecessors, the *Australopitecins* (EQ 2). This brain volume and the EQ has been acquired very rapidly through evolution in the last 2–4 million years, from the *Australopitecins* (500 g brain wieght) to the *Homo habilis* (600–700 g), *Homo erectus* (900–1000 g), and then *Homo sapiens* (1300 g), and *Homo sapiens sapiens* (1400 g) (Holloway 2008; Tobias 1997, 1995). However, the size and weight of the brain of present man does not just constitute a quantitative and global increase of the number of neurons arranged in the same pre-existing pattern of the brain of a primitive primate, but represents also a reorganization of this brain with selective increases and decreases in specific areas (Holloway 2008). More specifically, the brain during those 2–4 millions of years of evolution suffered a profound reorganization with reductions of certain areas (area 17 of Brodmann) expansions (area 10 of Brodmann—six times larger than the corresponding area of a chimpanzee—or area 7). Also, asynchronic reorganization of some key brain areas such as the orbitofrontal cortex and asymmetrical hemispheric reorganizations (right versus left brains) occurred during this period of time (Holloway 2008).

The macro and micro internal reorganization (progression and regressions) of specific brain areas (Brodmann areas) occurred during that long period of evolution (Changeux and Chavallion 1995) is particularly relevant when considering the last reorganization, selective and profound, of the prefrontal and orbitofrontal cortex with its key significance for understanding human behavior (Holloway 2008).

The prefrontal cortex, together with the parietal and temporal areas (those parts of the brain generically called association areas), seems to have developed in parallel to the acquisition of mental capacities. During evolution, those areas of the cerebral cortex have selectively increased their volume and the number of neurons of which they comprise as well as enduring a large reorganization when compared to their predecessors. As an estimation, it has been suggested that the chimpanzees have accumulated 3.4 billion neurons compared to the 8.5 billion cumulative neurons in the brain of the *Homo sapiens sapiens* (Jerison 1976).

Evolution has followed multiple and unknown routes to form the present species, including man (Jerison 1973). It is possible that many neural circuits in the brain have changed their function throughout evolution. Functions of parts of the brain that first appeared with a specific survival value and specifically determined by certain environmental conditions at some time could well have changed their functions at other times, re-adapting them to new conditions of the environment (Gisolfi and Mora 2000). However, it should be pointed out that our current knowledge about the evolution of the human brain is very fragmentary; many pieces of the puzzle remain missing. The newly discovered *Homo georgicus* points to a probably reevaluation of the paleoantropological map (Lordkipanidze et al. 2013). Uncovering how the different parts of the brain were developed, shaped, and reorganized in relation to specific environmental determinants could help to further understand the codes orchestrating the higher functions of the human brain (Gisolfi and Mora 2000).

1.4 **Neuroculture**

The very rapid progress in our understanding about how the brain works is giving rise to a new way of thinking, changing knowledge about established concepts in philosophy, ethics, law, sociology, economics, art, and religion. New ways of thinking have already begun, with the "neuro" discourses and practices found in neurophilosophy, neuroethics, neuroeducation, neuroeconomics, neuroaesthetics, and neurotheology. This is neuroculture (Mora 2007; Rolls 2012), a culture based on the sciences of the brain which brings with it, a reassessment of the humanities, a bridge via which those two large bodies of knowledge, science and humanities, will join. Neuroculture offers a slow process leading to the re-evaluation of our conception of man. Kandel and Mack (2003) wrote: "[w]hile Sciences and Humanities will continue to have their own separate concerns, we should, in the decades ahead of us, realize more and more that both are generated through a common computer design: the human brain."

The epicenter of neuroculture rests not solely on neuroscience or cognitive neuroscience as such, but on the recognition, once again, that human existence is the product of a long process of chance and necessities (Monod 1971) that has lasted millions of years. Revealing the neural mechanisms through which man reached his current neuroanatomical organization and its corresponding functional organization will provide a better understanding not just of man in society but also of the world around him. This new perception will relocate man in a renewed position in that same world. Jose Ortega y Gasset (1983) once said "el pensamiento humano no descubre el universo sino que lo construye" ("human thought does not discover the universe but creates the universe"). Julian Huxley (1964) stated that "science does not discover things previously existing but creates something new." The workings of the brain are orchestrated through ancestral codes anchored and hidden in its depth, generation after generation. It is out of these that it constructs something new.

Neuroculture will give scientific support to those words of Steiner (1999) who stated that all cultures are mortal; adding the idea that now all cultures and cultural diversities are the result of the functioning of the brain. It may well be too ambitious a prediction, but I seek to propose that this new culture, which we are naming neuroculture, will eventually lead to that unified theory of knowledge proposed by Wilson fifteen years ago (Wilson 1998). Neuroculture may in this way produce an impact on human thinking similar to that produced by Darwinism over 100 years ago. It is possible that, for many, neuroculture will represent a backward step, a way of losing that spiritual essence of what it has been so far conceived as the true essence of human nature. However, for many others neuroculture will be an advance, reaching towards a truth about human beings that has previously been hidden, a conception of man, and the greatness of man, based this time on scientific evidence and critical and analytical thinking which provides a re-evaluation of those grand conceptions of human nature—education, privacy, dignity, equality, politics, love, friendship, good and evil, truth, freedom, justice, dignity, and nobility.

Neuroculture presupposes that nothing occurs and nothing exists in the human world that has not been filtered (back and forth, culture to brain, brain back to culture) through that sieve we call the brain. These may be world events, a beautiful work of art, the successful development of a complex mathematical formula, or the deep feeling of having experienced the existence of God. All this frames the concept of neuroculture (Wexler 2006; Mora 2007; Vidal 2009; Frazzetto and Anker 2009; Rolls 2012).

1.5 Neurophilosophy, consciousness, and the mind–brain problem

Neuroculture offers a revaluation of humanities. At its core it is perhaps "neurophilosophy", as stated clearly by Patricia S. Churchland, arising from the recognition that brain science and the technologies used in its investigations is advanced enough now to have achieved a real breakthrough in our understanding of the brain–mind problem (memory, learning, consciousness, free will). "This predicts that if philosophy of mind is developed without an understanding of neurons and the brain is highly possible to become sterile" (Churchland 1990, 2008). Certainly one of the central themes in neurophilosophy is that of the so-called brain–mind problem (Llinás and Churchland 1996; Llinás 2001).

For at least 2000 years, and certainly since Plato, we have had a clear separation of two worlds, that of the brain or matter, occupying space and time, and that of the mind or spirit, occupying time and only individually experienced and therefore unique (Beakley and Ludlow 1992; Mora 1995). This perspective has been changed with the advent of neuroscience. Only a few decades of research have been necessary to start attempting to bridge those two worlds, man as product of Darwinian evolution and the mind as a series of processes carried out by the brain. In fact, mind does not exist in the sense of being a real entity or global concept capturing a static or permanent entity. What really exist are mind processes; that is to say, events characterized in terms of the time-varying activity of interconnected and distributed neuronal circuits in the brain. Time, not space, is the crucial dimension of the mind (Mora 2008). Therefore, for most neuroscientists, mind is not a spiritual, non-material entity, nor a product emerging or caused by the brain and different from the brain. Mind is the activity of the brain itself.

The difficulty then arises when considering how feelings, thoughts, and even consciousness (intimate and subjective events—non material–which are difficult to grasp using scientific tools) are in fact the activity of molecules and neurons in circuits of the brain. Apart from some important considerations, when talking about consciousness, most neuroscientists believe that there is a link between brain processes and mind processes, and language contributes in a meaningful way to make the two worlds of brain and mind appear to be different (Mora 2008). An example might suffice as explanation. For instance, when we suffer a burn on a finger and describe in neurobiological terms all the events occurring in the spinal cord and brain—that is, sensory receptors, synapses, neurotransmitters, gene activity, electrical potentials, and activation of neuronal circuits in distributed systems and networks of the limbic system, thalamus, and cerebral cortex (cognition)—we are already describing the emotional reaction and the cognitive component of pain but in a different language to that of psychology. In other words, the neurobiological events are not producing something different expressed at psychological or mental level, called "pain". All the events taking place in the brain *are* "pain"; that is, events that are first described in neurobiological terms at one level and then a person feeling it, consciously perceiving it, and verbally expressing it, described in cognitive or psychological common terms (Mora 2000). As stated more radically by Kandel in his Nobel lecture, given in 2000, "the new biology suggests that not only the body but the mind and the specific proteins that participate in those mental processes—self-consciousness, and the consciousness of others, the past and the future—evolved from our predecessors the hominids. This new biology postulates that consciousness and mental processes are biological processes that, on due time, will be explained in terms of signalling molecular pathways used by nerve circuits that interact among them."

This is similar to Patricia Churchland's humanistic and poetic description:

> . . . bit by experimental bit, neuroscience is morphing our conception of what we are. The weight of evidence now implies that it is the brain, rather than some nonphysical stuff, that feels, thinks and decides. That means there is no soul to fall in love. We do still fall in love, certainly, and passion is as real as it ever was. The difference is that now we understand those important feelings to be events happening in the physical brain. It means that there is no soul to spend its post-mortem eternity blissful in Heaven or miserable in Hell. Stranger yet, it means that the introspective inside—one's own subjectivity—is itself a brain-dependent way of making sense of neural events. In addition, it means that the brain's knowledge that this is so is likewise brain-based business (Churchland 2002)

At the central core of any neurobiological approach to the processes of mind lies consciousness and consequently, neuroculture. I thought, therefore, that a brief reflection on the neurobiology of consciousness would be relevant here.

Consciousness has been addressed from different philosophical or cognitive points of view (Llinás and Churchland 1996). Although these approaches are important, many neuroscientists feel that they will never be sufficient to fully explain what consciousness is unless other studies or theories appear to anchor them at a neurobiological level. For some, consciousness is no longer a philosophical problem but is instead a scientific problem (Zeki 1993; Crick 1994; Edelman 2006). In recent years reports have appeared trying to tackle this problem from the perspective of the brain (Mora 2000, 2009).

The problem of consciousness and its definition presents a considerable challenge. There are important difficulties in reaching a formal definition. Crick and Koch (1990) admitted this in an initial approach to defining consciousness, and they recognized that the most useful way to advance is to use the descriptions of psychologists and cognitive scientists. With such an approach, together with other observations, they reached the conclusion that consciousness "is crucially dependent on some form of short-term memory and also some form of serial attentional mechanisms." Tononi and Edelman (1998), on the other hand, consider that "conscious experience is a process that is unified and private, that is extremely differentiated and that evolves on a time scale of hundreds of milliseconds." See Searle (1998) and how he addresses the problem of the definition of consciouness.

Consciousness or awareness of the external world (sensory perception) has received a great deal of attention from the neurobiological point of view (Llinás and Churchland 1996; Mora 2000, 2009). Of relevance here is the so-called "binding problem" which refers to that neural activity required to bind together all properties of an object (form, orientation, color, movement, depth, and so forth), and this has been postulated to be produced by the synchronous activity (neuron firing) of all those neurons relevant to the analysis of these properties of the object. This synchronous neural activity has been suggested to occur in bursts at frequencies in the 40–60 Hz range (Llinas 2001). In addition, the role of attention and re-entrant links to the areas activated has been proposed as two fundamental ingredients for consciousness to take place (Tononi and Edelman 1998). Thus attention seems to play a crucial role in binding, although the exact location and mechanisms of attention are open questions despite the parietal lobe mechanisms that have been considered relevant in this context (Crick 1996). To complicate matters, we know now that attention is not just a simple neural mechanism involving just one single set of neural circuits but is instead several differentiated neural processes according to the different types of attention that have been so far described (Raz and Buhle 2006). There have been many other different approaches to consciousness from cognitive neuroscience and neuropsychology, including studies of patients with blind-sight syndrome, memory deficits, split brain, unilateral and contralateral neglect, or patients with specific brain lesions (Damasio 1994; Damasio and Damasio 1990). The new techniques for cognitive experimentation such as positron emission tomography (PET), magnetic resonance imaging (MRI), and the most recent recording technique magneto-electro-encefalography (MEG) with a resolution capable of correlating electrical activity of specific brain areas and cognitive events measured in real time, have all contributed considerably to this development (Llinás 2001).

A brief neurobiological formulation of the most relevant theories of cognition and consciousness, those of Crick and Koch (1990), Tononi and Edelman (1998, 2004), and Llinás and Churchland (1996, 2004), will be of use here. Crick and Koch (1990) proposed that despite the many different forms of consciousness could take at any moment, this could be related to one or a few basic neurobiological mechanisms having at their core "a particular type of activity in a transient set of neurons that use a subset of a much larger set of potential candidates." This speculation led Crick and Koch (1990) to ask which and where these neurons are, and how they are special (connections, pattern of firing, etc.). Tononi and Edelman (1998) criticized this approach after considering that "conferring this property on neurons seems to contribute to a category error, in the sense of ascribing to things properties they cannot have." In the light of classical lesion and stimulation studies and neurophysiologic studies, they propose what they call the "dynamic core hypothesis of consciousness":

> We propose that a large cluster of neuronal groups that together constitute, on a time scale of hundreds of milliseconds, a unified neural process of high complexity be termed the "dynamic core," in order to emphasize both its integration and its constantly changing activity patterns. The dynamic core is a functional cluster: its participating neuronal groups are much more strongly interactive among themselves than with the rest of the brain. The dynamic core must also have high complexity: its global activity patterns must be selected within less than a second out of a very large repertoire.

Perhaps one of the original formulations of this hypothesis is that the dynamic core is conceived not as an invariant set of neurons or circuits distributed throughout the brain, but as a set of circuits that may change over time and also that at times could underlie conscious experience while other may be involved in unconscious processes. An important component of the neuroanatomical substrates of this dynamic core could include the thalamocortical connections, both posterior and anterior regions (Tononi and Edelman 1998).

Finally, Llinás and Paré (1996; Llinás 2001) have proposed that, in fact, the thalamo-cortical system, which comprises a collection of specific thalamic regions (the ventrobasal complex and the non-specific, middle-line nuclei, and a set of cortical counterparts that create resonant functional states), is a central component of the cognitive substrates of the brain. They maintain that the 40 Hz oscillations that can be recorded in many areas of the cerebral cortex are the resonance of neuronal oscillations in specific thalamic nuclei. These oscillations, which exhibit a 12–13 msec phase shift from the frontal to the posterior part of the brain, could serve to produce a temporal binding of sensory stimuli. During that unit of time (12–13 msec) the brain would process a single or quantum cognitive event. Cognition, therefore, would be a collection of periods of time. They concluded that "integration of sensory events into a larger computational state that underlies cognition is a function of its temporal relation to ongoing oscillatory activities of the brain" (Llinás 2004).

Highlighting those neurobiological hypotheses is important because they were "hypotheses"—that is to say, they comprise "philosophical thinking"—and, consequently, they constitute extraordinary examples of the integration between science and humanities which is the essence of neurophilosophy and therefore is neuroculture itself (Churchland 1990). In fact, when talking about consciousness, philosophy may still be permitted to have the last word. As Thomas Nagel (1991) wrote: "Science cannot address consciousness unless modifying radically its methodology to identify and analyze the elements of subjective experience."

1.6 Knowledge, pleasure, beauty, and neuroculture

Painting, music, literature, sculpture; all embellish human life. All cultures have studied, described, or discussed these subjects albeit from a humanistic, subjective, and descriptive perspective. Neuroscience permits a new perspective which we call "neuroaesthetics". In its essence, neuroaesthetics is that discipline which is starting to provide some neuronal insights into our understanding of art and beauty (Cela-Conde et al. 2011; Ishizu and Zeki 2011). Thus defined, neuroaesthetics is, along with the brain–mind problem, at the heart of neuroculture.

What is art? What is beauty? Art is a uniquely human activity associated with emotion as well as abstract and symbolic cognition. Art is in fact knowledge and pleasure; pleasure never consummated, never satisfied, that kind of special pleasure which as noted by Immanuel Kant is of "the finest nature because it tolerates enjoyment without satiety or exhaustion" (Kant 1960; Mora 2006). Neuroaesthetics provides a way, using neurobiological tools (neuronal recordings and imaging techniques such as magnetic resonance or magneto-encephalography) to investigate and dissect the bases in the brain (brain areas, nodes, and distributed neuronal systems) of both knowledge and pleasure.

Today we have hypotheses based on scientific evidence (neuronal single recordings) about how the brain may construct the abstractions which are the building blocks of knowledge (Mora 2009). Thus neurophysiologic studies performed in monkeys have provided evidence for neurons in the visual areas responding to a single perspective of an object (form, orientation, and depth), but not responding to other perspectives. We also know that there are neurons that respond to the presentation of the entire object irrespective of the perspective presented, giving probable indication of these latter neurons synthesizing all views from the previous neurons. Some of these neurons respond not just to the specific object presented but to different objects of similar shapes and colors. It has been suggested that these last neurons are probably arranged in circuits able to construct an abstraction or "ideal" object that may accommodate many different objects within that same category. There is speculation that these last neuronal circuits are the bases for the process of abstraction (Zeki 1999, 2001; Mora 2009). Thus, when nature shows us hundreds of birds of all shapes and sizes, movements and colors, songs and different behaviors, the brain is capable of creating the concept, the idea, of a bird that would summarize all birds in the world. This "universal" bird is an abstraction created by the brain, an idealization of a bird that neither exists nor *could* exist in reality, thus making the "pure and immutable essence of the bird" as Plato might have posited. That is how the human brain works; by categorizing and classifying the sensory world though concepts and ideas. Using this brain-based process of abstraction, man attained the basic principles for thought, language, and communication. Art, in fact, is abstract and symbolic cognition, emotion, and communication beyond common emotional or symbolic language (Mora 2006).

Perhaps these networks for abstraction are the result of codes acquired through evolution, as has been suggested for language (Chomsky 1965). According to Chomsky, our environment is highly variable and indeterminate and thus cannot be used to offer a complete explanation of the remarkable ability all young children have to learn complex concepts. From this follows that children may already be born with an innate grammar. The same could be applied for the process of abstraction since it shares with language the neural principles of being common to all human beings and being acquired very rapidly in children. In any case, the contents of these networks permit their holders to perform logical operations whose products are thought, language, art, and mathematics, and also possibly to possess the capacity to create ethical values (Agnati et al. 2007). Indeed, it has been speculated that all these operations may be a function of cerebral dominance, based on the existence of associative neural networks distributed among different areas of the cerebral cortex, thalamus, and brainstem, including the prefrontal cortex (dorsolateral prefrontal cortex, medial and lateral sectors of the orbitofrontal cortex), the anterior cingulate cortex, and regions of the superior temporal cortex (Cela-Conde et al. 2004; Agnati et al. 2007; Ishizu and Zeki 2011).

Beauty (Kawabata and Zeki 2003; Ishizu and Zeki 2011) requires, along with the process of abstraction, something of an "emotional plus." Today we know in part how the flow of sensory information (from sensory receptors to thalamus and cerebral cortex) reaches the limbic system (emotional brain, amygdala, and orbitofrontal cortex) where it is labeled with reward or punishment, pleasure or pain (Rolls and Grabenhorst 2008; Pessoa 2008). It is after this emotional labeling of sensory information has been performed that cognitive processes are elaborated in the association areas of the cerebral cortex (Pessoa 2008). Therefore the brain, and particularly the brain of an artist, already works with ideas and abstractions that are emotionally colored (Mora 2009). This should be very relevant to understand the act of creation and the capacity of a work of art to evoke beauty. But what seems more interesting when talking about beauty, and especially in relation to individuals contemplating a work of art, whether this painting, architecture, or sculpture, is the hypothesis suggesting that beauty is not something inherent to the artistic work itself but is in fact created by the observer. In other words, the beholder would actively create in his own brain his own conception of beauty which would involve both, in abstract and in the particular, emotion and pleasure. Furthermore it appears that when that creative personal process of beauty is produced in many people in front of a specific masterpiece, it is then that the work is recognized as universally sublime (Zeki 2001; Mora 2007).

Creativity, like ethics, beauty, decision-making, or the ability to think or reason, is not located anywhere in the brain. To create means having the ability to get excited and think, learn, remember, imagine, and all these are brain functions whose processing is distributed in areas, neural systems, and specific networks functioning within codes of time, not space (Bowden and Jund-Beeman 2003; Kounios et al. 2008; Mora 2009). Creativity is an important ingredient in any artistic activity. It is the capacity to do new things, therefore, that enables the artist to capture on canvas, wood, rock, or holograms feelings, abstracts, and ideals. It has been speculated that the artistic creativity may emerge from that clash between the abstractions created by the brain, with its emotional component, unique and personal, and the concrete sensory objects of the world from which precisely those same abstractions have been constructed (Zeki 2001; Mora 2009). Think of this as something like an unconscious dissatisfaction caused by a mismatch between mental objects and real sensory objects, between ideas of the objects created by the brain and the real objects themselves. something like a divorce between the "tree" the artist sees, which is "real," and the "tree" that the artist constructs unconsciously on the screen of his mind. From that conflict, an emotional reaction arises, leading to that dissatisfaction (frustration). It could be precisely this frustration that acts as the starter of the creative process. The creative act would then stand as the need of the artist to find a personal satisfaction through his work of art, with this work no longer reflecting the "objective reality" but his "mental reality" (Zeki 2001, 2004a, 2004b; Mora 2009).

In summary, neuroaesthetics offers an explanation about how the appreciation of art and beauty is linked to the activity of multiple brain regions, pathways, and networks, and how this activity is tied to the evolution of *Homo sapiens*, and the development of language and hierarchical social grouping, and further modified by cultural exposure (Zeidel 2010; Cela-Conde et al. 2011). Neuroaesthetics also helps develop neuroscience and neuroculture through the study of paintings and artists since these provide hints on the impact on culture and the function of the brain (Conway 2012).

1.7 **Neuroculture and the unity of knowledge**

Neuroculture, as I stated above, would be a new cultural event trying to bringing together all knowledge into a single conception of man, a unified theory of knowledge

(Wilson 1998). Over the past few thousand years, man has analysed every aspect of the universe, included himself, dividing this universe into pieces as it is done now. Every group of scientists in every discipline, using the available tools, has analysed their field of study by dividing and subdividing it, and then, at their level of analysis, tried to understand and describe the meaning of that little piece of reality. We are now starting to become aware that because everything is a continuum in the universe, an acceptable understanding of a single piece of reality can never be achieved since its full meaning only exists when linked to another fragment of that same reality, sometimes to a different level of organization (Mora 2007).

At each level of study, science in general, and neuroscience in particular, has created its own tools of analysis which include language of description and understanding specific to it alone. The science of man is like a tower of Babel with many people in it, speaking many different languages and, therefore, not understanding each other. The time has come for neuroscience and cognitive neuroscience to try to combine and build bridges between levels of analysis, to approximate languages in order to provide a better understanding of the unique reality which is man. That is, as I see it, one of the major tasks of neuroculture (Mora 2007).

One of the key events in such an approach is to understand that from one level of analysis to the next, there is a jump which implies the emergence of new properties in the new level which are not reducible to those of the preceding level. These new emergent properties, however, are tightly bound and completely dependent upon the elements of the preceding level. The most simple and classical example of this is water. Water is made up of oxygen and hydrogen atoms. However, we know that oxygen by itself and hydrogen by itself are not water. That particular molecule, called water, with all its peculiar components, oxygen and hydrogen, are acquired when a particular link between the atoms of oxygen and hydrogen is formed. We know that the laws governing the molecule of water are different from those governing the atoms of oxygen and hydrogen. We also know that the properties of water are not reducible to those of hydrogen and oxygen when separated. But when oxygen and hydrogen are linked together, an emergent new thing, the molecule of water, appears, and with it new laws governing it (Mora 2000).

In neuroscience, considerable effort is devoted to stitching together the different levels—genes, molecules, organelles, microcircuits, neuronal compartments, nerve cells, specific regions, nodes of distribution and distributed systems up to performance, mental operations, and cognitive systems (Churchland and Sejnowsky 1992)—which may in time shed light on those complete and complex cognitive processes, mind and consciousness. Thus, it may provide a solid basis for the emerging science of human nature (Fowler and Screiber 2008; Edelman 2006), and with this the consolidation of neuroculture.

1.8 Conclusion

Neuroculture may offer a new vision of humanity and society based on knowledge about how the brain works with help from the perspective of evolution. It might give rise to a new way of thinking, changing our understanding of established concepts in philosophy, ethics, law, sociology, economics, art, and religion. Neuroculture may eventually produce a reassessment of the humanities and may provide a bridge by which science and humanities can join, thus creating a unified theory of man and society.

Acknowledgments

An early draft of this work was written during my stay at the Department of Molecular Physiology and Biophysics at the Carver College of Medicine, University of Iowa where I was appointed Helen C. Lewitt Professor. I would like to express my gratitude to the Helen C. Lewitt Foundation for the professorship award received. I thank Dr Ana María Sangunetti for her critical comments. Also I should like to express my thanks to Angela Amores for reading and commenting on early stages of the manuscript.

References

Agnati, L.F., Agnati, A., Mora, F., et al. (2007). Does the human brain have unique genetically determined networks coding logical and ethical principles and aesthetics? From Plato to novel mirror networks. *Brain Research Review* **55**, 68–77.

Amunts, K., Lepage, C., Borgeat, L., et al. (2013). Big Brain: an ultrahigh-resolution 3D human brain model. *Science* **340,** 1472–5.

Beakley, B. and Ludlow, P. (1992). *The philosophy of mind. Classical problems/contemporary issues.* Cambridge, MA: MIT Press.

Beggs, J.M. (2008). The criticality hypothesis. How local networks might optimize information processing. *Philosophical Transactions of the Royal Society A* **366**, 329–43.

Bowden, E.M. and Jung-Beeman, M. (2013). Aha! Insight experience correlates with solution in the right hemisphere. *Psychonomic Bulletin Review* **10**, 730–7.

Bullmore, E. and Sporns, O. (2009). Complex brain networks: graph theoretical analysis of structural and functional systems. *Nature Reviews Neuroscience* **10**, 186–98.

Cela-Conde, C.J., Agnati, L., Huston, J.P., et al. (2011). The neural foundations of aesthetic appreciation. *Progress in Neurobiology* **94**, 39–48.

Cela-Conde, C.J., Marty, G., Maesta, F., et al. (2004). Activation of prefrontal cortex in the human visual aesthetic perception. *Proceedings of the National Academy of Sciences of the USA* **101**, 6321–5.

Changeux, J.P. and Chavallion, J. (eds) (1995). *Origins of the human brain.* Oxford: Clarendon Press.

Chialvo, D.R. (2010). Emergent complex neural dynamics. *Nature Physics* **6**, 744–50.

Chomsky, N. (1965). *Aspects of the theory of syntax.* Cambridge, MA: MIT Press

Churchland, P.S. (1990). *Neurophilosophy.* Cambridge, MA: MIT Press.

Churchland, P.S. (2002). *Brain-Wise.* Cambridge, MA: MIT Press.

Churchland, P.S. (2008). The impact of neuroscience on philosophy. *Neuron* **60**, 409–11.

Churchland, P.S. and Sejnowsky, T.S. (1992). *The computational brain.* Cambridge, MA: MIT Press.

Conway, B.R. (2012). Color consilience: color through the lens of art practice, history philosophy and neuroscience. *Annals of the New York Academy of Science* **1251**, 77–94.

Cortina, A. (2012). *Neuroética y neuropolítica. Sugerencias para la educación moral.* Madrid: Tecnos.

Crick, F. (1994). *The astonishing hypothesis.* New York, NY: Scribner's and Sons.

Crick, F. (1996). Visual perception. Rivalry and consciousness. *Nature* **379**, 495–86.

Crick, F. and Koch, P. (1990). Towards a neurobiological theory of consciousness. *Seminars in the Neurosciences* **2**, 263–75.

Damasio, A.R. (1994). *Descartes's error*. London: Picador.

Damasio, H. and Damasio, A.R. (1990). The neural basis of memory, language and behavioural guidance: advances with the lesion method in humans. *Seminars in the Neurosciences* **2**, 277–86.

Dobzhansky, T. (1973). Nothing in biology makes sense except in the light of evolution. *American Biology Teacher* **35**, 125–9.

Edelman, G.M. (2004). *Wilder than the sky: The phenomenal gift of consciousness*. New Haven, CT: Yale University Press.

Edelman, G.M. (2006). *Second nature: Brain science and human knowledge*. New Haven, CT: Yale University Press.

Edelman, G.M. and Mountcastle, V.B. (1978). *The mindful brain*. Cambridge, MA: MIT Press.

Fowler, J.H. and Screiber, D. (2008). Biology, politics and the emerging science of human nature. *Science* **322**, 912–4.

Frazzetto, G. and Anker, S. (2009). Neuroculture. *Nature Reviews Neuroscience* **10**, 815–21.

Gisolfi, C.V. and Mora, F. (2000). *The hot brain: Survival, temperature and the human body*. Cambridge, MA: MIT Press.

Halmovichi, A., Tagliazzucchi, E., Balenzuela, P., et al. (2013). Brain organization into resting state networks emerges at criticality on a model of the human connectome. *Physical Review Letters* **110**, 178101.

Holloway, R.L. (2008). The human brain evolving: a personal retrospective. *Annual Review of Anthropology* **37**, 1–19.

Huxley, J. (1964). *Essays of a humanist*. New York, NY: Harper and Row.

Ishizu, T. and Zeki, S. (2011). Toward a brain-based theory of beauty. *PLoS One* 6, e21852. doi: 10.1371/ Journal.pone. 0021852.

Jenkins, W.M., Merzenich, M.M., Ochs, M.T., et al. (1990). Functional reorganization of primary somatosensory cortex in adult owl monkeys after behaviorally controlled tactile stimulation. *Journal of Neurophysiology* **63**, 82–104.

Jerison, H.J. (1973). *Evolution of the brain and intelligence*. New York, NY: Academic Press.

Jerison, H.J. (1976). Paleoneurology and the evolution of mind. *Scientific American* **234**, 90–101.

Kandel, E.R. (2003). *The molecular biology of memory storage: a dialog between genes and synapses*. Nobel lecture 2000. H. Jörnwal (ed.). Singapore: World Scientific Publishing Co.

Kandell, E.R. and Mack, J. (2003). A parallel between radical reductionism in science and in art. *Annals of the New York Academy of Science* **1001**, 272–94.

Kant, I. (1960). *Observations on the feeling of the beautiful and sublime*. Oakland, CA: University of California Press.

Kawabata, H. and Zeki, S. (2003). Neural correlates of beauty. *Journal of Neuroscience* **91**, 1699–705.

Kounios, S., Fleck, J.L., Green, D.L., et al. (2008). The origins of insight in resting-state brain activity. *Neuropsychologia* **46**, 281–91.

Kroeber, A.L. and Kluckhohn, C. (1952). Culture: A critical review of concepts and definitions. *Harvard University Peabody Museum of American Archeology and Ethnology Papers* **47**, 643–4.

Levallois, C., Clithero, J.A., Wouters, P., et al. (2012). Translating upwards: linking the neural and social sciences via neuroeconomics. *Nature Review Neuroscience* **13**, 789–97.

Llinás, R.R. (2001). *I of the vortex*. Cambridge, MA: MIT Press.

Llinás, R.R. (2004). La percepción como un mensaje seudo-onírico modulado por los sentidos. In F. Mora (ed.), *Esplendores y miserias del cerebro*. Madrid: FSCH, pp. 145–66.

Llinás, R.R. and Churchland, P.S. (1996). *The mind-brain continuum*. Cambridge, MA: MIT Press.

Llinás, R.R. and Pare, D. (1996). The brain as a closed system modulated by the senses. In R. Llinás and P.S. Churchland (eds), *The mind-brain continuum*. Cambridge, MA: MIT Press, pp. 1–118.

Lordkipanidze, D., Ponce de León, M.S., Margvelashvili, A., et al. (2013). A complete skull from Dmanisi, Georgia, and the evolutionary biology of early homo. *Science* **342**, 326–31.

Moll, J., De Oliveira-Souza, R., and Zahn, R. (2008). The neural basis of moral cognition: sentiments, concepts, and values. *Annals of the New York Academy of Science* **1124**, 161–80.

Moll, J., Zahn, R., De Oliveira-Souza, R., et al. (2005). Opinion: the neural basis of human moral cognition. *Nature Review Neuroscience* **6**, 799–809.

Monod, J. (1971). *El azar y la necesidad*. Barcelona: Barral.

Mora, F. (ed.) (1995). *El problema cerebro-mente*. Madrid: Alianza Editorial.

Mora, F. (2000). The brain and the mind. In M.G. Gelder, J.J. Lopez-Ibor, Jr, and N.C. Andreasen (eds), *New Oxford Textbook of Psychiatry*. Oxford: Oxford University Press, pp. 153–58.

Mora, F. (2006). *Los laberintos del placer en el cerebro humano*. Madrid: Alianza Editorial.

Mora, F. (2007). *Neurocultura*. Madrid: Alianza Editorial.

Mora, F. (2008). *El reloj de la sabiduría*. Madrid: Alianza Editorial.

Mora, F. (2009). *Como funciona el cerebro*. Madrid: Alianza Editorial.

Mora, F. (2011). *El Dios de cada uno. Por qué la neurociencia niega la existencia de un Dios universal*. Madrid: Alianza Editorial.

Mora, F. (2012a). *¿Está nuestro cerebro diseñado para la felicidad?* Madrid: Alianza Editorial.

Mora, F. (2012b). Dio e il cervello: alcune riflessioni dal punto di vista della neuroscieza. In L. Agnati and L. Alfieri (eds), *La comunicazione nel cervello delluomo e nella societa*. Fano: Aras Edizioni, pp. 145–61.

Mora, F., Segovia, G., and Del Arco, A. (2007). Aging, plasticity and environmental enrichment: structural changes and neurotransmitter dynamics in several areas of the brain. *Brain Research Review* **55**, 78–88.

Mora, F., Segovia, G., Del Arco, A., et al. (2012b). Stress neurotransmitters, corticosterone and body-brain integration. *Brain Research* **1475**, 71–82.

Moretti, P. and Muñoz, M.A. (2013). Griffiths phases and the stretching of criticality in brain networks. *Nature Communications* 4, 2521. doi: 10.1038/ncomms 3521.

Nagel, T. (1991). *Mortal questions*. Cambridge: Cambridge University Press.

Ortega y Gasset, J. (1983). *Obras completas*. Madrid: Alianza Editorial.

Pessoa, L. (2008). On the relationship between emotion and cognition. *Nature Reviews Neuroscience* **9**, 148–58.

Raz, A. and Buhle, J. (2006). Typologies of attentional networks. *Nature Reviews Neuroscience* 7, 367–79.

Rolls, E.T. (2012). *Neuroculture*. Oxford: Oxford University Press.

Rolls, E.T. and Grabenhorst, F. (2008). The orbitofrontalcortex and beyond: from affect to decision-making. *Progress in Neurobiology* **86**, 216–44.

Searle, J.R. (1998). How to study consciousness scientifically. *Brain Research* **26**, 379–87.

Shepherd, G.M., Mirsky, J.S., Healy, M.D., et al. (1998). The human brain project: neuroinformatics tools to integrating searching and modelling multidisciplinary neuroscience data. *Trends in Neurosciences* **21**, 460–8.

Steiner, G. (1999). *Conferencias Internationais de Filosofia e Epistemologia do Instituto Piaget: "Da condicao Humana."* Viseu, Portugal.

The Blue Brain Project. <http://bluebrain.epfl.ch>

The Human Connectome Project. <http://www.Humanconnectomeproject.org>

Tobias, P.V. (1995). The brain of the first hominids. In J.P. Changeux and J. Chavaillon (eds), *Origins of the human brain*. Oxford: Clarendon Press, pp. 61–83.

Tobias, P.V. (1997). Evolution of brain size, morphological structuring and longevity in early hominids. In S.U. Dani, A. Hori, and G.F. Walters (eds), *Principles of neural aging*. Amsterdam: Elsevier, pp. 153–74.

Tononi, M. and Edelman, G. (1998). Consciousness and complexity. *Science* **282**, 1846–51.

Vidal, F. (2009). Brainhood, anthropological figure of modernity. History of the Human. *Science* **22**, 5–36.

Wexler, B.E. (2006). *Brain and culture*. Cambridge, MA: MIT Press.

Wilson, E.O. (1998). *Consilience*. New York, NY: Alfred A. Knopf.

Wilson, E.O. (2003). *The future of life*. New York, NY: Knopf Publishing Group.

Zaidel, D.W. (2010). Art and the brain: insights from neuropsychology, biology and evolution. *Journal of Anatomy* **216**, 177–83.

Zeki, S. (1993). *A vision of the brain*. Oxford: Blackwell.

Zeki, S. (1999). *Inner vision. An exploration of art and the brain*. Oxford: Oxford University Press.

Zeki, S. (2001). Artistic creativity and the brain. *Science* **293**, 51–2.

Zeki, S. (2004a). Esplendores y miserias del cerebro. In F. Mora (ed.), *Esplendores y miserias del cerebro*. Madrid: FSCH, pp. 20–60.

Zeki, S. (2004b). Formación del concepto neural del arte: Dante, Miguel Angel y Wagner. In F. Mora (ed.). *Esplendores y miserias del cerebro*. Madrid: FSCH, pp. 392–428.

Chapter 2

Art, meaning, and aesthetics: The case for a cognitive neuroscience of art

William P. Seeley

> It is important to note that I am not suggesting that we should directly import the results of empirical psychology to aesthetics. The direct application of empirical results in aesthetics can, and very often does, go terribly wrong. What I suggest is that aesthetics should take some new paradigms of philosophy of perception seriously. The specific paradigm I am interested in here, the paradigm of multimodality, is based on a large body of empirical research. However, my aim is not to urge an empirical turn in aesthetics, but to urge a turn in aesthetics towards philosophy of perception, and this sometimes entails a turn towards empirically informed philosophy of perception.
> *(Nanay 2012, pp. 353–54)*

2.1 Introduction

The philosophy and psychology of art often behave as estranged bedfellows, fighting tooth and nail over absent common ground that neither concedes could exist. Empirical aesthetics is predicated on the claim that philosophical studies of art are methodologically suspect. The latter field has been defined as a *speculative aesthetics* grounded in deductive inferences drawn from the subjective judgments of experts. On this account philosophy of art, at best, reflects the subjective tastes of critics and theorists (Berlyne 1971; Fechner 1876; Martindale 1990). Philosophers behave no better. They claim that it is a category mistake, an error in logic, to ground a theory of art in behavioral studies—questions about the nature of art are questions about the structure and validity of artistic conventions, questions that can be no more answered by studying the preferences and physiological responses of large groups of naïve subjects than analogous questions about the validity of moral beliefs (Dickie 1962; Wittgenstein 1966; but see also Carroll et al. 2012; Carroll and Seeley 2013b). My suggestion, borrowing the letter, but perhaps not the intent, of

Bence Nanay's comment above, is that there is something to be learned in a methodological rapprochement on both sides of this great divide. Philosophical questions about the nature of artistic salience and the role evaluative judgments play in our interaction with artworks can be modeled as questions about the influence of cognitive and affective processing in perception. Philosophers would do well to pay attention to psychological theories that address the ways these variables influence our engagement with artworks. Likewise, psychology of art attributes a role to aesthetic appraisal in our engagement with artworks that is not consistent with contemporary artistic practices. Empirical aesthetics would, therefore, do well to pay attention to the range of ways philosophers of art model cognitive, affective, and aesthetic influences on judgments of artistic salience. In what follows I propose a model for a cognitive neuroscience of art that bridges the gap between philosophy of art and empirical aesthetics and discuss how the model resolves two standard philosophical objections to empirical aesthetics.

2.2 Art, aesthetics, and psychology

The history of mistrust between philosophy of art and empirical aesthetics can be traced to a disagreement about the nature of artistic salience. Think of it this way. Why would the subjective judgments and intuitions of art critics be of any interest at all to philosophers of art? This seems like odd evidence on which to build an objective, general theory of art! Well. Ontological questions about art are to a large extent questions about the conventions governing our categorical (and evaluative, but more on that later) judgments about a range of artifacts. If we want to know about these conventions, who should we ask—or better, who's behavior should we be interested in—experts or non-experts? It seems like the place to start is with people who understand the language, so to speak (Wittgenstein 1958; Dickie 1962). The ability to evaluate the judgments of non-experts in these contexts depends on a prior understanding of the structure of these conventions, conventions that are embedded in the behaviors of art experts. I suppose the worry expressed by empirical aesthetics is that this might render artistic judgments subjectively arbitrary and elitist. However, this does not necessarily follow. The claim is not that the judgments of experts arbitrarily determine the artistic conventions of a community. It is rather that expert judgments codify the shared practices that define an artistic community and in so doing reflect the conventions against which artistic judgments and behaviors are evaluated within that community.

Research in empirical aesthetics replaces the explicit judgments of expert consumers with physiological measures of arousal constitutive of (at least) the feeling (and in some theoretical contexts the content as well) of aesthetic experience. These physiological measures are treated as objective measures of *aesthetic interest and appraisal* that generalize across art experts and ordinary consumers and so putatively count as markers of the *artistic salience* of the formal, expressive, and semantic features of artworks within a community. The thought is that artworks are a natural kind of sorts defined by a common capacity to trigger aesthetic responses in consumers, however that might work. This kind of view is what philosophers call *an aesthetic theory of art*. Aesthetic theories of art collapse the concept of artistic salience into the concept of *aesthetic salience*, or reduce the former to the latter, if you prefer. Empirical aesthetics associates the aesthetic salience of artworks with a capacity to produce a robust, objectively measurable, unified class of physiological responses in consumers. This, in turn, putatively warrants replacing the judgments and intuitions of experts about the nature of art with objective measures of the physiological responses of groups of ordinary art consumers.

Perhaps no-one really holds the kind of strong non-cognitive aesthetic thesis expressed here. Nonetheless, the objective measures of arousal used to evaluate judgments of artistic salience in empirical aesthetics are aesthetic measures (see Berlyne 1974; Chatterjee 2012; Silvia 2012). So, on the surface of things, it looks as if empirical aesthetics relies on an aesthetic theory of art. The trouble is that a broad range of contemporary artworks are explicitly non-aesthetic, and, in the case of aesthetic works of art, there is no guarantee that the aesthetic responses of average members of an artistic community will track the correct use of the conventions governing artistic judgments and behaviors within that community.

2.2.1 Aesthetics and the philosophy of art

The terms *aesthetics* and *philosophy of art* are each used in ordinary contexts to refer generically to philosophical studies of art. However, these terms are not synonymous and the equivocation in this casual, non-reflective use of language is significant. The category of aesthetics is broader than the category of art. It includes snowy mountain vistas, stormy coastal landscapes, cozy capes and cottages with tidy lawns, industrial machinery, automotive design, and even the coffee pot on your kitchen counter. It also includes a range of artworks. The category "art" is broader than the extent of this latter, limited range of aesthetic artifacts. It includes formalist aesthetic Abstract Expressionist paintings and aesthetically striking Flemish landscapes from the seventeenth century. But it also includes a full range of non-aesthetic conceptual works like Robert Morris' ephemeral installations (e.g. *Threadwaste*, 1968[1], and *Steam*, 1968/9[2]), Robert Rauschenberg's "combines" (e.g. *Monogram*, 1955–59),[3] Jasper John's paintings of numbers and letters (e.g. *0 through 9*, 1961),[4] John Baldessari's paintings composed of text and definitions (e.g. *Tips for Artists Who Want to Sell*, 1966–68),[5] and Sol LeWitt's *Location Drawings* (e.g. *Wall Drawing 305*, 1977).[6] The concepts "art" and "aesthetics" therefore track different dimensions of the range of artifacts that we categorize as artworks, and philosophy of art represents a broader, more inclusive definition of art. This is critical. It is certainly the case that some artworks are constructed for the purpose of displaying their aesthetic properties. However, many more, and most likely the majority, of contemporary artworks are not like this. They are rather conceptual works. They express a concept or idea, challenge our conception of art, or perhaps reveal something to us about underlying biases within our communities (e.g. the works of John Baldessari, Joseph Kossuth, Robert Barry, Robert Morris, Wesley Meuris, The Judson Dance Theater Workshop, Barbara Krueger, Jenni Holtzer, or Adrian Piper). The trouble is that these works are not intentionally designed to trigger aesthetic experiences in consumers. In fact many of them are expressly anti-aesthetic, designed to draw attention to themselves as the ordinary, non-aesthetic objects, events, or actions that they are. Aesthetic theories of art would exclude these works from the category "art." Therefore, if empirical aesthetics depends upon a tacit aesthetic theory of art, it represents a narrow view of art that is inconsistent with contemporary practice.

A similar difficulty arises for the classification of aesthetic artworks themselves. Recall that the range of aesthetic objects, scenes, and etc. is broader than the range of artifacts we classify as artworks. What then distinguishes artistic from non-artistic aesthetic objects? Consider Danto's discussion of Warhol's *Brillo Boxes* as a limiting case (see Danto 2000). The original Brillo box is an aesthetically rich marketing device designed by a second generation Abstract Expressionist painter named Steve Harvey. Harvey created the canonical Brillo box while moonlighting in commercial design to make a living. We can perceptually discriminate Warhol's facsimile's from the original by their relative size (Warhol's are larger) and material (Warhol's are made of plywood). But these visible differences don't alone suffice to disentangle which is the artwork. We could, to lapse into philosophical fantasy, imagine a Martian, or even a pigeon with similar perceptual capacities to humans (see Danto 2003), mistaking Harvey's for a Warhol—for that matter I suppose that we could imagine a naïve consumer with an adequate understanding of Pop Art but limited knowledge of Warhol's works within this category making the same mistake. Further, given that Warhol's *Brillo Boxes* reproduce the original visual design fairly accurately, they share the same range of (or at least exhibit similar) dynamic, colorful aesthetic properties as Harvey's. The two types of artifacts aren't just visibly indiscernible in this regard; they are arguably also aesthetically indiscernible. What differentiates them as exemplars of different categories of artifacts, one an artwork and the other a commercial design object, is therefore neither their visible nor their aesthetic properties. Danto argued that what enables us to disambiguate these two sets of artifacts to different categories is the semantic salience of their visual design, or how these formal aesthetic properties were used to convey the content of the works in either case. Harvey's was a red, white, and blue visual celebration of the joy of Brillo intentionally designed to entice us to purchase a cleaning product. Warhol's was a Pop Art conceptual challenge to the dominant, formalist, Abstract Expressionist aesthetic of the time. Or, the artistic salience of the formal aesthetic properties of Warhol's *Brillo Boxes* lies in what their hard-edged formal design revealed about the then dominant conventions governing artistic practices within the New York avant garde art community. The aesthetic features of Harvey's Brillo box, in contrast, simply don't have artistic salience.

Warhol's work is conceptual, an expressly anti-aesthetic example of Pop Art. Someone might argue, consequently, that this is a bad test case. However, the model Danto proposes applies equally to works of pure abstraction as well. Consider, for instance, Jackson Pollock's *One: Number 31, 1950*.[7] Ordinary consumers are struck by the energy of Pollock's monumental canvases, by their chaotic construction, perhaps exemplified in their failure to resolve perceptually into stable geometric patterns. But the standard interpretation of these works locates their artistic salience in the trace of the artist's gesture visible in Pollock's brush stroke, in the expressive activity of the artist and the way this aspect of the paintings self-consciously reflects both Clement Greenberg's reductive formalist aesthetics (Greenberg 1960) and Harold Rosenberg's concept of action painting (Rosenberg 1952). The artistic salience of these works therefore lies in the way they exemplify Abstract Expressionism as a mid-twentieth century theoretical position about the nature of art, not their dynamic formal structure *per se*.

The discussion of *Brillo Boxes* and *One: Number 31, 1950* reveals something important about the nature of artworks: they are communicative devices. We typically engage artworks as intentional objects with cognitive content, as the product of a range of formal and compositional choices made by the artist, choices that carry information about their content. We are interested in how they were constructed, or perhaps better, in why they were constructed the way they were—even when they are canonically aesthetic artworks. It can, as a result, be argued that the artistic salience of the formal-compositional features of a work lies in their semantic salience, in how they have been used to express the meaning or content of the work. Of course, artists often use aesthetic devices as part of the array of formal-compositional strategies they use to convey the content of their work. However, the artistic salience of these features lies in their semantic salience, in what they contribute to the content or meaning of the work. This is true of the broad range of artifacts that count as artworks, from anti-aesthetic conceptual works to those limited cases where the content of a work simply involves the expression of some set of artistic conventions for constructing aesthetic artworks. For instance, the artistic salience of landscape paintings lies, at least in part, in what the stylistic choices of their painters reveals about the aesthetic conventions of their time (e.g. the way the skewed scale of Hudson River School paintings reveals mid-nineteenth century American beliefs about the relationship between man and nature and the current politics of American expansionism).

A standard philosophical objection to aesthetic theories of art can be used to illustrate this point. Aesthetic theories of art define artworks relative to a capacity for producing aesthetic experience. This entails that where an artwork doesn't succeed in triggering an aesthetic experience in consumers, it isn't an artwork. If sound this would mean that that there could be no failed artworks. This, in turn, would mean that generations of folks who have toiled away in studios, struggling in vain with the creative process, were in fact not engaged in art-making at all. They were doing something else altogether. Of course, philosophical arguments with absurd conclusions like this one are rarely intended to be taken literally. The point of a philosophical *reductio ad absurdum* is more often than not to suggest that there is something wrong with some aspect of an argument that deserves attention and modification.

"Art" is a success term. Its ordinary usage carries some degree of normative, or evaluative force. Some artifacts manage to "achieve" arthood, and we value them to the degree that they do. The question is, what are the criteria that govern our judgments in this regard? Aesthetic theories of art argue that the candidate criteria is the capacity of an artifact to trigger an aesthetic experience. However, this won't do. Failed aesthetic artworks are, nonetheless, artworks, and we judge them to be so despite the fact that they fail to produce aesthetic experiences in consumers. How is it, then, that we come to categorize members of this class of artifacts as artworks; we readily see that they were designed with an intention to function as aesthetic objects, that they are fit to some set of formal-compositional conventions for producing works in some aesthetic category of art, for instance nineteenth-century Hudson River School painting as opposed to mid-twentieth century Minimalist sculpture or late twentieth-century Installation Art. What matters in categorizing an artifact as an artwork is the fit between the intended content of the work and the conventions governing categorial judgments about the class of artworks it belongs to. A failed aesthetic work, then, is an artwork regardless of whether it ever succeeds in triggering any actual aesthetic responses in consumers.

There is nothing particularly special about art in this regard. Artifacts achieve "sawhood" (or any sort of toolishness) when they meet the categorial criteria for being a saw (or any sort of tool). Likewise we value them for their success in achieving this status. What makes something a saw is that it has a blade suited for tearing some material. A saw blade may fail to realize this function. Toothless friction saws designed for precision metal cuts are particularly bad at cutting wood. Similarly, cross-cut saws are ill suited for making rip cuts. But we still categorize them as saws. A length of rope coated in an experimental abrasive designed to cut plastics remains a saw, by virtue of its design as a blade for tearing material, even if the novel abrasive wears away too quickly to be effective. Likewise, a Claus Oldenburg soft saw sculpture is a saw of sorts even though it would fail to realize this function in any context.

2.2.2 Art, aesthetics, and appreciation

We might dub the issue discussed in the previous section *the problem of artistic salience.* The problem of artistic salience points towards a second philosophical challenge to empirically minded theories of art. Some philosophers have argued that questions about artistic appreciation, evaluative question about whether a work fits the conventions governing artistic practice, and if so whether it has been done well or poorly, are the critical questions that ground an understanding of art. This seems like too strong a position. Philosophers have always been interested in more general ontological and epistemological questions about the nature of artworks and our interactions with them. Aristotle, for instance, wrote on the narrative structure of tragedy and the nature of our psychological engagement with characters in *The Poetics.* There are similarly a range of interesting questions about the nature of installation art that are independent of questions about the value we assign particular installations (Irvin 2013). What are the identity conditions governing an installation? Is an installation, like a musical score, a repeatable recipe? How should the installation crew implement the relationship between a particular installation and the particular space of the gallery? What if the artist doesn't leave explicit instructions, is vague about them, or simply isn't around to query for advice? Should the work stay as true as possible to relevantly similar prior installations? If so, which ones? If not what sort of constraints govern the creative judgments of the curator? These aren't evaluative questions. They are questions about the way artworks work, questions about the conventions governing artistic production, understanding, and related psychological processes. Nonetheless, evaluative questions are among the questions a philosophical or psychological theory of art should answer. This is a place where philosophy of art and empirical aesthetics agree. Aesthetic appraisals play a critical role in empirical aesthetics. The objective physiological measures of arousal that ground research in the latter field are, in keeping with an aesthetic theory of art, explicit measures of artistic success, evaluative measures of aesthetic interest and appraisal.

Evaluative questions germane to artistic appreciation are *normative* questions, questions about the fit between an artwork and standard conventions governing artistic practice, or whether it was done well or poorly. The strong philosophical claim in this context is that causal explanations of behavior—explanations of behavior in terms of the causal-psychological processes that link stimuli to behavior—are irrelevant to normative questions. They apply equally whether or not that behavior is an appropriate response to the target situation or not (whether the arousal measures indicative of artistic interest and aesthetic responsiveness are appropriate to the work, are the product of contextually appropriate conventions and cognitive processes or not). For instance, we can imagine a consumer having a physiologically appropriate aesthetic response to a Jackson Pollock because it reminds them of the innocent exploratory creativity of childrens' finger-painting (perhaps their own child's finger-painting).

Explanations of aesthetic appraisals in experimental psychology are causal explanations. Therefore, it is argued, they are not sensitive to the appreciative dimension of artistic behavior (Wittgenstein 1966). Imagine, for instance, a study in which participants are asked to match musical compositions to a range of adjectives that expert music scholars and critics have identified as appropriate to those works (Dickie 1962). Appealing to agreement among subjects here would be like appealing to agreement among young children about the meaning of nonsense words, or appealing to agreement among toddlers about the syntactic well-formedness of sentences. What really matters in these contexts is something about the conventions governing linguistic practices and how apt the stimuli are to those conventions. This information would already be in place at the start of this kind of study, rendering its results otiose at best. Likewise with preference judgments. The preferences of uneducated novices are not the right kind of evidence to use to evaluate a work. The right place to look is the reflective judgments of experts about the fit between the work and the conventions governing those practices, not the averaged behavioral response of large samples of novice and expert consumers.

Appreciative judgments about artworks are grounded in comparisons between those artifacts and the conventions governing artistic practices for a particular category of art. The success of these judgments rests upon an understanding of that category of art, including knowledge of the normative conventions that determine whether an artwork of that type was done well or poorly (Levinson 1992; Rollins 2004). Philosophical questions germane to artistic appreciation are questions about the content of these conventions and why they have normative force, or how they constrain our evaluative judgments. These are not questions that can be answered by averaging the behavioral responses of ordinary untrained participants loosely associated with an artistic community. Consider the analogous case of morality. Ethical questions about moral behavior, questions about how one *ought* to behave in order to comport oneself as a genuinely moral person, cannot be answered by observing the ordinary day-to-day behavior of members of a community or by polling their preferences. These kinds of studies may reveal how groups of people *do* behave in a community, but the question about how one *ought* to behave is a different question (see Moore 1903). Objective measures of aesthetic interest and appraisal might similarly reveal what kinds of artistic judgments groups of people make. However, these are objective measures of a participant's subjective response to an artwork. They do not differentiate between judgments that reflect an appropriate understanding of particular artworks and those that do not. The argument, therefore, is that it is a methodological error, a category mistake, to use behavioral methods and results to address questions about the structure and function of artistic conventions.

2.2.3 A model for a cognitive neuroscience of art

I have suggested elsewhere that philosophical skepticism about the utility of empirical research to any understanding of art is misplaced (Carroll et al. 2012; Seeley 2010, 2011a, 2011b; Kozbelt and Seeley 2007). Cognitive science, in the broadest sense of the term, refers to any of a range of interdisciplinary approaches to understanding how an organism acquires, represents, manipulates, and uses information in the production of behavior. Research in neuroscience of art and empirical aesthetics falls squarely within cognitive science under this interpretation. Artworks are stimuli intentionally designed to trigger a range of affective, perceptual, and semantic responses in consumers constitutive of their artistically salient expressive, formal-compositional, aesthetic, and cognitive content. Empirical aesthetics and neuroscience of art are methodological tools that are regularly used to model the way consumers acquire, represent, manipulate, and use information carried in the formal-compositional structure of painting, films, dances, musical performances, novels, and poems (to name a few) in the processes underwriting canonical cognitive, expressive, and aesthetic artistic responses to artworks. These data can, in principle, be used to adjudicate between competing philosophical theories about the nature of art and our engagement with artworks, and to answer thorny questions the problems of artistic salience and artistic appreciation raise for empirical aesthetics more generally.

A sketch for a general model for a rapprochement between philosophy of art and empirical aesthetics emerges from some of the platitudes grounding research methods in neuroaesthetics and a fairly standard story about artists' productive practices. The local environment is replete with information about the structure, function, and dynamics of objects and events, only a small fraction of which is germane to our current goals and activities at any given time. Perceptual systems, on the other hand, are limited capacity cognitive systems. Selectivity is therefore a critical component of perception—perceptual systems must enable an agent to select task-salient perceptual features and discard the vast range of distracting information that clutters his or her perceptual world. Recent research suggests that one means to solve this problem is to develop visual routines and other types of perceptual strategies for directing attention to sets of features diagnostic of the identities, affordances, and locations of task-salient objects in the local environment (see Hayhoe and Ballard 2005; Rollins 2004; Schyns 1998).[8] Perceptual systems can, in this regard, be interpreted as evolved mechanisms for identifying and selecting task-salient information from a dynamic flux of sensory inputs and discarding distractors in the local surroundings (Zeki 1999; see also Pessoa et al. 2002; Stein et al. 2006). Artists have developed systematic methods for selecting sets of diagnostic features from ordinary perceptual experience, rendering them in a medium, and evaluating their salience as productive strategies, e.g. color studies for paintings or story boards and dailies for film. The resulting formal-compositional devices work as communicative strategies because they are directed at the range of perceptual strategies that cognitive systems employ in ordinary contexts to cope smoothly with a dynamic environment replete with sensory information on the fly. Therefore we can hypothesize that we ought to find a close coupling between artists' productive strategies and the operations of perceptual systems.

A significant amount of research in neuroaesthetics has been devoted to uncovering, tracking, and evaluating the range of ways artists' formal-productive strategies map to the operations of perceptual systems (e.g. Richard Latto's discussion of the relationship between irradiation and lateral inhibition in the retina, Beatrice Calvo-Merino and Corrine Jola's explorations of the role played by pre-motor areas in the action-observation-network in audience engagement with dance, and Nicole Speer's research on the role played by motor simulation in narrative understanding in text and film) (Calvo-Merino et al. 2008; Jola et al. 2012; Latto 1995; Speer et al. 2009). However, correlations between the neurophysiological operations of perceptual systems and artists' formal-compositional strategies do not alone suffice to explain the artistic salience of the formal, expressive, or semantic features of an artwork. Irradiation works to enhance figure ground segregation and the perception of depth in Monet's paintings because lateral inhibition produces half-shadows everywhere in our perceptual environment that play an analogous role in ordinary depth perception and object recognition (Latto 1995). Motor simulation, likewise, works as a mechanism to support the role of *kinetic transfer* in dance communication and narrative understanding in film because it plays an analogous role in action recognition and the perception of biological motion in ordinary contexts (see Carroll and Seeley 2013a, 2013b). Similarly, motor simulation plays a role in narrative understanding in fiction because (or to the degree that) it plays a critical role in the semantic comprehension of action sentences (Fischer and Zwaan 2008; Speer et al. 2009). What one needs is a further argument connecting the way artworks work generally as cognitive stimuli to the way consumers recognize and understand the artistic salience of their formal, expressive, and semantic content. The notion of an aesthetic appraisal is often called on to do the heavy lifting here in empirical aesthetics. However, as discussed above, the preponderance of non-art aesthetic objects, events, and experiences raises a question about whether aesthetic measures are appropriate measures of artistic salience.

A solution to this problem emerges from the observation that there is no ideal formal-compositional strategy for artistic production in a medium. Rather, artists must choose how to render their subject matter from a wide range of alternatives. Categories of art in which productive practices are directed towards shared subject matter are the easiest place to see this (for example, realistic landscape painting and Hollywood movies), but any exploration of formal stylistic differences would ultimately do. The stylistic variance in what has passed as realistic depiction in this context demonstrates that any of a range of formal-compositional strategies suffice to render these environmental features realistically (see Gombrich 1960; Lopes 1995). We might tell a similar story about the range of variants of common stories in movie and theater production, for example, Pierre Choderlos de Lacos' eighteenth-century novel *Dangerous Liasons*, Christopher Hampton's theatrical version of the same story, or the three film versions by Stephen Frear, Jin-ho Hur, and Milos Foreman. What are the constraints governing the development of those formal-compositional strategies that define the artistic styles of different schools, eras, and artists along with the surface structure of different artworks? The intended content of the work, the expressive, formal aesthetic, and cognitive effects an artist intends the work to have on consumers. What are the constraints on the artistic content of a work? The conventions governing productive and appreciative practices for a category of art. Therefore, the formal-compositional structure of an artwork naturally carries information about the conventions governing artistic practices, information that cognitivist models within the philosophy of art argue determines the artistic salience of the expressive, formal aesthetic, and semantic features of a particular work. This entails that artworks are *attentional engines*, or cognitive stimuli intentionally designed to direct attention those aspects of their formal-compositional structure that carry information about their content, including information about the category of art a work belongs to (Carroll and Seeley 2013a; Levinson 1992). Categorizing an artwork as belonging to a particular school, era, or artistic style provides a recipe for how to approach it—this instructs a consumer as to which sets of artistic conventions are appropriate to his or her cognitive and appreciative engagement with the work (e.g. as a satire or a dramatic novel, an Impressionist or a Hudson River School painting, or a Modernist or a Postmodernist dance).

A core set of critical questions loosely defines this research model for philosophy and neuroscience of art (Seeley 2013). What are the range of formal-compositional strategies that define productive practices within a category of art? How do these strategies function to carry and communicate diagnostic information about the content of these artworks? What accounts for the artistic salience of the expressive, formal aesthetic, and semantic features constitutive of the content of these works? A solution to the problem of artistic appreciation follows naturally from the answer to the last question. The artistic salience of the expressive, formal-aesthetic, and semantic features of an artwork lies in the way they reflect the range of artistic conventions that govern artistic practices for the category of art to which it belongs. Danto's discussion of Warhol's *Brillo Boxes* is a canonical illustration of this position. The artistic salience of Warhol's *Brillo Boxes* lies in the fact that they violate, and so cause us to reflect on, the conventions governing productive practices within, and appreciative judgments about, Abstract Expressionist artworks. Appreciative judgments about Warhol's work involve judgments about the validity of this project in the context of conventions governing avant-garde art, which are conventions governing the reflective use of artistic practices to evaluate the validity of artistic conventions. Warhol chose to approach this problem using a visually pleasing design. But ultimately his strategy was to highlight the identity of the work as what it was, an exemplar of an ordinary household object whose salience was tied to its everyday use, not a metaphysically charged, abstract, high-art aesthetic object. Other examples of this strategy are Jasper Johns' *Painted Bronze* (1960)[9] and *Painted Bronze (Ballantine Ale)* (1960)[10] and Steve Paxton's untitled dance in which he sat on a bench and ate a sandwich (see Banes and Caroll 2006). We might argue that the visually striking features of the Brillo design are, in this regard, the least successful aspect of Warhol's work! Alternatively we might argue that he employed a visually striking design to draw attention to its irrelevance to the salience of the artifact as an artwork.

2.3 Hierarchical models and attentional engines

We need a cognitive neuroscience of art because we need a model for how we use semantic content to disambiguate artworks from similarly structured non-art artifacts, for example, works of commercial design. This is transparently obvious in the case of non-aesthetic conceptual art, but it is also true of abstract aesthetic works at the opposite end of the spectrum. The artistically salient content of a work emerges from the way its formal-compositional structure reflects the conventions governing artistic practice within the appropriate category of art. The canonical example used to illustrate this fact within philosophy of art comes from Kendall Walton (1970).[11] Imagine two perceptually similar categories of art, Picasso's *Guernica*[12] and a hypothetical, imaginary category of bas reliefs that share its formal-compositional structure called *Guernicas*. The artistically salient dynamic features of *Guernica* are a product of its fractured, dynamic, black and white cubist composition (which Picasso used to express something about the horror of war for the communities within which they are fought). *Guernicas* are, in contrast, colorless. Their artistically salient dynamic features emerge from the way figure-ground relations give way to a depiction of the chaos of war in shallow relief, in their three-dimensional structure.

Guernica is flat. Therefore, relative to the conventions governing bas reliefs it is static rather than dynamic. A analogous criticism can be directed at the *Guernicas*. They are colorless. Therefore, relative to the conventions governing artistic production in painting they lack an articulated structure sufficient to support the dynamics of *Guernica*. We need a cognitive neuroscience of art because we need a theory that is sensitive to the way these kinds of conventions govern artistic practice within a category of art, constrain the contents of artworks, and influence our engagement with them. We need a theory that can explain how artworks work as attentional engines to focus attention on formal-compositional cues diagnostic for the appropriate category of art, *Guernica* and not *Guernicas*, Pop Art and not commercial design, or Abstract Expressionism and not finger painting.[13]

Biased competition theories of selective attention (Desimone and Duncan 1995; Kastner 2004; Pessoa et al. 2002) suggest a model for how artworks function as attentional engines. Recall the problem of selectivity. How do perceptual systems solve this problem? How does an organism flexibly orient its attention to diagnostic information on the fly in the kinds of novel contexts it confronts in everyday behavior? The simplest solution is that these systems are fine-tuned to perceptual salience. Some environmental features are simply brighter, more vividly colorful, or move differently than their neighbors. We easily orient our attention to them because they stand out in a crowd. The trouble is that the vast majority of behaviorally salient features in the environment simply are not perceptually salient in this way—certainly none of the keys on my keyboard, the papers on my desk, or the baking products in my cupboard sing out their relative salience to my current needs like a klaxon. Therefore perceptual systems require an independent mechanism for assigning salience to diagnostic features—for tagging the salience of objects and their features with relevance to me rather than their relative difference from their surroundings.

Biased competition models suggest that we employ top-down, fronto-parietal, and cortico-fugal attentional networks to solve this problem in everyday perceptual contexts. These networks bias perception by priming the firing rates of populations of neurons in sensory systems to the expectation of diagnostic features at particular locations. A fast first sweep through the perceptual system is sufficient to match minimal diagnostic features to semantic knowledge of the structure, function, and affordances of object and event types stored in declarative memory (Pessoa and Adolphs 2010). This information, in turn, can be used as a scaffold on which to fix perception—to direct attention and bias perceptual processing to task-salient features, objects, or object parts at expected locations within the visual field.

We can tell a similar story about artworks. Some of the features of an artwork are perceptually salient, and so naturally draw our attention. For instance, Winslow Homer rendered the trigger hand of the Union sniper brighter than any other feature in his wood engraving (for Harper's Weekly) *The Army of the Potomac—A Sharp-Shooter on Picket Duty* (1862).[14] The morality of the practice of using sharpshooters during the American Civil War was questionable. These soldiers never faced the enemy they engaged and so it was thought that the use of sharpshooters dehumanized the practice of war. We can't see anything of the face of the sharpshooter in the print beyond a shadowed eye trained on his target down the barrel of his rifle. Our full attention is drawn to his hand on the trigger (which hides his face). The sharpshooter is, thereby, depicted as a dehumanized mechanism of destruction. Likewise, the raw stroke of yellow oil crayon John Singer Sargent used to depict the fishing rod in *Val d'Aosta: A Man Fishing* (1907)[15] and the red flowers of the poppies depicted in Monet's *Wild Poppies Near Argenteuil* (1873)[16] are perceptually salient features. However, the artistic salience of these works does not lie in either the coarse-structured yellowishness of the fishing rod or the bright, rough redness of the poppies. Rather, these features function to draw our attention to more general structural features of the works, the dynamic interplay of light on the surface of the pond beside the fishermen in the Sargent and the generically greenish-brown field of grass waving in the wind that dominates the lower half of the Monet. In fact, the whole composition of the Sargent painting—the orientation and posture of the prone figures lying beside the pond, the gaze of the fisherman, and the orientation of the fishing rod—drives the consumer's attention to the coarse, expressive brushwork constitutive of the dynamic depiction of the water. These structural features are diagnostic cues to the identities of the two works as exemplars of an Impressionist style of painting. These cues, in turn, let us know how to engage the work -- in this case to pay attention to the way the brushstrokes are used to depict the dynamic interplay of light, form, and movement in the visual field.

What is the mechanism in all of this? Biased competition models of selective attention identify a broad network of fronto-parietal and cortico-fugal circuits that facilitate the cross-modal integration of sensorimotor and unimodal sensory information in perception. This network includes projections from prefrontal areas involved in spatial working memory, for example, dorsolateral prefrontal cortex (dlPFC) to the frontal eye fields (FEF) which are associated with endogenous shifts of attention, and projections from both FEF and dlPFC to superior colliculus (SC), pulvinar, and the lateral geniculate nucleus (LGN) in the thalamus (Kastner 2004). These subcortical areas are hypothesized to participate in processes critical for neural synchronization (LGN and pulvinar), directing eye movements (SC), and the cross-modal integration of visual, auditory, and somatosensory information (SC). LGN is the primary relay station between the retina and the visual cortex. Therefore, attentional priming reaches back to the earliest stages of visual processing (Beck and Kastner 2009; Stein et al. 2004; Pessoa and Adolphs 2010). These feedback projections from prefrontal and parietal areas that are associated with object recognition, spatial working memory, and attention to topologically encoded cortical and subcortical perceptual areas facilitate the integration of sensory information into coherent representations of task-salient aspects of the distal environment, and they account for the influence of categorization processing in perception. Feed-forward projections from perceptual processing areas to these same prefrontal and parietal areas, in turn, function to reinforce perceptual representations and instantiate working memory in a distributed multi-modal network.

The trick is that our engagement with paintings depends on a capacity to reconstruct their depictive and representational content from diagnostic information encoded in the abstract sets of marks constitutive of their two-dimensional formal-compositional structure—to recognize objects and scenes in a painting and comprehend what it would have meant for an artist to render them in the particular way they have been painted (e.g. the coarse brushstrokes used to depict the pattern of reflected light that gives the water its dynamic appearance in Sargent's *Val d'Aosta: A Man Fishing*). This requires matching abstract diagnostic cues both to category knowledge in object recognition and knowledge of the stylistic conventions regulating artistic production in a category of art—or, recognizing an artwork both as a visual representation of what it depicts and as a member of a category of art. The suggestion is that cortico-fugal attentional circuits (in conjunction with a range of further recurrent connections, including projections from areas TE and TEO to visual areas V4 and MT associated with the perception of color, abstract patterns, motion and depth) play a critical role in this capacity.

Hierarchical theories within empirical aesthetics can be used to articulate the range of processes that might influence cognitive and appreciative judgments about the artistic salience of the content of a work. These theories model artistic production and our engagement with artworks as a multi-stage process that includes sensory processes, perceptual analysis, explicit classification judgments, decision procedures, and affective responses that underwrite our capacity to recognize the formal-compositional and depictive content of a work, understand its semantic-representational content, and appraise its artistic salience (Chatterjee 2003, 2012; Kozbelt and Seeley 2007; Leder et al. 2004; Nadal et al. 2008). For instance, Leder and his colleagues differentiate between two distinct kinds of aesthetic responses to artworks, aesthetic judgments and aesthetic emotions. Aesthetic judgments are a product of what they call cognitive mastering, or the explicit classification and subsequent interpretation of the content of a work relative to expert knowledge of stylistic conventions that define artistic schools, movement, eras and the works of individual artists, for example, Rothko and Motherwell's Abstract Expressionists works, or differences between Sargent's portraits and his Impressionistic landscapes. Explicit classification is the key in these processes. It specifies the set of artistic conventions that constrain our cognitive engagement with a work. The output of perceptual analyses can then be meshed with salient expert art critical knowledge to yield an interpretation of the meaning of the work in cognitive mastering. A reduction of ambiguity, or uncertainty (see Berlyne 1971), via cognitive mastering yields an aesthetic evaluation of the success of a work. Success at the explicit classification, cognitive, and evaluative stages are experienced as changes in affective state that yield a state of pleasure or satisfaction Leder defines as an aesthetic emotion.

Leder argues that content responsiveness will lie on a continuum from naïve to expert consumers (see also Cela-Conde et al. 2002). Naïve consumers who lack expert art-critical or art-historical knowledge will respond to the depictive content of works, evaluating them relative to their success as either representational works or abstract constructions on this model.[17] Expert consumers, on the other hand, will exhibit sensitivity to the artistically salient formal, expressive, and semantic features of the work and the way these are used to generate its art-critically salient content. This is precisely what our model would predict. Loosely, explicit classification and cognitive mastering are processes that involve the implementation of judgments about the relative diagnosticity of features of the artwork for its content. These processes in turn, bias processing in fronto-parietal and cortico-fugal attentional circuits, driving our perceptual, affective, and cognitive interaction with the work (see also Chatterjee 2003, 2012).

2.4 Artistic salience, artistic appreciation, affective and embodied responses to artworks

We can now return to the questions about artistic salience and appreciation that have driven philosophical skepticism about empirical aesthetics. The first thing to note is that hierarchical theories have resources ready to hand to tackle these issues. These theories model our engagement with artworks as a multi-stage process that employs expert art critical knowledge to integrate perceptual, cognitive, and affective aspects of our engagement with artworks. There are other theories within empirical aesthetics that model our engagement with artworks as a cognitive process, for example, prototype preference (Martindale and Moore 1988) and processing fluency theories (Reber 2012). The virtue of hierarchical theories over these other approaches is that nested sets of hierarchically related recurrent circuits enable them to model explicitly the way the formal-compositional features of a work achieve artistic salience—the way they carry the information about salient artistic conventions constitutive of the formal aesthetic, expressive, and semantic content of a work. Nonetheless, and this is the real rub, all of these theories model the influence of cognitive variables in our engagement with artworks as a part of a process of aesthetic appraisal—they, like aesthetic theories of art more generally, treat artworks as aesthetic engines. For instance, Chatterjee (2003, 2012) argues that *aesthetically* salient perceptual, expressive, and semantic features of artworks drive top-down processing and attentional circuits in our engagement with artworks. This, as discussed earlier, narrows the potential explanatory scope of these models and opens the door to the problems of artistic salience and appreciation. However, if we drop the aesthetic desiderata and treat artworks as attention engines instead, these problems fall away as spurious worries.

The problem of artistic salience is a question of what differentiates artworks from non-art artifacts. This is a philosophically important question. A minimum desiderata for a theory of art is that it explain how we sort the range of artifacts that fall under that category, how we discriminate artworks from non-artworks and sort them into salient categories of art. Theories in neuroaesthetics, and empirical aesthetics more generally, treat artworks as cognitive stimuli like any other and model our engagement with art as a product of everyday cognitive processes. The challenge for any model like this that treats artworks as ordinary cognitive stimuli is to explain then how we manage to sort the artworks from the other everyday perceptual stimuli. One common strategy in neuroaesthetics is to simply dodge this question and study the way a set of standard, successful, formal aesthetic strategies work, for example, discussion of irradiation and McKay effects in Monet's paintings (Latto 1995; Livingstone 2002). In these contexts researchers define artistic salience relative to aesthetic salience, take the aesthetic salience of a well-accepted formal strategy as a given in a particular context, and then explain associated aesthetic effects as products of our perceptual engagement with the work. However, as discussed, the artistic salience of the aesthetic features of a work of art is not a perceptual given, but rather depends on the category of art to which we assign the work, to the way we differentiate it from other non-art artifacts and sort it from its other artistic cousins.

The diagnostic recognition framework suggested by biased competition theories of selective attention models our engagement with artworks differently. In ordinary perceptual contexts minimal sets of diagnostic features suffice to enable an organism to recognize the identities and affordances of objects and events perceptually (Kosslyn and Koenig 1995; Schyns 1998). How might this work? A fast first sweep through unimodal perceptual systems might suffice to match sparse sets of coarse-coded sensory stimuli to declarative knowledge of the structure and function of categories of objects or events stored in semantic or episodic memory. This knowledge can, in turn, be used to direct attention and prime the firing rates of populations of neurons across the full range of perceptual modalities to other environmental features diagnostic for the task at hand. These processes enhance the encoding of task-salient perceptual information, inhibit the encoding of potential distracters, and thereby facilitate the smooth and efficient selection and perception of behaviorally salient information. Categorical knowledge of the structure and function of object and event types therefore serves as a strong determinative constraint that shapes sensory processing, influencing not only what we see, but how we see it (Pessoa et al. 2002; Schyns 1998).

Artworks can be sorted into unique categories of objects and events defined by media (sculpture vs painting), productive practices (dance vs performance), content (e.g. portraits vs landscapes; formalist vs conceptual), and formal-compositional stylistic features (e.g. Thomas Cole's Hudson River School paintings, Diego Rivera's Social Realism, Thomas Hart Benton's Regionalism, or Robert Bechtle's Hyperrealism). These categories of art, in turn, articulate sets of artistic conventions that constrain our engagement with artworks. The identities of individual artworks may cut across categories of art—or, categories of art may not sort into functionally discrete sets, but rather form an interlocking web, and integrated network that defines an artworld at a time. For instance, Bill T. Jones borrows elements of performance to articulate the narrative content of his choreographed works (e.g. his work about the life of Abraham Lincoln, *Fondly Do We Hope . . . Fervently Do We Pray*)[18] and Judy Pfaff borrows from Cubism and the painterly drawing style of Jackson Pollock in her sculptural installations (e.g. *NYC, BQE*, 1987,[19] or *Deep Water*, 1980).[20] Recognizing the productive role these shared categorical conventions play in the content of the work is critical to understanding them. A capacity to categorize the formal-compositional content of the work in this way guides attention, facilitates the recovery of diagnostic features, and, via the same attentional processes operative in ordinary perceptual categorization tasks, determines the artistic salience of what we perceive and experience in our engagement with artworks. Therefore, treating artworks as attentional engines and acknowledging the productive role of categorization processing in attention and perception transparently resolves the problem of artistic salience for empirical studies of the arts.

A solution to the problem of artistic appreciation falls out for free on this view of how we recover the artistic salience of the features and content of an artwork. Recall that aesthetic appraisals and other appreciative judgments about artworks are normative judgments about whether a work was done well or poorly, about whether it is suited to a context. Normative judgments of this sort involve comparisons between the formal, expressive, and semantic content of the work and conventions that govern salient artistic practices, for example, conventions constitutive of artistic practices within the category of art to which a work belongs. Psychological explanations are causal explanations, or explanations of the causal-psychological interactions between a work and a consumer. The initial skeptical claim is that causal interactions between a consumer and an artwork are immune to the influences of these conventions. This claim is taken to entail that psychological explanations of our interactions with artworks are not sensitive to the right kinds of variables to stand as explanations of the success or failure of appreciative judgments—they can't alone discriminate between normative judgments that are appropriate and those that are not. This simply isn't true of the model articulated here. From the perspective of a diagnostic recognition framework, our perceptual engagement with artworks is driven by attentional processes that are themselves governed by knowledge of the range of conventions that define different categories of art. These conventions are, as philosophers say, constitutive of way we experience the work. They are productive constraints that determine what we perceive in a work and how we perceive it. Where the work doesn't match the range of conventions appropriate to a category of art, our perceptual expectations will be unsatisfied, our attentional strategies will misfire and remain unrealized. Our experience of the work will consequently be degenerate. It will fail to meet the cognitive goals that drive it and we will appraise it as ineffective.

The model I have proposed here, like hierarchical models generally, has the resources to resolve the problem of artistic appreciation because the salient artistic conventions are built into the perceptual processes that enable a consumer to recover the content of a work. Two caveats are in order here. First, a consumer's expert art-critical knowledge can be incomplete, incorrect, or non-existent. In these kinds of cases we would expect that the appraisals they generate for artworks in those categories would not adequately track their artistic salience. But this is not a problem for the model. The fit between a work and categorially relevant artistic conventions is the marker of artistic success on this account, not the physiological responses or preference judgments of consumers. We, therefore, have the resources ready to evaluate inappropriate appraisals. Nonetheless, the skeptic may yet find this approach wanting. Everything turns on how one interprets the claim that questions about art are questions about the conventions governing appreciative judgments about artworks. The kinds of questions answered here are *how questions* about the role artistic conventions play in our engagement with artworks. They are not *why questions* about the reasons we find value these conventions in the first place. The short answer to the second kind of question is, ". . . because they define the objects, events, and practices constitutive of our categories of art." But this answer won't be particularly satisfying to anyone who takes why questions seriously. My intuition is that we can appeal to dopaminergic reward systems to model the emergence of artistic value in our engagement with artworks as a how question. It is possible that anthropological and evolutionary accounts of art could give us purchase on related why questions, as might studies in cultural criticism at the humanities end of research on the arts. These are hard questions for future directions and another day, though.

Notes

1 Morris, R. (1968). *Threadwaste* [Installation]. Museum of Modern Art, New York, NY. Retrieved from: <http://www.moma.org/collection/object.php?object_id=81515>.

2 Morris, R. (1968/9). *Steam* [Installation]. Museum of Modern Art, New York, NY. Retrieved from: <http://www.moma.org/collection/artist.php?artist_id=4103>.

3 Rauschenberg, R. (1955-59). *Monogram* [Sculpture]. Moderna Museet, Stockholm. Retrieved from: <http://www.modernamuseet.se/en/Moderna-Museet/About/Research/Robert-Rauschenberg/>.

4 Johns, J. (1961). *O through 9* [Painting]. Tate Modern, London. Retrieved from: <http://www.tate.org.uk/art/artworks/johns-0-through-9-t00454>.

5 Baldessari, J. (1966-68). *Tips for Artist's Who Want to Sell* [Painting]. Marion Goodman Gallery, New York, NY. Retrieved from: <http://www.art21.org/images/john-baldessari/tips-for-artists-who-want-to-sell-1966-68>.

6 LeWitt, S. (1977). *Wall Drawing, 305* [Installation]. Mass MoCA, North Adams, MA. Retrieved from: <http://www.massmoca.org/lewitt/walldrawing.php?id=305>.

7 Pollock, J. (1950). *One: Number 31* [Painting]. Museum of Modern Art. New York, NY. Retrieved from: <http://www.moma.org/collection/object.php?object_id=78386>.

8 An affordance is a relation between an organism and some aspect of their environment that "affords" the opportunity for that organism to perform an action. For instance, chairs are artifacts that afford sitting for *Homo sapiens*. In this context, representational paintings afford the perception of dynamic scenes in depth, figurative sculpture affords the perception of the dynamics of animate figures, and narrative artworks afford empathetic engagement with characters by virtue of replicating critical stimulus features of our ordinary physical/social environment.

9 Johns, J. (1960). *Painted Bronze* [Sculpture]. Collection of the artist. See Varnedoe (1960), 185.

10 Johns, J. (1961). *Painted Bronze (Ballantine Ale)* [Sculpture]. Kunstmuseum Basel, Basel. See Varnedoe (1960), 184.

11 See Seeley (2014) for a more detailed discussion of these points.

12 Picasso, P. (1937). *Guernica* [Painting]. Museo Reina Sofia, Madrid. Retrieved from: <http://www.museoreinasofia.es/en/collection/artwork/guernica>.

13 It is critical in this context to disentangle the notion of content from the related concept of interpretation. Interpretation is a tricky business. I take it that contextual factors influence the process of interpretation and constrain the validity of a particular interpretation on an occasion—consider, for instance, differences in the interpretation of politically charged conceptual artworks as contemporary vehicles for social change as opposed to historical artifacts. The content of these kinds of works does not change across historical contexts. But an interpretation of the work as a live expression of current normative social issues may well be strikingly different than retrospective historical interpretations of the same work as an expression of a political movement embedded within the broader social structure of its time. The same might be said for the relationship between an artwork and an artistic movement—surely the broad lens of history yields a different interpretation of action paintings like Jackson Pollock's than did the polemical lens of mid-century art criticism current to their time. This is not to say that interpretation is a subjective matter. Rather, interpretation must be flexibly sensitive to changes in social and epistemic context in a way that categorical judgments that constrain our understanding of the content of an artwork are not (see Danto 2001). Pollock's work is an expression of aspects of Greenbergian Modernism, facts about the flatness of its composition and the transparent roughness of his brushstrokes reflect Greenberg's attention to the formal aspects of medium and support, whatever it might have meant then or means to us now for someone to have painted that way.

14 Homer, W. (1862). *The Army of the Potomac—A Sharp-Shooter on Picket Duty* [Lithograph]. Metropolitan Museum of Art, New York, NY. Retrieved from: <http://www.metmuseum.org/collection/the-collection-online/search/349139>.

15 Sargent, J.S. (1907). *Val d'Aosta: A Man Fishing* [painting]. Addison Gallery of American Art, Andover, MA. Retrieved from: <http://accessaddison.andover.edu/CUS.18.zoomobject._923?sid=25782&x=638408&x=638409>.

16 Monet, C. (1873). *Wild Poppies Near Argenteuil* [Painting]. Musée D'Orsay, Paris. Retrieved from: <http://www.musee-orsay.fr/en/collections/works-in-focus/painting/commentaire_id/poppy-field-8951.html?cHash=1e5e9216e3>.

17 In the case of abstract works naïve consumers are responding to the depiction of space and dynamic formal relationships.

18 Jones, B.T. (2012). *Fondly Do We Hope . . . Fervently Do We Pray* [Dance]. See *A Good Man*. Documentary video produced by Bob Hercules and Gordon Quinn for the PBS American Masters Series, 2011.

19 Pfaff, J. (1987). *NYC, BQE* [Installation]. See Dandler (2003), 26.

20 Pfaff, J. (1980). *Deepwater* [Installation]. See Dandler (2003), 11.

References

Banes, S. and Carroll, N. (2006). Cunningham, Balanchine, and postmodern dance. *Dance Chronicle* **29**(1), 49–68.

Beck, D.M. and Kastner, S. (2009). Top-down and bottom-up mechanisms in biasing competition in the human brain. *Vision Research* **49**(10), 1154–65.

Berlyne, D.E. (1971). *Aesthetics and psychobiology*. New York, NY: Appleton Century Crofts.

Berlyne, D.E. (1974). The new experimental aesthetics. In D.E. Berlyne (ed.), *Studies in the new experimental aesthetics*. New York, NY: John Wiley and Sons, pp. 1–26.

Calvo-Merino, B., Jola, C., Glaser, D.E., et al. (2008). Toward a sensorimotor aesthetics of performing art. *Consciousness and Cognition* **17**, 911–22.

Carroll, N., Moore, M., and Seeley, W.P. (2012). The philosophy of art and aesthetics, psychology, and neuroscience: studies in literature, music, and visual arts. In A.P. Shimamura and S.E. Palmer (eds), *Aesthetic science: connecting minds, brains, and experience*New York, NY: Oxford University Press, pp. 31–62.

Carroll, N. and Seeley, W.P. (2013a). Cognitivism, psychology and neuroscience: movies as attentional engines. In A. Shimamura (ed.), *Psychocinematics: exploring cognition at the movies*. New York, NY: Oxford University Press, pp. 53–75.

Carroll, N. and Seeley, W.P. (2013b). Kinesthetic understanding and appreciation. *Journal of Aesthetics and Art Criticism* **71**(2), 177–86.

Cela-Conde, C.L., Marty, G., Munar, E., et al. (2002). The "style scheme" ground perception of paintings. *Perceptual and Motor Skills* **95**(1), 91–100.

Chatterjee, A. (2003). Prospects for a cognitive neuroscience of visual aesthetics. *Bulletin of Psychology of the Arts* **4**(2), 55–60.

Chatterjee, A. (2012). Neuroaesthetics: growing pains of a new discipline. In A.P. Shimamura and S.E. Palmer (eds), *Aesthetic science: connecting minds, brains, and experience*. New York, NY: Oxford University Press, pp. 299–317.

Dandler, A. (2003). *Judy Pfaff*. Easthampton, MA: Hudson Hills Press.

Danto, A.C. (2000). Art and meaning. In N. Carroll (ed.), *Theories of art today*. Madison, WI: University of Wisconsin Press, pp. 130–40.

Danto, A.C. (2001). The work of art and the historical future. Reprinted in A.C. Danto, *The Madonna of the Future*. Berkeley, CA: University of California Press, pp. 416–31.

Danto, A.C. (2003). Seeing and showing. *Journal of Aesthetics and Art Criticism* **59**(1), 1–9.

Desimone, R. and Duncan, J. (1995). Neural mechanisms of selective visual attention. *Annual Review of Neuroscience* **18**, 193–222.

Dickie, G. (1962). Is psychology relevant to aesthetics? *The Philosophical Review* **71**(3), 285–302.

Fechner, G. (1876). *Vorschule der aesthetik*. Leipzig: Breitkopf and Hartel.

Fischer, M.H. and Zwaan, R.A. (2008). Embodied language: a review of the role of the motor system in language comprehension. *Quarterly Journal of Experimental Psychology* **61**(6), 825–50.

Gombrich, E.H. (1960). *Art and illusion*. Princeton, NJ: Princeton University Press.

Greenberg, C. (1960). Modernist Painting. In J. O'Brien (ed.), *Clement Greenberg: The Collected Essays and Criticism, Modernism with a Vengeance, 1957–1969*. Chicago, IL: University of Chicago Press, pp. 85–93.

Hayhoe, M. and Ballard, D. (2005). Eye movements in natural behavior. *Trends in Cognitive Sciences* **9**(4), 188–93.

Irvin, S. (2013). Installation art and performance: a shared ontology. In C. Mag Uidhir (ed.), *Art and abstract objects*. New York, NY: Oxford University Press, pp. 242–262.

Jola, C., Abedian-Amiri, A., Kuppuswammy, A., et al. (2012). Motor simulation without motor expertise: enhanced corticospinal excitability in visually experienced dance spectators. *PLoS One* **7**(3), 1–11.

Kastner, S. (2004). Attentional response modulation in the human visual system. In M.I. Posner (ed.), *The Cognitive Neuroscience of Attention*. New York, NY: The Guilford Press, pp. 144–56.

Kosslyn, S.M. and Koenig, O. (1995). *Wet Mind: The New Cognitive Neuroscience*. New York, NY: Free Press.

Kozbelt, A. and Seeley, W.P. (2007). Integrating art historical, psychological, and neuroscientific explanations of artists' advantages in drawing and perception. *Psychology of Aesthetics, Creativity, and the Arts* **1**(2), 80–90.

Latto, R. (1995). The brain of the beholder. In R. Gregory, J. Harris, P. Heard, et al. (eds), *The artful eye*. New York, NY: Oxford University Press, pp. 66–94.

Livingstone, M. (2002). *Vision and art: the biology of seeing*. New York, NY: Harry N. Abrams.

Leder, H., Belke, B., Oeberst, A., et al. (2004). A model of aesthetic appreciation and aesthetic judgments? *British Journal of Psychology* **95**, 489–508.

Levinson, J. (1992/1996). Intention and interpretation in literature. Reprinted in J. Levinson, *The pleasure of aesthetics*. Ithaca, NY: Cornell University Press, pp. 175–213.

Livingstone, M.S. (2002). *Vision and art: the biology of seeing*. New York: Henry M. Abrahms.

Lopes, D.M. (1995). Pictorial realism. *Journal of Aesthetics and Art Criticism* **53**(3), 277–85.

Martindale, C. (1990). *The clockwork muse*. New York, NY: Basic Books.

Martindale, C. and Moore, K. (1988). Priming, prototypicality and preference. *Journal of Experimental Psychology: Human perception and Performance* **14**(4), 661–70.

Moore, G.E. (1903/2000). *Principia ethica*. Cambridge: Cambridge University Press.

Nadal, M., Munar, E., Capo, M.A., et al. (2008). Towards framework for the study of the neural correlates of aesthetic preference. *Spatial Vision* **21**(3–5), 379–96.

Nanay, B. (2012). The multimodal experience of art. *British Journal of Aesthetics* **52**(4), 353–63.

Pessoa, L. and Adolphs, R. (2010). Emotion processing and the amygdala: from a "low road" to "many roads" of evaluating biological significance. *Nature Reviews Neuroscience* **11**, 773–82.

Pessoa, L., Kastner, S., and Ungerleider, L.G. (2002). Attentional control of the processing of neutral and emotional stimuli. *Cognitive Brain Research* **15**(1), 31–45.

Reber, R. (2012). Processing fluency, aesthetic pleasure, and culturally shared taste. In A.P. Shimamura and S.E. Palmer (eds), *Aesthetic science: connecting minds, brains, and experience*. New York, NY: Oxford University Press, pp. 223–49.

Rollins, M. (2004). What Monet meant: intention and attention in understanding art. *The Journal of Aesthetics and Art Criticism* **62**(2), 175–88.

Rosenberg, H. (1952). American Action Painters. *ArtNEWS* **51**(8), 22–3; 48–50.

Schimmel, P. (2005). *Robert Rauschenberg*: Combines. Los Angeles: The Museum of Contemporary Art, Los Angeles / Stedl.

Schyns, P.G. (1998). Diagnostic recognition: task constraints, object information, and their interactions. *Cognition* **67**(2), 147–79.

Seeley, W.P. (2010). Imagining crawling home: a case study in cognitive science and aesthetics. *Review of Philosophy and Psychology* **1**(2), 407–26.

Seeley, W.P. (2011a). What is the cognitive neuroscience of art . . . and why should we care? *The American Society for Aesthetics Newsletter* **31**(2), 1–4.

Seeley, W.P. (2011b). The Science of Art Is as Relevant to the Philosophy of Art as Artistic Representations Are to Science: A Reply to Roger Seamon. *American Society of Aesthetics Newsletter* **31**(3), Winter: 6–7.

Seeley, W.P. (2013). Movement, gesture, and meaning: a sensorimotor model for audience engagement with dance. In H. de Preester (ed.), *Moving Imagination: Explorations of Gesture and Inner Movement in the Arts*. Amsterdam: John Benjamins Publishing Company, pp. 51–68.

Seeley, W.P. (2014). Philosophy of art and empirical aesthetics: resistance and rapprochement. In P.P.L. Tinio and J.K. Smith (eds), *Cambridge Handbook of the Psychology of Aesthetics and the Arts*. New York, NY: Cambridge University Press, pp. 35–59.

Silvia, P.J. (2012). Human emotions and aesthetic experience: an overview of empirical aesthetics. In A.P. Shimamura and S.E. Palmer (eds), *Aesthetic science: connecting minds, brains, and experience*. New York, NY: Oxford University Press, pp. 250–75.

Speer, N.K., Reynolds, J.R., Swallow, K.M., et al. (2009). Reading stories activates neural representations of visual and motor experiences. *Psychological Science* **20**(8), 989–99.

Stein, B.E., Stanford, T.R., Wallace, M.T., et al. (2006). Crossmodal spatial interactions in subcortical and cortical circuits. In C. Spence and J. Driver (eds), *Crossmodal space and crossmodal attention*. New York, NY: Oxford University Press, pp. 25–50.

Varnedoe, K. (1960). *Jasper Johns: A Retrospective*. New York, NY: The Museum of Modern Art / Harry N. Abrams.

Walton, K.L. (1970). Categories of art. *Philosophical Review* **79**, 334–67.

Wittgenstein, L. (1958). *Philosophical investigations*. New York, NY: MacMillan Publishing Company.

Wittgenstein, L. (1966). *Lectures and conversations on aesthetics, psychology, and religious belief*. Berkeley, CA: University of California Press.

Zeki, S. (1999). *Inner vision*. New York, NY: Oxford University Press.

Chapter 3

States, people, and contexts: Three psychological challenges for the neuroscience of aesthetics

Kirill Fayn and Paul J. Silvia

The psychology of aesthetics is one of psychology's oldest fields, dating back to Fechner (1876) and the early experimental psychologists. The early decades saw a small but interesting stream of research (for reviews, see Gordon 1909; Valentine 1913), but psychological aesthetics didn't catch fire until the emergence of Daniel Berlyne in the 1960s. Berlyne and his research group situated aesthetic problems within the science of experimental psychology, which both introduced aesthetics to a mainstream research community and inspired researchers to apply traditional paradigms to aesthetic questions.

Although best remembered for his many contributions to psychological aesthetics, Berlyne viewed his research as essentially neurological (Silvia 2005b, 2006b). Throughout his writings, one sees a tough-minded attempt to anchor aesthetics in the neuroscience of reward. His early model (Berlyne 1960) linked aesthetic preference to changes in arousal, which were generated by the reticular activating system in response to stimulus properties such as complexity, novelty, and conflict. Based on the emerging field of affective neuroscience (Olds 1962, 1973), Berlyne's (1971, 1973, 1974) later model proposed a set of interacting neurological reward and aversion systems that created affect.

Psychological aesthetics and neuroaesthetics thus have a long and shared history. Research since Berlyne's time has been primarily behavioral rather than biological, but like Berlyne, we see these as complementary approaches. The recent rise of neuroaesthetics is an exciting development, one that promises to fulfill Berlyne's vision of a model of aesthetic experience anchored in modern neuroscience.

Our goal in this chapter, rather than review the vast literature on psychological aesthetics, is to isolate a handful of challenges that psychological aesthetics has grappled with since Berlyne's time. By considering what psychological aesthetics has learned, neuroaesthetics can leverage past knowledge, avoid making similar mistakes, and lend insight into some problems not easily understood with behavioral methods. We will focus on the three challenges that we think are most pressing for modern research: the diversity of emotional reactions to the arts; the surprising individual variability in aesthetic responses to the same art objects; and the range of contexts where aesthetic experience can occur.

3.1 States: the diversity of emotional response to the arts

As any teenager will tell you, feelings are complicated. The first major challenge for neuroaesthetics—and an ongoing challenge for psychological aesthetics—is to illuminate the wide range of emotional states that are experienced in response to the arts. Psychological aesthetics has historically concerned itself with ratings of low-intensity positive states such as liking, beauty, and pleasure (Silvia 2009). Due to the dominance of liking in psychological studies, it is no surprise that the neuroscience of aesthetics has similarly focused on measures of mild positive states as the defining "aesthetic response." Theories of emotion, however, propose a wide array of emotions—happiness, pride, interest, confusion, anger, sadness, shame, awe, and disgust, among others (Izard 1977; Scherer 2001)—and all emotions are relevant to something as profoundly human as the arts.

Within neuroaesthetics, it has been suggested that judgments of beauty are the cornerstone of aesthetic evaluations (Jacobsen et al. 2004). Accordingly, many studies have used a single positive–negative dimension to represent aesthetic experience when analysing how neural activation correlates with subjective ratings, such as dichotomous decisions of beauty (Cela-Conde et al. 2004), scales that range from *appealing* to *not appealing* (Kirk et al. 2009), and scales ranging from *ugly* to *beautiful* (Kawabata and Zeki 2004). Beauty and its related positive states are obviously important to aesthetics. Beauty is central to definitions of aesthetic experience (Gordon 1909) and to how people evaluate art objects (Augustin et al. 2012; Jacobsen et al. 2004), but aesthetics research should look beyond the cluster of positive feelings to consider other aesthetic states, which we highlight below.

3.1.1 Some curious aesthetic emotions

Interest and the knowledge emotions

Interest, an emotion associated with learning and exploring (Silvia 2006b, 2008b), was one of the two states emphasized in Berlyne's research. Many studies found that self-report scales, behavioral measures of preference and exploration, and physiological outcomes formed two broad factors: a *pleasingness* factor and an *interestingness* factor (e.g. Berlyne 1974). The interestingness factor was hard to fit into Berlyne's neurological model, which emphasized pleasingness, but it remains one of the central aesthetic responses to the arts. Finding something interesting promotes attention, exploration, and learning (Izard 1977; Tomkins 1962), and interest might be a precursor to more powerful aesthetic emotions (Markovic 2012). Interest comes about when people appraise something as both novel and complex but comprehensible (Silvia 2005a, 2005c).

Recent work has placed interest within a family of *knowledge emotions*, which include surprise, confusion, and awe (Silvia 2010). The knowledge emotions are essentially meta-cognitive—they come about from people's appraisals of their own knowledge and expectations—and they motivate learning, viewed broadly. Several studies show that people commonly express surprise (Ludden and Schifferstein 2007; Ludden et al. 2008, 2009), confusion (Silvia 2010, 2013; Silvia and Berg 2011), and awe (Keltner and Haidt 2003) in response to the arts.

Beauty

Armstrong and Detweiler-Bedell (2008) sought to redefine the aesthetic experience of beauty. In their view, beauty is "the exhilarating feeling that something complex, perhaps to the point of being profound, might yield an understanding" (p. 312), not a mild pleasant state that people experience upon viewing something pretty. They theorize that this powerful aesthetic experience is rooted in prolonged, effortful processing and a goal of understanding something particularly challenging. Understanding is acquired in stages, which increases pleasure as the beholder gets closer to comprehension. They distinguish the beautiful from the merely pretty based on cognitive effort: prettiness is facilitated by fluent processing (Reber 2012), but stimuli that elicit beauty thwart conceptual understanding and yet make understanding seem within reach. Beauty is conceptualized via a tripartite framework of cognition (appraisal), affect (valence and arousal), and motivation. Importantly, Armstrong and Detweiler-Bedell emphasize a motivational *promotion focus*, which occurs when the perceiver is interested in expanding their schema, knowledge, and experience. Although not much research has examined this conception of beauty, it is clearly far from the usual use of the concept of beauty in aesthetic research—an item used to measure mild positive experiences.

Awe

Awe, a powerful emotional experience that is fundamental to the arts, has received surprisingly little mention in psychological aesthetics (Nusbaum and Silvia 2014). Keltner and Haidt (2003) described awe as a highly arousing emotion that is felt in response to vastness and the need for accommodation. Vastness comes from situations and objects that are either physically larger than the self or experientially larger than normal everyday experiences; accommodation is the organism's ability to come to terms with the vastness of the experience. A study analysing experiences of awe found that it was mostly felt in response to information-rich environments of nature, art, and music (Shiota et al. 2007).

Chills

Chills—a feeling of shivers, thrills, or goosebumps, particularly on the scalp, neck, and back—is an intriguing aesthetic response. It can be evoked by anything but seems especially common when listening to music. Curiously, some people report never experiencing chills, and others report experiencing chills daily (Nusbaum and Silvia 2014). Unlike the other states discussed here, much of the work on chills comes from neuroscience, such as studies of chills while listening to music (Blood and Zatorre 2001; Salimpoor and Zatorre 2013) and studies of the autonomic nervous system's activity during the chills response (Benedek and Kaernbach 2011; Grewe et al. 2007).

Negative emotions

Many interesting aesthetic experiences aren't positive at all (Cooper and Silvia 2009; Silvia and Brown 2007). Anger, for example, is a common reaction to art that strikes viewers as transgressive. Much art does deliberately try to offend and provoke the audience, and audiences do become angry—sometimes to the point of defacing and censoring art—when they think someone is intentionally trying to go against their values. Disgust, another negative emotion, is a common response to art that seeks to unease the audience. The experience of disgust can be both literal (e.g. from depictions of revolting things) and metaphorical (e.g. disgust from ideas, behaviors, and values that strike viewers as morally repulsive). Shame and embarrassment—two negative emotions in the family of "self-conscious emotions" (Tracy and Robins 2007)—seem relevant to the arts, but essentially no research has applied them to aesthetic experience.

3.1.2 Cognitive appraisals and emotional differentiation

How can we recognize the diversity of aesthetic emotions without becoming bewildered by it? How can we understand the differences between these many states? One approach to studying diverse emotions that has been particularly successful is the cognitive appraisal theory of emotions (Lazarus 1991; Scherer 2001). Appraisal theories suggest that emotions emerge from cognitive evaluations of the circumstances that people find themselves in, particularly evaluations of how what is occurring relates to their needs, preferences, goals, and abilities (Ellsworth and Scherer 2003). These evaluations, called *appraisals*, are subjective and therefore able to account for the variations in responses to the same situation (Siemer et al. 2007). The appraisal approach has been applied to the study of aesthetic emotions and forms the basis of several theories (Armstrong and Detweiler-Bedell 2008; Silvia 2005b).

Within aesthetics research, the appraisal approach has been most often applied to the emotion of interest. Two appraisals that have consistently predicted interest in aesthetic stimuli are *novelty-complexity* and *coping potential*. The novelty-complexity appraisal facilitates attention to possibly important changes in the environment (Ellsworth and Scherer 2003), and the coping-understanding appraisal estimates our capacity to cope with a given situation (Lazarus 1991). In the context of the arts, the novelty-complexity appraisal reflects how unfamiliar, complicated, and surprising the work strikes someone. This is followed by the coping-understanding appraisal, which reflects the perceiver's beliefs about his or her ability to come to understand the new and complex stimuli. This appraisal structure has been replicated for a wide range of aesthetic objects—poems, film, and visual art—indicating stability across different types of stimuli (Silvia 2005a, 2005c, 2008a; Silvia and Berg 2011; Silvia et al. 2009).

The value of the appraisal approach in differentiating emotions in response to the arts has been applied to differentiate knowledge and negative emotions. Confusion has been shown to be a result of a high novelty-complexity appraisal and a low coping-understanding appraisal (Silvia 2010). Surprise has been shown to rely primarily on the appraisal of novelty-complexity (Scherer 2001). Anger and disgust have both been shown to rely on the appraisal of value congruence (seeing something as consistent or contrary to one's values), but they are differentiated by the intention appraisal for anger (seeing something as deliberately caused by someone), and the pleasantness appraisal (seeing something as inherently aversive) for disgust (Silvia and Brown 2007).

Another important finding using appraisals is the dissociation between interest and pleasure (Turner and Silvia 2006). When people viewed a set of pleasant and disturbing paintings, ratings of pleasantness and interest were unrelated. Pleasantness and interest may overlap in many—and perhaps most—situations, but pleasantness isn't necessary for something to be interesting. A person may thus find an artwork to be displeasing but still be aesthetically engaged with it. The distinction between pleasantness and the other emotions we have discussed is less subtle but just as important. Things needn't be pleasant or beautiful for most emotions to occur, so judgments of beauty alone aren't enough to capture and differentiate aesthetic experiences.

3.1.3 What does this mean for neuroaesthetics?

The key message—one that applies equally to the psychology and neuroscience of aesthetics—is that aesthetic experiences are much broader than judgments of beauty and preference. People seek and experience powerful and diverse feelings. Such states can be both positive and negative, and importantly, they can be independent of preference and beauty. Yet they are united by making the perceiver feel moved, creating a special subject–object relationship that may be the quintessential aesthetic experience.

Evidence for the benefits of a more expansive approach to aesthetic emotions comes from a recent study in neuroaesthetics, in which a novel MRI assessment was used to find step-like increases in activity in frontal regions of the brain, including the default mode network (Vessel et al. 2012). Participants were asked to rate a series of images based on how much it moved them, and were told to focus on the powerful, pleasing, and profound. The step-like activation in the default mode network for artworks that yielded only the highest rating is of particular interest because this region is associated with self-referential mentation, particularly memory retrieval and envisioning the future (Buckner et al. 2008). This finding might indicate the special subject–object relationship where art elicits introspection and contemplation. Further, it was found that post-scanning ratings of awe were related to unique activation, differentiating awe from other types of emotional experiences (Vessel et al. 2012). This study is a good model for neuroaesthetics research that attempts to look at more powerful, personal, and enduring aesthetic emotions.

The appraisal approach has proved useful in distinguishing aesthetic states and could easily be implemented in neuroaesthetics research. Measuring key appraisals along with emotion ratings could identify activation associated with these evaluative judgments and clarify some of the inconsistencies in findings based on preference ratings. It is important to note that some neuroaesthetic studies have chosen stimuli that were above the mean in complexity (Cela-Conde et al. 2004, 2009). These ratings, however, were done by a separate sample, and as we have seen in our research, complexity and other appraisals are in the eye of the beholder, and therefore, should be measured individually whenever possible.

3.2 People: the challenge of individual differences in aesthetic experiences

Individual differences in emotional responses are the rule, not the exception; few things evoke the same feelings in everyone (Siemer et al. 2007). For the arts, then, we would expect massive variability in how people respond to the same things, and we do see this in everyday life. The challenge, then, is to understand the basis behind these differences, to identify what explains why people experience art so differently. The theoretical and practical value of combining experimental and differential methods has been stressed for a long time (Cronbach 1957) and recently been reiterated, particularly for the neurosciences (Kanai and Rees 2011). The idea of moving away from the general in lieu of the particular has also been raised as a challenge to neuroaesthetics (Nadal and Pearce 2011). Given the small sample sizes that are inherent in neuroaesthetic research, researchers should use the literature on individual differences to select, study, and compare thoughtfully chosen participants.

Many traits predict variability in aesthetic experience, but two—openness to experience and expertise in the arts—are by far the most important. Different standings on these variables predict different intensities and kinds of aesthetic experience. If studying powerful aesthetic emotions is a goal of neuroaesthetics, then recruiting participants who are more likely to experience such emotions becomes important. Additionally, an individual differences approach to neuroaesthetics can also give insight into these constructs by illuminating the different ways aesthetics is processed neurologically by different people.

3.2.1 Openness to experience

One of the dimensions in the Five Factor Model of personality (McCrae and Costa 1999), openness to experience is fundamental to aesthetic appreciation and creativity (Feist 1998; Silvia 2012). Openness includes the subtraits of fantasy (having a vivid imagination), aesthetic sensitivity (appreciating art, nature, and beauty), feelings (being sensitive to the subtleties of inner emotional states), actions (being open to trying new things and visiting new places), ideas (having high intellectual curiosity), and values (being willing to re-examine values).

Openness is the most heterogeneous of the five factors, and disagreements about its nature might be the most contentious (DeYoung et al. 2012; McCrae 1994; McCrae and Sutin 2009). One recent model of openness to experience—the O/I model—divides it into two facets: *openness* and *intellect* (DeYoung et al. 2007). The openness facet is theoretically related to aesthetic sensitivity, creativity, and imaginativeness. The intellect facet, in contrast, is linked to fluid intelligence, vocabulary knowledge, and an intellectual approach to life. Several studies have found evidence for this carving of openness to experience. For example, openness and intellect diverge in how they predict performance on creativity and intelligence tasks (Nusbaum and Silvia 2011a), implicit and explicit learning (Kaufman et al. 2010), and major creative achievements (Kaufman 2013).

Openness to experience has consistently predicted aesthetic appreciation and engagement in the arts. Openness was found to predict engagement in aesthetic activities such as reading, painting, visiting art galleries, and valuing the arts (McManus and Furnham 2006). Additionally, it has been found to predict higher liking and interest in several different forms of visual art, with particularly strong associations with abstract forms of art (Fayn and Tiliopoulos 2012; Feist and Brady 2004; Rawlings 2003). Finally, in a large-scale population study ($n = 91\,692$) openness was found to be the strongest predictor of art appreciation and artistic activities (Chamorro-Premuzic et al. 2009). These findings indicate that people higher in openness enjoy their experiences with art more, have more positive attitudes towards the arts, and engage in behaviors that expose them to the arts.

Apart from liking and enjoying art more, people high in openness are more likely to have powerful and unusual emotional experiences in response to the arts. One of the items on the openness scale of the NEO Personality Inventory asks participants about their tendencies to experience chills in response to poetry and the arts. A cross-cultural study by McCrae (2007) found that in 50 out of 51 cultures, the item asking about chills had one of the highest item-total correlations with the total openness to experience scale; for the whole sample that combined all 51 cultures, the chills item had the highest item-total correlation of all the items. McCrae concluded that the experience of aesthetic chills was a universal and central marker of the openness trait.

In our own work, we have found that openness to experience is the most common predictor of how often people reported experiencing chills from listening to music (Nusbaum and Silvia 2011b). People high in openness were more deeply engaged with music in many ways (e.g. going to concerts, playing an instrument), and this deeper engagement mediated the effect of openness on chills. In other work, we split openness to experience into the O and I facets, and when people were asked to report how often they got chills in responses to the arts, openness but not intellect emerged as an important predictor (Silvia and Nusbaum 2011).

Moving beyond chills, openness to experience predicts other strong aesthetic experiences. For example, there is a strong association between openness and self-reported experiences of awe in everyday life (Shiota et al. 2006). When people rated how often they had a range of aesthetic experiences in response to whatever domain of the arts they encountered most often, the openness O/I facet strongly predicted how often people reported experiences such as wanting to cry and feeling moved, touched, and absorbed (Silvia and Nusbaum 2011). Finally, those scoring higher on the openness O/I facet experience a stronger relationship between interest and aesthetic fascination, indicating greater emotional reactivity to interesting art (Fayn and Tiliopoulos 2013).

3.2.2 Knowledge, expertise, and training

Expertise in the arts is the second big individual-difference variable in aesthetic experience. Some people know an incredible amount about the arts; others have a more casual knowledge of art; and many people know little. Expertise—knowledge in a domain of the arts gained through training and formal study—moderates virtually everything, and it is currently a hot topic in psychological aesthetics. Compared to novices, experts in the arts evaluate their subject using different dimensions (Hekkert and van Wieringen 1996), think about art using concepts that are more formal and abstract (Parsons 1987), and view visual images differently, according to eye-tracking studies (Locher 2006).

Most relevant to our purposes, however, are emotional differences: experts and novices have different aesthetic responses to the arts. Expert–novice differences in emotional reactions to the arts are robust effects that are easy to find (e.g. Leder et al. 2012; Locher et al. 2001; Millis 2001). One common finding is that experts, relative to novices, enjoy work that is more complex, abstract, and non-representational (e.g. Hekkert and van Wieringen 1996). Experts are also less likely to find complex works confusing (Silvia 2013; Silvia and Berg 2011) and are more likely to find them interesting (Axelsson 2007; Silvia 2006a).

Expertise is measured in a handful of straightforward ways. One approach is to classify people as experts or novices based on tangible achievements and attainments, such as working in the arts, having a college major or degree in the arts, or having taken classes in the arts (e.g. Haanstra et al. 2013). Some research combines attainments with interest in the arts: people must have both extensive attainments and indicate an elevated interest in the arts to qualify (e.g. Locher et al. 2001). Another approach measures people's knowledge about the arts. Smith and Smith's (2006) aesthetic fluency scale, for example, presents people with a list of figures and topics in art history. For each one, people are asked to rate how much they know about it using a scale that ranges from 0 (*I have never heard of this artist or term*) to 4 (*I can talk intelligently about this artist or idea in art*). The scale is based on the simple notion that experts know more than novices, so asking for simple knowledge judgments will catch most of the variance associated with expertise. This scale has been used in many studies and works quite well (DeWall et al. 2011; Silvia 2013; Silvia and Barona 2009; Silvia and Nusbaum 2011). The basic method lends itself well to modified versions, such as extensions to the domains of film, writing, and the decorative arts (Silvia 2007; Silvia and Berg 2011).

It should come as no surprise that expertise and openness to experience overlap. Experts in the arts are much more open to experience than novices, consistent with the role of openness in aesthetic sensitivity and creativity. In one study of college students, scores from the aesthetic fluency scale strongly predicted openness to experience ($\beta = 0.55$; Silvia 2007). Openness (a personality trait) and expertise (acquired crystallized knowledge) are clearly different, but being high in openness to experience provides the interest and motivation needed to develop expertise in the arts.

3.2.3 What does this mean for neuroaesthetics?

Individual differences influence the intensity and quality of emotional responses to the arts. Openness and expertise are particularly involved in more powerful and transcendent states that define aesthetic experience. The mechanisms for these differences are not totally clear, but early evidence suggests that people high in openness and art expertise are more likely to approach art with the view of expanding their lives, are less likely to experience confusion in the face of complexity, and are more likely to have powerful emotions in response to the arts.

One clear implication for neuroaesthetics concerns sampling. Neuroscience research has much smaller samples than behavioral research, and individual differences can wreak havoc in a small sample. If nothing else, it is important to measure personality and expertise to describe what the sample is like. More fruitful, however, would be to select participants based on individual differences. One approach is to select the people for whom intense aesthetic reactions are common, such as people high in openness and expertise. Another is to select extreme groups based on these variables, such as picking people low and high in openness. By expanding the variance between people, this design increases statistical power, which is always handy for small-sample methods. Regardless of the specific implications, paying more attention to individual differences can help with the problem of de-contextualizing participants and the differences that exist between them (Nadal and Pearce 2011).

3.3 Contexts: the diversity of aesthetic objects and environments

One of Berlyne's legacies is an emphasis on controlled laboratory experiments with carefully selected stimuli. The contexts of aesthetic experience—the things people encounter and the places in which they encounter them—are thus largely ignored. We know a great deal about aesthetic experience as it happens in an unfamiliar room in response to unfamiliar images that someone else picked; we know much less about what aesthetic experience involves in people's natural environments. We thus see the problem of context as having two sides. The first is the diversity of aesthetic objects: things that are easy to study experimentally—such as visual images—have been widely studied, whereas things that are complicated and inconvenient—such as narrative text, film, and theater—have not. The second is the diversity of aesthetic contexts: people experience aesthetic objects essentially everywhere—at home, in museums, in their cars, on smartphones, and anywhere people come into contact with ideas and objects with an aesthetic purpose. Few studies, however, have taken arts research out of the lab and into the real world.

3.3.1 Varieties of aesthetic objects

Some kinds of art are more convenient to study. Visual art and music—perhaps the two most widely studied domains—lend themselves well to lab paradigms. They can be presented using inexpensive equipment, researchers don't need special facilities or personnel, aspects of the presentation can be precisely determined, and it doesn't take much time in an experiment to show pictures or play music. For neuroimaging research, visual art and music are the most scanner-friendly, given current technologies. It is easy to show pictures or play music—presenting a lengthy text, short movie, or a stage play present more or less obvious challenges.

Our sense is that psychological aesthetics has taken a more expansive approach in recent years, which has seen more interest in a wide range of art domains. In narrative text, recent years have seen many new papers along with a new journal, *Scientific Studies of Literature*. Likewise, many interesting papers have appeared recently on literature, theater, music, and film in *Empirical Studies of the Arts* and *Psychology of Aesthetics, Creativity, and the Arts*, the field's two leading journals. Being researchers who focus on lab-based studies of visual art and music, we can appreciate the effort involved in extending the field to new areas.

Some researchers are moving outside of the traditional fine arts entirely. One notable program of research examines the aesthetic experience of designed products in both the lab and in natural environments (Demir et al. 2009; Ludden et al. 2008, 2009). In a similar vein, Whitfield has argued for the study of *mundane aesthetics* (Whitfield and de Destefani 2011). He has used designed objects as contexts for evaluating competing models of aesthetic experience (Whitfield 1983; Whitfield and Slatter 1979), and recent work has focused on how people make common aesthetic decisions, such as choosing paint colors for their interior walls (Whitfield and de Destefani 2011). An historic line of work by Berlyne and his collaborators examined aesthetic responses to architecture and the built environment (Oostendorp and Berlyne 1978a, 1978b), and an emerging line of research explores the aesthetic experience of images of space (L.F. Smith 2014).

3.3.2 Varieties of aesthetic locations

People encounter aesthetic objects in so many places: they listen to music at concerts, view art in museums, experience architecture as they amble through the built environment, experience nature as they wander through the natural environment, and interact with designed products—from clothes to watches to cars—everywhere. The diversity of aesthetic environments will be harder to analyse: research is much easier to do in a lab, especially when it involves massive magnetic machines.

Nevertheless, researchers have made some intriguing progress in studying aesthetic experience in the world that exists beyond the laboratory doors. An emerging body of work has applied experience sampling methods (ESM) to aesthetic experience. Such methods assess people repeatedly in their natural environments. By randomly sampling people's everyday experience, researchers can understand how often things actually happen, capture experiences as they occur, and study things that can't be studied using static, one-shot assessments, such as the variability and volatility of people's experiences (Conner et al. 2009; Hektner et al. 2007).

ESM research has been most commonly applied to music. Randomly sampling a person's day is a good way to estimate how often people listen to music and what kinds of emotions they typically experience when doing so (e.g. Juslin et al. 2008; Sloboda et al. 2001). Recently, our group has used ESM to study how often people get chills when listening to music and what kinds of contexts promote chills (Nusbaum et al. 2014) as well as why some people only rarely listen to and enjoy music (Nusbaum et al. 2015). In another project (Beaty et al. 2013), we have examined how often people hear "inner music"—imagined music when none is playing—and whether it is typically an aversive experience, as stereotypes about "earworms" and stuck-song syndrome would suggest (Kellaris 2008; Levitin 2006).

In psychological aesthetics, there is a small but excellent body of work that studies the experience of art within museums (e.g. Höge 2003; Mastandrea et al. 2007; J.K. Smith 2014). These studies are both correlational and experimental, and they have examined a range of interesting problems, such as how long people spend looking at art (Smith and Smith 2001), what people tend to look at (Heidenreich and Turano 2011), and how the experience of a reproduction differs from the experience of an original (Locher et al. 2001).

The natural extension, as technology advances, is to add biological measures to assess experience in natural environments. Ambulatory brain imaging won't be practical and widespread for some time, but ambulatory measures of the autonomic nervous system are relatively inexpensive and easy to use. One outstanding example of the use of ambulatory autonomic physiology is a project by Tschacher and colleagues (2012). In this research, participants wore a glove device during a visit to a fine art museum. The device tracked their movements through the museum and provided markers of sympathetic and parasympathetic activity. The researchers could thus link physiological activity to aspects of the art (e.g. a painting's fame or intensity) and to participants' self-reports (Tröndle and Tschacher 2012).

Other options include ambulatory impedance cardiography (Cybulski 2011), which uses electrodes placed on the chest and back to measure the electrical and mechanical activity of the heart (Thayer et al. 2008). From this, it can assess the influence of the sympathetic and parasympathetic branches of the autonomic nervous system by calculating measures of the heart's rate, force of contraction (e.g. the cardiac pre-ejection period, a measure of sympathetic influence; Kelsey 2012), and variability (e.g. heart-rate variability, a measure of parasympathetic influence; Segerstrom et al. 2012). Ambulatory impedance devices are small, comfortable, and discreet, so they hold great potential for measuring biological aspects of aesthetics in everyday public environments.

3.3.3 What does this mean for neuroaesthetics?

The challenge of contexts presents the greatest difficulty for both neurological and behavioral research on aesthetics. For many neuroscience methods, visual imagery and music are the only practical aesthetic domains. Certain methodological requirements, such as precise time-linking of events and measurements and the practicality of what people can perceive while lying in a scanner, mean that many kinds of dynamic domains are out of reach for now. Likewise, ambulatory neuroscience methods are in their infancy, so extending neuroaesthetics into museums and recital halls is far off. At the same time, some exciting options for ambulatory central imaging are looming on the horizon for basic research, and options for ambulatory autonomic nervous system assessment are practical and relatively affordable (Cybulski 2011). For now, then, the challenge of context is something still at the back of the field's mind, something to take up as the field finds new methods, technologies, and insights.

3.4 Conclusion

These are exciting times for the scientific study of the arts. Based on who we see at conferences and what we see published in journals, the field is attracting a diverse pool of young talent and an intriguing range of topics and methods. The emergence of neuroaesthetics is one sign that the science of aesthetics is maturing. In this chapter, we offered some food for thought based on the experience of psychological aesthetics. The past decades of research using behavioral methods has revealed a few core issues—the diversity of aesthetic states, the diversity of individual responses, and the diversity of aesthetic contexts. These are challenging problems that neuroscience approaches to aesthetics should both confront and tackle. In some cases, neuroaesthetics can learn from the mistakes of behavioral work, such as the perils of defining aesthetic experience solely in terms of positive states. In other cases, we expect that neuroaesthetics can solve some problems that are hard to crack with psychological methods, such as the nature of obscure states like chills and awe. In either case, we are curious to see what the future holds for the fledgling field of neuroaesthetics.

References

Armstrong, T. and Detweiler-Bedell, B. (2008). Beauty as an emotion: The exhilarating prospect of mastering a challenging world. *Review of General Psychology* **12**, 305–29.

Augustin, M.D., Wagemans, J., and Carbon, C.C. (2012). All is beautiful? Generality vs. specificity of word usage in visual aesthetics. *Acta Psychologica* **139**, 187–201.

Axelsson, Ö. (2007). Individual differences in preferences for photographs. *Psychology of Aesthetics Creativity, and the Arts* **1**, 61–72.

Beaty, R.E., Burgin, C.J., Nusbaum, E.C., et al. (2013). Music to the inner ears: Exploring individual differences in musical imagery. *Consciousness and Cognition* **22**, 1163–73.

Benedek, M. and Kaernbach, C. (2011). Physiological correlates and emotional specificity of human piloerection. *Biological Psychology* **86**, 320–9.

Berlyne, D.E. (1960). *Conflict, arousal, and curiosity*. New York, NY: McGraw-Hill.

Berlyne, D.E. (1971). *Aesthetics and psychobiology*. New York, NY: Appleton-Century-Crofts.

Berlyne, D.E. (1973). The vicissitudes of aplopathematic and thelematoscopic pneumatology (or the hydrography of hedonism). In D.E. Berlyne and K.B. Madsen (eds), *Pleasure, reward, preference*. New York, NY: Academic Press, pp. 1–33.

Berlyne, D.E. (1974). The new experimental aesthetics. In D.E. Berlyne (ed.), *Studies in the new experimental aesthetics*. Washington, DC: Hemisphere, pp. 1–25.

Blood, A.J. and Zatorre, R.J. (2001). Intensely pleasurable responses to music correlate with activity in brain regions implicated in reward and emotion. *Proceedings of the National Academy of Sciences of the USA* **98**, 11818–23.

Buckner, R., Andrews-Hanna, J., and Schacter, D. (2008). The brain's default network: Anatomy, function and relevance to disease. *Annals of the New York Academy of Sciences* **1124**, 1–38.

Cela-Conde, C.J., Ayala, F.J., Munar, E., et al. (2009). Sex-related similarities and differences in the neural correlates of beauty. *Proceedings of the National Academy of Sciences of the USA* **106**, 3847–52.

Cela-Conde, C.J., Marty, G., Maestu, F., et al. (2004). Activation of the prefrontal cortex in the human visual aesthetic perception. *Proceedings of the National Academy of Sciences of the USA* **101**, 6321–5.

Chamorro-Premuzic, T., Reimers, S., Hsu, A., et al. (2009). Who art thou? Personality predictors of artistic preferences in a large UK sample: The importance of openness. *British Journal of Psychology* **100**, 501–16.

Conner, T.S., Tennen, H., Fleeson, W., et al. (2009). Experience sampling methods: A modern idiographic approach to personality research. *Social and Personality Psychology Compass* **3**, 292–313.

Cooper, J.M. and Silvia, P.J. (2009). Opposing art: Rejection as an action tendency of hostile aesthetic emotions. *Empirical Studies of the Arts* **27**, 111–28.

Cronbach, L.J. (1957). Two disciplines of scientific psychology. *American Psychologist* **12**, 671–84.

Cybulski, G. (2011). *Ambulatory impedance cardiography: The systems and their applications*. Heidelberg: Springer.

Demir, E., Desmet, P.M.A., and Hekkert, P. (2009). Appraisal patterns of emotions in human–product interaction. *International Journal of Design* **3**, 41–51.

DeWall, C.N., Silvia, P.J., Schurtz, D.R., et al. (2011). Taste sensitivity and aesthetic preferences: Is taste only a metaphor? *Empirical Studies of the Arts* **29**, 171–89.

DeYoung, C.G., Grazioplene, R.G., and Peterson, J.B. (2012). From madness to genius: The Openness/Intellect trait domain as a paradoxical simplex. *Journal of Research in Personality* **46**, 63–78.

DeYoung, C.G., Quilty, L.C., and Peterson, J.B. (2007). Between facets and domains: 10 aspects of the Big Five. *Journal of Personality and Social Psychology* **93**, 880–96.

Ellsworth, P.C. and Scherer, K.R. (2003). Appraisal processes in emotion. In R.J. Davidson, K.R. Scherer, and H.H. Goldsmith (eds), *Handbook of affective sciences*. New York, NY: Oxford University Press, pp. 572–95.

Fayn, K. and Tiliopoulos, N. (2012). Open to not understanding and grateful for the experience: A process approach to studying the association between aesthetic appreciation and personality. *Proceedings of the 11th Australian Conference on Personality and Individual Differences*.

Fayn, K. and Tiliopoulos, N. (2013). Gratefully curious: A multilevel analysis of individual differences in aesthetic appreciation. *Proceedings of the 1st World Conference on Personality*.

Fechner, G.T. (1876). *Vorschule der Ästhetik*. Leipzig: Breitkopf & Härtel.

Feist, G.J. (1998). A meta-analysis of personality in scientific and artistic creativity. *Personality and Social Psychology Review* **2**, 290–309.

Feist, G.J. and Brady, T.R. (2004). Openness to experience, non-conformity, and the preference for abstract art. *Empirical Studies of the Arts* **22**, 77–89.

Gordon, K. (1909). *Esthetics*. New York, NY: Henry Holt.

Grewe, O., Nagel, F., Kopiez, R., et al. (2007). Listening to music as a re-creative process: Physiological, psychological, and psychoacoustical correlates of chills and strong emotions. *Music Perception* **24**, 297–314.

Haanstra, F., Damen, M.L., and Van Hoorn, M. (2013). Interestingness and pleasingness of drawings from different age and expertise groups. *Empirical Studies of the Arts* 31, 173–94.

Heidenreich, S.M. and Turano, K.A. (2011). Where does one look when viewing artwork in a museum? *Empirical Studies of the Arts* **29**, 51–72.

Hekkert, P. and van Wieringen, P.C.W. (1996). Beauty in the eye of expert and nonexpert beholders: A study in the appraisal of art. *American Journal of Psychology* **109**, 389–407.

Hektner, J.M., Schmidt, J.A., and Csikszentmihalyi, M. (2007). *Experience sampling method: Measuring the quality of everyday life*. Thousand Oaks, CA: Sage.

Höge, H. (2003). A museum experience: Empathy and cognitive restoration. *Empirical Studies of the Arts* **21**, 155–64.

Izard, C.E. (1977). *Human emotions*. New York, NY: Plenum.

Jacobsen, T., Buchta, K., Köhler, M., et al. (2004). The primacy of beauty in judging the aesthetics of objects. *Psychological Reports* **94**, 1253–60.

Juslin, P. N., Liljeström, S., Västfjäll, D., et al. (2008). An experience sampling study of emotional reactions to music: Listener, music, and situation. *Emotion* **8**, 668–83.

Kanai, R. and Rees, G. (2011). The structural basis of inter-individual differences in human behaviour and cognition. *Nature Reviews Neuroscience* **12**(4), 231–42.

Kaufman, S.B. (2013). Opening up openness to experience: A four-factor model and relations to creative achievement in the arts and sciences. *Journal of Creative Behavior* **47**, 233–55.

Kaufman, S.B., DeYoung, C.G., Gray, J.R., et al. (2010). Implicit learning as an ability. *Cognition* **116**, 321–40

Kawabata, H. and Zeki, S. (2004). Neural correlates of beauty. *Journal of Neurophysiology* **91**, 1699–705.

Kellaris, J.J. (2008). Music and consumers. In C.P. Haugtvedt, P. Herr, and F.R. Kardes (eds), *Handbook of consumer psychology*. New York, NY: Taylor and Francis, pp. 837–56.

Kelsey, R.M. (2012). Beta-adrenergic cardiovascular reactivity and adaptation to stress: The cardiac pre-ejection period as an index of effort. In R.A. Wright and G H.E. Gendolla (eds), *How motivation affects cardiovascular response: Mechanisms and applications*. Washington, DC: American Psychological Association, pp. 43–60.

Keltner, D. and Haidt, J. (2003). Approaching awe, a moral, spiritual, and aesthetic emotion. *Cognition and Emotion* **17**, 297–314.

Kirk, U., Skov, M., Christensen, M.S., et al. (2009). Brain correlates of aesthetic expertise: A parametric fMRI study. *Brain and Cognition* **69**, 306–15.

Lazarus, R.S. (1991). *Emotion and adaptation*. New York, NY: Oxford University Press.

Leder, H., Gerger, G., Dressler, S.G., et al. (2012). How art is appreciated. *Psychology of Aesthetics, Creativity, and the Arts* **6**, 2–10.

Levitin, D.J. (2006). *This is your brain on music: The science of a human obsession*. New York, NY: Plume.

Locher, P.J. (2006). Experimental scrutiny of the role of balance in the visual arts. In P. Locher, C. Martindale, and L. Dorfman (eds), *New directions in aesthetics, creativity, and the arts*. Amityville, NY: Baywood, pp. 19–32.

Locher, P.J., Smith, J.K., and Smith, L.F. (2001). The influence of presentation format and viewer training in the visual arts on the perception of pictorial and aesthetic qualities of paintings. *Perception* **30**, 449–65.

Ludden, G.D.S. and Schifferstein, H.N.J. (2007). Effects of visual-auditory incongruity on product expression and surprise. *International Journal of Design* **1**(3), 29–39.

Ludden, G.D.S., Schifferstein, H.N.J., and Hekkert, P. (2008). Surprise as a design strategy. *Design Issues* **24**(2), 28–38.

Ludden, G.D.S., Schifferstein, H.N.J., and Hekkert, P. (2009). Visual–tactual incongruities in products as sources of surprise. *Empirical Studies of the Arts* **27**, 61–87.

Marković, S. (2012). Components of aesthetic experience: Aesthetic fascination, aesthetic appraisal and aesthetic emotion. *i-Perception* **3**, 1–17.

Mastandrea, S., Bartoli, G., and Bove, G. (2007). Learning through ancient art and experiencing emotions with contemporary art: Comparing visits in two different museums. *Empirical Studies of the Arts* **25**, 173–91.

McCrae, R.R. (1994). Openness to experience: Expanding the boundaries of Factor V. *European Journal of Personality* **8**, 251–72.

McCrae, R.R. (2007). Aesthetic chills as a universal marker of openness to experience. *Motivation and Emotion* **31**, 5–11.

McCrae, R.R. and Costa, P.T., Jr. (1999). A five-factor theory of personality. In L.A. Pervin and O.P. John (eds), *Handbook of personality* (2nd edn). New York, NY: Guilford, pp. 139–53.

McCrae, R.R. and Sutin, A.R. (2009). Openness to experience. In M.R. Leary and R.H. Hoyle (eds), *Handbook of individual differences in social behavior.* New York, NY: Guilford, pp. 257–73.

McManus, I.C. and Furnham, A. (2006). Aesthetic activities and aesthetic attitudes: Influences of education, background and personality on interest and involvement in the arts. *British Journal of Psychology* **97**(4), 555–87.

Millis, K. (2001). Making meaning brings pleasure: The influence of titles on aesthetic experience. *Emotion* **1**, 320–9.

Nadal, M. and Pearce, M.T. (2011). The Copenhagen Neuroaesthetics Conference: Prospects and pitfalls for an emerging field. *Brain and Cognition* **76**, 172–83.

Nusbaum, E.C. and Silvia, P.J. (2011a). Are openness and intellect distinct aspects of openness to experience? A test of the O/I model. *Personality and Individual Differences* **51**, 571–4.

Nusbaum, E.C. and Silvia, P.J. (2011b). Shivers and timbres: Personality and the experience of chills from music. *Social Psychological and Personality Science* **2**, 199–204.

Nusbaum, E.C. and Silvia, P.J. (2014). Unusual aesthetic states. In P.P.L. Tinio and J.K. Smith (eds), *Cambridge handbook of the psychology of aesthetics and the arts.* Cambridge: Cambridge University Press, pp. 519–39.

Nusbaum, E.C., Silvia, P.J., Beaty, R.E., et al. (2014). Listening between the notes: Aesthetic chills in everyday music listening. *Psychology of Aesthetics, Creativity, and the Arts* **8**, 104–9.

Nusbaum, E.C., Silvia, P.J., Beaty, R.E., et al. (2015). *Turn that racket down! Physical anhedonia and diminished pleasure from music.* Manuscript under review.

Olds, J. (1962). Hypothalamic substrates of reward. *Physiological Review* **42**, 554–604.

Olds, J. (1973). Brain mechanisms of reinforcement learning. In D.E. Berlyne and K.B. Madsen (eds), *Pleasure, reward, preference: Their nature, determinants, and role in behavior.* New York, NY: Academic Press, pp. 35–63.

Oostendorp, A. and Berlyne, D.E. (1978a). Dimensions in the perception of architecture: I. Identification and interpretation of dimensions of similarity. *Scandinavian Journal of Psychology* **19**, 73–82.

Oostendorp, A. and Berlyne, D.E. (1978b). Dimensions in the perception of architecture: II. Measures of exploratory behavior. *Scandinavian Journal of Psychology* **19**, 83–9.

Parsons, M.J. (1987). *How we understand art: A cognitive developmental account of aesthetic experience.* Cambridge: Cambridge University Press.

Rawlings, D. (2003). Personality correlates of liking for "unpleasant" paintings and photographs. *Personality and Individual Differences* **34**, 395–410.

Reber, R. (2012). Processing fluency, aesthetic pleasure, and culturally shared taste. In A. Shimamura and S.E. Palmer (eds), *Aesthetic science: Connecting minds, brains, and experience.* New York, NY: Oxford University Press, pp. 221–49.

Salimpoor, V.N., and Zatorre, R.J. (2013). Neural interactions that give rise to musical pleasure. *Psychology of Aesthetics, Creativity, and the Arts* **7**, 62–75.

Scherer, K.R. (2001). Appraisal considered as a process of multilevel sequential checking. In K.R. Scherer, A. Schorr, and T. Johnstone (eds), *Appraisal processes in emotion: Theory, methods, research.* New York, NY: Oxford University Press, pp. 92–120.

Segerstrom, S.C., Hardy, J.K., Evans, D.R., et al. (2012). Pause and plan: Self-regulation and the heart. In R.A. Wright and G.H.E. Gendolla (eds), *How motivation affects cardiovascular response: Mechanisms and applications.* Washington, DC: American Psychological Association, pp. 181–98.

Shiota, M.N., Keltner, D., and John, O.P. (2006). Positive emotion dispositions differentially associated with Big Five personality and attachment style. *Journal of Positive Psychology* **1**, 61–71.

Shiota, M.N., Keltner, D., and Mossman, A. (2007). The nature of awe: Elicitors, appraisals, and effects on self-concept. *Cognition and Emotion* **21**, 944–63.

Siemer, M., Mauss, I., and Gross, J.J. (2007). Same situation—different emotions: How appraisals shape our emotions. *Emotion* **7**, 592–600.

Silvia, P.J. (2005a). Cognitive appraisals and interest in visual art: Exploring an appraisal theory of aesthetic emotions. *Empirical Studies of the Arts* **23**, 119–33.

Silvia, P.J. (2005b). Emotional responses to art: From collation and arousal to cognition and emotion. *Review of General Psychology* **9**, 342–57.

Silvia, P.J. (2005c). What is interesting? Exploring the appraisal structure of interest. *Emotion* **5**, 89–102.

Silvia, P.J. (2006a). Artistic training and interest in visual art: Applying the appraisal model of aesthetic emotions. *Empirical Studies of the Arts* **24**, 139–61.

Silvia, P.J. (2006b). *Exploring the psychology of interest*. New York, NY: Oxford University Press.

Silvia, P.J. (2007). Knowledge-based assessment of expertise in the arts: Exploring aesthetic fluency. *Psychology of Aesthetics, Creativity, and the Arts* **1**, 247–9.

Silvia, P.J. (2008a). Appraisal components and emotion traits: Examining the appraisal basis of trait curiosity. *Cognition and Emotion* **22**, 94–113.

Silvia, P.J. (2008b). Interest—The curious emotion. *Current Directions in Psychological Science* **17**, 57–60.

Silvia, P.J. (2009). Looking past pleasure: Anger, confusion, disgust, pride, surprise, and other unusual aesthetic emotions. *Psychology of Aesthetics, Creativity, and the Arts* **3**, 48–51.

Silvia, P.J. (2010). Confusion and interest: The role of knowledge emotions in aesthetic experience *Psychology of Aesthetics, Creativity, and the Arts* **4**, 75–80.

Silvia, P.J. (2012). Curiosity and motivation. In R.M. Ryan (ed.), *The Oxford handbook of human motivation*. New York, NY: Oxford University Press, pp. 157–66.

Silvia, P.J. (2013). Interested experts, confused novices: Art expertise and the knowledge emotions. *Empirical Studies of the Arts* **31**, 107–16.

Silvia, P.J. and Barona, C.M. (2009). Do people prefer curved objects? Angularity, expertise, and aesthetic preference. *Empirical Studies of the Arts* **27**, 25–42.

Silvia, P.J. and Berg, C. (2011). Finding movies interesting: How expertise and appraisals influence the aesthetic experience of film. *Empirical Studies of the Arts* **29**, 73–88.

Silvia, P.J. and Brown, E.M. (2007). Anger, disgust, and the negative aesthetic emotions: Expanding an appraisal model of aesthetic experience. *Psychology of Aesthetics, Creativity, and the Arts* **1**, 100–6.

Silvia, P.J., Henson, R.A., and Templin, J.L. (2009). Are the sources of interest the same for everyone? Using multilevel mixture models to explore individual differences in appraisal structures. *Cognition and Emotion* **23**, 1389–406.

Silvia, P.J. and Nusbaum, E.C. (2011). On personality and piloerection: Individual differences in aesthetic chills and other unusual aesthetic experiences. *Psychology of Aesthetics, Creativity, and the Arts* **5**, 208–14.

Sloboda, J.A., O'Neill, S.A., and Ivaldi, A. (2001). Functions of music in everyday life: An exploratory study using the experience sampling method. *Musicae Scientiae* **5**, 9–32.

Smith, J.K. (2014). *The museum effect: How museums, libraries, and cultural institutions educate and civilize society*. Lanham, MD: Rowman & Littlefield.

Smith, L.F. (2014). The science and aesthetics of astronomical images. *Psychology of Aesthetics, Creativity, and the Arts* **8**, 506–13.

Smith, J.K. and Smith, L.F. (2001). Spending time on art. *Empirical Studies of the Arts* **19**, 229–36.

Smith, L.F. and Smith, J.K. (2006). The nature and growth of aesthetic fluency. In P. Locher, C. Martindale, and L. Dorfman (eds), *New directions in aesthetics, creativity, and the arts*. Amityville, NY: Baywood, pp. 47–58.

Thayer, J.F., Hansen, A.L., and Johnsen, B.H. (2008). Noninvasive assessment of autonomic influences on the heart: Impedance cardiography and heart rate variability. In L.J. Luecken and L.C. Gallo (eds), *Handbook of physiological research methods in health psychology*. Los Angeles, CA: Sage, pp. 183–209.

Tomkins, S.S. (1962). *Affect, imagery, consciousness: Vol. 1, The positive affects*. New York, NY: Springer.

Tracy, J.L. and Robins, R.W. (2007). The self in self-conscious emotions: A cognitive appraisal approach. In J.L. Tracy, R.W. Robins, and J.P. Tangney (eds), *The self-conscious emotions: Theory and research*. New York, NY: Guilford, pp. 3–20.

Tröndle, M. and Tschacher, W. (2012). The physiology of phenomenology: The effects of artworks. *Empirical Studies of the Arts* **30**, 75–113.

Tschacher, W., Greenwood, S., Kirchberg, V., et al. (2012). Physiological correlates of aesthetic perception of artworks in a museum. *Psychology of Aesthetics, Creativity, and the Arts* **6**, 96–103.

Turner, S.A., Jr. and Silvia, P.J. (2006). Must interesting things be pleasant? A test of competing appraisal structures. *Emotion* **6**, 670–4.

Valentine, C.W. (1913). The aesthetic appreciation of musical intervals among school children and adults. *British Journal of Psychology* **6**, 190–216.

Vessel, E.A., Starr, G.G., and Rubin, N. (2012). The brain on art: Intense aesthetic experience activates the default mode network. *Frontiers in Human Neuroscience* **6**, 1–17.

Whitfield, T.W.A. (1983). Predicting preference for familiar, everyday objects: An experimental confrontation between two theories of aesthetic behavior. *Journal of Environmental Psychology* **3**, 221–37.

Whitfield, T.W.A. and de Destefani, L.R. (2011). Mundane aesthetics. *Psychology of Aesthetics, Creativity, and the Arts* **5**, 291–9.

Whitfield, T.W.A. and Slatter, P.E. (1979). The effect of categorization and prototypicality on aesthetic choice in a furniture selection task. *British Journal of Psychology* **70**, 65–75.

Chapter 4

Aesthetic appreciation: Convergence from experimental aesthetics and physiology

Helmut Leder, Gernot Gerger, and David Brieber

4.1 Introduction—strange behaviors?

Among the puzzling things that humans do which at first sight have no apparent biological use are many of those activities we subsume under the concept of "culture". For example, humans spend hours in theaters and concert halls, and they also visit art exhibitions, which often require traveling and other costs. In more abstract terms, regarding art, people spend time in quiet spaces and look at human-made two- or three-dimensional structures that remain unchanged, delivering a stable flow of rather invariable sensory stimulation. In the present chapter we discuss the study of processes involved in such behaviors, specifically in art appreciation, and how researchers can measure psychological processes and states of aesthetic appreciation by using behavioral and psychophysiological methods.

The object of our studies, the consumption of art by mankind, is quite old. Available evidence suggests that art is a unique human feature (Davies 2012). Examples of prehistoric art range back at least 80 000 years. Thus, along with weapons and tools, artworks are among the oldest human artifacts. Although we know little about how the early artworks were perceived, and we have no evidence for the particular kinds of receptor perspectives involved (Dutton 2009), artistry is deeply embedded in our evolutionary past. For many people today, at least in many Western cultures, going to a museum is a common behavior, for some even a frequent behavior. Nevertheless, the concept of a museum as we know it today is a relatively new invention (Shiner 2003). The Capitolian museum that opened in 1471 in Rome can be regarded as the first modern museum. Independently of whether cave paintings are seen as ancestors of the modern concept of museums, or whether museums are classified as a recent development in history, the existence of art raises the question of what kind of behavior the perception of art is, and what kind of experiences it yields. Art's appeal might be due to various distinct possibilities. Psychology of the arts has focused on the quality of experiences that explain why this behavior is demonstrated (Leder et al. 2004). In this vein, artworks in exhibitions represent an unquestioned aesthetic value to visitors, and promise exciting, enlightening, or uplifting experiences. This relates to the most-often assumed explanation of aesthetic behavior: that art affords hedonic states of positive experiences of beauty, pleasure, or the sublime.

However, there are various examples of contemporary art exhibitions for which such simple, straightforward effects can hardly be true. For example, consider Teresa Margolles' contribution to the Mexican Pavillion at the Venice Biennale in 2009, named *What else can we talk about?*. In Margolles' installation family members of homicide victims of the Mexican drugs war cleaned the floors of a Venetian palace with water that was taken from the burying grounds of the victims. This could not be said to offer the viewer a first-hand pleasurable experience. Thus, at least when it comes to contemporary art, alternative explanations of the appeal of art should be discussed: art can promise interesting, arousing, sometimes challenging experiences, by presenting new, innovative, and even provoking artworks, which do not explicitly aim to display the beautiful or the sublime. However, these artworks might receive their appeal from situations that require some resolution—that is, perceivers are trying to understand and make sense of what is displayed by the artwork and intended by the artist, similar to the solution of a cognitive puzzle or the resolution of visual ambiguity. Such processes might be self-rewarding (Leder et al. 2004).

This also relates to explanations that stress social functions of art. Artists adopt a certain view of the world; they often express a specific attitude. Consequently, early psychological approaches emphasized that the perceiver has to be able to empathize with the artwork—or the artist (see Currie 2011, for a comprehensive discussion). Thus, art also allows for practising playfully in learning to perceive, interpret, and exchange viewpoints and opinions (Dissanayake 2007). Related to this explanation, evolutionary approaches argue that "art making and appreciation serves an important ritualistic function that enhances social cohesion" (Chatterjee 2011, p. 56). Psychological analysis of art is usually directed towards an inner perspective, such as the psychological states and processes during perception of art.

In this chapter we aim to address two issues. First, we discuss how psychology understands the behavior of perceiving and enjoying art. Second, we deconstruct the experience of art into different components and discuss how experimental aesthetics, and particularly its use of psychophysiological measures, provides insight into the complex processes during aesthetic receptions of visual art.

4.2 How do psychologists understand aesthetic experiences of art?

Psychology aims to describe, explain, and, through scientific understanding, predict human behavior and experience. The focus of most psychological theories is a kind of inner view of a person (James 1890). In this respect, psychology of aesthetics aims to determine the processes that take place in a perceiver, in order to explain what he or she likes aesthetically. Fechner (1876) in his *Vorschule der Ästhetik* expressed this in the following way:

> In considering the aesthetic effect of anything upon us, we do not only think about the concepts under which it can be subsumed, and its categorization within such a system, although such questions need to be asked. However, the most interesting and important question has to be: *Why is something liked or disliked?* . . . The answer to this question can only be found in the specific laws

> relating to liking and disliking, and in the laws of obligation, in just the same way that the particular movements of physical objects are explained by the laws of motion, and by the reasons and purposes of those movements. Movements cannot be explained alone by naming concepts or categorizing into different types of movement, but require precise laws of motion, whilst the purposes of the movements are also taken into account. Conceptual explanations in aesthetics will therefore be nothing but an empty framework, until they are confirmed by proper explanatory laws. [Translation by Leder and Pleines, see Leder (2014, p. 117).]

Fechner founded the field of empirical aesthetics with his *Vorschule der Ästhetik* (1876). In this approach he distinguished bottom-up, stimulus-driven, from top-down, cognitive-driven, approaches and anticipated the interplay of these approaches during aesthetic receptions. This is much in accordance with what psychology today has identified as happening within a person when he or she experiences a work of art (Leder et al. 2004). Fechner's empirical studies were clearly concerned with aesthetic reception—not production. He presented the first laboratory studies regarding the principle of the Golden Section in rectangles, according to which a ratio of 1 to 1.618 should be particularly beautiful, and conducted a first field study in a museum by comparing the appreciation of two versions of Holbein's *Madonna*, exhibited in the Semper Gallerie in Dresden. Interestingly, although Fechner made aesthetics one of his two main research topics (in addition to psychophysics), the progress of empirical aesthetics throughout the twentieth century was surprisingly slow. Mainstream psychology had a stronghold in perceptual studies, but other topics of clinical, social, and basic processes in perception, like attention and thinking, became more dominant themes. There were exceptions, such as Karl Bühler's (1913) approach to Gestalt psychology, which very explicitly starts with a reference to art history. However, only after the influence of behaviorism subsided did experimental aesthetics experience a revival, owing mainly to Berlyne's biological arousal approach in the 1970s. Since the cognitive revolution, the information processing concept has guided the study and understanding of human behavior for half a century. As a result, cognition, thinking, and internal representations are common to many psychological theories.

The information-processing model developed by Leder and colleagues (2004), and Chatterjee's (2003) neuropsychological model, are two examples of this. In both models, the aesthetic experience is defined as a sequence of and interaction between information processing stages. Figure 4.1 shows Vartanian and Nadal's (2007) integration of both models. Chatterjee (2003) distinguished aesthetic decisions and emotional responses as outputs of his model. Both are the result of processing stages of early and intermediate vision, a representational domain, and are modulated through attention. Leder and colleagues' (2004) model aims to explain similar outputs, aesthetic judgments and aesthetic emotions. It proposed five information-processing stages that explain how an aesthetic experience develops—from perception of an artwork to its interpretation and evaluation.

Most of this model is concerned with processes that are not directly visible; they represent inner-states, and constitute what informally would be called thinking. According to the information-processing assumption, cognitive processes form representations and change psychological and physiological states in the person. For example, the combination

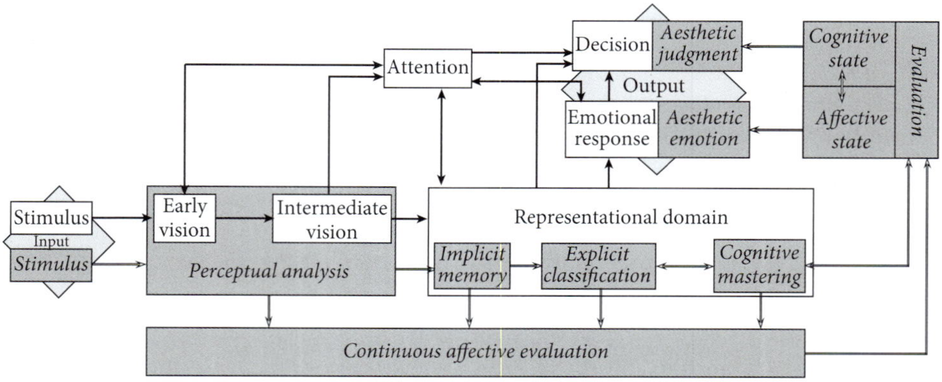

Figure 4.1 Combination of models of aesthetic experience. (Adapted from Vartanian and Nadal 2007.)

of text and image in Magritte's "*La trahison des images*,"[1] an image of a pipe, and the text "*ceci nest pas une pipe*" ("this is not a pipe"), creates a perceptual representation of image and words that are instantly seen as contradictory. However, by means of interpretation the perceiver might resolve the seeming contradiction by realizing that the pipe is not a pipe, but an image of a pipe, and that images can represent an object, but images aren't objects. This way, due to cognitive processes, a state of resolved ambiguity replaces a state of ambiguity. In the model developed by Leder and colleagues (2004) these cognitive processes are also seen as eliciting changes in emotional states, mainly as a consequence or byproduct of the various cognitive activities (see also Leder 2013). Thus, an art experience is a complex episode, consisting of an interaction between an object and a perceiver, and the interplay of cognitive as well as emotional processes (see Figure 4.3).

The focus on the recipient's individual states might be due to the proximity and relatedness of the research programs with the field of perception. From the beginning, perception was a focus for the new discipline of empirical psychology. The early works of Helmholtz or Wundt developed research approaches to human perception with high experimental vigor. If the goal of the methods subsumed under the heading *psychophysics* was to understand scientifically the relation between the external world and its mental representation, then perception was the natural place to start. Throughout the twentieth century, expertise and knowledge about perception increased alongside technological developments in image presentation and manipulation. With these developments the complexity of research programs in psychology increased by the consideration of more complex visual arrays, such as artworks. As a consequence, today's psychology demonstrates an increased interest in the arts. Moreover, the scope has also broadened through the integration of various methods, comprising experimental, cognitive-behavioral studies as well as psychophysiology and brain research. The combination of an outer psychophysics—how object features affect inner states, and the relation between inner states and physiological changes, termed *inner psychophysics* by Fechner—pretty much determines the current program of empirical aesthetics.

4.2.1 Psychological methods

Empirical psychology faces the problem that most of the cognitive and emotional processes it posits are not directly observable, and they take place in very fast succession. These successive processes are usually not even directly accessible to consciousness, thus they are difficult to report. However, based on theoretical considerations, a systematic variation of independent variables and measurement of the outcome by deploying one or more dependent variables, allows us to infer these processes from observations and reporting. In experimental aesthetics, most commonly only behavioral measurements are used as dependent variables, but due to technical advances in recent decades these measurements are being complemented by psychophysiological methods.

Regarding behavioral measurements, aesthetic evaluations are very often measured by preference decisions. These can be given using forced-choice preference tasks or Likert scales measuring liking or preference. The scale "how much do you like this" is frequently used, but must be seen as a somewhat vague and very general measure of aesthetic evaluations (Leder et al. 2005). The questions about how much an object is preferred, how strongly it evokes an aesthetic experience, or how beautiful it is, can also be posed. Interestingly, different object classes differ due to the dimensions that are associated with their aesthetic evaluation (Augustin et al. 2012). This should be considered in experiments.

Psychophysiology provides an interesting improvement to and extension of the behavioral approach, as it allows testing hypotheses associated with physiological correlates. Physiological responses are closely linked to inner states. Some of the psychophysiological methods (e.g. EEG, electroencephalography; MEG, magnetoencephalography) allow the tracking of changes in processes in "real time" as they have very high temporal resolution. Additionally, psychophysiological measures of the peripheral nervous system function, such as electrodermal activity (EDA), heart rate (HR), facial electromyography (facial EMG), and pupillometry complement the experimental repertoire for psychologists and can contribute to a deeper understanding of psychological processes (see Figure 4.2).

Regarding EDA, one commonly used measure is skin conductance, which records how well the skin conduces current flow in between a pair of skin electrodes. As sweat secretion increases, due to sympathetic activation, so does skin conductance. Both slow (tonic) changes in skin conductance level (SCL) and fast event-related (phasic) skin conductance responses (SCR) can be interpreted as physiological indexes of arousal. SCR increases with stimulus intensity, regardless of whether it has positive or negative valence (e.g. Bradley et al. 1990).

In contrast to EDA, facial EMG can be used to measure responses associated with emotional valence because facial muscle activity is crucial to the expression of emotions (Ekman et al. 1980). Facial EMG can detect subtle changes in facial muscle activity related to emotional processing, even if no overt changes in facial expression are visible. The *musculus zygomaticus major* which raises the corner of the mouth and causes smiling, and the *musculus corrugator supercilii* which draws the eyebrows downwards while frowning, are reliable indicators of affective processes. Whereas affectively positive stimuli elicit an

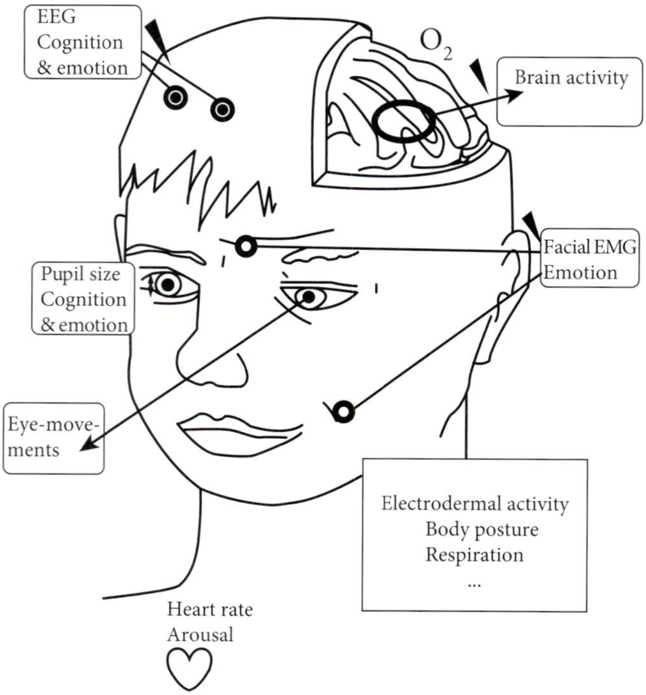

Figure 4.2 Illustration of the various methods that can be used to find correlates of the inner states of a person.

increase in zygomaticus muscle activity, affectively negative stimuli increase corrugator muscle activity (Bradley et al. 2001; Dimberg and Karlsson 1997). These changes are usually in accordance with self-reports of subjectively felt emotions (Dimberg 1988). One should note that facial EMG is also sensitive to changes in cognitive processing (Scherer and Ellgring 2007). For example, a higher cognitive load leads to stronger *musculus corrugator supercilii* activation (Lishner et al. 2008).

Pupillometry, the measurement of pupil dilation and contraction, is another psychophysiological measure that can be applied to understand processes in aesthetics. Pupil size is controlled by the iris sphincter and the dilator muscles. The former is innervated by the parasympathetic nervous system, the latter by the sympathetic nervous system. Although pupil size variation mainly owes to changes in luminance (the pupillary light reflex), it is also associated with changes in arousal (Bradley et al. 2008) and cognitive load (e.g. Laeng et al. 2012). Highly arousing pictures elicit an increase in pupil dilation regardless of stimulus valence (Bradley et al. 2008). High cognitive loads while processing visual and auditory stimuli increases pupil diameter and decreases light reflex (Kahneman and Beatty 1966; Steinhauer et al. 2000). Pupillometry is usually measured with eye-tracking systems. Such systems also allow inferences about spatial (where?) and temporal (when and how long?) eye movement characteristics of information intake during visual exploration.

During the last decade, brain activity underlying aesthetic experiences was examined using methods such as electroencephalography (EEG), magnetoencephalography (MEG), and functional magnetic resonance imaging (fMRI). Electrical potentials of neurons produce electric and magnetic fields that can be measured from outside the skull. EEG electrodes attached to the scalp record the electrical activity; the highly sensitive magnetometers used in MEG measure the magnetic fields. EEG can be used to measure different frequency bands of neural oscillations associated with specific psychological states. Alternatively, more specific psychological processes can be analysed with the event-related potential (ERP) technique, which averages EEG responses for a specific time-locked event. In comparison to other brain imaging techniques, EEG and MEG have the advantage of a high temporal resolution. Functional MRI, on the other hand, has a high spatial resolution. The demand for oxygen due to the increase in the activity of specific brain regions raises the regional cerebral blood flow. Functional MRI detects these changes in blood oxygenation level and thus provides an indirect measure of brain activity. These techniques are not restricted to local measures, thereby enabling the identification of networks of brain regions involved in aesthetic processing.

Taken together, fMRI is suitable for testing assumptions regarding the location and involvement of brain processes, while EEG and MEG shed light on temporal characteristics within a fraction of a second. Facial EMG, measures of EDA, and pupillometry are especially useful to test overt and subtle affective responses. Studies using these psychophysiological methods have corroborated results from behavioral studies and have revealed new insights regarding the components and processes involved in art experience.

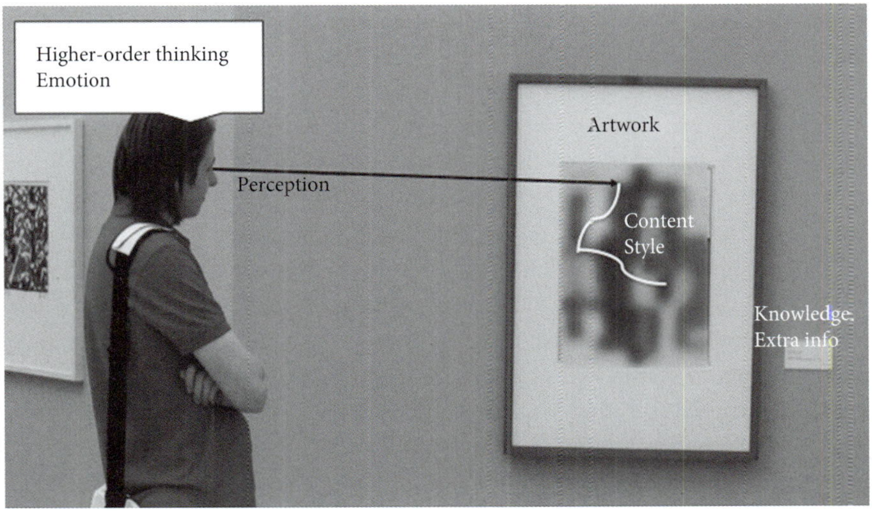

Figure 4.3 Components of an aesthetic experience.

4.3 Components of aesthetic experience of art—behavioral and physiological approaches

Psychological theories of aesthetic experience disentangle the experience into distinctive processing stages, as noted earlier. Hereafter, we use examples concerned with some of the components in order to illustrate how their role in aesthetic episodes and art appreciation can be addressed using behavioral and psychophysiological methods. Figure 4.3 illustrates an aesthetic episode and the components that we consider in the next paragraphs.

We discuss how contexts enable aesthetic processing, we study early perceptual processes including the important stages of style or content processing, and how knowledge and expertise, as well as emotion, shape aesthetic experiences from art. Regarding these components, we present findings about the processes involved by putting the emphasis on scientific studies that combined behavioral and physiological methods in their investigations.

4.3.1 The role of context

The aesthetic experience is often associated with a specific context. Such contexts can be established by pre-classifying an object as art. This is accomplished by displaying the object in a museum or by being asked to evaluate it aesthetically in the laboratory. An aesthetic context encourages the perceiver to adopt what Cupchik and Laszlo (1992) called an aesthetic attitude, and it provides a necessary but not sufficient pre-condition for aesthetic experiences (Leder et al. 2004). Aesthetic experiences are then viewed as a special state of mind qualitatively different from those summoned by everyday experiences (Markovic 2012). Thus, context is a defining feature of aesthetic experiences.

Many famous artists played with such contextual frames by placing everyday objects in an aesthetic context—see, for example, Marcel Duchamp's *Fountain* (1917, which is a urinal),[2] Andy Warhol's *Brillo Boxes* (1964),[3] or Beuys' blocks of fat (1968)[4] displayed in art museums. Indeed, seeing a urinal as an exhibit in a museum might lead to a different experience to seeing a urinal when visiting the rest room, but how do contextual cues change our aesthetic experiences?

Contexts exert influence on our experience via several different routes. Critically, aesthetic experiences usually take place within rather safe environments (Frijda and Schram 1995) which pose no risk to the health and well-being to the individual. In such a context, everyday concerns can be suppressed (Cupchik and Winston 1996), allowing the individual to focus attention on the object, and to engage in an evaluative processing style of thought (Leder et al. 2004). Artists and art curators seem to be well aware of this; O'Doherty (1976) characterized galleries as "*white spaces*" where the "*outside world must not come in.*" Everyday stimulus processing, on the contrary, is pragmatic and goal oriented. In order to secure well-being, perception usually ensures successful identification of and appropriate reaction to various objects in the visual field. In aesthetic processing this immediacy of generating unambiguous representations and goal-directed actions is far less pronounced. Given that aesthetic contexts reduce pragmatic and motivational considerations, the perceiver is allowed to enjoy a rather ambiguous state when seeing an artwork (Jakesch and Leder 2009). Indeed, artists produce works of art not just for the sake of mere representation of an object; they also exploit stylistic and formal properties in their artworks to provide the viewer with an engaging viewing experience. Pablo Picasso's cubist paintings, Rene Magritte's often semantically ambiguous paintings, or in the case of abstract art, paintings devoid of any representational content, are all examples of artworks exploiting such stylistic and formal properties. In addition, an aesthetic context might be marked by certain expectations, promising aesthetically pleasing and cognitively challenging experiences when looking at art (Leder et al. 2004).

It is assumed that an aesthetic context encourages an evaluative processing style (Cupchik and Winston 1996; Leder et al. 2004). In a series of experiments, Jacobsen and colleagues directly tested this assumption. Participants evaluated abstract patterns in either an aesthetic task ("Is this stimulus beautiful?"), or in a pragmatic—or at least non-aesthetic—task ("Is this stimulus symmetrical?"), while physiological brain responses were recorded employing EEG (Jacobsen and Höfel 2003) or fMRI (Jacobsen et al. 2006). An aesthetic context led to a strong right hemisphere lateralization of an ERP component called the late positive potential, LPP. The authors interpret this lateralization as reflecting an evaluative style of processing in aesthetic contexts, because the LPP is increased for stimuli of positive and negative valence (Cuthbert et al. 2000) and is associated with enhanced attentional processing for motivationally salient stimuli (Hajcak et al. 2012). Jacobsen and colleagues (2006) used fMRI to show that an aesthetic context elicited stronger activity in frontal networks including the ventral prefrontal cortex (BA45/47) and the frontomedian cortex (BA10/9). These brain regions are linked to evaluative judgments of persons and actions.

Activation of a similar region of the BA10 was also reported by Cupchik and colleagues (2009) in a study where artworks had to be contemplated in either an aesthetic ("focus on colors, tones, composition, and feelings elicited by the artwork") or pragmatic ("focus on what is depicted and follow the visual narrative of the artwork") context. Activation of BA10 is consistent with its role in top-down control of cognition, shifting processing towards more evaluative processing. Specifically, these areas are linked to the introspective evaluation of internal mental states (i.e. one's own thoughts and feelings), and might indicate self-referential processes in aesthetic evaluations (Cupchik et al. 2009).

In another study, Kirk, Skov, Hulme, Christensen, and Zeki (2009) showed that top-down classification of stimuli by defining them as genuine artworks, and hence framing an aesthetic context, or as computer-generated, changed evaluative stimulus processing. Participants gave higher aesthetic ratings for the same material presented as art as compared with non-art. Analyses of the brain states revealed that this was accompanied by specific changes in brain activations in the medial orbitofrontal cortex, which indicates hedonic value. Additionally, the prefrontal cortex was more strongly activated. The authors argue that this pattern of activation is consistent with subjects' expectations about the hedonic value of aesthetic experiences leading to an evaluative processing of the artwork.

These studies also provided evidence for the prediction of enhanced stylistic and formal processing in aesthetic evaluations. For example, Cupchik and colleagues (2009) reported that painting style influenced emotional ratings only in an aesthetic context. This was corroborated by a specific neuronal activation pattern. Perceptual visual areas associated with analyses of these structures, like the left superior parietal lobule (BA 7), were more strongly activated in the aesthetic compared to pragmatic context. Additionally, Kirk, Skove, Hulme, and colleagues (2009) reported increased activations in the primary visual cortex during aesthetic evaluations. Enhanced activation in these areas is attributable to a greater cognitive load placed on the visual systems through systematic analyses of stylistic aspects and formal (i.e. visual narrative) aspects of the paintings. Moreover, the entorhinal cortex and the temporal pole (see also Jacobsen et al. 2006) were activated more strongly. Such activations are linked with declarative memory processes (Kroll et al. 1997; Nakamura and Kubota 1995) and, according to the authors, they reflect recollection of contextually appropriate art-related and cultural information—such as recalling the specific style of the painting.

Noguchi and Murota (2013) studied the combined effects of two factors, a contextual manipulation (defining artworks, Renaissance sculptures of human bodies, as "genuine" or "fake") and, unknown by the participants, a stylistic manipulation (visually altering the proportions of the sculptures resulting in likenesses of deformed bodies) while recording EEG. As in the study by Kirk, Skov, Hulme, and colleagues (2009), participants provided higher aesthetic ratings for artworks defined as genuine compared to those defined as non-genuine. Additionally, participants showed marginally higher aesthetic ratings for the original compared to the altered versions. Both manipulations also produced significant differences in the ERPs. The P200 component showed a more positive going waveform for genuine as compared to non-genuine artworks, and for original as compared to altered artworks. This suggests that both kinds of manipulations were integrated and processed in the brain very rapidly (after between 200 ms and 300 ms). The authors also conducted source analyses of the EEG data, which permits estimation of the locations in the brain where this activity was generated. The parietal and medial prefrontal cortices were the sources of activation when a genuine versus fake context or when the original versus altered stimuli were compared. Additionally, for the stylistic comparison the occipital cortex was more strongly activated. These findings are consistent with the studies of Cupchik and colleagues (2009), Jacobsen and his team (2006), and Kirk and co-workers (2009).

Together, these studies confirm that context plays an important role in aesthetic experiences. It changes processing to a more evaluative style comprising activation changes in a widespread network of lower- and higher-order cognitive and emotional areas (Nadal et al. 2008; Chatterjee 2003; Vessel et al. 2012). It is a long-standing issue in the arts as to whether changes in emotional processing reflect genuine emotional reactions or whether these are specific to aesthetic experiences (Frijda 1988; Scherer 2005). In the next section we will discuss how emotional experience is affected by aesthetic processing.

4.3.2 Where physiology plays the decisive role: aesthetic emotions?

Philosophers and psychologists have discussed whether emotions triggered by the arts are different from common everyday emotions (Beardsley 1958; Frijda 1988; Scherer 2005). Scherer sees a crucial difference between aesthetic and everyday emotions, which he calls "utilitarian". Utilitarian emotions serve adaptive and homeostatic purposes to secure the well-being of an individual in significant events in life. Besides appraisals of novelty and intrinsic pleasantness, they comprise motivational appraisals like approach or withdrawal, action appraisals like fight or flight, and appraisals of coping potential, all designed to fulfil current and future needs. Aesthetic emotions, on the other hand, lack this kind of necessary immediacy of adapting to current events. They are instead produced by intrinsic appraisals of the aesthetic qualities of the object. Thus, motivational and action appraisals and the need to cope with the current situation should be less pronounced, or even absent, in aesthetic emotions (Scherer 2005). Similarly, Frijda (1988) assumes that the intensity of emotional reactions is reduced when a person appraises an event as not realistic or not directly relevant to ones' current or future needs. An aesthetic context, signaling a safe experience detached from reality (Beardsley 1958), provides the frame for such appraisals. For example, one rarely runs away from a cinema during a frightening horror movie, although fear-related responses will be triggered, including changes in facial expressions, increased heartbeat, and reported fear. Alternatively, some authors (Goldstein 2009; Mar and Oatley 2008) argue that because an aesthetic context lacks real-world consequences, it allows the individual to indulge in emotional reactions which are similar to or even stronger than those experienced in everyday life. Experiencing such emotions allows for risk-free practice for later life events (Dissanayake 2007; Tooby and Cosmides 2001), if and when similar circumstances require them.

Only a few empirical studies tested whether and how aesthetic and everyday emotions differ (especially in the domain of visual art). Several studies employed a contextual manipulation, which is very similar to the distinction between aesthetic versus realistic processing, by telling participants that stimuli were either fictional (aesthetic processing has somehow a fictional character; Tan 2000) or realistic. Goldstein (2009) asked participants to watch films evoking sadness or anger under these two contexts and rated the intensity of their emotional experience. Additionally, participants rated the intensity of anger and sadness in personal events in real life. Overall, intensity ratings did not differ much. Participants reported equal amounts of anger and sadness when watching a film either in a realistic or fictional context. In relation to personal life events, participants reported more intense anger reactions but not more intense sadness reactions. This means that, at the very least, consciously reported intensity of emotional reactions is very similar under fictional and real life events.

However, Mocaiber and colleagues (2010, 2011) reported that a fictional context attenuates components of the emotional reactions to negative stimuli as indicated by psychophysiological and behavioral parameters. A fictional, as compared to a realistic, context led to reduced amplitude of the LPP (Mocaiber et al. 2010), which is indicative of reduced emotional processing. Also, specific changes in heart rate—a lower defensive bradycardia reaction—were observed (Mocaiber et al. 2011) and reaction times slowed down. These results imply that emotional reactions to negative stimuli are attenuated within a fictional context, as a stimulus might not be judged to be relevant for oneself (Frijda 1988).

Critically, these studies used a very specific contextual manipulation by framing a fictional context (which is related to an aesthetic context, but is not the same; see discussion in section 4.3.1) and they did not use visual artworks. Conversely, in the study by Gerger, Leder, and Kremer (2014), participants evaluated positively and negatively valenced visual contemporary artworks and pictures from the International Affective Picture System (IAPS; Lang et al. 1999) under an aesthetic ("These stimuli are artworks") or a realistic context ("These stimuli are press photographs depicting real scenes"). To track changes of subjects' experiences, positive reactions were measured by levels of liking, of joy, and recording of the *musculus zygomaticus major* activity, and negative reactions by measuring anger, disgust, shame, sorrow, sadness, fear ratings, and *musculus corrugator supercilii* activity. Similarly to Goldstein (2009), they did not find any changes due to context for negative reactions, suggesting a very similar emotional experience for negative aspects. However, they found some changes for the positive reactions. Specifically, in the aesthetic compared to realistic context, negatively valenced artworks received higher liking and joy ratings, which was accompanied by a stronger reaction of the *musculus zygomaticus major*. For the more realistic-looking IAPS pictures, only the joy ratings were influenced. Emotionally negative pictures were perceived less negatively (in accordance with Mocaiber's results), but also emotionally positive IAPS pictures were perceived less positively. These results are consistent with the following conclusions. An aesthetic context enables a more positive perception of negatively valenced stimuli, especially when they are artworks. At the same time, the context does not erase negative reactions to emotionally negative stimuli. Thus, positive and negative aspects of the emotional experience are differentially affected by an aesthetic context. This is consistent with emotional models which posit that positive and negative aspects of emotional experiences are processed somewhat independently in different subsystems (Cacioppo et al. 1997; Norris et al. 2010). The results also imply that it is important to consider both positive and negative aspects of the emotional experience in order to answer the question whether and how an aesthetic context changes our emotions. We have seen that an aesthetic context changes our experiences on a cognitive as well as emotional level. In the next section we take a closer look at the early stage of information processing, from behavioral and physiological studies.

4.3.3 Early processes: changes in perceptual areas

As was shown in Figure 4.1, at the beginning of each visual aesthetic experience is a stage of perceptual processing. All psychological theories of art emphasize these kinds of processes. Perception, however, comprises a fairly complex set of stages and processes and involves the transformation of sensory information into visual experiences which to some extent elicit aesthetic pleasure. Chatterjee (2003) summarized this as early and intermediate visual processing.

Several studies have shown that aesthetic processing results in enhanced activities in the early and intermediate areas of visual processing. Fairhall and Ishai (2008) compared the perception of representational, ambiguous, and abstract artworks with scrambled control stimuli. The former showed stronger activation of the inferior occipital gyrus and gyrus fusiformis, where object form is processed, and the dorsal occipital cortex, which processes spatial properties of visual objects. Additionally, Cupchik and colleagues (2009) found indications for stronger visuo-spatial analyses in an aesthetic context compared to a pragmatic context, and Kirk and his team (2009) reported stronger activations in early visual areas (primary visual cortex) under an aesthetic context (see section 4.3.1). Additionally, increased activations in these visual areas correlate with subjective aesthetic evaluations. Vartanian and Goel (2004), employing abstract and representational paintings, showed that activation in the occipital gyrus, including the fusiform gyrus, linearly increased as a function of subjective preference ratings. Similarly, Vessel and colleagues (2012) used abstract and representational paintings as well, and reported linear increases of activity in visual areas in the occipito-temporal cortex as a function of subjectively reported aesthetic value.

However, activation patterns in these early and intermediate visual processing areas often show little overlap in different studies (e.g. Nadal et al. 2008). They depend critically on task demands, instructions, procedures, and stimulus materials used. For example, aesthetic processing of representational as compared to abstract artworks engages networks of object perception more strongly (Fairhall and Ishai 2008; Vartanian and Goel 2004). Overall, empirical studies provided some evidence that early and intermediate visual processing areas are consistently more strongly activated during aesthetic processing. This stronger activation might reflect a top-down enhancement due to attentional and/or emotional processes (Vuilleumier and Driver 2007). Further research is required in order to clarify the exact nature of these processes.

The stage of perceptual analyses (Figure 4.1) considers the effects of a number of variables that influence a given image's aesthetic appeal. Among these variables are complexity, contrast, symmetry, order, and grouping. The combination of psychophysiological and behavioral methods can be fruitful in clarifying the contribution of such perceptual factors. For example, according to Berlyne (1971), visual complexity affects aesthetic responses through changes in arousal. However, attempts to test this prediction by using abstract patterns and psychophysiological measures of EDA failed to detect any effect of visual complexity on physiological arousal (Berlyne and Lawrence 1964). Krupinski and Locher (1988) manipulated complexity with different levels of symmetry. Their participants rated asymmetrical, single-, and double-axes symmetrical patterns according to their hedonic value while SCR was recorded. In contrast to Berlyne and colleagues (1963), they found a linear increase in SCR and hedonic ratings with an increase in complexity. These diverging results regarding the relationship between visual complexity and physiological arousal during the aesthetic evaluation of abstract patterns resemble results concerning the effects of visual complexity on art appreciation.

Nadal and colleagues (2010) sought to resolve the ambiguity concerning the results of complexity in art appreciation—ambiguous, because some studies found that simplicity was preferred. Nadal and colleagues (2010) combined various methods. They employed a large set of abstract and representational artworks as well as non-artistic images which were selected to represent three levels of complexity, high, medium, or low. When participants rated these stimuli for perceived beauty, there was no clear pattern of influences of complexity on aesthetics. Interestingly, Nadal and colleagues (2010) classified their stimuli according to several dimensions that in the literature had been discussed as determinants of complexity, such as the number and variety of elements, an image's three-dimensionality, the general appeal of symmetry, etc. When they analysed the effects according to these dimensions, none of them had a general effect on beauty. The authors concluded that the perceived complexity might have—more or less—individual causes in different dimensions and that these variations might conceal clear effects of complexity on appreciation of artworks. However, the nature of the processes that determine how complexity affects the perception of artworks still remains an interesting topic for future studies. In Leder and colleagues' (2004) approach, the specificity of art is not so much related to formal features such as complexity, but rather it is assigned to the distinction between content and style.

4.3.3.1 Representations: content and style? Art specific?

Kuchinke, Trapp, Jacobs, and Leder (2009) showed how physiology can inform the processes concerned with explicit content recognition. They tested the hypothesis that at the moment of insight when content of an artwork is recognized, changes in emotional responses can be observed. They looked at Cubist paintings, whose content takes the viewer time to recognize (Hekkert and van Wieringen 1996). Pablo Picasso and Georges Braque developed a style which depicts the content in simultaneously overlaid or combined perspective views. This style slows recognition because the constituting elements cannot easily be assigned to one view. In some cases, recognition of the depicted object exceeds several seconds. In everyday life, however, object recognition is usually very fast and takes only fractions of a second. Kuchinke and colleagues (2009) measured pupillary responses and showed that the moment of content recognition was indeed accompanied by an emotional response (Figure 4.1). Thus, at exactly the same time as an artwork's subject was identified, participants showed a pupillary response that could be interpreted as an elicited emotion of pleasurable insight.

Style is considered an important feature in the processing of art (Leder et al. 2004). Regarding the processes underlying the recognition of style and content Augustin, Leder, Hutzler, and Carbon (2008) studied the time it takes for style and content to affect the perception of artworks by presenting images for very short times. Two artworks—sometimes the same, sometimes different, in content, style, by the same or a different artist—had to be evaluated according to their perceived similarity. With a presentation time of 10 ms, viewers could only distinguish matters of content. However, perceived similarity also was affected by painting style with 50 ms. Thus, temporally, style follows content. In a later study, Augustin, Defranceschi, Fuchs, Carbon, and Hutzler (2011) employed ERP to study the neural time-course underlying the perception of content or style. Participants viewed pictures that systematically varied in style and content and had to perform a go/no-go dual-choice task. Results again supported the hypotheses that in the processing of art, style follows content. Style-related information was available later than content-related information as indicated by certain ERP components (lateralized readiness potential—LRP, and N200). In the future, physiological measures will serve to improve our understanding of the aesthetic experiences as they are experienced over time, and as we become able to measure short-term effects and dynamic changes continuously.

4.3.3.2 Art expertise and knowledge

Experts are experts because they know more about a given subject than most individuals, and their knowledge is deeper. Knowledge about an artist's work and life, specific art styles and movements, and the process of artistic production aids stylistic and conceptual interpretation (Leder et al. 2004). The increased ability to interpret art, in turn, facilitates understanding and art appreciation. Psychological research on art expertise mainly used two approaches: (i) studying the effect of art-specific information on laypeople's art experience, or (ii) contrasting laypeople and art experts (people with formal training in the arts or art professionals). Both approaches revealed how knowledge and art expertise affects processing on a cognitive, perceptual, and affective level.

Art-specific knowledge provided by additional information shapes processing on a higher-order cognitive level such as finding meaning, resolving ambiguity, and developing understanding. Due to their lack of pre-existing knowledge, laypeople are particularly suitable for projects which systematically manipulated art-specific knowledge. Several behavioral studies using this approach showed that additional information positively affects art appreciation and understanding (Leder et al. 2006; Millis 2001; Russell 2003; Swami 2013). However, this depends on relative expertise level and art style. Especially with regard to abstract art, additional information helps laypeople adopt style-focused processing opens new perspectives for interpretation, and therefore increases their understanding of what they are viewing. Swami (2013) emphasized that increased understanding mediates the positive effect of additional information on art appreciation. However, the impact of additional information on art appreciation also relies on the pre-existing (relative) level of expertise. For example, Belke, Leder, and Augustin (2006) presented abstract artworks with or without stylistic information to participants differing in their relative level of art expertise. Participants with low art expertise showed an increased liking for abstract art when style-related information was given; participants with high expertise, in contrast, showed a decrease in liking. This negative effect of information on the appreciation of art was presumably due to its interference with pre-existing knowledge of the experts

Only a few studies examined the effect of additional information on art appreciation by employing psychophysiological measures. Lengger, Fischmeister, Leder, and Bauer (2007) studied the effect of stylistic information on the evaluation of abstract and representational art. Understanding increased when stylistic information was provided. However, on a neuronal level, artworks presented with stylistic information compared to artworks without stylistic information evoked a weaker activation mainly in left frontal and parietal networks. This could indicate that art-specific information, by increasing knowledge and understanding, helps to process artworks more efficiently on a neuronal level (Haier et al.

1988; Lengger et al. 2007). Taken together, these studies show that specific information and knowledge significantly affect understanding of art and its appreciation.

Choosing another approach to investigate the effect of knowledge on evaluation of art, Pang, Nadal, Müller-Paul, Rosenberg, and Klein (2013) compared the neuronal response of laypeople and art experts while evaluating art and a set of control stimuli. Consistent with the study by Lengger and co-workers (2007), Pang and colleagues found that knowledge leads to more efficient processing on a neuronal level—ERP amplitudes of the P3b and LPP, which are linked to knowledge retrieval, were reduced for art experts compared to laypeople. Interestingly, in the study by Pang and colleagues, these expertise effects were not limited to artworks; they also extended to control stimuli (filtered artworks and simple color-field stimuli). This suggests that art experts in general have a different style of visual processing.

This is partly supported by a study by Long, Peng, Chen, Jin, and Yao (2011). They showed that early perceptual processes like color processing differs between art experts (artists) and laypeople on a neuronal level. First, MRI scans revealed structural differences between artists and laypeople indicated by a higher density of gray matter in brain areas involved in color processing (V4) for artists. Secondly, during a color-naming task, artists showed higher activity in V4 than laypeople. This contradicts the neural-efficiency hypothesis, in which we would expect a decreased activity for artists. These diverging results imply that the relation between expertise and efficient neural processing may be limited to specific tasks, brain regions, or subtypes of art expertise (but see Neubauer and Fink 2009). Nevertheless, a concurrent finding is that the level of art expertise affects visual processing on a neural level.

On a behavioral level, experts and laypeople differ in how they visually explore artworks. Cupchik and Laszlo (1992) assume that art experts have learned to focus on stylistic and conceptual features whereas laypeople adopt an object-oriented style of processing typical for everyday perception. Several eye-tracking studies support this hypothesis. Laypeople directed their gaze more often to pictorial content, but art experts attended more to structural and compositional features (Nodine et al. 2009; Vogt and Magnussen 2007).

Art expertise also influences affective components of aesthetic experiences. According to Leder and colleagues (2004), understanding and successful interpretation, both core factors of art expertise, are self-rewarding and affectively positive. In line with this, Kirk, Skov, Christensen, and Nygaard (2009) showed that architecture experts exhibited higher activity in brain areas associated with reward processing (medial orbitofrontal cortex and subcallosal cingulate gyrus) compared to non-experts when they evaluated pictures of buildings. If a high level of art expertise generally increases positive affective experiences of artworks, then art experts compared to laypeople should also feel more positive even when they look at negatively valenced art. Leder, Gerger, Brieber, and Schwarz (2014) addressed this issue by measuring emotional responses when art experts and laypeople looked at artworks and non-artworks (IAPS pictures) with affectively positive and negative content. In general, positive stimuli led to a higher zygomaticus activation and negative stimuli to a higher corrugator activation, and positive stimuli were liked more than negative ones, but this effect was modulated by expertise. In line with the argumentation that expertise fosters a detached (Beardsley 1958) style of processing, experts exhibited attenuated emotional responses as indicated by less extreme corrugator supercilii activations and less extreme valence ratings. Additionally, there was a specific effect of liking—experts liked negative artworks more compared to laypeople. Thus, experts' reduced emotional reactivity might have fostered the positive aesthetic evaluation of even emotionally negative art.

In addition, art experts' aesthetic evaluations are less susceptible to biasing information. In an fMRI study by Kirk, Harvey, and Montague (2011) participants evaluated artworks which were unobtrusively presented with either of two company logos. One of them was supposed to be from the company sponsoring the study which paid $300 for participation. Laypeople showed higher liking ratings for artworks which were presented with a sponsor logo; art experts, in contrast, did not show this sponsorship effect. For both groups, activity in the ventromedial prefrontal cortex (VMPFC) increased with liking, but laypeople showed an even higher increase for sponsored artworks. Interestingly, for art experts, dorsolateral prefrontal cortex (DLPFC) activity correlated negatively with sponsorship effect and showed a stronger connectivity with VMPFC. Activity in the DLPFC has been associated with top-down executive control and emotion regulation, which could explain why the sponsor logo had no effect on art experts' aesthetic judgments. Hence, art experts were less susceptible to external biasing information and, thus, more independent in forming their aesthetic judgment.

In sum, knowledge as a core feature of art expertise seems to facilitate independent aesthetic judgments and enables us to master viewing artworks more extensively on a higher-order cognitive level. Moreover, art knowledge and frequent encounters with art not only improves our ability to understand and interpret art, but also changes how we process art on perceptual and affective levels.

4.4 Conclusion

In this chapter we have covered a set of research topics from empirical aesthetics to show how they can be studied by employing psychophysiological methods. This overview was selective rather than comprehensive. Nevertheless, regarding most components involved in art experience, the combination of behavioral and physiological measures is very fruitful. Where affective states are concerned, facial EMG apparently can be a good method to adopt, but in the future, fMRI may prove more productive when looking at the involvement of brain regions. Undoubtedly, EEG will remain a valuable tool in the future, particularly with regard to studying the sequence of processes.

However, there are also important restrictions. Our selection of methods is incomplete and future developments are already apparent. Kapoula and colleagues (2011) studied whole-body movements and the shifting of body weight towards or away from an artwork. Moreover, peripheral measures like heart-rate variability and electro-dermal activity might play a role in uncovering the nature of arousal and activation during art perception. Finally, much is anticipated from more network-based analyses of scanned activity, and from combining genetic research and brain function to uncover individual differences, and the field of eye-movement research could be a rich source of data, only briefly covered here.

Empirical aesthetics was often restricted to laboratory studies with distinct limitations: lab set-ups are not representative of the diverse experiences one has in a museum and they might miss essential features. Technical limitations, for example, in the resolution of artwork reproductions, or their size, can obstruct research. Moreover, there is still a huge bias towards reception. Production of artworks, although hard to imagine in the high arts as a topic for cognitive experimenters, will be a rich field of future research.

Finally, the objects of our inquiries—the arts—are in constant development. Today, the acknowledgment of contemporary world art (Vogel 2013) challenges the traditional Eurocentric concepts of high art, and it requires flexible, self-critical, and globalized views that are often not compatible with what psychologists currently understand as "art".

Acknowledgment

This chapter was supported by a grant from WWTF CS11-023-LedRos and a FWF grant P23538 to Helmut Leder.

Notes

1 Magritte, R. (1929). *La trahison des images [The treachery of images]*. Retrieved from <http://collections.lacma.org/node/239578>.

2 Duchamp, M. (1917, replica 1964). *Fountain*. Retrieved from <http://www.tate.org.uk/art/artworks/duchamp-fountain-t07573>.

3 Warhol, A. (1964). *Brillo Boxes*. Philadelphia Museum of Art, Philadelphia, PA. Retrieved from <http://www.philamuseum.org/collections/permanent/89204.html>.

4 For images, see: <https://www.google.co.uk/search?q=beuys+blocks+of+fat&rlz=1C1OPRB_enGB566GB566&espv=2&biw=1280&bih=699&tbm=isch&tbo=u&source=univ&sa=X&ei=aYOmVKmEDsPJOamxgNAP&ved=0CCgQsAQ>.

References

Augustin, M.D., Defranceschi, B., Fuchs, H.K., et al. (2011). The neural time course of art perception: An ERP study on the processing of style versus content in art. *Neuropsychologia* **49**, 2071–81. doi: 10.1016/j.neuropsychologia.2011.03.038.

Augustin, M.D., Leder, H., Hutzler, F., et al. (2008). Style follows content: On the microgenesis of art perception. *Acta Psychologica* **128**, 127–38.

Augustin, M.D., Wagemans, J., and Carbon, C.-C. (2012). All is beautiful? Generality vs. specificity of word usage in visual aesthetics. *Acta Psychologica* **139**, 187–201. doi: 10.1016/j.actpsy.2011.10.004.

Beardsley, M.C. (1958). *Aesthetics: Problems in the philosophy and criticism*. New York, NY: Harcourt.

Belke, B., Leder, H., and Augustin, M.D. (2006). Mastering style—Effects of explicit style-related information, art knowledge and affective state on appreciation of abstract paintings. *Psychology Science* **48**, 115–34.

Berlyne, D.E. (1971). *Aesthetics and psychobiology*. New York, NY: Appleton Century Crofts.

Berlyne, D.E., Craw, M., Salapatek, P.H., et al. (1963). Novelty, complexity, incongruity, extrinsic motivation, and the GSR. *Journal of Experimental Psychology* **66**, 560–7.

Berlyne, D.E. and Lawrence, G. (1964). Effects of complexity and incongruity variables on GSR, investigatory behavior and verbally expressed preference. *Journal of General Psychology* **71**, 21–45.

Bradley, M.M., Codispoti, M., Cuthbert, B.N., et al. (2001). Emotion and motivation I: Defensive and appetitive reactions in picture processing. *Emotion* **1**, 276–98. doi:10.1037//1528-3542.1.3.276.

Bradley, M.M., Cuthbert, B.N., and Lang, P.J. (1990). Startle reflex modification: Emotion or attention? *Psychophysiology* **27**, 513–22.

Bradley, M.M., Miccoli, L., Escrig, M., et al. (2008). The pupil as a measure of emotional arousal and autonomic activation. *Psychophysiology* **45**, 602–7. doi:10.1111/j.1469–8986.2008.00654.x.

Bühler, K. (1913). *Die Gestaltwahrnehmungen. Experimentelle Untersuchungen zur psychologischen und ästhetischen Analyse der Raum- und Zeitanschauung*. Stuttgart: Spemann.

Cacioppo, J.T., Gardner, W.L., and Berntson, G.G. (1997). Beyond bipolar conceptualizations and measures: the case of attitudes and evaluative space. *Personality and Social Psychology Review* **1**, 3–25. doi: 10.1207/s15327957pspr0101_2.

Chatterjee, A. (2003). Prospects for a cognitive neuroscience of visual aesthetics. *Bulletin of Psychology and the Arts* **4**, 55–60.

Chatterjee, A. (2011). Neuroaesthetics: A coming of age story. *Journal of Cognitive Neuroscience* **23**, 53–62.

Cupchik, G.C. and Laszlo, J. (1992). *Emerging visions of the aesthetic process. Psychology, semiology and philosophy*. New York, NY: Cambridge University Press.

Cupchik, G.C., Vartanian, O., Crawley, A., et al. (2009). Viewing artworks: Contributions of cognitive control and perceptual facilitation to aesthetic experience. *Brain and Cognition* **70**, 84–91. doi: 10.1016/j.bandc.2009.01.003.

Cupchik, G.C. and Winston, A.C. (1996). Confluence and divergence in empirical aesthetics philosophy and mainstream psychology. In E.T. Carterette and M.P. Friedman (eds), *Handbook of Perception & Cognition: Cognitive Ecology*. San Diego, CA: Academic Press, pp. 62–85.

Currie, G. (2011). Empathy for objects. In A. Coplan and P. Goldie (eds), *Empathy: Philosophical and psychological perspectives*. Oxford: Oxford University Press, pp. 82–97.

Cuthbert, B.N., Schupp, H.T., Bradley, M.M., et al. (2000). Brain potentials in affective picture processing: Covariation with autonomic arousal and affective report. *Biological Psychology* **52**, 95–111. doi: 10.1016/s0301-0511(99)00044-7.

Davies, S. (2012). *The Artful Species: Aesthetics, Art, and Evolution*. Oxford: Oxford University Press.

Dimberg, U. (1988). Facial electromyography and the experience of emotion. *Journal of Psychophysiology* **2**, 277–82.

Dimberg, U. and Karlsson, B. (1997). Facial reactions to different emotionally relevant stimuli. *Scandinavian Journal of Psychology* **38**, 297–303. doi:10.1111/1467-9450.00039.

Dissanayake, E. (2007). What art is and what art does: An overview on contemporary evolutionary hypothesis. In C. Martindale, P.J. Locher, and V. Petrov (eds) *Evolutionary and Neurocognitive Approaches to Aesthetics, Creativity, and the Arts*. Amityville, NY: Baywood, pp. 1–14.

Dutton, D. (2009). *The Art Instinct: Beauty, Pleasure and Human Evolution*. New York, NY: Bloomsbury Press, 2009.

Ekman, P., Friesen, W.V., and Ancoli, S. (1980). Facial signs of emotional experience. *Journal of Personality and Social Psychology* **39**, 1125–34. doi: 10.1037/h0077722.

Fairhall, S.L., and Ishai, A. (2008). Neural correlates of object indeterminacy in art compositions. *Consciousness and Cognition* **17**, 923–32. doi: 10.1016/j.concog.2007.07.005.

Fechner, G.T. (1876). *Vorschule der Ästhetik. Erster Teil*. Leipzig: Breitkopf & Haertel.

Frijda, N.H. (1988). The laws of emotion. *American Psychologist* **43**, 349–58. doi: 10.1037//0003-066x.43.5.349.

Frijda, N. and Schram, D. (1995). Emotions and cultural products: Psychological reaction to the arts and literature. *Poetics* **23**, 1–6. doi: 10.1016/0304-422X(95)90009-W.

Gerger, G., Leder, H., and Kremer, A. (2014). Context effects on emotional and aesthetic evaluations of artworks and IAPS pictures. *Acta Psychologica* 151, 174–83. doi: 10.1016/j.actpsy.2014.06.008.

Goldstein, T.R. (2009). The pleasure of unadulterated sadness: experiencing sorrow in fiction, nonfiction, and "in person." *Psychology of Aesthetics Creativity and the Arts* **3**, 232–7. doi: 10.1037/a0015343.

Haier, R.J., Siegel Jr., B.V., Nuechterlein, K.H., et al. (1988). Cortical glucose metabolic rate correlates of abstract reasoning and attention studied with positron emission tomography. *Intelligence* **12**, 199–217.

Hajcak, G., Weinberg, A., MacNamara, A., et al. (2012). ERPs and the study of emotion. In S.J. Luck and E.S. Kappenmann (eds), *Oxford Handbook of ERP components*. New York, NY: Oxford University Press, pp. 441–74.

Hekkert, P. and van Wieringen, P.C.W. (1996). The impact of level of expertise on the evaluation of original and altered versions of post-impressionistic paintings. *Acta Psychologica* **94**, 117–31. doi: 10.1016/0001-6918(95)00055-0.

Jacobsen, T. and Hoefel, L. (2003). Descriptive and evaluative judgment processes: Behavioral and electrophysiological indices of processing symmetry and aesthetics. *Cognitive Affective and Behavioral Neuroscience* **3**, 289–99. doi: 10.3758/cabn.3.4.289.

Jacobsen, T., Schubotz, R.I., Höfel, L., et al. (2006). Brain correlates of aesthetic judgment of beauty. *Neuroimage* **29**, 276–85. doi: 10.1016/j.neuroimage.2005.07.010.

Jakesch, M. and Leder, H. (2009). Finding meaning in art: Preferred levels of ambiguity in art appreciation. *Quarterly Journal of Experimental Psychology* **62**, 2105–12. doi: 10.1080/17470210903038974.

James, W. (1890). *The principles of psychology*. New York, NY: Holt, Rinehart & Winston.

Kahneman, D. and Beatty, J. (1966). Pupil diameter and load on memory. *Science* **154**, 1583–5.

Kapoula, Z., Adenis, M.-S., Thanh-Thuan, L., et al. (2011). Pictorial depth increases body sway. *Psychology of Aesthetics Creativity and the Arts* **5**, 186–93. doi: 10.1037/a0022087.

Kirk, U., Harvey, A., and Montague, P.R. (2011). Domain expertise insulates against judgment bias by monetary favors through a modulation of ventromedial prefrontal cortex. *Proceedings of the National Academy of Science of the USA* **108**, 10332–6. doi:10.1073/pnas.1019332108/-/DCSupplemental. www.pnas.org/cgi/doi/10.1073/pnas.1019332108.

Kirk, U., Skov, M., Christensen, M.S., et al. (2009). Brain correlates of aesthetic expertise: A parametric fMRI study. *Brain and Cognition* **69**, 306–15. doi:10.1016/j.bandc.2008.08.004.

Kirk, U., Skov, M., Hulme, O., et al. (2009). Modulation of aesthetic value by semantic context: An fMRI study. *NeuroImage* **44**, 1125–32. doi: 10.1016/j.neuroimage.2008.10.009.

Kroll, N.E.A., Markowitsch, H.J., Knight, R.T., et al. (1997). Retrieval of old memories: the temporofrontal hypothesis. *Brain* **120**, 1377–99. doi: 10.1093/brain/120.8.1377.

Krupinski, E. and Locher, P.J. (1988). Skin conductance and aesthetic evaluative responses to nonrepresentational works of art varying in symmetry. *Bulletin of the Psychonomic Society* **26**, 355–8.

Kuchinke, L., Trapp, S., Jacobs, A.M., et al. (2009). Pupillary responses in art appreciation: Effects of aesthetic emotions. *Psychology of Aesthetics Creativity and the Arts* **3**, 156–63. doi: 10.1037/a0014464.

Laeng, B., Sirois, S., and Gredeback, G. (2012). Pupillometry: A window to the preconscious? *Perspectives on Psychological Science* **7**, 18–27. doi:10.1177/1745691611427305.

Lang, P.J., Bradley, M.M., and Cuthbert, B.N. (1999). *International affective picture system: Technical manual and affective ratings*: Gainesville, FL: The Center for Research in Psychophysiology, University of Florida.

Leder, H. (2013). Next steps in neuroaesthetics: which processes and processing stages to study? *Psychology of Aesthetics Creativity and the Arts* **7**, 27–37. doi: 10.1037/a0031585.

Leder, H. (2014). Beyond perception—Information processing approaches of art appreciation. In P.P.L. Tinio and J. Smith (eds), *Cambridge Handbook of Aesthetics and the Arts*. Cambridge, MA: Cambridge University Press, pp. 115–38.

Leder, H., Augustin, D., and Belke, B. (2005). Art and cognition! Consequences for experimental aesthetics. *Bulletin of Psychology and the Arts* **2**, 11–20.

Leder, H., Belke, B., Oeberst, A., et al. (2004). A model of aesthetic appreciation and aesthetic judgments. *British Journal of Psychology* **95**, 489–508. doi: 10.1348/0007126042369811.

Leder, H., Carbon, C.-C., and Ripsas, A.-L. (2006). Entitling art: Influence of title information on understanding and appreciation of paintings. *Acta Psychologica* **121**, 176–98. doi:10.1016/j.actpsy.2005.08.005.

Leder H., Gerger G., Brieber, D., et al. (2014). What makes an art expert? Emotion and evaluation in art appreciation. *Cognition and Emotion* **28**, 1137–47. doi: 10.1080/02699931.2013.870132.

Lengger, P.G., Fischmeister, F.P.S., Leder, H., et al. (2007). Functional neuroanatomy of the perception of modern art: A DC-EEG study on the influence of stylistic information on aesthetic experience. *Brain Research* **1158**, 93–102. doi:10.1016/j.brainres.2007.05.001.

Lishner, D.A., Cooter, A.B., and Zald, D.H. (2008). Rapid emotional contagion and expressive congruence under strong test conditions. *Journal of Nonverbal Behavior* **32**, 225–39. doi: 10.1007/s10919-008-0053-y.

Long, Z., Peng, D., Chen, K., et al. (2011). Neural substrates in color processing: A comparison between painting majors and non-majors. *Neuroscience Letters* **487**, 191–5. doi:10.1016/j.neulet.2010.10.020.

Mar, R.A. and Oatley, K. (2008). The function of fiction is the abstraction and simulation of social experience. *Perspectives on Psychological Science* **3**, 173–92. doi: 10 1111/j.1745-6924.2008.00073.x.

Markovic, S. (2012). Components of aesthetic experience: aesthetic fascination, aesthetic appraisal, and aesthetic emotion. *i-Perception* **3**, 1–17. doi: 10.1068/i0450aap.

Millis, K. (2001). Making meaning brings pleasure: The influence of titles on aesthetic experiences. *Emotion* **1**, 320–9. doi:10.1037//1528-3542.1.3.320.

Mocaiber, I., Perakakis, P., Pereira, M., et al. (2011). Stimulus appraisal modulates cardiac reactivity to briefly presented mutilation pictures. *International Journal of Psychophysiology* **81**, 299–304. doi: 10.1016/j.ijpsycho.2011.07.014.

Mocaiber, I., Pereira, M., Erthal, F., et al. (2010). Fact or fiction? An event-related potential study of implicit emotion regulation. *Neuroscience Letters* **476**, 84–8. doi: 10.1016/j.neulet.2010.04.008.

Nadal, M., Munar, E., Capo, M.A., et al. (2008). Towards a framework for the study of the neural correlates of aesthetic preference. *Spatial Vision* **21**, 379–83.

Nadal, M., Munar, E., Marty, G., et al. (2010). Visual complexity and beauty appreciation. Explaining the diverging of results. *Empirical Studies of the Arts* **28**, 173–91.

Nakamura, K. and Kubota, K. (1995). Mnemonic firing of neurons in the monkey temporal pole during a visual recognition memory task. *Journal of Neurophysiology* **74**, 162–78.

Neubauer, A.C. and Fink, A. (2009). Intelligence and neural efficiency. *Neuroscience and Biobehavioral Reviews* **33**, 1004–23. doi:10.1016/j.neubiorev.2009.04.001.

Nodine, C.F., Locher, P.J., and Krupinski, E.A. (2009). The role of formal art training on perception an aesthetic judgement of art compositions. *Leonardo* **26**, 219–27.

Noguchi, Y. and Murota, M. (2013). Temporal dynamics of neural activity in an integration of visual and contextual information in an esthetic preference task. *Neuropsychologia* **51**, 1077–84. doi: 10.1016/j.neuropsychologia.2013.03.003.

Norris, C.J., Gollan, J., Berntson, G.G., et al. (2010). The current status of research on the structure of evaluative space. *Biological Psychology* **84**, 422–36. doi: 10.1016/j.biopsycho.2010.03.011.

O'Doherty, B. (1976). *Inside the white cube: The ideology of the gallery space*. San Francisco, CA: Lapis Press, First University of California.

Pang, C.Y., Nadal, M., Müller-Paul, J.S., et al. (2013). Electrophysiological correlates of looking at paintings and its association with art expertise. *Biological Psychology* **93**, 246–54. doi:10.1016/j.biopsycho.2012.10.013.

Russell, P.A. (2003). Effort after meaning and the hedonic value of paintings. *British Journal of Psychology* **94**, 99–110. doi:10.1348/000712603762842138.

Scherer, K.R. (2005). What are emotions? And how can they be measured? *Social Science Information* **44**, 695–729. doi: 10.1177/0539018405058216.

Scherer, K.R. and Ellgring, H. (2007). Are facial expressions of emotion produced by categorical affect programs or dynamically driven by appraisal? *Emotion* **7**, 113–30. doi: 10.1037/1528-3542.7.1.113.

Shiner, L. (2003). *The invention of art*. Chicago, IL: Chicago University Press.

Steinhauer, S.R., Condray, R., and Kasparek, A. (2000). Cognitive modulation of midbrain function: Task-induced reduction of the pupillary light reflex. *International Journal of Psychophysiology* **39**, 21–30.

Swami, V. (2013). Context matters: Investigating the impact of contextual information on aesthetic appreciation of paintings by Max Ernst and Pablo Picasso. *Psychology of Aesthetics, Creativity, and the Arts* **7**, 285–95. doi:10.1037/a0030965.

Tan, E.S. (2000). *Emotion, art and the humanities*. In M. Lewis and J.M. Haviland-Jones (eds), *Handbook of Emotions*. New York, NY: Guilford Press, pp. 116–36.

Tooby, J. and Cosmides, L. (2001). Does beauty build adapted minds? Toward an evolutionary theory of aesthetics, fiction and the arts. *Sub-Stance* (94–5), 6–27.

Vartanian, O. and Goel, V. (2004). Neuroanatomical correlates of aesthetic preference for paintings. *Neuroreport* **15**, 893–7. doi: 10.1097/01.wnr.0000118723.38067.d6.

Vartanian, O. and Nadal, M. (2007). A biological approach to a model of aesthetic experience. In L. Dorfman, C. Martindale, and V. Petrov (eds), *Aesthetics and Innovation*. Newcastle: Cambridge Scholars Publishing, pp. 429–44.

Vessel, E.A., Starr, G.G., and Rubin, N. (2012). The brain on art: Intense aesthetic experience activates the default mode network. *Frontiers in Human Neuroscience* **6**, 1–17. doi: 10.3389/fnhum.2012.00066.

Vogel, S.B. (2013). Globalkunst—Eine neue Weltordnung. *Kunstforum*. Bd. 220.

Vogt, S. and Magnussen, S. (2007). Expertise in pictorial perception: Eye-movement patterns and visual memory in artists and laymen. *Perception* **36**, 91–100. doi:10.1068/p5262.

Vuilleumier, P. and Driver, J. (2007). Modulation of visual processing by attention and emotion: windows on causal interactions between human brain regions. *Philosophical Transactions of the Royal Society B* **362**, 837–55. doi: 10.1098/rstb.2007.2092.

Chapter 5

The moving eye of the beholder: Eye tracking and the perception of paintings

Raphael Rosenberg and Christoph Klein

5.1 Gaze movements as a literary model in art history

Gaze movements have been an issue in the history of art long before they became a research topic in psychology. Early evidence dates back to the sixth century, when in 553 the Byzantine historian Procopius of Caesarea described how the interior space of the Hagia Sophia (Istanbul) overwhelms the beholder:

> each detail attracts the eye and draws it on irresistibly to itself. So the vision constantly shifts suddenly, for the beholder is utterly unable to select which particular detail he should admire more than all the others. But even so, though they turn their attention to every side and look with contracted brows upon every detail, observers are still unable to understand the skilful craftsmanship. (Procopius 1940, pp. 23–5)

In the course of early modern times, reflexions on gaze movements have significantly increased in writings about art, and they were repeatedly used to justify aesthetic qualities. In the 1460s the Florentine architect and sculptor Filarete explains the superiority of the round arch over the gothic pointed arch by referring to the eye movements it induces:

> It cannot be doubted that nothing which impedes the sight in any way is as beautiful as the one that leads the eye rather than restraining it. Such is the round arch. As you have noticed, your eye is not arrested in the least when you look at a half-circle arch . . . The pointed is not so, for the eye, or sight, pauses a little at the pointed part and does not run along as it does on the half circle. (Filarete 1972, pp. 230–1, translation by the authors)

Denis Diderot, the French philosopher who was also a founding father of art criticism, went significantly further. In 1767, comparing two altarpieces by the painters Gabriel François Doyen and Joseph-Marie Vien, he drew an explicit link between the composition of paintings and the gaze of the beholder. For Diderot, composition is an instruction to the eye, a path which the gaze follows in a certain order. He describes Vien's painting (Figure 5.1) as follows:

> Such is the path followed in perusing this composition, Religion, the angel, the saint, the women at his feet, the listeners in the background, and those in the left background, the two tall standing female figures, the elderly man leaning forward at their feet, and the two figures, one a man and the

> other a woman, seen from the back and placed squarely in the foreground, this path descending gently and meandering . . .; a line of liaison that clearly, crispy, and effortlessly links the composition's principal features. (Diderot 1995, p. 29)

For Diderot, Vien's painting is an example of a beautiful "line of liaison" whereas he finds Doyen's piece (Figure 5.2) very poor in this regard:

> In every composition there's a path, a line . . . If this line, which I call the line of liaison, bends, folds over onto itself, twists, is agitated, if its circumvolutions are diminutive, multiple, recti-linear, and angular, the composition will be ambiguous and obscure; the eye, wandering at random through a labyrinth, bewildered, will find it difficult to grasp the connections . . . If it is broken, the composition will have empty spots, holes . . . Doyen's Miracle of Saint Anthony's Fire is not above reproach in this respect: its line of liaison is fractured, bent, folded, and twisted. It is difficult to follow; sometimes it's ambiguous, or comes to a sudden halt, or requires considerable indulgence from the eye that's trying to follow its course. A well-ordered composition will always have but one true line of liaison; and it will serve as guide to anyone looking at it as well as to anyone attempting to describe it. (Diderot 1995, p. 152)

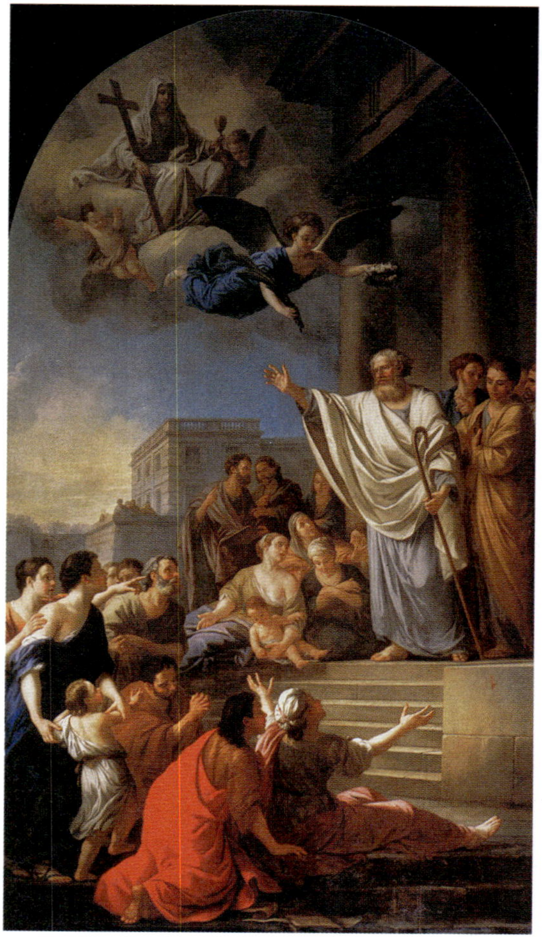

Figure 5.1 Joseph-Marie Vien, *St. Denis Preaching in Gaul*, 1767. © The Art Archive / Collection Dagli Orti.

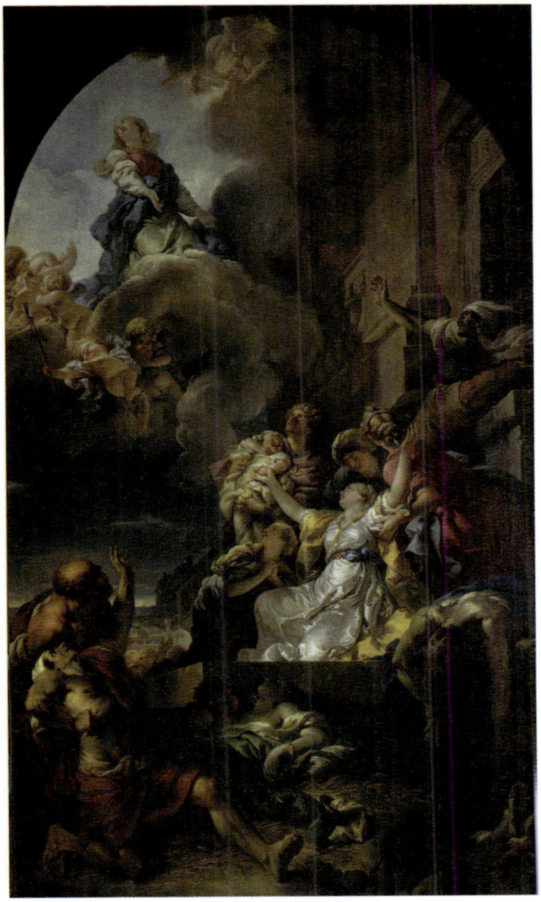

Figure 5.2 Gabriel François Doyen, *The Miracle of Saint Anthony's Fire*, 1767.

As an academic discipline, art history was established in the middle of the nineteenth century; the language it used was based on much older writings about art. The idea of the movement of the eye as related to the aesthetic and the structure of the work of art was to remain an important paradigm of this new discipline. Some art historians thought that stylistic differences between artworks of different times and places were due to changes in the behavior of the eye. Thus in 1899, Heinrich Wölfflin, one of the most influential art historians in the early twentieth century, explains the difference between the Italian arts of the early Renaissance—that is, the fifteenth century—and the High Renaissance of the early sixteenth century by the eye's need to relax:

> The Quattrocento placed unbelievable demands on the eye. The viewer not only has the greatest difficulty in picking out individual physiognomies from the serried ranks of heads but the figures are only discernible in fragmentary form. . . . By contrast, what great ocular satisfaction there is to be derived from the compositions of Raphael with the many figures. (Wölfflin 1899, p. 292, translation by the authors)

In 1912, Wilhelm Waetzoldt, who later became director-general of the State Museums in Berlin until the Nazis removed him from his post, explained the differences between the Italian and the German/Dutch style in Renaissance art as having a basis in national differences in the eye's aptitudes:

> Italians have an architectural-plastic talent accustoming the eye to trace the form of things, to see each individual figure in space and to ascertain the physicality of a thing by scanning it with the eye. . . . The Italian vision isolates, the vision of Dutch people and Germans connects; the former is used to the mobility of gaze, the latter to the quiet-looking eye. (Waetzoldt 1912, p. 211ff., translation by the authors)

Perception remained a central topic in art history throughout the twentieth century. On the one hand, the idea of the eye following the composition of artworks was sustained. On this basis, the art historian Kurt Badt (1961), for instance, developed a general method for the interpretation of paintings, insisting that the path of composition normally begins—and should begin—at the lower left-hand corner of pictures (Rosenberg 2014). On the other hand a more general, though sometimes rather metaphorical discussion about "cultures" of perception, also addressed as "scopic regimes," figured prominently not only in art history but also in visual culture studies, a new discipline which emerged from this very discussion in the 1990s (Foster 1988; Volkenandt 2011).

In speculating about the eye movements of virtual beholders, art critics and art historians touched upon a field that would later evolve as a central method in psychology: the recording of gaze movements to investigate cognitive processes. Psychological advances were made possible by the description and by recordings of the saccadic nature of eye movements by ophthalmologists since the end of the nineteenth century (see section 5.2), but art historians did not take note of it for over 100 years (first discussions appeared in Frangenberg 1990, p. 144ff; Baxandall 1994, p. 413; Clausberg 1999; Rosenberg 2000, p. 49ff; Giuliani 2003, p. 27ff.).

In the following sections we will introduce the measurement of gaze movements as a prominent psychological technique and outline its neural bases and relationships with visual attention before reviewing what we have learned so far about eye movements of the beholders of paintings from experiments.

5.2 Analysis of gaze movements as a psychological method

5.2.1 Overview

Gaze movements are basically alternations between periods of the eyes remaining relatively stationary, which we call "fixations," and shorter periods of greater mobility, which we call "saccades" (from the French word "saccade" meaning "jerk") (Figure 5.3). This was described for the first time in 1879 by Javal and Lamare (Javal 1879).

Two decades later, Erdmann and Dodge reported the first successful attempt to record and measure eye movements empirically (Erdmann and Dodge 1898). Not only did this pioneering technological work lay the foundations for eye-movement research in clinical populations (Klein and Ettinger 2008), but it also opened the field for experimental

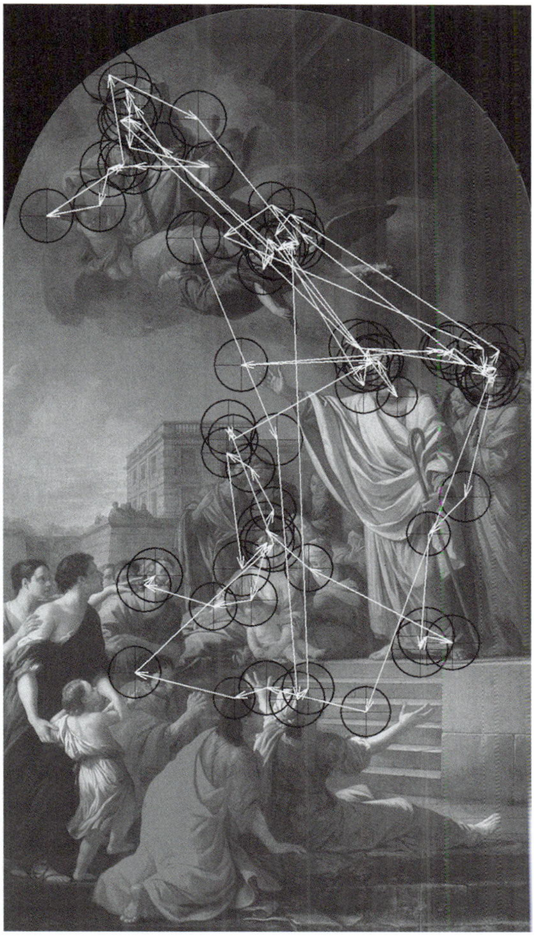

Figure 5.3 Saccades (white arrows) and fixations (black circles) of a beholder whilst viewing Vien's painting (*St. Denis Preaching in Gaul*, 1767) for 20 s. Adapted version: © Laboratory for Cognitive Research in Art History, University of Vienna.

psychological eye-movement research (Wade and Tatler 2005). Nowadays, the study of gaze movements figures prominently in basic and applied psychological research. In basic research, gaze movement analyses have been used to study interactions between covert and overt attention processes (e.g. Henderson et al. 1989; Hoffman and Subramaniam 1995), reading (Henderson and Fereira 1993), problem-solving processes (Grant and Spivey 2003; Knoblich et al. 2009), exploration of geometric patterns (Manor et al. 1995; Noton and Stark 1971c), the study of faces (Mertens et al. 1993), film watching (Smith 2014), visual processing of media (Bucher and Schumacher 2012), the analysis of visual scenes (Rayner et al. 2009), and in paintings. In applied research, gaze movements have been used to study, for example, product designs (Carbon et al. 2006), and print advertisements (Pieters et al. 1999), to construct human–computer interfaces (Jacob and Karn 2003), to investigate processing of social situations in autistic patients (Klin et al. 2009) and problem-solving strategies in patients with Parkinson's disease (Hodgson et al. 1999), and first-episode schizophrenics (Huddy et al. 2009).

5.2.2 Measurement of gaze movements as a distinguished method in psychology

There are at least three reasons for the popularity of eye movement recordings in psychological research: importance, immediacy, and versatility. Gaze movements are *important* to investigate because the ocular-motor system is doubtlessly the most used system among the sensory-motor systems in primates, supported by an extensively distributed functional neural system that is currently better understood than any other functional system of the primate brain. Due to the anatomy of the eye, this system's effects can be investigated with an unmatched degree of *immediacy*, both psychologically and technically. As there is only one small spot of highest visual acuity on the retina, the fovea centralis, both the exploration of the visual world and our responses to its changes require our gaze to move as a series of fixations and saccades which, together, provide the best available model of the structure of human consciousness unfolding in space and time. The final virtue of eye movement studies is the *versatility* of their application in all kinds of basic as well as applied scientific settings—from everyday situations (e.g. Hayhoe and Ballard 2005) to the narrow setting of a funcational magnetic resonance imaging (fMRI) scanner—and experimental paradigms.

5.2.3 Fixations and saccades

The analysis of gaze movements crucially hinges upon (a) the definition of fixations and saccades, and (b) the method of their grouping. We will explain the definitions in the following, and discuss methods of grouping in section 5.4.

Regarding the definition of fixations, if we assume that there is a *continuum* between "*fixation*," considered as periods of relatively stationary gaze that are characterized by the presence of micro-saccades (or "fixation saccades"; plus drifts), and "*exploration*," considered as periods of relative gaze mobility that are characterized by the presence of larger-amplitude saccades (and drifts) (Otero-Millan et al. 2013), the definition of fixations in terms of degrees of displacement of the eye is to some degree arbitrary. Similarly ambiguous is the question as to what minimum duration a so-defined fixation should have, as this parameter is likely to be different for different kinds of tasks and different individuals.

Accordingly, various definitions of fixations have been suggested in different domains of the literature, but these have been rarely compared empirically (see Manor and Gordon 2003). Typically, periods of relative gaze stationarity of at least 200 ms are considered as fixations. However, this definition has been based on early reading studies and may thus be inappropriate for other visual tasks that use visual stimuli that are more complex or less complex than words. Extracting the gist of a complex scene, for instance, takes less time (40–100 ms) than analysing it (with minimum fixation durations of 150 ms; Rayner et al. 2009). Fixations during reading also take at least 140 ms according to McConkie and colleagues (1992; but see Liechy et al. 2003). Westheimer (1954; but see Manor and Gordon 2003) reported intervals between successive saccades of 150 ms. Similar values, ranging between 100 ms and 200 ms have been reported in later studies (van Diepen et al. 1995; Harris et al. 1988; but see Manor and Gordon 2003). Russo and Rosen (1975) used minimum fixation durations of 200 ms to investigate gaze movement patterns during a reading and selection task. Even longer fixation durations, say, of 320 ms for the analysis of geometric stimuli (Manor et al. 1995) have been reported as well. This latter figure corresponds well with the average fixation duration of about 320 ms which we found for the participants of our study, irrespective of the painting they viewed (unpublished work).

Fixation durations, however, not only differ between situations but also between individuals, and within individuals. Furthermore, speaking about the painting reduced fixation durations by about 40–50 ms (Klein et al. 2014). And inter-individual fixation durations varied greatly between 200 ms and 800 ms, with a preference for 250–400 ms (Mertens et al. 1993).

One of the few studies that empirically investigated the impact of different fixation definitions on gaze movement parameters was published by Manor and Gordon (2003), who compared different fixation thresholds. Comparing 200 ms and 100 ms thresholds, this study reported significant increases in the number of fixations, total fixation duration, and scan path length (defined as a continuous line drawn through consecutive points of fixation), as well as significant decreases in average fixation duration for geometric and face stimuli. Particularly noteworthy here is the association of the shorter fixation threshold with increased scan path lengths, underlining the perceptual–cognitive significance of this relatively short (100 ms) fixation threshold. Also, schizophrenic patients could be more accurately discriminated on the basis of 100 ms as opposed to 200 ms fixation thresholds in showing less fixations and shorter scan paths, as well as longer overall fixation durations. Fixation parameters may differ according to stimulus types. Manor and Gordon (2003) compared face stimuli and geometric figures of the Rey–Osterrieth complex figure type and found more numerous and shorter fixations for face stimuli, independent of fixation threshold (100 ms or 200 ms).

Assuming again that fixation-related saccades and exploration-related saccades lie on a continuum and thus follow the same "main sequence" (Otero-Millan et al. 2008); that is, the positive function relating the amplitude of saccades to their peak velocities (which holds for saccades 0.03–50° in amplitude: Becker 1989), it is clear that the amplitude of saccades that are typically observed whilst participants are presented with various kinds of eye movement tasks on a computer screen are easily determined by their velocity curves and an arbitrary or calibration-based onset/offset criterion (e.g. 20° per s or 20 per cent of the amplitude of the velocity curve; Klein and Foerster 2001).

In such situations, saccades are easily recognizable as stepwise, discontinuous eye movements. It has been estimated that humans execute about 200 000 saccades per day (but see Fischer 1999). During these ballistic eye movements the eyeball reaches peak velocities of more than 400° per second (Fischer 1987; cf. Manor and Gordon 2003). Saccades are characterized by abrupt eye movement onsets with accelerations of up to 30 000°/s^2, peak velocities of 400–600°/s, and an almost instantaneous termination of the movement (Becker 1989). While at least seven different classes of saccades can be distinguished (see Becker 1989), *re-fixation saccades*, directed at objects that are selected beforehand from the environment, are of primary interest here. Such re-fixation saccades are frequently inaccurate and do not hit their target accurately; as a consequence, corrective eye movements, either glissades or corrective saccades, follow. The amplitude of the primary saccade may be too small ("undershoot") or too large ("overshoot"). Undershoots prevail in the execution of large amplitude saccades (Fioravanti et al. 1995) and are followed by secondary "corrective" saccades as the second part of a pre-programmed two-step sequence (Becker and Fuchs 1969).

5.2.4 Physiological and neural basis of fixations and saccades

The visual system of primates has been investigated very intensively. Light enters the eye through the pupil and reaches two types of photo-receptors on the retina, rods processing dim light and cones processing bright light and colors. These photoreceptors accomplish a transduction that translates light information from a spotlight into an electric signal. The central fovea, a roughly 1.5 mm small area located about 5° temporally to the visual axis, contains the highest density of cones and thus enables the most precise viewing. The axons of about a million ganglion cells leave the eye through the blind spot, form the optic disc, and transmit their neuronal information via the optic nerve, passing the optic chiasm and the optic tract to reach the lateral geniculate nucleus of the thalamus and from there the primary visual cortex (V1) and further the visual association cortices (V2–V5). From these association cortices the visual information spreads out to be processed by a complex system of cortical and sub-cortical areas. It is important to note that the central fovea covers only less than 1 per cent of the retinal area but projects to a large portion of the visual cortex to process visual information coming from the central 2° of the visual field. These mere anatomic conditions highlight two important functional aspects of the visual attention—its anatomically "embodied" selective nature and its close link with the continual alternation between fixations and saccades to (re)construct the visual outer world.

The brain structures involved in "fixations" and "saccades" overlap to a large degree. This overlap can be expected by *a priori* reasoning under the aforementioned assumption of a continuum between fixation-related and exploration-related saccades. These structures include: (a) the frontal eye fields (Bruce and Goldberg 1985; Bruce et al. 1985); (b) the supplementary eye field (Bon and Lucchetti 1990; Schall 1991); (c) parts of the parietal cortex (Robinson et al. 1978; Andersen 1989); (d) the substantia nigra of the basal ganglia (Hikosaka and Wurtz 1983); (e) the superior colliculus (Dorris and Munoz 1998; Schiller and Stryker 1972); and (f) regions in the brainstem (Paré and Guitton 1994; Bruce 1990).

While the functional system involved in gaze movements during the contemplation of visual art is already complex, it is by no means exhaustive with respect to the complexity of processes that are supposed (Leder et al. 2004) and can be experienced subjectively during the contemplation of paintings.

5.3 Visual attention and gaze movements

The (functional) anatomic and neural features of the visual system outlined in section 5.2 not only underline the fact that the visual system is a system of information selection; they also indicate that visual attention unfolds in space and time as a continual alternation between fixations and saccades (or "fixation-related" and "exploration-related" saccades), which we will henceforth call "gaze movements." Decades of experimental research with human and non-human primates have scrutinized these intricate relationships between visual attention and gaze movements (recently reviewed in Kowler 2011) in order to address—among others—two questions of fundamental importance: how do "top-down" and "bottom-up" factors interact; and what is the relative importance of low-level stimulus features (in particular, contrast, color, and motion) as opposed to higher-level stimulus structures (in particular, objects or gist; Einhaeuser et al. 2008)?

With regard to the importance of low-level features, Koch and Ullman (1985) have suggested that locations within the visual field that exhibit relatively high physical salience are more likely to be attended and fixated. According to these authors, visual scenes contain spatial distributions of local contrasts of luminance, color, or movement that can be described in "saliency maps" and are processed in early stages of visual information processing. On the one hand, the concept of saliency maps is certainly theoretically intriguing and straightforward in predicting the locations of fixations better than models that posit random fixation locations (Foulsham and Underwood 2008). On the other hand, empirical research has shown that many further factors co-determine gaze positions and that these factors are more "top-down" than "bottom-up." Such factors include strategic decisions to optimize task performance such as looking at locations that maximize the probability of finding a searched target (Najemnik and Geisler 2005), or looking near the center position in a scene to identify large portions of the scene with this single optimal viewing position (Tatler 2007). Another top-down factor is specific interests that direct the gaze to relevant areas of the visual field (e.g. looking at eyes and heads in social scenes to extract relevant social information; Birmingham et al. 2009). Similarly, interesting objects may attract the allocation of attention and fixations better than the perceptual saliency of an area of the visual field (Einhaeuser et al. 2008). Also, specific tasks can strongly impact which information participants select when looking at visual scenes; this includes the contemplation of paintings as well and will be discussed in the following section.

Such bottom-up and top-down factors, rather than being mutually exclusive, are likely to interact. This was demonstrated, for instance, in Cerf and colleagues' (2008) study, which reported an improved prediction of gaze positions when the combined influences of top-down face preference and bottom-up low-level saliency were considered. These examples thus show that visual attention is controlled both by exogenous bottom-up processes of perceptual saliency and endogenous top-down processes (Chica et al. 2013).

That visual attention unfolds in space and time suggests that it is the *sequencing of saccades* during the exploration of visual scenes (rather than an individual fixation or an individual saccade) that requires scientific investigation. In this regard, Kowler (2011) has pointed to two principles that seem to rule the sequential selection of saccadic goals. According to the "winner-takes-all" principle, the area of the visual field that momentarily exhibits the highest "strength" (e.g. saliency, interest value) attracts visual attention. According to the "inhibition of return" principle, however, the selected area quickly (within a few hundred milliseconds) loses its "strength" to other areas to be fixated as the new "winners." It is important to note that during such sequences of saccades to different fixation locations, the proper eye movements as manifestations of overt visual attention are *preceded* by movements of covert visual attention to the location that is to be fixated next (Henderson et al. 1989). While the movements of covert and overt visual attention are temporally staggered, they are spatially isomorphic as it has been shown that it is impossible to orient attention (covertly) to one location while moving the eyes to another (Hoffman and Subramaniam 1995).

Gaze movements as the sequence of fixations and saccades to new locations generate scan paths across a stimulus of interest (Liechty et al. 2003). Such scan paths were first described by Noton and Stark (1971a, 1971b, 1971c), who showed that when participants looked at visual patterns their gazes repeatedly created a fixed series of fixation locations, the "scan path," that was characteristic for an individual participant. Noton and Stark also found that different scan paths were created by different individuals for the same pattern, and different scan paths by the same individual for different patterns. Interestingly, these scan paths remained consistent across different presentations of the same stimulus to the same participant and were even replicated when a previously scanned stimulus was removed and had to be imagined. The apparent existence of an ocular-motor "memory" of an individual stimulus certainly stands as among the most compelling evidence of top-down factors in gaze movements.

Gaze movements as alternations of fixations and saccades seem to be governed by two different states of visual attention, according to Liechty and colleagues (2003). During states of *local visual attention* and under the presumed control of the inferior temporal cortex, information is extracted from specific and proximal locations by means of short saccades. Conversely, during states of global visual attention presumably influenced by the posterior parietal cortex, long saccades are employed to integrate information from distal locations. According to Liechty and colleagues (2003), the interaction between temporal/local and parietal/global states is controlled by the prefrontal cortex.

To summarize, the fact that only a very small area of the retina provides the best visual acuity requires visual attention to unfold as gaze movements. The dynamics of gaze movements are co-determined by perceptual bottom-up factors as well as cognitive or motivational top-down influences, governed by principles that regulate how a fixation location is selected and then abandoned as a new location is chosen, and may switch between local and global states of visual attention.

5.4 Gaze movements during the beholding of paintings

5.4.1 History of research

What do we do when we behold works of visual arts? Given the central role of the ocular system in this process and in the light of the emphasis of gaze movements in literature on art (section 5.1), eye tracking is logically among the foremost methods used to investigate this field. However, although the first comprehensive research dates back to the 1930s, the number of investigations into this area has remained rather limited.

The psychologist Guy T. Buswell (1891–1994) accomplished the first in-depth studies on how people look at paintings and published the only book to-date about eye tracking and art perception (Buswell 1935). He recorded gaze movements of children and adults, of art experts and lay persons, of American and "Oriental" subjects looking at reproductions of works of art (paintings, drawings, prints, sculptures, buildings, applied arts) various in cultural origin, color, design, and balance, and the amount of detail incorporated. He included silhouettes and pictures with outline drawings as well as partly unfinished paintings. Compared to present day eye trackers, Buswell's device was cumbersome for the experimental subjects and the data it provided were extremely time-consuming to analyse. However, he did manage to collect a total of 1877 records from 200 different individuals looking at different pictures for several tens of seconds at a rate of 30 measurements per second.

Buswell thoroughly addressed most of the questions that research in this area eversince investigated. His main chapters—"The Duration of Fixation Pauses," "Variations in Perception Related to Characteristics of the Picture," "Variations in Perception Related to Characteristics of Individuals," and "Variations in Perception Due to Directions for Looking at Pictures"— follow a systematic order that we have adopted for the headings of the present section. Buswell and his collaborators at the University of Chicago were technically innovative, thoughtful in the design of their studies, and very rigorous in their methodologies. However, their overall conclusions are rather sobering. A contemporary reviewer noticed in the *Burlington Magazine*: "Our admiration for the ingenuity of Professor Buswell's apparatus and the heroic laboriousness of his experiments and calculations may be a little tempered by our disappointment at noticing that nothing of the slightest importance to the sciences of aesthetics or psychology seems to result from this research" (Thouless 1936, p. 58). In retrospect one may say that Buswell's rather flawed general results were due to technical limitations. Buswell (1935, p. 90) would have loved to investigate longer periods of beholding, especially with experts, but for technical reasons this was not feasible. In addition, without electronic data processing his capacity for a comparative analysis of the huge amount of recordings was very restricted. Longer measurements of gaze movements when viewing images were first carried out by the Russian psychologist Alfred L. Yarbus (1914–1986). Some of the stimuli used by Yarbus were works of art. However, the topic of his seminal book was "Eye Movements and Vision" in a broader sense. He did not focus solely on art as Buswell did.

Due to the very nature of art, the essence of which withstands any bold attempts at manipulation without being destroyed, and to the difficulty of offering aesthetic experiences in the cool atmosphere of a laboratory, scant proper experimental research has been accomplished regarding viewing visual art. Approaches that do not manipulate the painting to be viewed and which measure paintings in their original contexts are the most appropriate, but most studies were governed by the technical feasibility of eye trackers. Only a few studies have as yet been conducted in museums rather than in labs, and the limitations imposed on the subjects by wearing eye trackers are not negligible (Heidenreich and Turano 2011; Kapoula and Lestocart 2006). Worth bearing in mind too is that the time taken to look at artworks used to be quite short. The average time taken in most experimental settings was less than the mean time of 27.2 and median of 17.0 seconds, a period which general visitors were found to spend looking at paintings at the Metropolitan

Museum (Smith and Smith 2001), and was much less than the time art experts would consider as sufficient to have an aesthetic experience (Massaro and colleagues (2012) is a recent example using a viewing time of 3 s). We assume that it can take several minutes to enjoy a work of art fully and to experience the aesthetic values of a painting that one does not know. The viewing time of experts such as artists or art historians, especially when speaking in groups with each other, can often last for several minutes, sometimes more than an hour, including alternating phases of silent contemplation and discussions about the artwork under consideration. In an interdisciplinary project between departments of art history and psychology we sought to overcome these limitations as far as technically possible. In a first study at the University of Heidelberg we recorded the gaze movements of 99 participants who were presented four facsimiles of paintings, looking at each for 15 minutes. The paintings had been reproduced in (almost) original size with the best photographic techniques currently available, were mounted with appropriate frames, and hung on the wall of an otherwise "neutral" room that, in terms of art exhibitions, would be defined as a "white cube." We used a head-mounted eye tracker with an electromagnetic-positioning system (SMI® IViewX HED-HT) that allows the participant to move within a radius of about 120 cm (Klein et al. 2014). In a later series of recordings at the University of Vienna, subjects looked at high-quality digital reproductions of paintings on a high-resolution screen (Apple® 30" 2560 × 1600 pixel) at a distance of about 90 cm, each for two minutes. Their gaze movements were recorded without any physical contact between subjects and the device (remote eye tracker SMI® IViewX RED 120). From the binocular recordings we processed the data recorded of the dominant eye of every subject (Brinkmann et al. 2014).

5.4.2 Variations of gaze movements during the time lapse of beholding

Eye tracking enables a precise description of the time lapse during the viewing of paintings, and this was always a central issue in eye-tracking studies about art. One of the chief findings of Buswell (1935) was that the duration of fixations tends to increase during the course of viewing. He differentiates two "general patterns of perception" within the first 10–15 seconds—a first phase of general survey with shorter fixations followed by a phase of more detailed study with longer fixation durations and (shorter) saccades concentrated over small areas of the picture. The major distinction between these patterns of perception has been confirmed and elaborated several times, among others by Berlyne (1971), Antes (1974), and more recently by Liechty and colleagues (2003) (see section 5.3), Locher (2006), and Locher and colleagues (2007) in the light of cognitive theories. Molnar (1981) accepted this distinction as well but tried to also distinguish a third "kind of exploration" with distinct patterns of the duration of fixations and of the length of saccades. He thought that these patterns also depend on the complexity and style of the painting, but he failed to give sufficient experimental evidence for his supposition. Our own studies, which for the first time consider much longer periods of time, endorse Buswell's findings and underline that there is a significant inter-individual variability in the duration of fixations, in the length of saccades, and in the sequence of different patterns of perception. The most significant average changes to the duration of fixations and the length of saccades occur in the first seconds of viewing. However, repeated alternations of patterns of perception, oscillating between general survey and detail study continue for at least 15 minutes. Within the first few seconds, the average duration of fixations significantly increases. In the following minutes the slope of the curve becomes rather flat but the increase in duration of fixations endures. We interpret this as indicating a deeper perception that can be arguably related to aesthetic experience. This would confirm a study by Molnar and Ratsikas (1987), where a group of beholders that were asked to judge the aesthetic qualities of the picture had clearly longer fixations than another group of beholders who were simply asked to describe what they saw.

5.4.3 Variations of gaze movements related to paintings

Buswell (1935) demonstrated that gaze movements meet common assumptions made in art literature (such as those reported in section 5.1) only in some respects. His numerous eye-tracking graphs make clear that the eye hardly ever moves systematically along a composition line from one end to the other. Nevertheless, Buswell's work provided important indications as to the relationships between gaze movements and pictures. He showed that every picture has specific centers of interest that attract a higher density of fixations than other areas (Buswell 1935, pp. 18–24). Regarding gaze movements, his results were very restricted and his analyses limited to the two initial saccades and to three images showing mainly vertical lines (interior of a gothic church), mainly horizontal lines (Library of the Oriental Institute of the University of Chicago), and lines following the direction of a wave (Hokusai, *The Wave*). He states that his results were "quite in accord" with the general assumption "that the eye will follow the direction of the principal lines in a picture" (Buswell 1935, pp. 79 and 82). Yarbus (1967, pp. 171–96) published single recordings of subjects viewing reproductions of landscape paintings for up to 30 minutes. Probably due to technical reasons, those visualizations only show a portion of the total fixations and saccades. However, they are revealing since they make clear that while beholding the same painting for many minutes the subject's eyes repeat the same movements—Yarbus talks about individually specific "cyclical patterns." Molnar (1981, 1992) came to similar conclusions, and described these visualizations using statistical methods as follows: the transition probability between the areas of interest of two paintings (by Titian and Manet) reached a stable state after just 20 or 25 saccades. Like the "scan paths" described by Noton and Stark (1971a, 1971b, 1971c, and see section 5.3; but see Mannan et al. 1997; Krieger et al. 2000), those findings were related to single spectators.

In our studies (Rosenberg et al. 2008; Engelbrecht et al. 2010; Brinkmann et al. 2014; Klein et al. 2014; Rosenberg 2014) we have meanwhile shown 57 paintings from different epochs (fifteenth to the twentieth centuries) and different genres (history paintings, portraits, landscapes, still life, abstract art). The viewing time was either 2 or 15 minutes and the number of participants varied between 10 and 99. Taken together, our studies confirm Buswell's conclusion that most paintings have specific centers of interest with significantly higher densities of fixations (Figure 5.4). This is hardly the case for paintings such as

(a) (b)

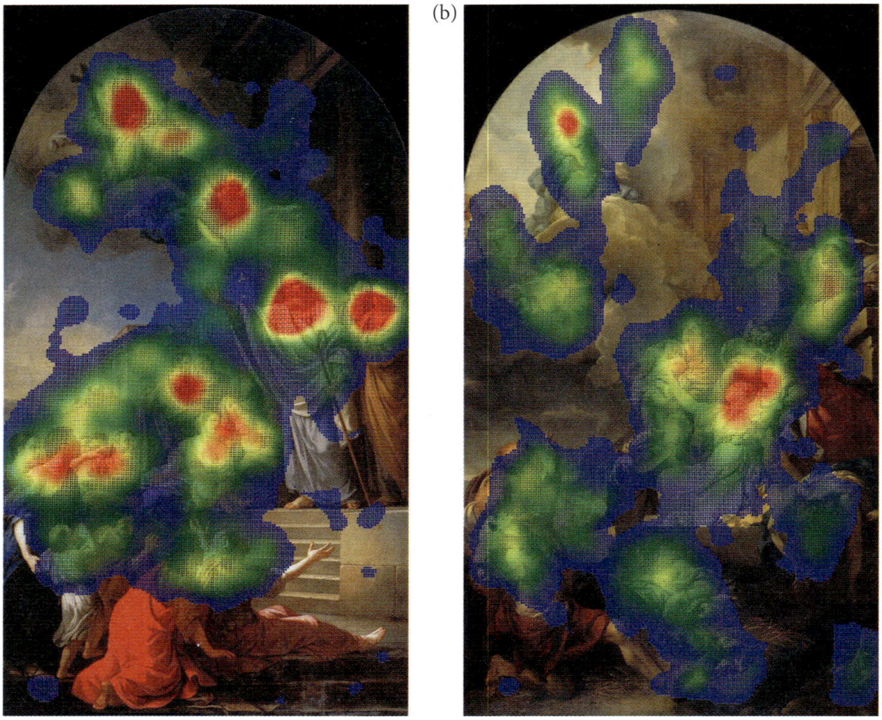

Figure 5.4 Heat maps of fixations of 40 viewers (20 art experts and 20 non-experts) looking at Vien's (a, *St. Denis Preaching in Gaul*, 1767) and Doyen's (b, *The Miracle of Saint Anthony's Fire*, 1767) paintings for 2 mins each. Adapted versions © Laboratory for Cognitive Research in Art History, University of Vienna.

Jackson Pollock's drippings (Figure 5.5), however, where the artist intentionally avoided a hierarchical composition, and denied as much as possible any distinction between figure and ground, resulting in no single zone with enough fixations to be red in the heat map.

For a significant number of paintings and despite major differences between subjects, not only fixations but also saccades build patterns that are specific to each painting: Beholders tend to reiterate particular paths with their eyes. This is evident when the saccades of different viewers are superimposed (see Figure 5.6a and 5.6b, with a reduction to 10 per cent of all saccades to reduce confusion). Hence we conclude that the "cyclical patterns" observed by Yarbus (1967) do not only occur for single subjects but are very similar for different subjects viewing the same painting as long as they do so for longer stretches of time.

We have developed different methods to determine whether and how saccades repeat specific paths: "Frequent cluster transitions" show the spots (clusters) with the highest density of fixations and the often-repeated saccadic transitions. We thereby define clusters of fixations as circular regions of a given diameter (here 90 pixels) containing more than a certain amount of fixations as per minute rate (for an alternative definition of fixation clusters see also Santella and DeCarlo 2004). An algorithm calculates the

(a)
(b)
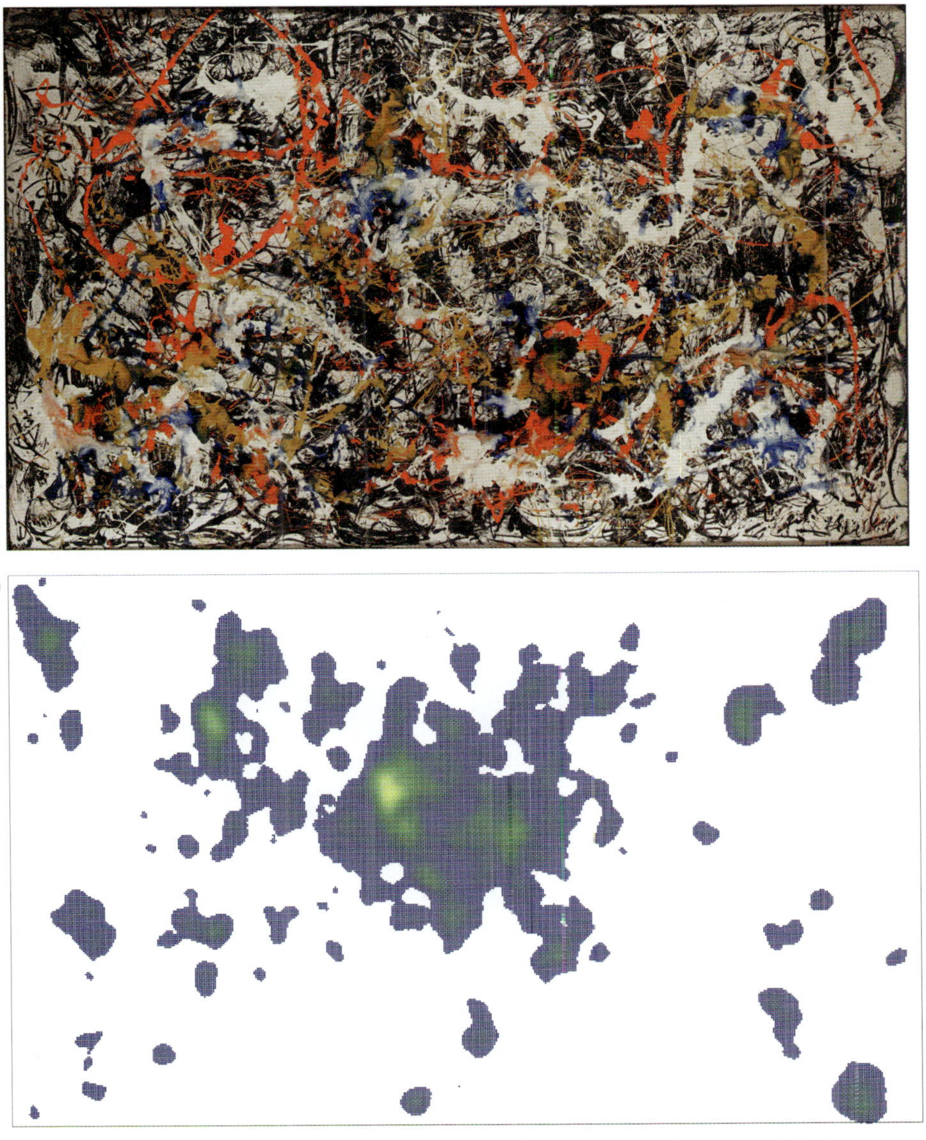

Figure 5.5 (a) Jackson Pollock's *Convergence*, 1952, and (b) heat map of fixations of 40 viewers (20 art experts and 20 non-experts) viewing this painting for 2 mins each. Albright Knox Art Gallery/Art Resource, NY/Scala, Florence. © The Pollock-Krasner Foundation ARS, NY and DACS, London 2015. Part (b) © Laboratory for Cognitive Research in Art History, University of Vienna.

amount of saccadic transitions between the fixations of those clusters. In Figure 5.7a and 5.7b we visualize all transitions repeated at least 0.5 times per minute. The width of the lines between the clusters encodes the frequency of saccades, with more frequent transitions resulting in greater width. Color can be used to encode the direction of saccades.

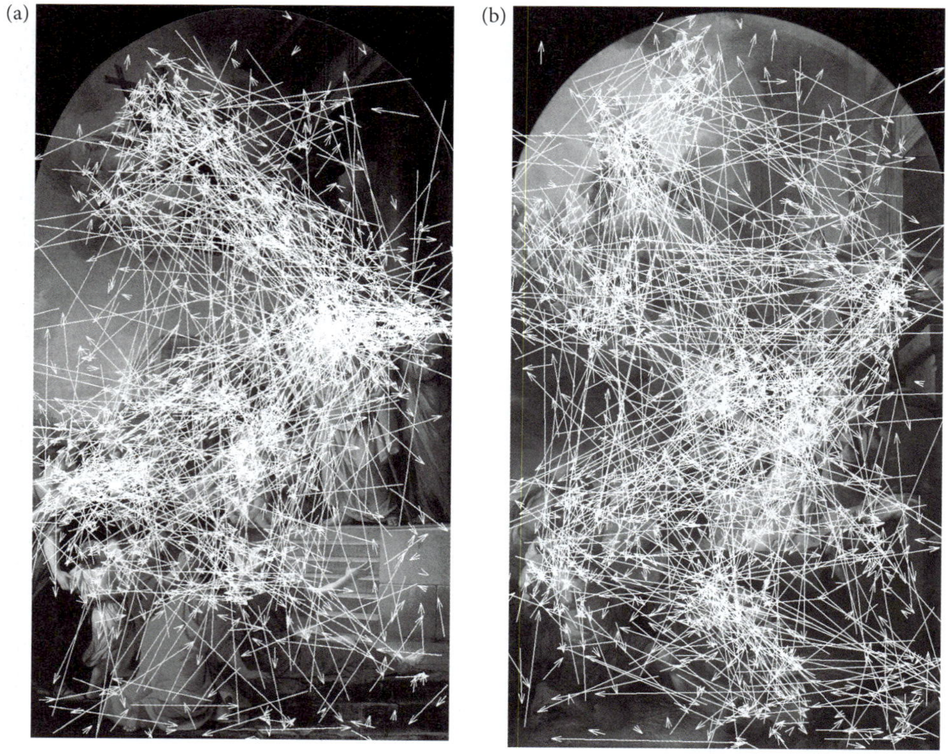

Figure 5.6 10 per cent of all saccades of 40 viewers (20 art experts and 20 non-experts) beholding Vien's (a) and Doyen's (b) paintings for 2 mins each. Adapted versions: © Laboratory for Cognitive Research in Art History, University of Vienna.

The visualizations of gaze movements (see Figures 5.3, 5.6a and 5.6b) show that Diderot's analyses (section 5.1)—as well as similar claims recurring in the art historical literature—do not match the real dynamic of the eye. The gaze jumps from fixation to fixation, moving forth and back. Eyes do not follow any line of composition in a continuous manner, nor do beholders scan paintings from top to bottom or left to right continuously. In addition, the general assumption of art historians (Badt 1961; Wölfflin 1941) that viewers' eyes predominantly move from left to right—as when reading—when looking at European paintings is empirically incorrect since the average direction of saccades between most clusters of fixations tends to be balanced in all the paintings we have studied to date.

However, the analysis of average eye movements, both for fixations (Figure 5.4a and 5.4b) and saccades (Figure 5.7a and 5.7b), reveals that several aspects of Diderot's text are indeed correct: viewers concentrate much longer on certain parts of Vien's painting (greater amount of red areas in the heat map), whereas their fixations are more scattered when looking at Doyen's piece. This corresponds with Diderot's description of the differences between those altarpieces. His description of a line of composition in Vien's altarpiece is correct as long as we consider the frequently repeated saccades and not the actual

(a) (b)

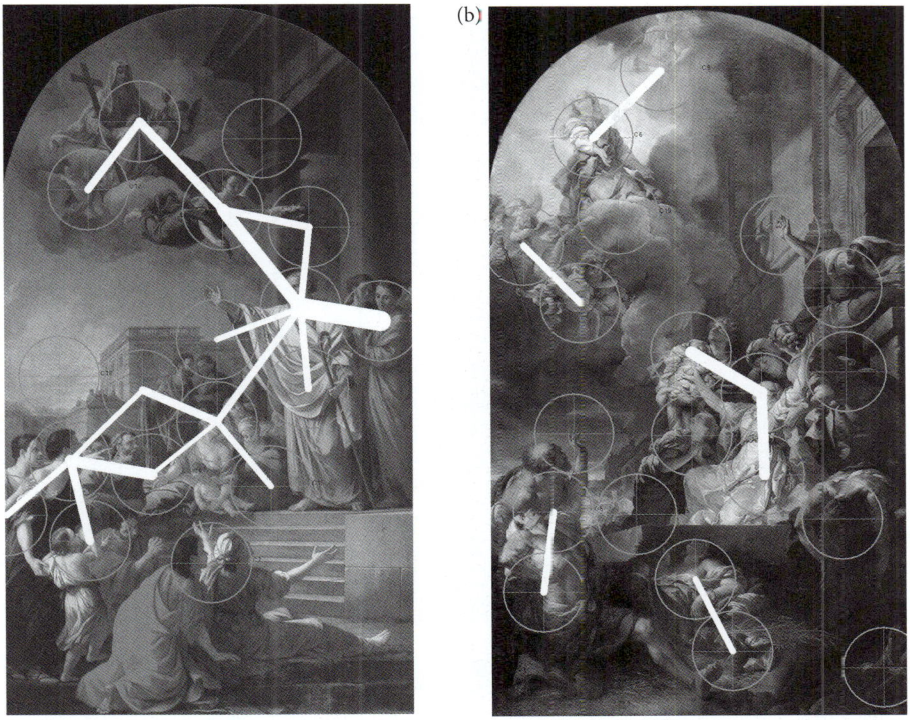

Figure 5.7 Frequent saccadic transitions between fixations clusters for Vien's (a) and Doyen's (b) paintings (average of 40 viewers, 20 art experts and 20 non-experts, whilst viewing for 2 mins each). Adapted versions: © Laboratory for Cognitive Research in Art History, University of Vienna.

course of the movement of the eye. The line he describes matches the graph of the most frequent saccadic transitions between clusters of fixations (Figure 5.7a). Correspondingly, it is clear that the saccadic movements executed when viewing Doyen's painting (Figure 5.7b) connect the figures within the four groups, but they do not connect those groups with each other quite as much. In contrast, fixations are almost equally distributed when looking at Pollock's canvas (Figure 5.4), and this corresponds to the "allover", a term commonly used to describe his canvases covered with paint from edge to edge, without preferences.

Another method to illustrate and evaluate the paths repeated by the beholder's eyes is the visualization of "similar saccades". We developed an algorithm merging all saccades that are more or less close to each other and run more or less in parallel to each other. This method is useful for paintings that do not have clear centers of interest like human figures and especially faces; for example, landscapes, abstract images, etc. (Figure 5.8a and 5.8b).

We have chosen the altarpieces by Vien and Doyen because Diderot's description of (hypothetical) eye movements during the contemplation of these paintings is prominent in the history of art literature. However, those altarpieces stand for a wide range of paintings where the fixations and saccades of almost all beholders repeat

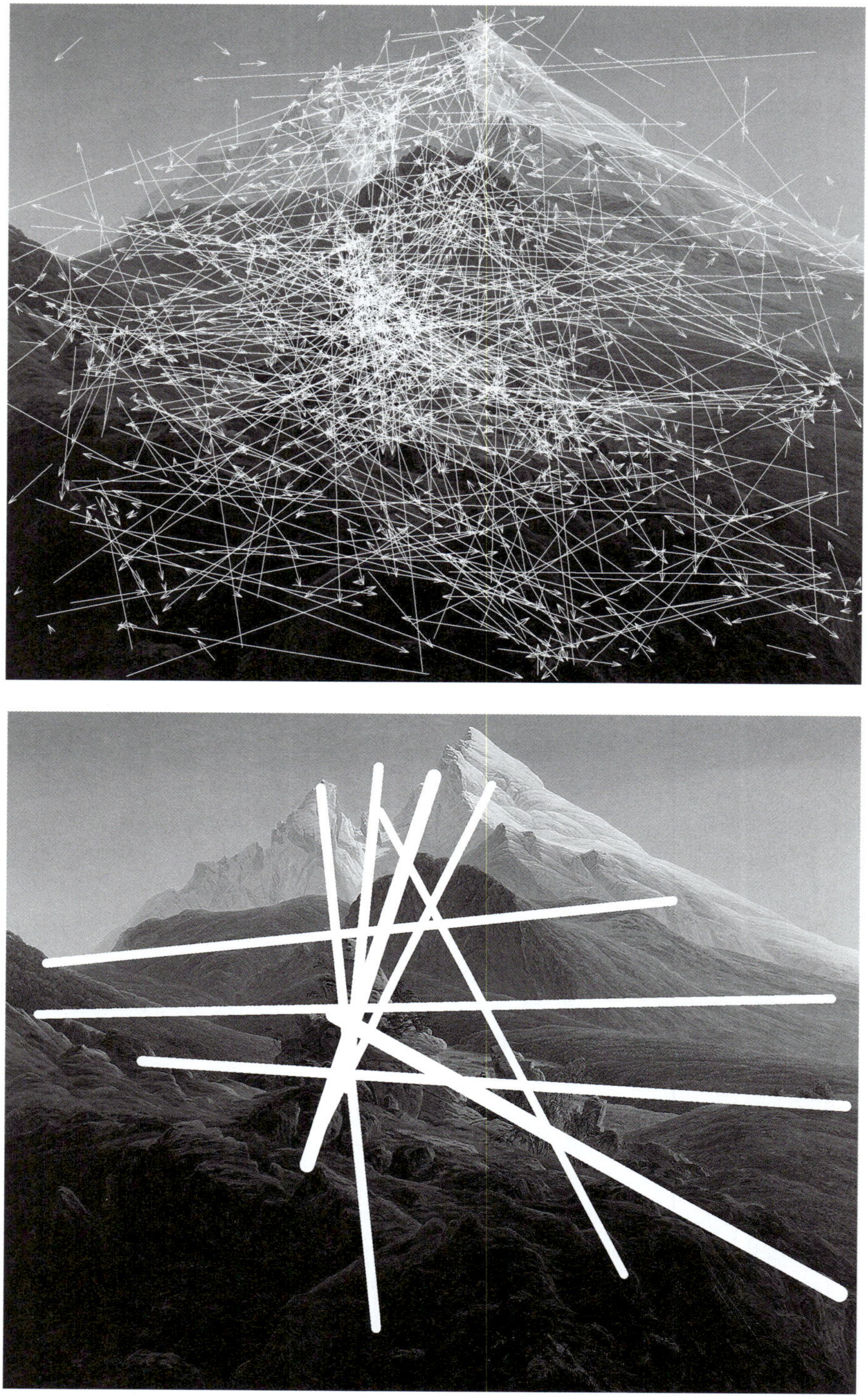

Figure 5.8 20 per cent of all saccades and visualization of similar saccades of 20 viewers (10 art experts and 10 non-experts) whilst looking at C.D. Friedrich, *Der Watzmann*, 1824/25, for 2 mins each. © akg-images. Adapted versions: © Laboratory for Cognitive Research in Art History, University of Vienna.

patterns that are specific to the composition of each painting. Those patterns often match the analysis of the paintings given by art critics over the centuries, and they are very similar to diagrams of composition of those paintings, as they have been sketched by artists and art historians since the second half of the eighteenth century (Rosenberg 2008). We thus conclude that the eyes of beholders do indeed often reconstruct the structure of paintings, and that the repetitions of the structures of works of art with our eyes might be a basis for aesthetic experience. As with Diderot's text, subjective analysis seems often a kind of general summary of real-life gaze movements; an abstraction of factual gaze movements highly influenced by cognitive processes, such as the awareness about the content (iconography) of the painting. Knowledge about the content and the history represented in a painting affects the order in which we arrange the pieces of this painting in our minds and hence the order in which we describe it, regardless of the fact that our eyes perceive these parts again and again in alternating sequences.

However, for some types of paintings the patterns of gaze movements do not mirror their structure. This is particularly the case for paintings where faces are dominant, as in portraits, since the beholder's eyes mainly concentrate on eyes, nose, and mouth, and very few other details (Figure 5.9), although any art critic and most art viewers will normally describe many more aspects of the painting than those facial details. Another limitation is given by the fact that gaze patterns only appear if the painting has points of reference, saliencies in regard to form and or content. This is less the case for abstract than for

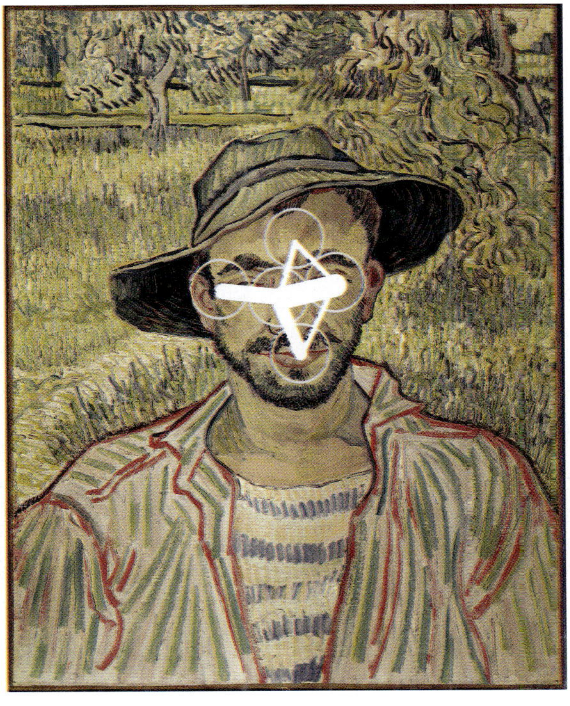

Figure 5.9 Frequent saccadic transitions between fixation clusters of 40 viewers (20 art experts and 20 non-experts) of Vincent van Gogh, *Portrait of a Peasant*, 1889, whilst beholding for 2 mins each.

representational art (Brinkmann et al. 2014), and of course even less for color-field or even monochrome paintings where fixations and saccades do not form any patterns.

5.4.4 Variations of gaze movements related to beholders

There is literally no psychologically relevant feature in which individuals do *not* differ greatly. Given the complexity and general openness of art to varying interpretations (Eco 1962), individual differences can be assumed to be particularly relevant with regard to the way people look at paintings. Indeed, pronounced individual differences in gaze movement patterns while looking at paintings had already been noticed by Buswell (1935). Three decades later, Yarbus (1967, p. 192) concluded that "individual observers differ in the way they think and, therefore, differ also to some extent in the way they look at things." Locher (1996) analysed individual subjects' fixation distributions and reported individual differences in identifying the visual center of a picture (an individual's mean eye position).

In one of our studies (Klein et al. 2014), given its large sample size and long recording duration, we also looked at various eye movement parameters, grouped according to factor analysis into saccades, fixations, and transitions between fixations clusters. We found individual differences in all of these factors to be highly stable across the viewing time of 10 minutes (unpublished work); these individual differences were, furthermore, somewhat consistent across the four paintings of this study. Overall, our results thus suggest that individual differences in elementary facets of gaze movement control during the contemplation of paintings are stable across stretches of several minutes and are consistent across different types of painting. Hence, we can assume major personality-related contributions, the nature of which is yet to be determined, to the way people look at paintings.

Beyond inter-individual variances, group-specific differences are one of the most interesting topics of eye-tracking research about the perception of images—both for psychology and art history. Buswell (1935) asked about differences due to expertise and culture but he was not able to discern any. Kristjanson and Antes (1989) reported differences in gaze movement patterns between artists and non-artists that seemed to correlate with the familiarity of the paintings. Artists showed longer fixations during the contemplation of known paintings and shorter ones with unknown paintings, whilst the opposite held for non-artists. The importance of art expertise for the way people look at paintings has also been emphasized by Nodine and colleagues (1993). They presented original and manipulated paintings for 12 seconds to artists and non-artists and found that untrained viewers failed to recognize the perceptual organizing function of pictorial elements. Furthermore, Zangemeister and colleagues (1995) compared the gaze movements of art experts and laypersons when viewing representational or abstract paintings. While fixation durations did not differentiate paintings or participant groups, art experts were found to explore paintings with longer saccades than non-experts. This group difference was more pronounced for abstract as compared to representational paintings. Vogt and Magnussen (2007) as well as Pihko and colleagues (2011) also describe differences in viewing strategies influenced by expertise.

In our studies, for the first time we compared experts (students of art history above degree level) and non-experts (students from the same university not interested in art) for viewing periods longer than one minute per painting. We could not detect any differences between the two groups for non-complex paintings. By contrast, we found that for complex paintings these groups differed in their eye movement patterns, but only at the beginning of the contemplation for periods of time of several tens of seconds (Rosenberg 2014). In the case of Vien's altarpiece, for instance, such group differences lasted for approximately 30 seconds (Figure 5.10). One possible explanation is that non-experts might need more time to understand the structure—the intrinsic meaning of the painting—conversely, experts are faster in mapping the structure of a painting, in particular when it is complex.

Given the fact that historical changes can be described as cultural changes, a crucial question for art historians is whether and to what extent the cultural imprint may influence looking at artworks. Art historians have discussed whether stylistic changes in art might be due to changes of cognitive modes (Baxandall 1972), and whether the perception of artworks has changed during centuries (Rosenberg 2000). We are currently carrying out a project on the "cultural eye" (by doctoral student Hanna Brinkmann) focusing on

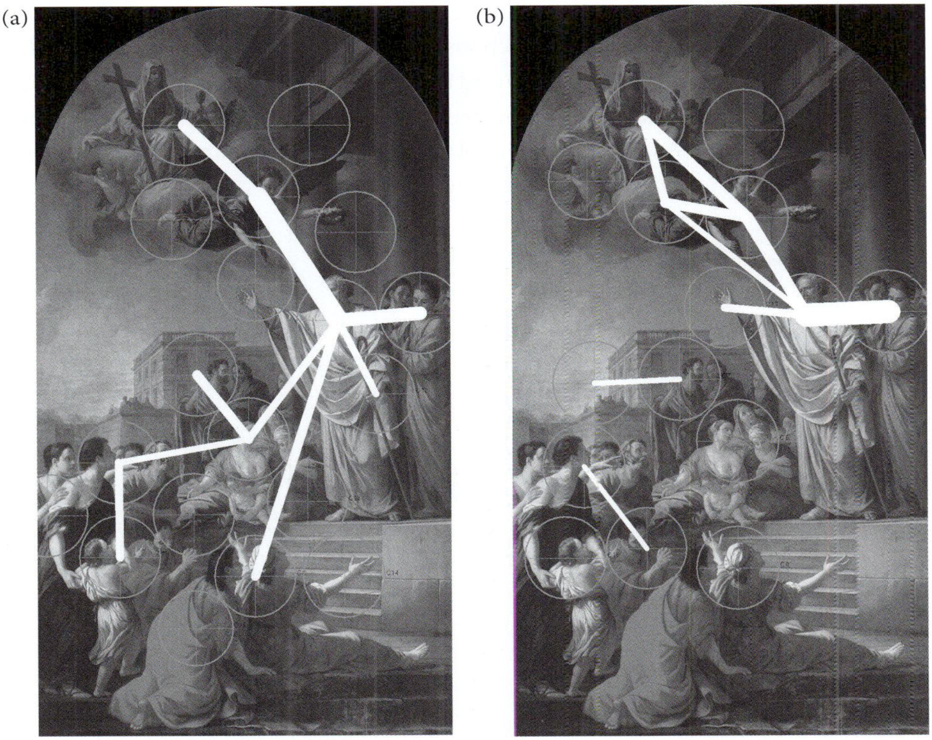

Figure 5.10 Frequent saccadic transitions between fixations clusters of 20 art experts (a), and 20 non-experts (b) whilst looking at Vien's painting for 30 s each. Adapted versions: © Laboratory for Cognitive Research in Art History, University of Vienna.

differences between Japanese and Austrian beholders. In a preliminary study with 6 Japanese and 6 Austrian test people looking at ten paintings for two minutes each, we found significant differences. The Japanese concentrated on the background of the paintings (compare Boland et al. 2008; Nisbett et al. 2001; Nisbett 2003), and made fewer horizontal and more vertical saccades, compared to the Austrian individuals (Figure 5.10). This might be due to reading habits. Finally, the Japanese executed longer fixations. This could be due to the fact that all paintings in this test were European, hence might have been unfamiliar to the Japanese viewers.

5.4.5 Variations of gaze movements related to task and context

Buswell (1935, pp. 136–41, 144) demonstrated that the "directions given prior to looking at pictures have a marked influence upon the character of perception." Yarbus recorded the eye movements of a beholder viewing the reproduction of a painting by I. Repin, with seven different instructions such as "give the ages of the people," or "remember the position of the people and objects in the room." The page illustrating those recordings (Yarbus 1967, p. 174) has been quoted regularly, and the experiment was replicated with 17 observers (DeAngelus and Pelz 2009). There can be no doubt that the variation of task has a stronger influence on gaze movements than differences between beholders, and it is interesting to use paintings to test such differences and hence demonstrate the top-down control of gaze movements. However, instructions as used in the Yarbus experiment are alien to the purpose of paintings as works of art and to the normal conditions of their beholding. More relevant to the understanding of aesthetic processes is an experiment conducted by Molnar and Ratsikas (1987). They compared gaze movements of participants looking at

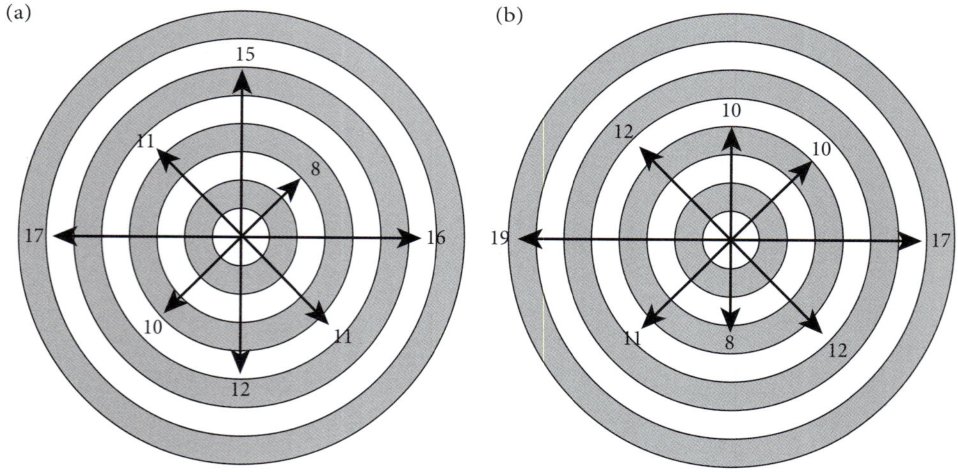

Figure 5.11 Average direction of saccades (percentage) of six Japanese (a) versus six Austrians (b) non-experts test individuals looking at C.D. Friedrich, *Der Watzmann* (see Figure 5.8), for 2 mins each (eight segments of 45°, arrows are segment bisectors). © Laboratory for Cognitive Research in Art History, University of Vienna.

reproductions of paintings knowing that they were expected to report verbally later on either the semantic content of the paintings or their aesthetic qualities. The group with the aesthetic instruction had significantly longer fixations.

Until the eighteenth century, paintings normally did not have titles. Yet, in the course of the nineteenth century, titles became an integral part of works of visual art. Viewers nowadays often look at a painting after they have read its title. The title might hence be regarded as an element which guides the eye of the spectator, and artists and art historians are thus aware of the importance of titles for the spectator (Welchman 1997). Kapoula and colleagues (2009) demonstrated that at least in the case of Fernand Léger's Cubist paintings, which include rather unclear motifs, knowledge of different titles has a significant influence on gaze movements.

A normal "task" for the spectator of art is to speak about the observed works. Since the Italian Renaissance, talking about artworks in front of them was a common practice not only among artists but also among the growing group of interested laypersons: connoisseurs and, later, art historians. There are extensive sources on discourses and descriptions of paintings that were originally spoken in front of the works of art (Welzel 1997), and the practice of describing works of art remains fundamental in museums and art history courses. In order to understand whether and how speaking about the work influences the processing of beholding we designed a study in which participants contemplated paintings, selected to cover different genres, for 10 minutes, followed either by a period of up to 5 minutes during which participants were required to answer several open questions regarding the painting or by a period of 5 minutes of continued silent viewing. The results suggest how striking an effect speaking has on the way people look at paintings: consistent with a shift towards the global state of attention (that is, not focusing on details; see section 5.3), participants employed shorter fixations and longer saccades while they were speaking compared to silent beholding. In addition, the number of transitions between clusters of fixations significantly increased. We therefore suggest that speaking about known visual artworks activates a mnemonic representation of a particular painting, built up during the previous period of viewing that guides gaze movement control during a global state of attention and in partial disregard as to the details (Klein et al. 2014).

We also assume that context affects the reception of art in general and gaze movements of spectators in particular. As long as eye trackers could be used only in laboratories, it was not possible to test contextual influences. Thanks to new devices which enable eye tracking of mobile spectators, some studies have begun to investigate the influence of the context where paintings are exhibited (Brieber et al. 2014).

5.5 Conclusion and directions for future research

The mutual interest between art history and psychology can be traced back through over a century of research. Fechner (1876), one of the founding fathers of psychology, was also a pioneer of experimental investigation of aesthetics just as Wölfflin (1886) and Riegl (1901), two leading figures in art history during their own lives, developed psychological approaches in order to gain a basic understanding of the historical evolution of art. Since the 1950s publications of the psychologist Rudolf Arnheim were read by artists and art historians, as those by the art historian Ernst Gombrich were by psychologists. In terms of major recent developments in psychophysiology, eye tracking is probably the most promising technique, opening new horizons for collaborative research on the psychology of visual arts. The scope of this collaborative research is as broad and versatile as the many disciplines involved. It requires decisions about what should be some fundamental questions and it offers the possibility of addressing a range of relevant topics.

Among the fundamental questions are the following. First, what is the nature of the setting in which the appreciation of art is to be studied? While the vast majority of the limited number of studies in the psychology of visual arts have been conducted in the laboratory, ambulatory assessment nowadays uses versatile equipment which can be used to study the viewing and appreciation of art in environments that are more valid than the laboratory, such as museum, gallery, or the home. These studies would not be confined to subjective reports (e.g. using mobile devices which could be used to prompt responses, such as cell phones) or the measurement of gaze movements. They could investigate a wealth of psychophysiological processes supposedly associated with the appreciation of art (e.g. pupil diameter, electro-encephalography, heart rate, skin conductance response). Here, as in all domains of ambulatory assessment, the gain in ecological validity comes at the expense of "softened" cause–effect relationships ("internal validity") which still can be ascertained best in the laboratory. Secondly, what kind of co-variation should be investigated? Should it be "natural" co-variation such as the variation of aesthetic judgments across the paintings exhibited in a gallery, or "experimental" co-variation such as the changes in brain responses to a painting as a consequence of varying beholding instructions? Here, the fundamental question is whether to vary or manipulate the painting or the beholder (or both). Varying paintings—for example, by comparing abstract versus representational paintings or by comparing paintings from different epochs—leaves the object of study "intact" but renders the analysis of its effects ambiguous due to the many differences between paintings. Conversely, manipulating paintings—for example, by applying graphical filters to alter colors, contours, and the like, or relocating its elements to alter symmetry or composition—may yield at best clear "cause and effect relationships," but possibly ones that border on art-historical irrelevance due to the "artificial" nature of the manipulation. Varying (by selection) or "manipulating" (by instruction) the beholder, by contrast, tells us little about how a piece of art influences our perception of it. This individual-centered approach instead emphasizes the subject of the process of constructing "beauty in the eye of the beholder" and offers a variety of different approaches such as the following. We can study the impact of individual-related variables such as the degree of art expertise, age, gender, and cultural background (cross-cultural and sub-cultural) as well as their interaction by keeping the object artwork constant. This approach can be easily combined both with the art-centered approaches outlined earlier and with individual-centered manipulations of viewing instructions. Such instructions can focus on aspects of the paintings (e.g. its colorfulness, contents, etc.) or the processes in the beholder (e.g. remembering, evaluating, etc.), or both. All these approaches have been employed in the pertinent literature so far, albeit often not in combination with the recording of gaze movements. Due their close coupling with visual attention, detailed in section 5.3, gaze movements should be more systematically recorded in these endeavours.

Irrespective of the selection of approaches chosen, there are a number of topics that should be addressed in future research of a psychology of visual art. First, given the importance of hypothetical eye movements for the analysis of visual art (see section 5.1), more empirical work using gaze movement recordings is required to revise and possibly to re-evaluate and extend this kind of theorizing on the basis of firm empirical data. Secondly, given the inherently subjective nature of aesthetic experience and the openness of any kind of art, the systematic study of "personality" (in the broadest sense of the term) factors in looking at artworks is of utmost importance for this field of research. This does not only refer to the influences of gender, age, expertise, general education, general intelligence, or culture (with its pertinence for historical studies, see section 5.4.4), but it also touches upon the influences of "situational" factors (gallery versus lab), or the stability of individual differences. Thirdly, the observed gaze movement patterns should be validated against psychologically meaningful parameters (e.g. does fixation duration predict encoding success? Is the understanding of an artwork's complexity related to the complexity of the gaze movement patterns?). Fourthly, assuming that art is essential to the education systems of many cultures, how can we utilize the measurement of gaze movements during art contemplation to foster the understanding of art at different ages across the life span? Furthermore, eye-tracking studies have been used already in the field of conservation. Maisey and colleagues (2011) measured the reaction of beholders to discover how to re-integrate a highly damaged painting by John Martin at Tate Britain. We assume that in the future eye tracking will have significant implications in decisions about art display (museum and exhibition design) and for art-educational programs. Fifthly, while almost all research on the psychology of visual art has focused on the perception of art, experimental research on art production (as Miall and Tchalenko 2001) still remains a greater challenge. However, it is conceivable that even in this field eye tracking—for instance of a painter at work—will give us clues in the future about the nature of the creative act.

Acknowledgments

We thank Johanna Aufreiter, Hanna Brinkmann, Laura Commare, Caroline Fuchs, Martin Hirschbühl, Tanja Jenni, and Chiara Pompermaier for their help in recording and evaluating eye-tracking data at the Laboratory for Cognitive Research in Art History of the University of Vienna. The work was supported by grants from the DFG (RO 2281/3-1), the Institute for Advanced Study Berlin (Fellowship 2007-08), the WWTF (CS11-023), the ÖAW (Doc Team 68), and the FWF (P 25821-G21).

References

Andersen, R.A. (1989). Visual and eye movement functions of the posterior parietal cortex. *Annual Review of Neurosciences* **12**, 377–403.

Antes, J.R. (1974). The time course of picture viewing. *Journal of Experimental Psychology* **103**(1), 62–70.

Badt, K. (1961). *Modell und Maler von Jan Vermeer. Probleme der Interpretation; eine Streitschrift gegen Hans Sedlmayr*. Köln: DuMont Schauberg.

Baxandall, M. (1972). *Painting and Experience in Fifteenth Century Italy. A Primer in the Social History of Pictorial Style*. Oxford: Clarendon Press.

Baxandall, M. (1994). Fixation and distraction: the nail in Braque's Violin and Pitcher (1910). In J. Onians (ed.), *Sight and Insight. Essays on Art and Culture in Honour of E.H. Gomrich at 85*. London: Phaidon, pp. 399–415.

Becker, W. (1989). Metrics. In R.H. Wurtz and M.E. Goldberg (eds), *The Neurobiology of Saccadic Eye Movements*. Amsterdam: Elsevier, pp. 13–67.

Becker, W. and Fuchs, A.F. (1969). Further properties of the human saccadic system: eye movements and correction saccades with and without visual fixation points. *Vision Research* **9**, 1247–58.

Berlyne, D.E. (1971). *Aesthetics and Psychobiology*. New York, NY: Appleton-Century-Crofts.

Birmingham, E., Bischof, W.F., and Kingstone, A. (2009). Saliency does not account for fixations to eyes within social scenes. *Vision Research* **49**, 2992–3000.

Boland, J.E., Chua, H.F., and Nisbett, R.E. (2008). How we see it: culturally different eye movement patterns over visual scenes. In K. Rayner (ed.), *Cognitive and Cultural Influences on Eye Movements*. Tianjin: People's Publishing House, pp. 363–78.

Bon, L. and Lucchetti, C. (1990). Neurons signalling the maintenance of attentive fixation in frontal area 6aβ of macaque monkey. *Experimental Brain Research* **82**, 231–3.

Brieber, D., Nadal, M., Leder, H., et al. (2014). Art in time and space: context modulates the telation between art experience and viewing time. *PLoS One* **9**(6), e99019, June 2014, doi:10.1371/journal.pone.0099019.

Brinkmann, H., Commare, L., Leder, H., et al. (2014). Abstract art as a universal language? *Leonardo* **47**(3), 256–7.

Bruce, C.J. (1990). Integration of sensory and motor signals in primate frontal eye fields. In G.M. Edelman, E.W. Gall, and W.M. Cowan (eds), *Signal and Sense*. New York, NY: Wiley-Liss, pp. 261–314.

Bruce, C.J. and Goldberg, M.E. (1985). Primate frontal eye fields. I. Single neurons discharging before saccades. *Journal of Neurophysiology* **53**, 603–35.

Bruce, C.J., Goldberg, M.E., Bushnell, C., et al. (1985). Primate frontal eye fields. II. Physiological and anatomical correlates of electrically evoked eye movements. *Journal of Neurophysiology* **54**, 714–34.

Bucher, H.-J. and Schumacher, P. (2012). *Interaktionale Rezeptionsforschung: Theorie und Methode der Blickaufzeichnung in der Medienforschung*. Wiesbaden: Springer VS.

Buswell, G.T. (1935). *How People Look at Pictures: A Study of the Psychology and Percpetion in Art*. Chicago, IL: University of Chicago Press.

Carbon, C.-C., Hutzler, F., and Minge, M. (2006). Innovativeness in design investigated by eye movements and pupillometry. *Psychology Science* **48**(2), 173–86.

Cerf, M., Harel, J., Einhaeuser, W., and Koch, C. (2008). Predicting human gaze using low-level saliency combined with face detection. *Advances in Neural Information Processing Systems* **20**, 241–58.

Chica, A.B., Bartolomeo, P., and Lupianez, J. (2013). Two cognitive and neural systems for endogenous and exogenous spatial attention. *Behavioural Brain Research* **237**, 107–23.

Clausberg, K. (1999). *Neuronale Kunstgeschichte. Selbstdarstellung als Gestaltungsprinzip*. Wien: Springer.

DeAngelus, M. and Pelz, J.B. (2009). Top-down control of eye movements: Yarbus revisited. *Visual Cognition* **17**(6/7), 790–811.

Diderot, D. (1995). *The Salon of 1767*. New Haven: Yale University Press.

Dorris, M.C. and Munoz, D.P. (1998). Saccadic probability influences motor preparation signals and time to saccadic initiation. *Journal of Neuroscience* **18**, 7015–26.

Eco, U. (1962). *Opera Aperta*. Milano: Bompiani.

Einhaeuser, W., Spain, M., and Perona, P. (2008). Objects predict fixations better than early saliency. *Journal of Vision* **8**, 1–26.

Engelbrecht, M., Betz, J., Klein, C., et al. (2010). Dem Auge auf der Spur: Eine historische und empirische Studie zur Blickbewegung beim Betrachten von Gemälden. *IMAGE – Zeitschrift für interdisziplinäre Bildwissenschaft* **11**, 29–41.

Erdmann, B. and Dodge, R. (1898). *Psychologische Untersuchungen über das Lesen auf experimenteller Grundlage*. Halle: Max Niemeyer.

Fechner, G.T. (1876). *Vorschule der Ästhetik*. Leipzig: Breitkopf & Härtel.

Filarete (1972). *Trattato di architettura*. (**Vol. 1**). Milano: Ed. Il Polifilo.

Fioravanti, F., Inchingolo, P., Pensiero, S., et al. (1995). Saccadic eye movement conjugation in children. *Vision Research* **35**(23-4), 3217–28.

Fischer, B. (1987). The preparation of visually guided saccades. *Reviews of Physiology, Biochemistry and Pharmacology* **106**, 1–35.

Fischer, B. (1999). *Blick-Punkte*. Bern: Verlag Hans Huber.

Foster, H. (1988). *Vision and visuality*. Seattle, WA: Bay Press.

Foulsham, T. and Underwood, G. (2008). What can saliency models predict about eye movements? Spatial and sequential aspects of fixations during encoding and recognition. *Journal of Vision* **8**, 1–17.

Frangenberg, T. (1990). *Der Betrachter. Studien zur florentinischen Kunstliteratur des 16. Jahrhunderts*. Berlin: Mann.

Giuliani, L. (2003). *Bild und Mythos. Geschichte der Bilderzählung in der griechischen Kunst*. München: Beck.

Grant, E.R. and Spivey, M.J. (2003). Eye movements and problem solving: guiding attention guides thoughts. *Psychological Science* **14**, 462–6.

Harris, C.M., Hainline, L., Abramov, I., et al. (1988). The distribution of fixation durations in infants and naive adults. *Vision Research* **28**, 419–32.

Hayhoe, M. and Ballard, D. (2005). Eye movements in natural behavior. *Trends in Cognitive Science* **9**, 188–94.

Heidenreich, S.M. and Turano, K.A. (2011). Where does one look when viewing artwork in a museum? *Empirical Studies of the Arts* **29**(1), 51–72.

Henderson, J.M. and Ferreira, F. (1993). Eye movement dontrol during reading: fixation measures reflect foveal but not parafoveal processing difficulty. *Canadian Journal of Experimental Psychology* **47**, 201–21.

Henderson, J.M., Pollatsek, A., and Rayner, K. (1989). Covert visual attention and extrafoveal information use during object identification. *Perception and Psychophysics* **45**, 196–208.

Hikosaka, O. and Wurtz, R.H. (1983). Visual and oculomotor functions of monkey substantia nigra pars reticulata. I. Relation of visual and auditory responses to saccades. *Journal of Neurophysiology* **49**, 1230–53.

Hodgson, T.L., Dittrich, W.H., Henderson, L., et al. (1999). Eye movements and spatial working memory in Parkinson's disease. *Neuropsychologia* **37**, 927–38.

Hoffman, J.E. and Subramanian, B. (1995). The role of visual attention in saccadic eye movements. *Perception and Psychophysics* **57**, 787–95.

Huddy, V.C., Aaron, A.R., Harrison, M., et al. (2009). Impaired conscious and preserved unconscious inhibitory processing in recent onset schizophrenia. *Psychological Medicine* **39**(6), 907–16.

Jacob, R.J.K. and Karn, K.S. (2003). Eye tracking in Human-Computer Interaction and usability research: Ready to deliver the promises. In J. Hyönä, R. Radach, and H. Deubel (eds), *The Mind's Eye: Cognitive and Applied Aspects of Eye Movement Research*. Oxford: Elsevier Science BV.

Javal, E. (1879). Essai sur la physiologie de la lecture. *Annales d'Oculistique* **82**, 242–53.

Kapoula, Z., Daunys, G., Herbez, O., et al. (2009). Effect of title on eye-movement exploration of cubist paintings by Fernand Léger. *Perception* **38**(4), 479–91.

Kapoula, Z. and Lestocart, L.-J. (2006). Perception de l'espace et du mouvement dans Study of a Dog (Étude de chien) de Francis Bacon. *Intellectica* **44**, 215–26.

Klein, C., Betz, J., Hirschbuehl, J., et al. (2014). Describing art – an interdisciplinary approach to the effects of speaking on gaze movements during the beholding of paintings. *PLoS One* **9**(12), 2014, e102439, doi:10.1371/journal.pone.0102439.

Klein, C. and Ettinger, U. (2008). A hundred years of eye movement research in psychiatry. *Brain and Cognition* **68**, 215–8.

Klein, C. and Foerster, F. (2001). Development of prosaccade and antisaccade task performance in participants aged 6 to 26 years. *Psychophysiology* **38**, 179–89.

Klin, A., Lin, D.J., Gorrindo, P., et al. (2009). Two-year-olds with autism orient to non-social contingencies rather than biological motion. *Nature* **459**, 257–61.

Knoblich, G., Ohlsson, S., and Raney, G.E. (2009). An eye movement study of insight problem solving. *Memory and Cognition* **29**, 1000–9.

Koch, C. and Ullman, S. (1985). Shifts in selective visual attention: towards the underlying neural circuitry. *Human Neurobiology* **4**, 219–27.

Kowler, E. (2011). Eye movements: the past 25 years. *Vision Research* **51**, 1457–83.

Krieger, G., Rentschler, I., Hauske, G., et al. (2000). Object and scene analysis by saccadic eye-movements: an investigation with higher-order statistics. *Spatial Vision* **13**(2-3), 201–14.

Kristjanson, A.F. and Antes, J.R. (1989). Eye movement analysis of artists and nonartists viewing paintings. *Visual Arts Research* **15**(2), 21–30.

Leder, H., Belke, B., Oeberst, B., et al. (2004). A model of aesthetic appreciation and aesthetic judgments. *British Journal of Psychology* **95**, 489–508.

Liechty, J., Pieters, R. and Wedel, M. (2003). Global and local covert visual attention: evidence from a Bayesian Hidden Markov Model. *Psychometrika* **68**, 519–41.

Locher, P.J. (1996). The contribution of eye-movement research to an understanding of the nature of pictorial balance perception: A review of literature. *Empirical Studies of the Arts* **14**(2), 143–63.

Locher, P.J. (2006). The usefulness of eye movement recordings to subject an aesthetic episode with visual art to empirical scrutiny. *Psychology Science* **48**(2), 106–14.

Locher, P.J., Krupinski, E.A., Mello-Thoms, C., et al. (2007). Visual interest in pictorial art during an aesthetic experience. *Spatial Vision* **21**, 55–77.

Maisey, S., Smithen, P., Vilaro-Soler, A., et al. (2011). Recovering from destruction: the conservation, reintegration and perceptual analysis of a flood-damaged painting by John Martin. *International Council of Museums: Committee for Conservation. Published Proceedings. Lisbon, Portugal, 19–23 September*. Available at: <http://www.bbk.ac.uk/psychology/our-staff/academic/tim-smith/documents/1320_459_MAISEY_paper_EN1.pdf>.

Mannan, S.K., Ruddock, K.H., and Wooding, D.S. (1997). Fixation patterns made during brief examination of two-dimensional images. *Perception* **26**(8), 1059–72.

Manor, B.R. and Gordon, E. (2003). Defining the temporal threshold for ocular fixation in free-viewing visuocognitive tasks. *Journal of Neuroscience Methods* **128**, 85–93.

Manor, B.R., Gordon, E., and Touyz, S.W. (1995). Consistency of the first fixation when viewing a standard geometric stimulus. *International Journal of Psychophysiology* **20**, 1–9.

Massaro, D., Savazzi, F., Di Dio, C., et al. (2012). When art moves the eyes: a behavioral and eye-tracking study. *PLoS One* **7**(5).

McConkie, G., Reddix, M.R., and Zola, D. (1992). Perception and cognition in reading: where is the meeting point? In K. Rayner (ed.), *Eye Movements and Vision Cognition: Scene Perception and Reading*. New York, NY: Springer, pp. 293–303.

Mertens, I., Siegmund, H., and Gruesser, O.J. (1993). Gaze motor asymmetries in the perception of faces during a memory task. *Neuropsychologia* **31**, 989–98.

Miall, R.C. and Tchalenko, J. (2001). A painter's eye movements: a study of eye and hand movement during portrait drawing. *Leonardo* **34**(1), 35–40.

Molnar, F. (1981). About the role of visual exploration in aesthetics. In H.I. Day (ed.), *Advances in Intrinsic Motivation*. New York, NY: Plenum, pp. 385–414.

Molnar, F. (1992). A science of vision for visual art. *Leonardo* **30**(3), 225–32.

Molnar, F. and Ratsikas, D. (1987). Some aesthetical aspects of visual exploration. In J.K. O'Regan and A. Levy-Schoen (eds), *Eye Movements: From Psychology to Cognition*. Amsterdam: Elsevier, pp. 353–74.

Najemnik, J. and Geisler, W.S. (2005). Optimal eye movement strategies in visual search. *Nature* **434**, 387–91.

Nisbett, R.E. (2003). *The geography of thought. How Asians and Westerners think differently . . . and why*. New York, NY: Free Press.

Nisbett, R.E., Peng, K., and Choi, I. (2001). Culture and systems of thought: holistic versus analytic cognition. *Psychological Review* **108**(2), 291–310.

Nodine, F., Locher, P.J., and Krupinski, E.A. (1993). The role of formal art training on perception and aesthetic judgement of art compositions. *Leonardo* **26**, 219–27.

Noton, D. and Stark, L. (1971a). Eye movements and visual perception. *Scientific American* **224**(6), 35–43.

Noton, D. and Stark, L. (1971b). Scanpaths in eye movements during pattern perception. *Science* **171**(968), 308–11.

Noton, D. and Stark, L. (1971c). Scanpaths in saccadic eye movements while viewing and recognizing patterns. *Vision Research* **11**, 929–42.

Otero-Millan, J., Troncoso X.G., Macknik, S.L., et al. (2008). Saccades and microsaccades during visual fixation, exploration, and search: Foundations for a common saccadic generator. *Journal of Vision* **8**(14). <http://www.journalofvision.org/content/8/14/21.long>.

Paré, M. and Guitton, D. (1994). The fixation area of the cat superior colliculus: effects of electrical stimulation and direct connection with brainstem omnipause neurons. *Experimental Brain Research* **101**, 109–22.

Pieters, R., Rosbergern, E., and Wedel, M. (1999). Visual attention to repeated print advertising: a test of scanpath theory. *Journal of Marketing Research* **36**, 424–38.

Pihko, E., Virtanen, A., Saarinen, V.-M., et al. (2011). Experiencing art: the influence of expertise and painting abstraction level. *Frontiers in Human Neuroscience* **5**. <http://www.ncbi.nlm.nih.gov/pmc/articles/PMC3170917>.

Procopius (1940). On Buildings (**Vol. 1**). Cambridge, MA: Harvard University Press.

Rayner, K., Smith, T.J., Malcolm, G.L., et al. (2009). Eye movements and visual encoding during scene perception. *Psychological Science* **20**, 6–10.

Riegl, A. (1901). *Die Spätrömische Kunstindustrie nach den Funden in Österreich-Ungarn*. Wien: Hof- und Staatsdruckerei.

Robinson, D.L., Goldberg, M.E., and Stanton, G.B. (1978). Parietal association cortex in the primate: sensory mechanisms and behavioral modulations. *Journal of Neurophysiology* **41**, 910–32.

Rosenberg, R. (2000). *Beschreibungen und Nachzeichnungen der Skulpturen Michelangelos. Eine Geschichte der Kunstbetrachtung*. München: Deutscher Kunstverlag.

Rosenberg, R. (2008). Le schéma de composition, outil et symptôme de la perception du tableau. In R. Recht, P. Sénéchal, C. Barbillon, et al. (eds), *Histoire de l'histoire de l'art en France au XIXe siècle*. Paris: La documentation Française, pp. 419–31, 523–4.

Rosenberg, R. (2014). Blicke Messen. Vorschläge für eine empirische Bildwissenschaft, *Jahrbuch der Bayerischen Akademie der Schönen Künste* **27**, 2013, 71–86

Rosenberg, R., Betz, J., and Klein, C. (2008). Augensprünge. *Bildwelten des Wissens. Kunsthistorisches Jahrbuch für Bildkritik* **6**(1), 127–9.

Russo, J.E. and Rosen, L.D. (1975). An eye fixation analysis of multialternative choice. *Memory and Cognition* **3**(3), 267–76.

Santella, A. and DeCarlo, D. (2004). Robust clustering of eye movement recordings for quantification of visual interest. *Eye Tracking Research and Application* **23**, 27–34.

Schall, J.D. (1991). Neuronal activity related to visually guided saccades in the frontal eye fields of rhesus monkeys: comparison with supplementary eye fields. *Journal of Neurophysiology* **66**, 559–79.

Schiller, P.H. and Stryker, M. (1972). Single-unit recordings and stimulation in superior colliculus of the alert monkey. *Journal of Neurophysiology* **35**, 915–24.

Smith, T.J. (2014). Audiovisual correspondences in Sergei Eisenstein's Alexander Nevsky: a case study in viewer attention. In P. Taberham and T. Nannicelli (eds), *Cognitive Media Theory (AFI Film Reader)*. <http://www.bbk.ac.uk/psychology/our-staff/academic/tim-smith/documents/Alex_Nev_CogMediaTheory_2014_timjsmith_figs.pdf>.

Smith, J.K. and Smith, L.F. (2001). Spending time on art. *Empirical Studies of the Arts* **19**(2), 229–36.

Tatler, B. W. (2007). The central fixation bias in scene viewing: selecting an optimal viewing position independently of motor biases and image feature distributions. *Journal of Vision* **7**, 1–17.

Thouless, R.H. (1936). How people look at pictures. *Burlington Magazine* **68**, 58.

van Diepen, P.M.J., De Graef, P., and d'Ydewalle, G. (1995). Chronometry of foveal information extraction during scene perception. In J.M. Findlay, R. Walker, and R.W. Kentridge (eds), *Eye Movement Research/Mechanisms, Processes and Applications*. Amsterdam: Elsevier, pp. 349–62.

Vogt, S. and Magnussen, S. (2007). Expertise in pictorial perception: eye-movement patterns and visual memory in artists and laymen. *Perception* **36**(1), 91–100.

Volkenandt, C. (2011). Art. "Visual Culture/Visual Studies." In: U. Pfisterer (ed.), *Metzler Lexikon Kunstwissenschaft. Ideen, Methoden, Begriffe* (2nd edn). Stuttgart: Metzler, pp. 462–6.

Wade, N.J. and Tatler, B.W. (2005). *The Moving Tablet of the Eye. The Origins of Modern Eye Movement Research*. Oxford: Oxford University Press.

Waetzoldt, W. (1912). *Einführung in die Bildenden Künste*. Leipzig: Hirt.

Welchman, J.C. (1997). *Invisible Colors. A Visual History of Titles*. New Haven, CT: Yale University Press.

Welzel, B. (1997). Galerien und Kunstkabinette als Orte des Gesprächs. In W. Adam (ed.), *Geselligkeit und Gesellschaft im Barockzeitalter*. Wiesbaden: Harrassowitz. Vol. 1, pp. 495–504.

Westheimer, G. (1954). Eye movement responses to a horizontally moving visual stimulus. *Archives of Ophthalmology* **52**, 313–26.

Wölfflin, H. (1886). *Prolegomena zu einer Psychologie der Architektur*. München: C. Wolf & Sohn.

Wölfflin, H. (1899). *Die klassische Kunst*. München: Bruckmann.

Wölfflin, H. (1941). Über das Rechts und Links im Bilde. In H. Wölfflin (ed.), *Gedanken zur Kunstgeschichte. Gedrucktes und Ungedrucktes* (2nd edn). Basel: Schwabe & Co, pp. 82–90.

Yarbus, A.L. (1967). *Eye Movement and Vision*. New York, NY: Plenum Press.

Zangemeister, W.H., Sherman, K., and Stark, L. (1995). Evidence for a global scanpath strategy in viewing abstract compared with realistic images. *Neuropsychologia* **33**(8), 1009–25.

Section 2

Cognitive neuroscience of visual aesthetics and art

Chapter 6

Neural mechanisms for evaluating the attractiveness of faces

Spas Getov and Joel S. Winston

6.1 Introduction

Faces are central to human social interaction and their aesthetic valuation has important and wide-ranging societal implications. For example, facial attractiveness is likely to be key to determining mate choice (Thornhill and Gangestad 1999), and it can predict performance across many domains (Langlois et al. 2000), including income and success in job interviews (Dipboye et al. 1977; Frieze et al. 1991). Comparisons across individuals as well as across different cultures show some variation in what is considered to be attractive (Bruce and Young 1998), yet there is also cross-cultural agreement (Jones and Hill 1993), as well as converging evidence from studies in infants that perception of attractiveness has a strong innate component (Langlois et al. 1987; Slater et al. 1998). Ultimately, it is likely that what is perceived as attractive in a face generalizes across race, gender, and age (Langlois et al. 1991).

Mechanisms for evaluating the attractiveness of other people's faces have been widely investigated in the fields of social and experimental psychology, and more recently there have been focused neuroscientific attempts to delineate the brain processes facilitating the evaluation of attractiveness. This chapter will focus on findings from a group of studies performed over the last 10–15 years with the aid of functional brain imaging technologies (including functional magnetic resonance imaging, fMRI; positron emission tomography, PET; electroencephalography, EEG; and magnetoencephalography, MEG), as well as studies of patients with focal brain lesions. Our aim has been to outline what is currently known about the neural mechanisms for evaluation of facial attractiveness, undertake a critical appraisal of the key publications in this field, and discuss how this endeavor contributes more generally to the study of neuroaesthetics.

We briefly preface our discussion of the neuroscientific data with a summary of the known physical determinants of attractiveness (section 6.2). Additionally, we outline several relevant theoretical frameworks that have informed the neuroscientific study of attractiveness (section 6.3), before moving on to discuss the relevant studies themselves (sections 6.4 and 6.5).

6.2 What makes a face attractive

There have been a number of theories regarding the physical determinants of facial attractiveness. Perhaps the oldest of these relate beauty to optimal proportions between various facial features. Plato and Leonardo da Vinci, amongst others, were proponents of these theories. A brief account is provided by Bruce and Young (1998, p. 131).

An evolutionarily driven, adaptationist perspective is that facial attractiveness judgments are based on phenotypic characteristics that accurately reflect an individual's health and fitness to their particular environment and its demands (for reviews see Johnston 2006; Rhodes 2006; Thornhill and Gangestad 1999). In other words, this account posits that facial beauty reflects the reproductive quality of the owner of the face (Senior 2003). There are features that reflect health fairly transparently, including clear eyes, smooth and even-toned skin, and an average body mass index, and clues to many of these can be found in the face. Three other physical properties of faces that have received special attention in terms of their impact on attractiveness are symmetry, averageness (or typicality), and sexual dimorphism. A detailed review of evidence relating to each of these areas is beyond the scope of this chapter. However, each will be covered briefly in the following paragraphs in the interest of providing an outline of the kinds of physical characteristics that may need to be encoded by neural processes for evaluating attractiveness.

Undoubtedly, there are other determinants of facial attractiveness among the physical features and configurations of human faces, besides those already discussed. In addition, other information, both inside the visual domain (e.g. shape of body parts, overall physical condition, and body movement) and beyond (e.g. smell, voice), certainly makes important contributions to attractiveness judgments in real-life situations. However, a broader discussion of these other determinants of human attractiveness is outside the scope of this chapter.

6.2.1 **Symmetry**

Certain facial traits, which are normally symmetrical, can become less so in some individuals. This phenomenon, termed *fluctuating asymmetry*, arguably reflects increased sensitivity to genetic and environmental stresses (Møller and Pomiankowski 1993). This idea has been offered as a basis for explaining a link between facial symmetry and attractiveness in evolutionary terms (Thornhill and Gangestad 1999); faces which are more symmetrical are found more attractive because, according to this account, they suggest the bearer is more robust in dealing with environmental stressors. A number of studies have shown that increasing facial symmetry using computer manipulations leads to increased attractiveness ratings (Perrett et al. 1999; Rhodes et al. 1998). It is important to note, however, that there is evidence that symmetry covaries with other facial features such as reduced face length and increased cheekbone prominence. These may also be important determinants of judgments about attractiveness, and therefore the amount of variance in attractiveness accounted for by symmetry alone may be small (Thornhill and Gangestad 1999). Investigation into the brain areas which process facial symmetry has been conducted using fMRI (Chen et al. 2007). Faces activated a number of brain regions in the well-established face processing network (Haxby et al. 2000; see section 6.3.1) including the fusiform face area, the occipital face area, the middle occipital gyri, and the superior temporal and intraoccipital sulci. Symmetrical versions of the faces caused differential activation only in the right occipital face area, suggesting a role for this region in facial symmetry processing (Chen et al. 2007).

6.2.2 Typicality

It can be argued from an evolutionary perspective that the most typical faces, which are close to the mean of the population, should be favored as they denote genetic heterozygosity (Thornhill and Gangestad 1999). Indeed, there is empirical evidence that computer-generated composite (averaged) faces are rated as more attractive than real-life individual faces (Langlois and Roggman 1990), and that this is independent of the increase in facial symmetry that such manipulations can produce (Rhodes et al. 1999). It has been claimed that infant preference for attractive faces may also in fact reflect a preference for typical faces (Rubenstein et al. 1999). However, averageness only increases attractiveness to a point, and specific non-average features can make faces even more attractive (Thornhill and Gangestad 1999). An alternative account to explain differential responses to attractiveness in terms of typicality posits that faces are indexed by their similarity to unfit or disfigured faces, and that disfigured faces thus anchor the lower end of the attractiveness dimension. This view places more importance on the negatively valenced half of the attractiveness scale and proposes that viewing a face as unattractive can be conceptualized as an overgeneralization of mechanisms for representing an unfit face (Zebrowitz and Montepare 2008).

6.2.3 Sexual dimorphism

One prevalent argument is that a strong contribution of secondary sexual characteristics to facial features (e.g. wider mandible and prominent eyebrow ridges in men, or enlarged lips and upper cheeks in women) has an asymmetric effect on judgments of attractiveness of faces of men and women. Specifically, feminized female faces are judged as more attractive, while masculinized male faces are perceived as more dominant and dishonest but less attractive, and feminized male faces are instead preferred (Perrett et al. 1998). A possible explanation for this finding is that feminized male faces predict more cooperative, honest, and caring behavior, and may thus indicate willingness to invest in a long-term and stable relationship. The preference for feminized male faces is by no means a universal finding, however. Further work has shown that in fact, women prefer feminized male faces only when in the less fertile phase of their menstrual cycle, and preference shifts to more masculine faces during the more fertile phase (Penton-Voak et al. 1999). In addition, the shift towards preference for masculinized male faces during the fertile phase of the menstrual cycle was abolished when women were instructed to select specifically long-term male partners (Penton-Voak et al. 1999). A number of other studies have found that sex-specific facial features are significantly predictive of attractiveness judgments (Cunningham 1986; Cunningham et al. 1990; Johnston and Franklin 1993) but caution must be exercised given the variability of judgments, particularly in women, depending on cyclical fertility or the specific purpose of face choice.

6.3 Conceptual frameworks for studying brain processing of facial attractiveness

6.3.1 Distributed neural systems for face perception

Faces are highly complex visually, as well as highly socially informative, and it is therefore no surprise that face perception has received a great deal of attention in the neurosciences. For some years, the most influential cognitive model of face perception has been that of Bruce and Young (1986) which emphasized distinct processing streams to deal with different kinds of information relevant to faces. A key distinction was made between processes supporting identity recognition and those supporting recognition of facial expressions and speech. The advent of fMRI allowed brain areas involved in facial processing to be determined with considerable anatomical specificity, leading to influential findings of activation by faces of an area in the fusiform gyrus (subsequently named the fusiform face area, FFA; Kanwisher et al. 1997) as well as an area in the superior temporal sulcus (STS; e.g. Puce et al. 1998), amongst others. Such findings have since been widely replicated and have fuelled the development of an influential neurobiological framework describing human face perception (Haxby et al. 2000). Haxby and colleagues extended Bruce and Young's model by proposing, with new anatomical specificity guided by neuroimaging findings in humans and neuronal recordings in non-human primates, a core system for face perception made up of the inferior occipital gyri, or occipital face area (OFA, facilitating visual analysis of facial features), FFA (facilitating face recognition, or analysis of invariant face information), and STS (facilitating processing of the dynamic aspects of faces such as expression and eye gaze). The model also describes interactions between this core system and an extended system that includes regions involved primarily in other cognitive functions, but which make important contributions to face perception, such as the amygdala (emotion evaluation) and anterior temporal cortex (retrieval of biographical information). Subsequent studies of face perception have largely adopted this framework and it anchors much of the research into the neural processing of facial attractiveness also. At the same time, it is important to note that while the overall concept of dual streams of core face processing (invariant and dynamic) has been very powerful in framing work in this field over the last 10–15 years, it is increasingly apparent that there is considerable cross-over in the roles of the two processing streams and they are far from completely dissociable. For example, there is evidence for the involvement of STS in representing facial identity (Winston et al. 2004) as well as involvement of the FFA in representing facial expression (Fox et al. 2009). A detailed critique of the separate face-processing pathways has previously been undertaken, along with proposals of alternative models (Calder and Young 2005). In particular, the role of STS is probably much broader than simply covering dynamic aspects of faces, and the region has been re-framed as being involved in evaluation of the social relevance of faces, especially in terms of inferring the goals and intentions of others (Allison et al. 2000), be this conveyed via dynamic or static signals.

6.3.2 Attractive faces as rewarding stimuli

Another common framework within which to place investigation of the neural processing of facial attractiveness conceptualizes faces as rewarding stimuli, or as positive "reinforcers" (i.e. stimuli which encourage behaviors that are associated with their perception). Examples of behavioral modulation by facial attractiveness lend support to this view. However, it has been pointed out that unlike drives towards food, water, or warmth, a drive towards attractiveness cannot arise from a deficit state (Aharon et al. 2001). In addition, it has been proposed that faces with an adaptive value (e.g. opposite-sex faces) are processed as rewarding whereas those without such value (e.g. beautiful same-sex faces) are instead processed as aesthetically pleasing, but not as rewarding (Senior 2003). The well-established "reward circuit" that guides approach and avoidance behaviors has the striatum (caudate, putamen, and nucleus accumbens) at its core, but it also includes important anatomical connections between the striatum and midbrain nuclei such as the ventral tegmental area, as well as orbitofrontal, ventromedial, and ventrolateral prefrontal cortex, and "limbic" structures such as the amygdala (Delgado 2007). The involvement of this circuit in the processing of facial attractiveness now has considerable empirical support from fMRI studies (Aharon et al. 2001; Liang et al. 2010; O'Doherty et al. 2003; Winston et al. 2007), and this is discussed in more detail in section 6.4.1. Whether attractiveness is always processed through reward circuitry in the brain, or this is only the case for extremely attractive faces (Mende-Siedlecki et al. 2012), and perhaps extremely unattractive faces (Liang et al. 2010), remains debatable. Besides extremes of attractiveness or unattractiveness, it has also been suggested that specifically making attractiveness evaluation a goal (Mende-Siedlecki et al. 2012) or the presence of other social cues, such as gaze direction, that may indicate a favorable or unfavorable engagement with an attractive face (Kampe et al. 2001), may contribute importantly to activation of reward systems.

6.3.3 Attractiveness as a trait; commonality with mechanisms of facial emotion recognition

Another way of conceptualizing facial attractiveness is as a trait exhibited by faces, in common with other traits such as trustworthiness, aggressiveness, and confidence. When people are asked to provide unconstrained descriptions of real-life face images, attractiveness is the most commonly used descriptor (Oosterhof and Todorov 2008), which presumably reflects the high evolutionary importance of this trait. On the other hand, Oosterhof and Todorov (2008) found that different trait descriptions of faces were highly intercorrelated and used principal components analysis to show that evaluation of faces based on such traits can be described by a two-dimensional conceptual space, with axes of dominance and trustworthiness (the latter approximating to valence). Todorov and colleagues have gone on to argue that there are common neural mechanisms subserving evaluation of different facial traits, as well as evaluation of facial emotional expressions (Said et al. 2011), which is consistent with emotion overgeneralization hypotheses (Zebrowitz 2004). Behaviorally, it has been shown that ratings of faces according to traits and ratings according to emotional expressions are frequently intercorrelated. For example, faces rated as highly dominant by one sample of participants are consistently rated as angry by an independent sample (Said et al. 2011). In addition, principal components analysis shows that emotion and trait judgments have very similar dimensional structures (Said et al. 2011). Finally, evidence for commonality of neural mechanisms for emotion and trait judgments comes from an adaptation study (Engell et al. 2010) demonstrating that neutral faces were judged as more trustworthy after adaptation to angry faces and less trustworthy after adaptation to happy faces, suggesting commonality in mechanisms for processing trustworthiness with those for anger and happiness.

These behaviorally demonstrated commonalities, while not direct evidence of common neural mechanisms, are at least suggestive of commonalities in the neural evaluation of facial traits and expressions. Further support for this claim can be found in neuroimaging findings, which have established a similarity in the brain regions active during evaluation of attractiveness and trustworthiness (Bzdok et al. 2010; Mende-Siedlecki et al. 2012; Winston et al. 2007, 2002) and during evaluation of untrustworthy and angry faces (Mende-Siedlecki et al. 2012). Commonality of mechanisms for facial expression and facial trait processing could clearly be a very useful framework for understanding facial attractiveness evaluation, particularly given the comparatively well-developed understanding of neural mechanisms for evaluation of emotional facial expressions (reviewed by Vuilleumier and Pourtois 2007). At the same time, some dissociation of mechanisms is highly likely, with some regions or networks being involved specifically in one aspect of social face processing only.

6.3.4 Distance from mean in valence-space

An alternative view, related to that discussed in section 6.3.3, again hinges on placing attractiveness within a two-dimensional space described by axes of valence and dominance (Oosterhof and Todorov 2008; Said et al. 2011). There is evidence to suggest that neural responses to faces, for example in the amygdala, may represent distance from typicality in valence-space, rather than evaluation of any specific social dimension (Said et al. 2010). Viewed within this framework, inconsistencies in findings from different studies of the neural correlates of attractiveness can be explored, and at times explained, through differences in the typicality of the face stimuli used. More specifically, this theory predicts that whether the stimulus sets used in a study have placed most typical faces in the center of the attractiveness dimension or at one extreme of it could have important implications for the results (Mende-Siedlecki et al. 2012; Said et al. 2011). For example, the use of naturalistic face photographs (e.g. O'Doherty et al. 2003) leads to the most typical faces being at the positive end of the attractiveness continuum, leading to a linear typicality function, and correspondingly, to mainly linear neural responses (Said et al. 2010). In contrast, synthetic face stimuli (e.g. Chatterjee et al. 2009; Kim et al. 2007) would tend to have a non-linear typicality function, with the most typical faces in the middle of the continuum, and thus result in non-linear neural responses (see Figure 6.1 for examples of naturalistic and computer-generated face images used in studies of facial attractiveness). Whether this is an entirely accurate claim remains open to debate, but it could certainly be useful in guiding attempts to interpret discrepancies between the results of different studies.

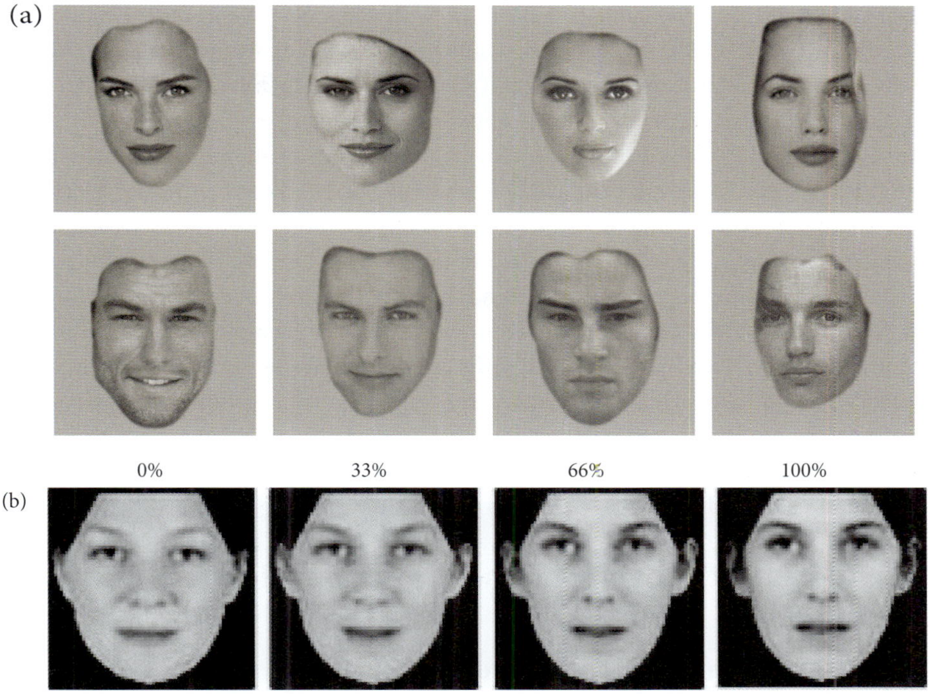

Figure 6.1 Examples of face stimuli used in studies of facial attractiveness evaluation. Panel (a) shows several highly attractive faces, half of which have a mildly smiling expression and the other half a neutral expression. Faces were obtained from photographs in magazines. Reprinted from *Neuropsychologia*, 41 (2), J. O'Doherty, J. Winston, H. Critchley, D. Perrett, D.M. Burt, and R.J. Dolan, Beauty in a smile: the role of medial orbitofrontal cortex in facial attractiveness, pp. 147–55, Copyright (2003), with permission from Elsevier. Panel (b) shows an example of computer-generated faces, with one identity on the left hand side, another on the right, and morphs (in percentage change from the left hand identity) in the middle. Participants were asked to judge how attractive each face was. Reproduced from A. Chatterjee, A. Thomas, S.E. Smith, and G.K. Aguirre, The neural response to facial attractiveness, *Neuropsychology*, 23 (2), pp. 135 © 2009, American Psychological Association, with permission.

6.3.5 Neuropsychological models of aesthetic preference

Neuroaesthetics, a term pioneered by Zeki (1999), has been defined as "the neural underpinnings of aesthetic experience of beauty, particularly in visual art" (Di Dio and Gallese 2009). As in other spheres of social life, faces have an important place in art. An understanding of the neural mechanisms for evaluating facial attractiveness will help explain specifically how we ascribe aesthetic value to faces, but it is also of much broader relevance to the study of neuroaesthetics, with the potential to provide clues to any neural mechanisms shared by all aesthetic experience. It is therefore important to consider how studies of the neural correlates of aesthetic (and/or attractiveness) evaluation of faces can be framed using neuropsychological models of aesthetic preference. A recent review tackling this issue (Nadal et al. 2008) evaluates models of the processing steps involved in determining the aesthetic value of a stimulus, with particular reference to a model proposed by Chatterjee (2003). Chatterjee posited three processing stages relevant to the hierarchy of perceptual processing in general (early processing to extract and analyse simple visual components, intermediate-level processes to segregate or group elements, and late processes where key aspects of the stimulus can be integrated with relevant memories and emotions). Such a framework can facilitate the integration of aesthetic theory with relevant neurobiological findings and reference will be made to this when discussing neurobiological findings in sections 6.4 and 6.5 of this chapter. Notwithstanding the utility of such models, it is important to bear in mind that they are rarely likely to decribe fully and accurately the interplay of all relevant neural processes. For example, it is well known that emotive aspects of faces influence both early and late processing components (Eimer and Holmes 2002; Sato et al. 2001), and an early component of emotional processing does not fit neatly into the cognitive model described here. As discussed in more detail in section 6.4.4 facial attractiveness also modulates both early and late components of neural processing (van Hooff et al. 2011; Werheid et al. 2007).

6.4 Studies of the neural mechanisms for evaluating facial attractiveness

6.4.1 Involvement of the brain's reward circuitry in evaluation of facial attractiveness

The earliest investigation into neural mechanisms for evaluating facial attractiveness made use of PET and demonstrated involvement of left frontal cortical regions. Increased regional cerebral blood flow (rCBF) in left anterior frontal cortex was correlated with judgment of faces as unattractive, and increased rCBF in the left fronto-temporal junction was correlated with judgments of faces as attractive (Nakamura et al. 1998). Subsequently, a number of studies have used fMRI to explore the neural correlates of attractiveness evaluation with greater anatomical specificity. Much of this early work has been guided by the framework that casts attractive faces as stimuli that engage the brain's reward-processing systems, as discussed earlier in section 6.3.2. Aharon and colleagues (2001) hypothesized that there are separable neural mechanisms for aesthetic assessment of attractiveness in faces (termed "liking") and for driving motivated behavior in relation to attractive faces (termed "wanting"). Separate groups of heterosexual males either rated the beauty of faces (a mix of neutral and attractive faces of each gender) or pressed a key to prolong viewing of the same set of faces. Participants rated both attractive males and females as such, but exerted effort specifically and only to prolong viewing of attractive females (Figure 6.2a–c). A third group of males underwent fMRI during passive viewing of the same set of faces. There was a significant interaction between attractiveness and face gender in the nucleus accumbens, an area strongly associated with reward processing (e.g. Bardo 1997; Knutson et al. 2001). This interaction was driven by stronger activations to beautiful female faces and average male faces than to beautiful male faces or average female faces (Figure 6.2d). Other regions of interest, including

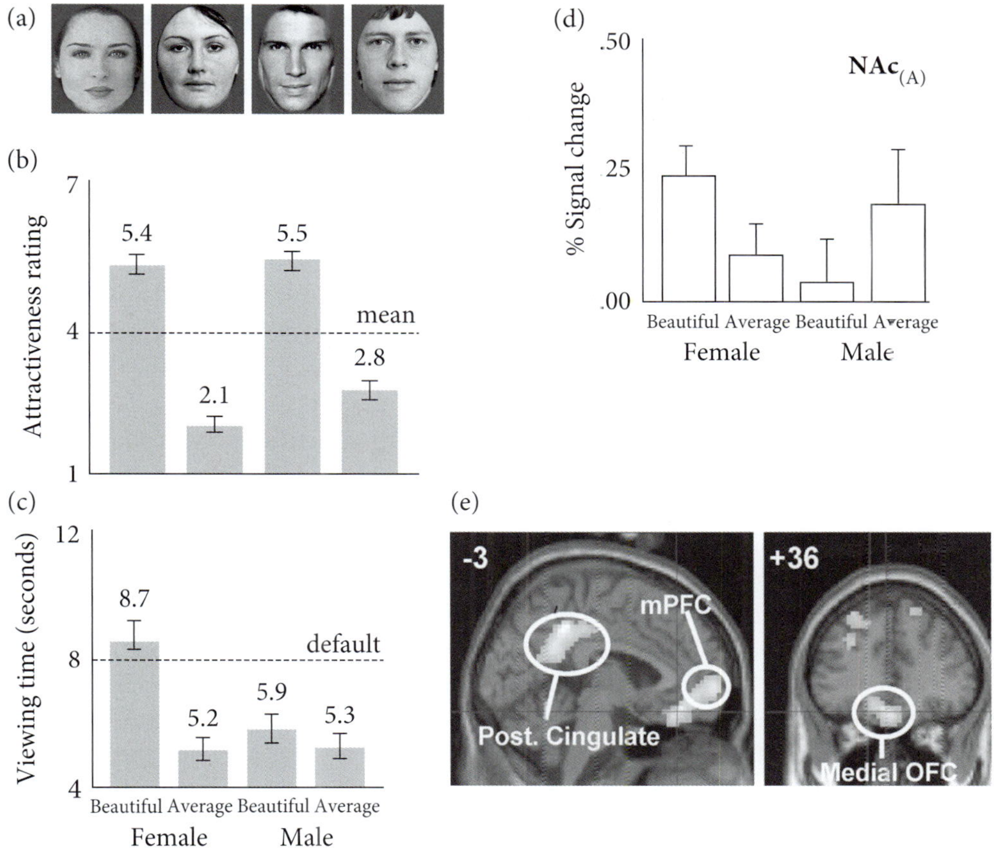

Figure 6.2 Stimulus-related neural activation, from studies highlighting brain regions that respond differentially to varying levels of facial attractiveness. Panel (a) gives an example of stimuli used, from left to right: beautiful female face, average female face, beautiful male face, average male face. Panel (b) shows the attractiveness ratings (on a scale of 1–7) given for each of the four categories of face from panel (a) by a sample of male participants; panel (c) shows average viewing time (increased by pressing a button repeatedly) achieved for each of the four face categories from panel (a) by a separate sample of male participants. Panel (d) shows blood-oxygen-level-dependent fMRI signal change in the nucleus accumbens (NAc) for each of the four face categories depicted in panel (a). Figures 6.2a–d are reprinted from *Neuron*, 32 (3), Itzhak Aharon, Nancy Etcoff, Dan Ariely, Christopher F Chabris, Ethan O'Connor, and Hans C Breiter, Beautiful Faces Have Variable Reward Value: fMRI and Behavioral Evidence, pp. 537–551, Copyright (2001), with permission from Elsevier. Panel (e) shows the main effect of attractiveness (high attractiveness > low attractiveness) in medial orbitofrontal cortex (OFC), posterior cingulate and medial prefrontal cortex (mPFC). Reprinted from *Neuropsychologia*, 41(2), J. O'Doherty, J. Winston, H. Critchley, D. Perrett, D.M. Burt, and R.J. Dolan, Beauty in a smile: the role of medial orbitofrontal cortex in facial attractiveness, pp. 147–155, Copyright (2003), with permission from Elsevier. Numbers in top-left corner of brain images show relevant coordinate (*x*-coordinate for sagittal images, *y*-coordinate for coronal images) of the displayed slice.

sublenticular extended amygdala (SLEA) and ventral tegmental area (VTA), also showed strong activations associated with viewing of attractive female faces, but no differential activation for average and attractive male faces. The pattern of activation in SLEA and VTA was noted to mirror closely results of the key-pressing behavioral task. In summary, the results suggest some dissociation between "liking" and "wanting" behaviorally (both attractive male and female faces are liked but only attractive female faces are wanted by heterosexual males). In terms of neural activation, such a dissociation is less clear. No brain areas mirrored the "liking" pattern of behavioral results, but areas important for reward processing did mirror the "wanting" pattern of behavioral results. The most significant and noteworthy findings were in the nucleus accumbens, suggesting that it may play an important role in distinguishing attractive and rewarding faces from attractive and unrewarding ones.

In order to resolve whether neural processes signaling attractiveness evaluation can be dissociated from those engaged by reward expectation, O'Doherty and colleagues (2003) performed an fMRI experiment with an event-related design.[1] The authors found activation in the medial orbitofrontal cortex (OFC) during viewing of attractive faces, while more lateral prefrontal regions, including lateral OFC, were responsive to unattractive faces (Figure 6.2e). These findings reflect implicit evaluation of attractiveness because participants were unaware that attractiveness was the critical experimental measure and were asked to determine the gender of each presented face, rather than to rate attractiveness directly. In addition, half of the stimulus set was made up of mildly smiling faces, which resulted in further enhancement of OFC activity. The key findings were thus that OFC is engaged automatically by attractive faces, and that this engagement is positively modulated by happy facial expression.

Results showing activation of lateral OFC by unattractive faces and/or medial OFC by attractive faces have since been replicated (Cloutier et al. 2008; Hampshire et al. 2012).

Further work has extended the conceptualization of attractive faces as rewards by establishing involvement of the amygdala, which plays an important role in evaluation of social stimuli (see Adolphs 2010 for a review). Winston and colleagues (2007) hypothesized that the amygdala would show non-linear changes in blood-oxygen-level-dependent (BOLD) fMRI responses to attractiveness, with higher responses to both attractive and non-attractive (relative to neutral) faces. This would explain why previous work, seeking only linear relationships between brain activity and attractiveness, found no activation in amygdala. The authors' findings were consistent with this hypothesis, additionally showing non-linear and linear responses to attractiveness in different regions of OFC. This study also compared neural responses during implicit attractiveness judgments (while evaluating gender) with those during explicit attractiveness judgments. Explicit judgment of attractiveness recruited a network including medial prefrontal and paracingulate cortices, posterior OFC, insula, and STS (Figure 6.3a).

Non-linear neural responses to facial attractiveness have been found in other regions of the established reward-processing network, including anterior cingulate cortex (ACC), lateral OFC, striatum (including nucleus accumbens, caudate, and putamen), and VTA

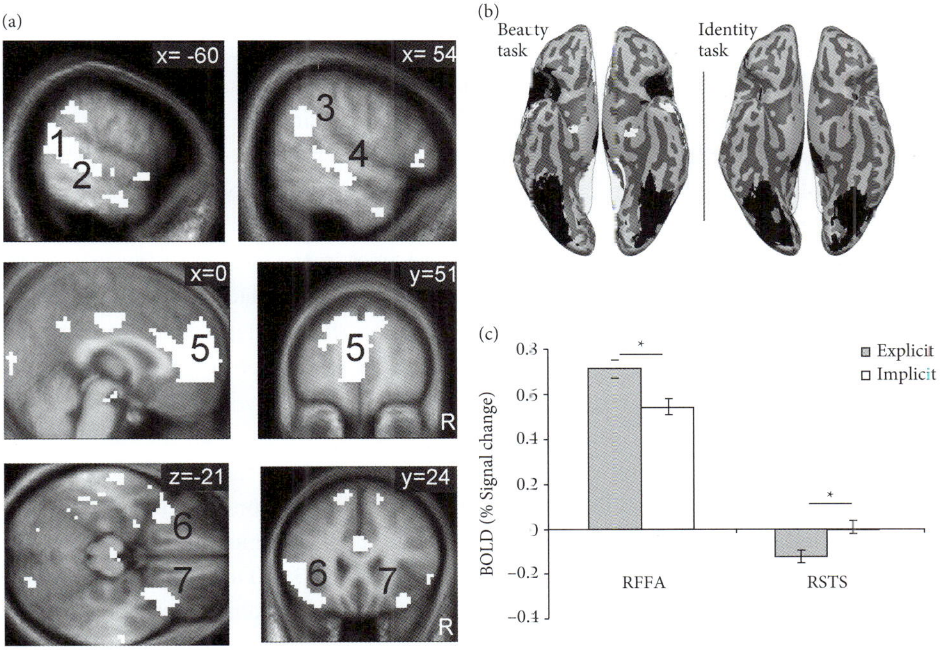

Figure 6.3 Task-related neural activation, from studies highlighting brain regions that respond differentially to attractiveness evaluation or an alternative control task. Panel (a) shows superior temporal sulcus (STS) and prefrontal responses for attractiveness judgments (contrasted with age judgmnts). There are four individual loci of activation in STS (labeled 1–4; two on each side of the brain), and three individual loci of activation in prefrontal areas (labeled 5–7). Reprinted from *Neuropsychologia*, 45 (1), Joel S. Winston, John O'Doherty, James M. Kilner, David I. Perrett, and Raymond J. Dolan, Brain systems for assessing facial attractiveness, pp. 195–206, Copyright (2007), with permission from Elsevier. Numbers in top right corner of brain images show relevant coordinate (*x*-coordinate for sagittal images, *y*-coordinate for coronal images, *z*-coordinate for axial images) of the displayed slice. Panel (b) shows neural responses to facial beauty (contrasted with facial identity, left), and to facial identity (contrasted with facial beauty, right). Black-colored areas represent increased responses to attractive faces and white colored areas represent increased responses to unattractive faces. Reprinted from *Neuropsychology*, 23 (2), Anjan Chatterjee, Amy Thomas, Sabrina E. Smith, and Geoffrey K. Aguirre, The neural response to facial attractiveness, pp. 135–43, doi: 10.1037/a0014430 Copyright (2009), with permission from the American Psychological Association. Panel (c) shows blood-oxygen-level-dependent signal change in right fusiform face area (RFFA) and right STS (RSTS) during evaluation of attractiveness (explicit condition) or a one-back stimulus identification task (implicit condition). Reprinted from *Journal of Neuroscience*, 155 (2), Giuseppe Iaria, Christopher J. Fox, Christopher T. Waite, Itzhak Aharon, and Jason J. S. Barton, The contribution of the fusiform gyrus and superior temporal sulcus in processing facial attractiveness: neuropsychological and neuroimaging evidence, pp. 409–22, figure 8, doi: 10.1016/j.neuroscience.2008.05.046 Copyright (2008), with permission from the Society for Neuroscience.

(Liang et al. 2010). Stimuli used by Liang and colleagues included extremely unattractive and disfigured faces, in line with the overgeneralization account for unattractiveness processing already described in section 6.2 (Zebrowitz and Montepare 2008). Besides supporting a view of attractive faces as rewarding, these findings go further by proposing that faces varying on the attractiveness dimension in either direction should be viewed as motivationally salient, and that there is widespread processing of both attractive and unattractive stimuli within the same regions.

Further support is lent to the view of attractive faces in the context of reward and reinforcement by a study which explores the capacity of attractive faces to influence behavioral preference (Bray and O'Doherty 2007). The authors show that it is possible to establish classical conditioning by pairing visual stimuli with subsequent presentation of attractive or unattractive faces. In a group of participants of both genders, changes in behavioral responses to the conditioned stimuli were observed specifically when they were paired with attractive female faces. When responses were modeled in terms of reward-prediction error, such error correlated with ventral striatal activity during conditioning with attractive faces. These findings closely parallel previous findings with monetary rewards or the provision of juice to thirsty participants (McClure et al. 2003; O'Doherty et al. 2004) and suggest that attractive faces may indeed be treated as other rewarding stimuli by the brain.

In sum, the hypothesis that attractive faces are rewarding has received support from a growing body of functional imaging evidence, which delineates a network of both cortical and subcortical structures. This network has most consistently included the nucleus accumbens and OFC. Some areas (e.g. STS and medial prefrontal cortex) are differentially activated by implicit and explicit evaluation of attractiveness. Some reward processing regions, including OFC, encode facial attractiveness in both a linear and non-linear fashion, while others, most notably the amygdala, appear to respond specifically in a non-linear manner. The conceptualization of attractive faces as rewarding stimuli has been a fruitful one, and some dissociation between neural determinants of pure aesthetic evaluation versus motivated behavior in relation to attractive faces has been demonstrated.

6.4.2 Processing of attractive faces as high-level visual stimuli

A number of neuroimaging studies have sought to explore evaluation of facial attractiveness by focusing on engagement of brain areas specialized for visual processing, rather than on areas for reward processing. While there is clearly overlap between the neural processes that underlie these two aspects of facial attractiveness evaluation, and neither is likely to proceed in complete isolation from the other, each has been investigated using distinct experimental questions and approaches.

In a study by Chatterjee and colleagues (2009), use was made of computer-generated face images, which were equated in terms of "distinctiveness" from an average face (Figure 6.1b). This is a notably different approach from the studies discussed in section 6.4.1, all of which used naturalistic stimuli. The authors found that explicit attractiveness judgments activated a network including anterior insula, dorsal posterior parietal cortex, and inferior and medial prefrontal regions (we note considerable overlap with the findings of Winston et al. 2007). In addition, ventral occipito-temporal areas, including FFA and lateral occipital cortex, were also widely activated (Figure 6.3b). Importantly, Chatterjee and colleagues (2009) were able to show that activity in many of these regions varied parametrically with degree of facial attractiveness. Only the ventral occipito-temporal areas remained active during implicit attractiveness evaluation (participants were judging facial identity instead; Figure 6.3b). The conclusion was that the outlined ventral occipito-temporal network is triggered automatically by attractiveness.

Iaria and colleagues (2008) asked a more specific question about which structures in the core system for face perception (traditionally held to incorporate parallel processing routes via FFA and STS, as described in detail in section 6.3.1; see also Haxby et al. 2000) are most important in the processing of attractiveness in faces. Given the strong social relevance of facial attractiveness, one might predict a key role for STS, although given that attractiveness is a largely invariant trait, one might instead argue that structural features, predominantly processed in FFA, would be more important. In a first experiment the authors investigated how attractiveness judgments and attractiveness-motivated behavior differed between eight patients with acquired prosopagnosia (a deficit in recognizing personally familiar faces caused by damage to occipito-temporal cortex, usually including FFA; Figure 6.4) and 19 healthy controls. Prosopagnosic patients were impaired in both types of task, and the degree of each individual's impairment correlated with severity of their face-recognition deficit. However, half of the patients showed some residual ability to appropriately rate attractiveness; two of these had bilateral anterior temporal lesions (one with clear sparing of right FFA and STS) and the other two had unilateral right occipito-temporal lesions (one with clear involvement of right FFA). The authors conclude that the correlation between severity of prosopagnosic deficit and degree of impairment of attractiveness evaluation supports the involvement of identity-related processing in attractiveness evaluation. The majority of patients scored either normally or near-normally on emotional expression recognition testing and only two patients had lesions clearly involving STS. It is therefore difficult to make any strong inferences about importance of STS from these data. More generally, difficulty in interpreting the results stems from the heterogeneous sample with respect to lesion location (Figure 6.4), which precludes the possibility of ascribing the attractiveness evaluation deficit to a single anatomical structure, or set of structures. In a separate fMRI study, healthy participants viewed faces varying in attractiveness and either performed explicit evaluation of attractiveness, or performed an unrelated task (in which case attractiveness evaluation was considered implicit). Notably the implicit task used here (a "one-back" task requiring matching of each face to the last presented face) differed substantially to those used in other studies discussed earlier, which required age or gender judgments (Winston et al. 2007, Chatterjee et al. 2009), and this may contribute to differences in the results. Iaria and colleagues (2008) found that while FFA activation was not correlated with degree of attractiveness, such activation did increase bilaterally during explicit (as compared to implicit) evaluation. STS showed greater activity during implicit than during explicit evaluation (Figure 6.3c). Moreover, activation was increased in a network including OFA, amygdala and OFC during explicit processing

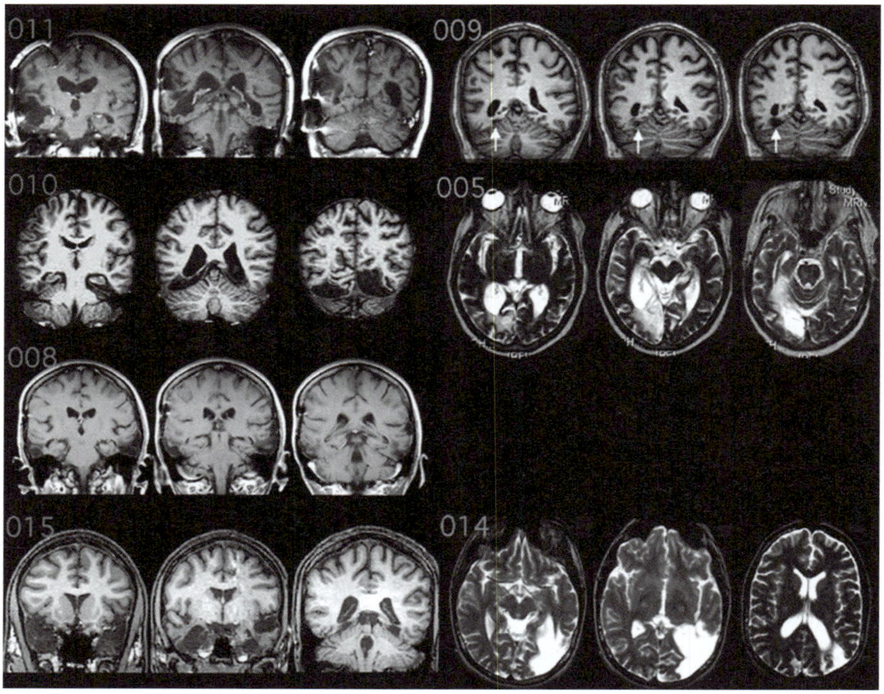

Figure 6.4 Example lesion study, which addressed the involvement of the distributed brain system for face perception in evaluation of facial attractiveness. Structural MR brain images from seven of the eight prosopagnosic patients tested by Iaria et al. (2008) are shown (the number at the top left of each set of images is an individual patient identifier). The heterogeneity of lesion location is clearly evident. Reprinted from *The Journal of Neuroscience*, 155 (2), Giuseppe Iaria, Christopher J. Fox, Christopher T. Waite, Itzhak Aharon, and Jason J.S. Barton, The contribution of the fusiform gyrus and superior temporal sulcus in processing facial attractiveness: neuropsychological and neuroimaging evidence, pp. 409–422, figure 1, doi: 10.1016/j.neuroscience.2008.05.046 Copyright (2008), with permission from the Society for Neuroscience.

of attractiveness. The findings from both the neuropsychological and fMRI experiments of Iaria and colleagues (2008) are consistent with a role for FFA in facial attractiveness evaluation, although it is difficult to draw more specific conclusions about relative importance of FFA and STS, particularly given the variability of patients' lesion extent and location. As discussed in section 6.3.1, there is substantial evidence of overlap between the function of the invariant (FFA-related) and dynamic (STS-related) streams of face processing even in healthy individuals, and such overlap may occur to an even greater extent as the result of remodeling in the wake of structural brain lesions.

Facial attractiveness processing has also been investigated in patients with congenital prosopagnosia, a condition of significantly impaired face recognition from birth (Behrmann and Avidan 2005). By definition, such individuals have no clear associated structural brain lesion visible on imaging, although recent studies have found more subtle regional differences in gray matter volume when compared to non-prosopagnosic brains (Dinkelacker et al. 2011; Garrido et al. 2009). Adding to the findings of Iaria and colleagues (2008), one study found that ability to make accurate attractiveness judgments was usually impaired in patients with congenital prosopagnosia (Le Grand et al. 2006). On the other hand, two studies have found that ability to rate attractiveness is preserved in at least some patients with congenital prosopagnosia, despite impairments in face recognition (Sadr et al. 2004), or rating of distinctiveness (Carbon et al. 2010). These findings would suggest that processing of facial identity and facial aesthetic value are dissociable at least in some cases. These results appear highly informative for cognitive models of attractiveness and facial processing more generally. It must be noted, however, that congenital prosopagnosia is a heterogeneous condition with respect to individual behavioral deficits, and this presumably also applies to heterogeneity in neural deficits.

The relative importance of FFA and STS (and other core brain regions involved in the perception of faces) in evaluation of facial attractiveness remains an intriguing question for further investigation. It is important to note that there are grounds to expect that regions for both "static" and "dynamic" aspects of face processing would be involved in attractiveness evaluation. First, it is clear that attractiveness is influenced by both structural properties (such as symmetry) and more dynamic signals (such as gaze direction or expression), which may be expected to be processed preferentially by one or the other of these systems, respectively. Secondly, it is increasingly appreciated that there is much more overlap between processing of structural face properties in the FFA and dynamic face properties in the STS than originally appreciated, and the roles of these two streams are not in fact entirely dissociable (Calder and Young 2005).

6.4.3 Meta-analyses of studies investigating neural processing of facial attractiveness

Two recent meta-analyses have combined and re-framed results from studies of the neural correlates of facial attractiveness. It is important to note that while meta-analyses of neuroimaging data provide a powerful means of aggregating findings across many studies, which can result in more accurate and robust results, there are important methodological limitations (for reviews see Kober and Wager 2010; Radua and Mataix-Cols 2012).

Bzdok and colleagues (2010) employed activation likelihood estimation analysis of 16 fMRI studies to investigate the hypothesis that facial attractiveness and facial trustworthiness judgments are processed by overlapping brain networks. In terms of attractiveness evaluation, most consistent neural activation was found in the amygdala (in particular the right amygdala), supporting the now widely postulated extended role of the amygdala in processing motivationally and socially relevant stimuli regardless of valence (Adolphs 2010). In addition, there was convergent activation of medial OFC and inferior frontal gyrus (IFG). The authors' interpretation is that the amygdala acts as a filter of evolutionarily relevant sensory information, including information with potential long-term payoff such as attractiveness. Meanwhile, the medial OFC is proposed to appraise beauty as indexed by reward value, while IFG processes semantic aspects of faces. A nucleus accumbens response to attractive faces was found in a number of neuroimaging studies and this structure was therefore also posited to play an important role in attractiveness evaluation.

A second meta-analysis by Mende-Siedlecki and colleagues (2012) again focused on studies that have used neuroimaging methods to investigate the neural correlates of attractiveness and/or trustworthiness evaluation, but using an alternative methodology (multilevel kernel density analysis). The authors included a larger number of studies than did Bzdok and colleagues (2010), with a total of 29 studies analysed (17 specifically focusing on the neural basis of attractiveness evaluation). While both meta-analyses report similar inclusion criteria, the work by Bzdok and colleagues (2010) excluded a number of studies based on unobtainable brain coordinates of results, restriction of findings to regions of interest defined *a priori*, or experiments that were deemed unsuitable for their analysis. Many of these studies were included in the analysis of Mende-Siedlecki and coworkers (2012), alongside a handful of studies published later than the meta-analysis of Bzdok and colleagues (2010). It is important to bear in mind these sampling differences and the differing analysis methodologies when comparing the results. Across both attractiveness and trustworthiness the findings of Mende-Siedlecki and colleagues (2012) were that activation patterns systematically vary as a function of face valence (bilateral amygdala for negative faces, and a network including left caudate and extending into the nucleus accumbens and medial OFC, as well as ventro-medial prefrontal cortex, peri-genual ACC, and right thalamus for positive faces). A key component of the analysis conducted by Mende-Siedlecki and colleagues (2012) concerned neural responses that vary with stimulus valence non-linearly (e.g. with stronger responses to both positively and negatively valenced faces compared to neutral). Such non-linear responses were found in the right amygdala, extending into the right putamen. Furthermore, the authors propose that results of their analysis support a functional fractionation of the right amygdala into a ventral portion with linear responses to negatively valenced faces and a dorsal portion with non-linear responses to socially relevant faces. Focusing on attractiveness specifically, no regions active during evaluation of unattractive faces were reported. However, attractiveness research has focused much more on the effect of positively valenced (attractive) faces, and evaluation of these is reported to activate a network including the left caudate, nucleus accumbens, ventromedial prefrontal cortex (vmPFC), OFC, and peri-genual ACC with significantly more consistency than for trustworthy faces.

Mende-Siedlecki and colleagues (2012) comment that the use of extremely attractive faces (e.g. culled from magazines) in some studies and not in others, could explain an important discrepancy found within the facial attractiveness literature. They argue that studies showing activation in reward-related areas including nucleus accumbens, caudate, and medial OFC, have made use of faces possessing extremely attractive features (e.g. O'Doherty et al. 2003), as well as extremely unattractive ones (Liang et al. 2010). In contrast, computer-generated faces, often used in studies that have failed to show involvement of reward-related regions (e.g. Chatterjee et al. 2009), tend to show variation of valence in both directions from an averagely attractive face. Furthermore, faces at the extremes of the attractiveness dimension tend not to be included in such studies. While it is possible that only extremely attractive or unattractive faces are treated as rewarding or aversive, and striatal reward systems are only engaged in such contexts, we suggest an alternative interpretation could be that computer-generated faces are often not placed in the same context as naturalistic faces. It seems intuitive that a naturalistic face could more easily be imagined to belong to a potential mate than a computer-generated face. Whether such a distinction holds is also likely to be dependent on the experimental task and overall context since reward regions may respond with the same strength to rewards of very different magnitude providing they are placed in different contexts (e.g. Tobler et al. 2005). A related possibility is that experimental paradigms that make use of participant samples from one gender only, and expose them specifically to face stimuli of the opposite gender, may automatically create a more mate-seeking context in which attractive faces of opposite gender become rewarding in a way that they are not in a purely evaluative context (Cloutier et al. 2008). These are intriguing possibilities, but they are yet to be evaluated empirically.

6.4.4 Neurophysiological studies of the processing of facial attractiveness

This chapter focuses primarily on evidence for neural processing of facial attractiveness obtained using fMRI. While such studies form by far the most numerous (as well as most highly cited) group of research outputs on this topic, it is important to remember that fMRI has a number of limitations as a method for studying neuronal function. Two of these limitations—namely, fMRI's modest temporal resolution, and the indirect relationship between fMRI's BOLD signal and neuronal activity in the brain—are well addressed by EEG and MEG, which provide far superior temporal resolution and a more direct measure of neuronal function. Findings from studies employing such neurophysiological methodologies are therefore often complementary to findings obtained using fMRI. In the light of this, we will briefly review the contribution of neurophysiological methodologies to the understanding of neural processing of facial attractiveness.

An early EEG study on this topic, undertaken over 15 years ago, made use of computer-generated faces varying in attractiveness, which were shown to heterosexual male participants (Johnston and Oliver-Rodriguez 1997). Event-related potential recordings (ERP, representing EEG changes time-locked to stimulus presentation) showed that the amplitude of a late positive component of the signal (around 400–600 ms) was correlated with attractiveness ratings given to female (but not male) faces and was also modulated by the modification of relevant facial features (particularly short lower jaws, full lips, and widely spaced eyes) that also enhanced behavioral ratings of attractiveness. Since the late positive component has been extensively linked with changes in affect (e.g. Farwell and Donchin 1991), this finding was interpreted as a time signature of affective components of attractiveness evaluation. The modulation of the late positive component by facial attractiveness has since been replicated in a study that extended previous findings by demonstrating that variations in facial attractiveness are also associated with modulation of an early posterior negativity (around 150–250 ms), which is temporally distinct from the late positive

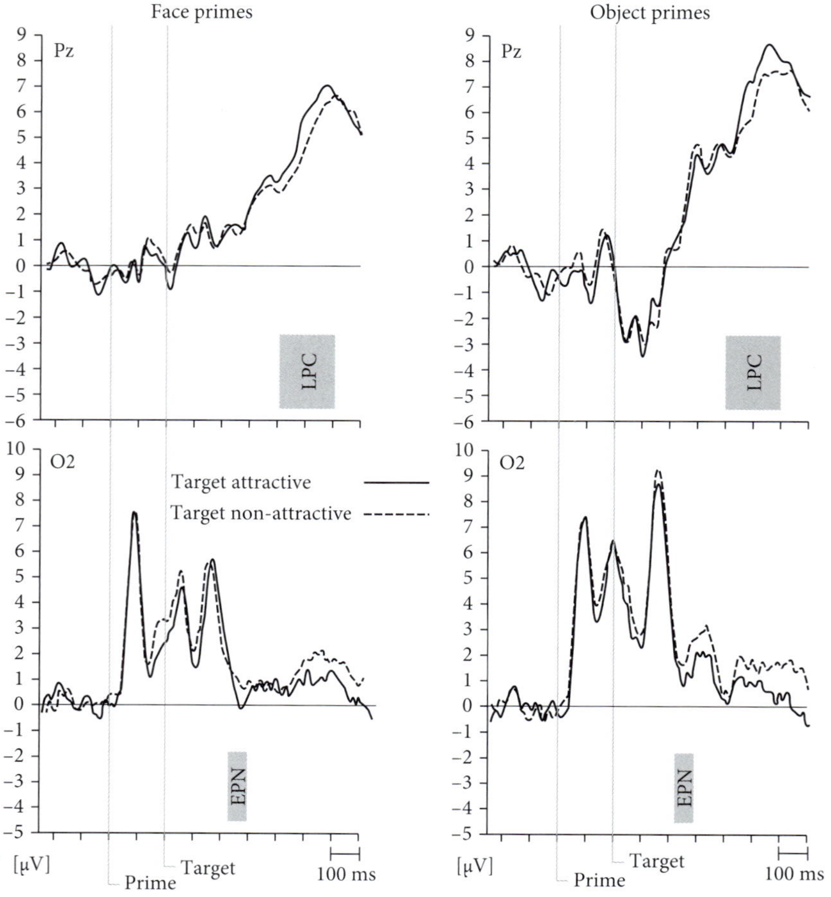

Figure 6.5 Event-related potential responses to attractive and non-attractive faces. Faces were preceded by prime images that were either also faces (left-side panels), or common everyday objects (right-side panels). The results show how facial attractiveness modulates both a late positive component (LPC) of the response (upper panels), and an early positive negativity (EPN) component of the response (lower panels); these modulations occur both when stimuli are primed with faces (left-sided panels) and with objects (right-sided panels). Pz and O2 refer to scalp electrode positions from the extended 10–20 system (midline parietal and right occipital, respectively). Reprinted from *Biological Psychology*, 79 (1–2), Katja Werheid, Annekathrin Schacht, and Werner Sommer, Facial attractiveness modulates early and late event-related brain potentials, pp. 100–8, Copyright (2007), with permission from Elsevier.

component, and argued to reflect evaluation of emotional significance that guides attentional capture (Werheid et al. 2007; Figure 6.5).

A more recent ERP study (van Hooff et al. 2011) employed an orienting task to explore attentional bias related to facial attractiveness. Faces were incidental to the experimental task (unlike in the experiments of Johnston and Oliver-Rodriguez 1997, and Werheid et al. 2007), enabling investigation of automatic stimulus-driven facial attractiveness processing. Faces at both ends of the attractiveness scale elicited larger early ERP components (at a similar time to the findings of Werheid et al. 2007). This finding was proposed to relate to implicit selective attention. A late slow-wave component of the response was more prominent for attractive opposite-sex faces in males but not in females, suggesting that males have a gender-specific mechanism for heightened motivated attention towards attractive female faces. These two ERP components, which occur at different latencies from stimulus onset, presumably reflect different aspects of attractiveness processing. The fact that the signal components show modulation by the attractiveness of task-irrelevant faces adds support to other evidence suggesting that aspects of attractiveness evaluation are automatic (Chatterjee et al. 2009; O'Doherty et al. 2003; Winston et al. 2007).

Other recent ERP studies have found similar temporal signatures for facial attractiveness evaluation, and in addition report further findings that shed light on attractiveness processing under limited cognitive resources (Rellecke et al. 2011), ERP changes corresponding to recognition of attractive faces (Zhang et al. 2011), and the interaction between attractiveness and cooperative behavior (Chen et al. 2012).

In sum, neurophysiological measures have played an important part in advancing our understanding of the neural mechanisms for facial attractiveness evaluation, largely by helping to establish temporally distinct components corresponding to various aspects of such processing. In addition, many of the principal findings, such as automaticity of processing for attractiveness, and engagement of systems specialized for evaluation of reward or emotion, converge with findings from the fMRI literature. On the other hand, there has been relatively little evidence from the neurophysiological literature to support the notion that attractiveness modulates mechanisms for structural encoding of faces. Future studies using methodologies with high temporal resolution will undoubtedly continue to make an important contribution to this area.

6.4.5 Evaluating facial attractiveness in a social context

As already discussed, faces play an important role in social interaction and evaluating their attractiveness has wide-ranging societal implications from guiding mate choice to influencing decisions about moral character and employability. While several studies discussed in other sections of this chapter address the socially relevant aspects of attractiveness evaluation (and will be again briefly mentioned here), a handful of studies not discussed elsewhere have also directly focused on social aspects of facial attractiveness evaluation.

Klucharev and colleagues (2009) showed that individual assessments of facial attractiveness are adjusted according to group opinion. Using fMRI, they found that the difference between an individual's decision and the group's decision about the atractiveness of faces was predicted by increased BOLD response in a rostral zone of cingulate cortex and decreased BOLD response in ventral striatum. The authors comment that the mechanisms found are analogous to prediction error postulated to underpin reinforcement learning. Besides placing social conformity in the same framework as reinforcement learning, the findings provide specific insight into the impact of group opinion on judgments of facial attractiveness.

More recently, a similar question was addressed by Zaki and colleagues (2011). Here the authors focused specifically on differentiating "private acceptance" of group opinion (true modification of an individual's beliefs) from "public compliance" (appearing to modify beliefs in order to meet public expectations). For this reason, instead of undergoing fMRI while making initial ratings of facial attractiveness and receiving feedback about group opinion of the same faces (as in Klucharev et al. 2009), in this case participants underwent fMRI 30 minutes later, at which time they were asked to rate all previously seen faces on an attractiveness scale for a second time. Stimuli that had been rated as more attractive by peers than by participants resulted in increased activation in the nucleus accumbens and the OFC (as compared to stimuli rated as less attractive by peers than participants). While the results of Klucharev and colleagues (2009) imply a role for reward systems in the experience of consensus with group opinion, those of Zaki and colleagues (2011) suggest that the actual neural representation of value associated with a stimulus converges towards that implied by other people's opinions.

Evaluation of attractiveness is relevant to social functioning in general, however, some studies have focused on the interaction between attractiveness and other specific socially relevant signals conveyed by faces. Kampe and colleagues (2001) showed that attractive faces lead to increased striatal activity when eye gaze is directed straight towards the observer (but not when it is slightly averted). They argue that this difference in gaze direction signals a willingness (versus a reluctance) to engage in social interaction.

Some elements of attractiveness evaluation may rely on inferences regarding goals and intentions of the observed face. Making these kinds of judgments about others has been termed "theory of mind" and converging evidence implicates brain regions including STS, temporo-parietal junction, and medial frontal cortex (MFC) in this complex set of processes (Amodio and Frith 2006; Frith and Frith 1999; Gallagher and Frith 2003; Saxe and Kanwisher 2003). STS and MFC activation has been observed in fMRI experiments involving attractiveness evaluation, specifically when judgments of attractiveness have explicitly been set as the experimental task (Winston et al. 2007). Such findings suggest that explicit evaluation of attractiveness may involve complex inferences about the goals and intentions of others, or perhaps self-referential processing. Whether this is the case, and if so, how such processing may interact with assessments of reward value or visual characteristics of attractive faces in the brain, is an intriguing question for future research.

6.4.6 Gender-specific mechanisms of attractiveness evaluation

Several studies have explored whether the neural responses to facial attractiveness are modulated by gender. Increased ACC activity in response to attractive male or female faces and accompanying changes in sympathetic arousal (as indexed by pupil dilation) have been observed in men only (Winston et al. 2007). Another study found increased medial OFC activity in response to attractive opposite-sex faces in men only (Cloutier et al. 2008). Liang and colleagues (2010) showed non-linear responses to both attractive and unattractive faces throughout the brain's reward circuitry (striatum, OFC, VTA) as well as linear effects in both medial and lateral OFC. Male participants exhibited stronger non-linear effects, whereas female participants exhibited stronger linear effects. One study (Kranz and Ishai 2006) has investigated how gender-specific effects of facial attractiveness are modulated by sexual orientation and found stronger activation in the thalamus and medial OFC specifically for sexually preferred faces. Attentional biases favoring attractive faces are more pronounced in men than in women (van Hooff et al. 2011).

These findings notwithstanding, we would argue that gender differences in the neural processing of attractiveness are surprisingly minor. A number of studies have found only limited gender differences in neural processing of attractiveness (Chatterjee et al. 2009; O'Doherty et al. 2003) and when found, differences have often mainly been down to specific activation patterns for male participants viewing opposite-sex faces (which have no direct correlate in women viewing opposite-sex faces); for example, in ACC (O'Doherty et al. 2003; Winston et al. 2007). Furthermore, it has been noted that gender-specific differences in processing for opposite-sex faces are rather small when compared to the main effects of attractiveness (Winston et al. 2007). Thus, while some gender-specific mechanisms of attractiveness evaluation can undoubtedly be demonstrated (and in particular, male observers seem to exhibit a special engagement of reward mechanisms when they are evaluating attractive opposite-sex faces) it appears that non-gender-specific ratings on an attractiveness dimension contribute much more strongly to attractiveness evaluation. Such observations argue against conceptualizing attractiveness evaluation chiefly in terms of reward, because the rewarding aspects of facial attractiveness may be expected to be highly gender-specific (although we note there is evidence to suggest that attraction occurs towards physically attractive strangers irrespective of gender e.g. Byrne et al. 1968). Instead, they support an account in which attractiveness is evaluated as a form of beauty. In such a case, one would predict some degree of commonality in the neural mechanisms underpinning evaluation of facial attractiveness and evaluation of other beautiful objects, such as works of art for example. This possibility will be explored in the discussion at section 6.5.

6.4.7 Role of the prefrontal cortex in neural processing of facial attractiveness

Another perspective, which implicates the prefrontal cortex in the processing of attractive faces, can be gained from the study of neural mechanisms of valuation. Attractive faces have both an immediate experienced value (i.e. they are intrinsically rewarding), and a more decision-related value (defined as value ascribed to an item to aid decision making, e.g. in an economic exchange task; Smith et al. 2010). Smith and colleagues (2010) made use of face stimuli varying in attractiveness and images of banknotes indicating monetary gain or loss. For both stimulus types, they showed that the ventromedial prefrontal cortex (VMPFC), a region known to be important in the representation of decision value (Knutson et al. 2007; Plassmann et al. 2007), could be functionally fractionated, with BOLD activity in anterior VMPFC being associated with experienced value and BOLD activity in posterior VMPFC being associated with decision value. Importantly, the decision value signals in posterior VMPFC were present even in the absence of an overt choice task.

6.5 Discussion

We now set out to integrate and discuss some of the key findings from imaging investigations of the neural mechanisms for evaluating attractiveness in faces, which have been outlined in this chapter. We make comments on possible fruitful future directions in the field, and place particular emphasis on discussing the relevance of this literature to the study of aesthetic preference more generally.

6.5.1 Summary and interpretation of findings from studies of the neural mechanisms for evaluating facial attractiveness

A number of studies have hypothesized that attractive faces are processed by the brain as rewarding stimuli. Findings seem to converge on increased BOLD fMRI activation in nucleus accumbens, medial OFC, and ACC when viewing attractive faces (Aharon et al. 2001; O'Doherty et al. 2003; Winston et al. 2007). In particular, there is convergent evidence to support a key role for the OFC. The results of several studies highlight a functional fractionation of this region, with medial OFC being responsive to attractive faces, and lateral OFC being responsive to unattractive faces (Cloutier et al. 2008; Hampshire et al. 2012; O'Doherty et al. 2003). Such findings sit well with more general frameworks implicating OFC in representing the affective and motivational value of stimuli (e.g. Rolls 2000).

Given the amygdala's responsiveness to both appetitive and aversive stimuli in a number of contexts, there has been a rationale for predicting non-linear BOLD responses to attractiveness in this region. Both linear and non-linear responses to attractiveness have been demonstrated not just in the amygdala and medial OFC (Winston et al. 2007), but also, when using highly unattractive disfigured faces at the negatively valenced end of the scale, in ACC, lateral OFC, and striatum (Liang et al. 2010). The right amygdala was the region most consistently associated with attractiveness evaluation in two recent meta-analyses (Bzdok et al. 2010; Mende-Siedlecki et al. 2012). The latter of these studies suggested a degree of functional fractionation, with linear BOLD responses in the ventral amygdala (responds most strongly to negative faces) and non-linear BOLD responses (to both ends of the attractiveness dimension) in the dorsal amygdala (Mende-Siedlecki et al. 2012). Mende-Siedlecki et al. (2012) argue that the ventral portion of the amygdala contributes to processing of valence, whereas the dorsal portion is involved in representing the salience or value of ambiguous facial information. This view dovetails neatly with the work of Whalen and colleagues (2001), which supports a similar functional fractionation of amygdala. Overall, these findings and their interpretation support an important role for the amygdala in facial attractiveness evaluation, consistent with both responding to negatively valenced faces (quite possibly to guide avoidance behavior), and indexing relevance of a particular face in a social context.

A handful of other studies have anchored hypotheses regarding the neural evaluation of attractiveness within popular frameworks for neural processing of faces (e.g. Haxby et al. 2000) rather than reward processing systems. Chatterjee and colleagues (2009) demonstrated increased BOLD activation in occipito-temporal areas (including FFA and STS) in association with increasing attractiveness of computer-generated faces. In patients with brain lesions resulting in prosopagnosia, Iaria and colleagues (2008) have shown that FFA, and possibly STS, are likely to play a part in processing facial attractiveness, though the data do not go as far as showing that either region is of critical importance. With regard to congenital, rather than acquired prosopagnosia, the evidence suggests that facial attractiveness evaluation is preserved in many, but not all individuals with the condition (Carbon et al. 2010; Le Grand et al. 2006; Sadr et al. 2004). Considerable evidence therefore exists to support the hypothesis that regions of the core system for face perception are involved in facial attractiveness evaluation, but there is a lack of clear evidence that any of them are of critical importance. It remains unclear how well evaluation of facial attractiveness fits in with the dual processing streams for invariant and dynamic/social aspects of faces proposed by Haxby and colleagues (2000). Senior (2003) presented an extension of the model of Haxby and colleagues (2000) to include neural processing of facial attractiveness. Particular focus was given to links between dynamic facial processing systems, centered on STS, and a proposed new part of the extended system for face processing, specialized for evaluation of facial attractiveness. Given more recent evidence, it seems likely that both invariant and dynamic streams of the model of Haxby and colleagues (2000) are involved in attractiveness evaluation. Whether certain aspects of facial attractiveness are processed primarily by the invariant stream and others primarily by the dynamic/social stream remains largely an open question. It is important also to bear in mind that considerable overlap between these two processing streams has been demonstrated.

One important distinction to make is between studies that have made attractiveness evaluation task-relevant and those that have investigated implicit evaluation of attractiveness (in which case participants are performing an unrelated task to facial attractiveness evaluation whilst viewing faces that vary in attractiveness). Implicit attractiveness evaluation has been shown to engage regions that form part of the reward network including the ventral striatum (Aharon et al. 2001) and OFC (O'Doherty et al. 2003), suggesting that processing of attractive faces as rewards occurs automatically. Winston and colleagues (2007) specifically addressed the distinction between implicit and explicit processing of facial attractiveness and therefore sought to determine how neural responses are modulated as a function of task (direct evaluation of attractiveness, versus determining the age of faces which also vary in attractiveness). Bilateral STS, bilateral posterior OFC/anterior insula, and medial prefrontal areas, all responded more strongly during explicit attractiveness evaluation, in most cases in a task-dependent rather than stimulus-dependent fashion. The involvement of STS and MFC, regions important for "theory-of-mind" functions such as inference about the goals and intentions of others, should prompt future empirical investigation of whether such inferences are necessary for direct and explicit attractiveness evaluation. Iaria and colleagues (2008) found that explicit attractiveness evaluation was correlated with increased BOLD activation in FFA. On the other hand a study by Chatterjee and colleagues (2009) found that ventral occipito-temporal areas were activated during both implicit and explicit attractiveness evaluation, whilst explicit attractiveness evaluation was selectively accompanied by activation in insula and inferior and medial prefrontal regions. Finally, an EEG study (van Hooff et al. 2011) found that both early and late ERP components related to attractiveness evaluation were present despite such evaluation being implicit. It seems that much of the motivationally relevant processing of attractive faces, as well as much of that pertaining to visual features of the face, can occur automatically, without an observer's explicit attempt to evaluate attractiveness (for a recent review on the topic of automaticity in face perception and social judgment, see Bargh et al. 2012). On the other hand, explicit attractiveness evaluation seems to involve activation of regions implicated in more complex processes of social cognition, which play a part in allowing us to infer the goals and intentions of others, or to compare the dispositions and traits of others to those of ourselves. Such explicit evaluation also results in activation of ventral parts of the visual system and preforntal regions that may support visual awareness (Rees 2007).

Two meta-analyses of studies of the neural processes underlying facial attractiveness evaluation have provided largely overlapping results, showing most consistent activation in amygdala, OFC, and striatum (Bzdok et al. 2010; Mende-Siedlecki et al. 2012). A handful of studies have used EEG to study the neural correlates of facial attractiveness evaluation, with findings converging towards the presence of both early and late ERP components of processing, postulated to correspond to attentional orienting and affective/motivational processing respectively (Johnston and Oliver-Rodriguez 1997; van Hooff et al. 2011; Werheid et al. 2007). Two studies have focused specifically on how neural correlates of attractiveness evaluation are modulated by group opinion, finding that activation of the ventral striatum decreases when an individual's opinion conflicts with the group's (Klucharev et al. 2009) and furthermore, that facial attractiveness-related representations in the ventral striatum and OFC are altered for more prolonged periods (at least 30 mins) to fit in with group opinion (Zaki et al. 2011). Aside from being modulated by social context, responses to facial attractiveness are affected by other socially relevant facial signals, such as direction of eye gaze (Kampe et al. 2001). Reward mechanisms seem to be engaged mostly for opposite-sex faces (Aharon et al. 2001), and more so for male observers (Cloutier et al. 2008). While there may also be other regions activated specifically for sexually preferred faces (Kranz and Ishai 2006) as well as some differences in the linearity/non-linearity of neural response to attractive faces dependent on the gender of the observer (Liang et al. 2010), overall gender-invariant effects of attractiveness appear to be more prominent (Winston et al. 2007). A handful of studies have explored other aspects of facial attractiveness processing, including interaction with memory formation systems (Tsukiura and Cabeza 2011), and the role of VMPFC in representation of both value and value-based decisions in relation to attractive faces (Smith et al. 2010).

In sum, investigations of the neural mechanisms of facial attractiveness evaluation have focused mainly on the role of reward systems and visual processing systems. A number of robust and replicable findings have enabled some key brain regions to be defined, and putative networks to be sketched out. However, a great many important details remain to be clarified. Our understanding of how some of the structural and configural facial determinants of attractiveness are processed in the brain remains limited. Connectivity analyses to establish how some of the regions involved in attractiveness evaluation interact, as well as experiments to establish the details of interplay between reward systems and visual systems during attractiveness evaluation, are as yet lacking. Some time ago, Senior (2003) has argued that the amygdala is likely to be a key link between reward and visual systems in the evaluation of facial attractiveness. How facial attractiveness evaluation interacts with processing systems for social cognition, memory formation, representation of value, and many other domains of cognitive processing, are also avenues that remain barely explored.

6.5.2 Do mechanisms for processing of facial attractiveness overlap with those for processing of beauty and aesthetic experience?

Neuroaesthetics, a term originating from the work of Zeki (1999), is concerned with the neural mechanisms governing beauty perception in art. In many instances, the scope can be broadened to encompass objects or scenes that would not traditionally be classed as art, but which nevertheless invoke intense (pleasurable) feelings in the observer. The study of neuroaesthetics has centered on the visual domain, although the principles are felt to apply equally to other domains, for example music or literature (Chatterjee 2010). Although the definition of "aesthetic experience" is broad, one can readily argue that both within the realm of visual art, and as part of real world visual scenes, the human face is often an object that invokes such experience in the observer. Certainly, it seems intuitive that attractive faces would also be aesthetically pleasing and indeed the two terms have often been used interchangeably in the social and evolutionary biological literature, but ultimately the distinction between these two constructs remains incompletely defined.

The neural mechanisms that underlie aesthetic evaluation and experience have come under scrutiny in recent years, and there are a small but growing number of neuroimaging studies aimed at this question. This line of work is tackled in more detail elsewhere in this volume and has previously been discussed in several detailed reviews and commentaries (Cela-Conde et al. 2011; Chatterjee 2010; Di Dio and Gallese 2009; Nadal et al. 2008; Ramachandran and Hirstein 1999).

To what extent does neural processing of facial attractiveness overlap with processes for evaluating beauty and aesthetic value outside the context of faces? The answer to this question would be best facilitated by studies that have made use of both attractive faces and other beautiful stimuli in the same set of experiments. Kawabata and Zeki (2004) investigated the neural processing associated with assessment of visual beauty irrespective of category of visual stimulus. Whilst undergoing fMRI, participants were asked to rate (in terms of their beauty) paintings depicting portraits (i.e. faces), landscapes, still-life scenes, and abstract compositions. Differentiation between beautiful and ugly images across all these categories was tracked by medial OFC activity. More recent work has shown that the degree of activation of the medial OFC correlates with beauty scores given not just to different classes of visual stimuli, but also to stimuli of different modalities (in this case, both visual and auditory stimuli; Ishizu and Zeki 2011). Such findings, converging on medial OFC, have led Zeki and colleagues to propose that this region represents the neural correlate of all beauty (Ishizu and Zeki 2011). Such a proposal places attractiveness evaluation in the OFC in a broad neuroaesthetic context. However, one may consider the role of the OFC even more broadly as a region mediating the rewarding effects of stimuli, be they beautiful or not. This broader view would be supported by findings that the medial OFC is activated by monetary reward, for example (Elliott et al. 2003; O'Doherty et al. 2001).

Another study employing fMRI to determine the neural correlates of aesthetic preference for paintings (Vartanian and Goel 2004) found a correlation between increased aesthetic preference and increased activity in cingulate and occipital cortices, as well as decreased aesthetic preference and decreased activity in right caudate. Again, the reported pattern overlaps with striatal and occipital areas activated during facial attractiveness evaluation. Cela-Conde and colleagues (2004) used MEG to investigate the neural correlates of aesthetic preference. Judging stimuli as beautiful was found to correlate with late signal (400–1000 ms) changes over left dorsolateral prefrontal cortex.

Nadal and colleagues (2008) comment on the surprising lack of correlation between aesthetic preference and neural activation in the amygdala in the above studies. One possible reason they suggest for this is that the stimuli used did not extend far enough into the negatively valenced end of the aesthetic value dimension. We would suggest that an alternative explanation is that the studies were not specifically designed to look for involvement of the amygdala. Activation in this brain region can be difficult to demonstrate using both fMRI and EEG/MEG, due to artifact from nearby air-filled sinuses, and the region's distance from the brain surface, respectively. In addition, as we have discussed, an important component of the amygdala's response to attractiveness (and potentially to beauty more generally) is non-linear (Mende-Siedlecki et al. 2012; Winston et al. 2007). By and large, the studies we describe in this section have not specifically looked for non-linear patterns of neural activation, and this may be an important reason for their lack of demonstrable amygdala activation.

One study that shows little concordance between neural networks involved in aesthetic preference and those we have outlined for facial attractiveness evaluation is that of Jacobsen and colleagues (2006). In this fMRI experiment, aesthetic judgments regarding visual stimuli (as compared to symmetry judgments) activated prefrontal, posterior cingulate, left temporal pole, and the temporoparietal-junctional regions. Importantly, rather than seeking out regions where activation correlates with degree of beauty ascribed to the stimulus (as undertaken in other studies investigating the neural correlates of aesthetic preference), this study contrasted aesthetic valuations with assessments regarding symmetry. This difference in experimental design may explain disparities between these findings and those of other studies addressing the neural correlates of aesthetic preference (Nadal et al. 2008). In addition, given that we know symmetry is an important determinant of facial attractiveness (Rhodes et al. 1998), as well as of beauty more generally (Enquist and Arak 1994), it seems likely that symmetry and beauty were confounded in this experiment, making interpretation of the results difficult.

While the results from these neuroimaging studies, which sought to clarify the neural mechanisms of aesthetic preference, seem quite diverse, they have recently been synthesized into unified conceptual frameworks (Cela-Conde et al. 2011; Chatterjee 2010). There appears to be a fair degree of overlap between the findings of fMRI studies of facial attractiveness evaluation and those of aesthetic preference. Findings from EEG also show such overlap; as reviewed by Nadal and colleagues (2008), EEG studies of aesthetic preference have led to the emergence of a two-stage processing model with early signal changes (around 300 ms) corresponding to "initial impression," and later signal changes (around 600 ms) corresponding to "a deeper aesthetic evaluation" (Jacobsen and Höfel 2003). Such a model bears clear similarity to the two temporal stages of processing described for evaluation of facial attractiveness in section 6.4.4.

In sum, there is likely to be considerable overlap between neural processing of aesthetic preference and facial attractiveness although in most cases this can currently only be inferred from similarities between findings of different studies and there is a paucity of direct comparison within the same study (with the exception of Kawabata and Zeki 2004). Future studies dedicated to addressing this question would be illuminating. In addition, investigations into the neural processing of facial attractiveness have been guided by highly developed conceptual frameworks that describe neural processing of faces (Bruce and Young 1986; Haxby et al. 2000), and rewarding stimuli (Delgado 2007), and have furthermore begun to generate unique frameworks of their own (e.g. Senior 2003). Such frameworks, besides those already built around understanding aesthetic evaluation (e.g. Chatterjee 2003), could usefully guide further investigation into the neural correlates of aesthetic preference. There is much scope for mutual benefit from continued interaction between these two fields.

6.6 Conclusion

A growing number of experimental studies, making use of neuroimaging methodologies as well as patients with brain lesions, have shown that there is an extended network of brain regions which are involved in evaluation of the attractiveness of faces. Attractiveness modulates fMRI-BOLD activity in regions important for visual processing, including those specialized for face processing (e.g. FFA), and for processing of social signals from faces as well as other sources (e.g. STS). Frequently, attractive (as well as unattractive) faces undergo processing within the brain's reward centers in the OFC and striatum. In addition, the amygdala plays an important role in indexing the social relevance of stimuli in the external environment, and in line with this, appears consistently to be involved in evaluation of attractiveness, a trait that carries important social implications. Some dissociation has been shown between regions activated during implicit attractiveness evaluation, and those activated during explicit evaluation. In particular, explicit attractiveness evaluation engages medial frontal and superior temporal regions which may reflect theory-of-mind or self-referential evaluations relevant to adapting attitudes and behavior in social contexts. A number of the neurobiological findings, not least the considerable involvement of reward systems, are potentially in keeping with the fundamentally important role of attractiveness evaluation in terms of reproduction and survival. In addition, there is growing evidence that some of the neural mechanisms for processing of facial attractiveness may be common to processing of beauty more generally, and thus the study of the neurobiology of facial attractiveness is of particular relevance to the study of neuroaesthetics. A great many questions remain open in this field, and we envisage substantial further development in coming years.

Note

1 In an event-related fMRI design, stimulus order can be randomized. This differs to a blocked design where stimuli must be presented in condition-specific groups. In the experiment under discussion, an event-related design meant that the attractiveness/reward value of a stimulus could not be predicted from its predecessors. In contrast, stimulus presentation in a blocked design is more predictable.

References

Adolphs, R. (2010). What does the amygdala contribute to social cognition? *Annals of the New York Academy of Sciences* **1191**, 42–61.

Aharon, I., Etcoff, N., Ariely, D., et al. (2001). Beautiful faces have variable reward value: fMRI and behavioral evidence. *Neuron* **32**, 537–51.

Allison, T., Puce, A., and McCarthy, G. (2000). Social perception from visual cues: role of the STS region. *Trends in Cognitive Sciences* **4**, 267–78.

Amodio, D.M. and Frith, C.D. (2006). Meeting of minds: the medial frontal cortex and social cognition. *Nature Reviews Neuroscience* **7**, 268–77.

Bardo, M. (1997). Neuropharmacological mechanisms of drug reward: beyond dopamine in the nucleus accumbens. *Critical Reviews in Neurobiology* **12**, 37–67.

Bargh, J.A., Schwader, K.L., Hailey, S.E., et al. (2012). Automaticity in social-cognitive processes. *Trends in Cognitive Sciences* **16**, 593–605.

Behrmann, M. and Avidan, G. (2005). Congenital prosopagnosia: face-blind from birth. *Trends in Cognitive Sciences* **9**, 180–7.

Bray, S. and O'Doherty, J. (2007). Neural coding of reward-prediction error signals during classical conditioning with attractive faces. *Journal of Neurophysiology* **97**, 3036–45.

Bruce, V. and Young, A. (1986). Understanding face recognition. *British Journal of Psychology* **77**, 305–27.

Bruce, V. and Young, A.W. (1998). *In the Eye of the Beholder: Science of Face Perception* (1st edn). Oxford: Oxford University Press.

Byrne, D., London, O., and Reeves, K. (1968). The effects of physical attractiveness, sex, and attitude similarity on interpersonal attraction. *Journal of Personality* **36**, 259–71.

Bzdok, D., Langner, R., Caspers, S., et al. (2010). ALE meta-analysis on facial judgments of trustworthiness and attractiveness. *Brain Structure and Function* **215**, 209–23.

Calder, A.J. and Young, A.W. (2005). Understanding the recognition of facial identity and facial expression. *Nature Reviews Neuroscience* **6**, 641–51.

Carbon, C.-C., Grüter, T., Grüter, M., et al. (2010). Dissociation of facial attractiveness and distinctiveness processing in congenital prosopagnosia. *Visual Cognition* **18**, 641–54.

Cela-Conde, C.J., Agnati, L., Huston, J.P., et al. (2011). The neural foundations of aesthetic appreciation. *Progress in Neurobiology* **94**, 39–48.

Cela-Conde, C.J., Marty, G., Maestú, F., et al. (2004). Activation of the prefrontal cortex in the human visual aesthetic perception. *Proceedings of the National Academy of Sciences of the USA* **101**, 6321–5.

Chatterjee, A. (2003). Prospects for a cognitive neuroscience of visual aesthetics. *Bulletin of Psychology and the Arts* **4**, 55–60.

Chatterjee, A. (2010). Neuroaesthetics: A coming of age story. *Journal of Cognitive Neuroscience* **23**, 53–62.

Chatterjee, A., Thomas, A., Smith, S.E., and Aguirre, G.K. (2009). The neural response to facial attractiveness. *Neuropsychology* **23**, 135.

Chen, C.-C., Kao, K.-L.C., and Tyler, C.W. (2007). Face configuration processing in the human brain: the role of symmetry. *Cerebral Cortex* **17**, 1423–32.

Chen, J., Zhong, J., Zhang, Y., et al. (2012). Electrophysiological correlates of processing facial attractiveness and its influence on cooperative behavior. *Neuroscience Letters* **517**, 65–70.

Cloutier, J., Heatherton, T.F., Whalen, P.J., et al. (2008). Are attractive people rewarding? Sex differences in the neural substrates of facial attractiveness. *Journal of Cognitive Neuroscience* **20**, 941–51.

Cunningham, M.R. (1986). Measuring the physical in physical attractiveness: quasi-experiments on the sociobiology of female facial beauty. *Journal of Personality and Social Psychology* **50**, 925–35.

Cunningham, M.R., Barbee, A.P., and Pike, C.L. (1990). What do women want? Facialmetric assessment of multiple motives in the perception of male facial physical attractiveness. *Journal of Personality and Social Psychology* **59**, 61–72.

Delgado, M.R. (2007). Reward-related responses in the human striatum. *Annals of the New York Academy of Sciences* **1104**, 70–88.

Di Dio, C. and Gallese, V. (2009). Neuroaesthetics: a review. *Current Opinion in Neurobiology* **19**, 682–7.

Dinkelacker, V., Grüter, M., Klaver, P., et al. (2011). Congenital prosopagnosia: multistage anatomical and functional deficits in face processing circuitry. *Journal Neurology* **258**, 770–82.

Dipboye, R.L., Arvey, R.D., and Terpstra, D.E. (1977). Sex and physical attractiveness of raters and applicants as determinants of resumé evaluations. *Journal of Applied Psychology* **62**, 288–94.

Eimer, M. and Holmes, A. (2002). An ERP study on the time course of emotional face processing. *Neuroreport* **13**, 427–31.

Elliott, R., Newman, J.L., Longe, O.A., et al. (2003). Differential response patterns in the striatum and orbitofrontal cortex to financial reward in humans: a parametric functional magnetic resonance imaging study. *Journal of Neuroscience* **23**, 303–7.

Engell, A.D., Todorov, A., and Haxby, J.V. (2010). Common neural mechanisms for the evaluation of facial trustworthiness and emotional expressions as revealed by behavioral adaptation. *Perception* **39**, 931–41.

Enquist, M. and Arak, A. (1994). Symmetry, beauty and evolution. *Nature* 372, 169–72.

Farwell, L.A. and Donchin, E. (1991). The truth will out: interrogative polygraphy ("lie detection") with event-related brain potentials. *Psychophysiology* **28**, 531–47.

Fox, C.J., Moon, S.Y., Iaria, G., et al. (2009). The correlates of subjective perception of identity and expression in the face network: an fMRI adaptation study. *NeuroImage* **44**, 569–80.

Frieze, I.H., Olson, J.E., and Russell, J. (1991). Attractiveness and income for men and women in management. *Journal of Applied Social Psychology* **21**, 1039–57.

Frith, C.D. and Frith, U. (1999). Interacting minds—A biological basis. *Science* **286**, 1692–5.

Gallagher, H.L. and Frith, C.D. (2003). Functional imaging of "theory of mind." *Trends in Cognitive Sciences* 7, 77–83.

Garrido, L., Furl, N., Draganski, B., et al. (2009). Voxel-based morphometry reveals reduced grey matter volume in the temporal cortex of developmental prosopagnosics. *Brain* **132**, 3443–55.

Hampshire, A., Chaudhry, A.M., Owen, A.M., et al. (2012). Dissociable roles for lateral orbitofrontal cortex and lateral prefrontal cortex during preference driven reversal learning. *NeuroImage* **59**, 4102–12.

Haxby, J.V., Hoffman, E.A., and Gobbini, M.I. (2000). The distributed human neural system for face perception. *Trends in Cognitive Sciences* **4**, 223–33.

Iaria, G., Fox, C.J., Waite, C.T., et al. (2008). The contribution of the fusiform gyrus and superior temporal sulcus in processing facial attractiveness: Neuropsychological and neuroimaging evidence. *Journal of Neuroscience* **155**, 409–22.

Ishizu, T. and Zeki, S. (2011). Toward a brain-based theory of beauty. *PLoS One* **6**, e21852.

Jacobsen, T. and Höfel, L. (2003). Descriptive and evaluative judgment processes: behavioral and electrophysiological indices of processing symmetry and aesthetics. *Cognitive, Affective, and Behavioral Neuroscience* **3**, 289–99.

Jacobsen, T., Schubotz, R.I., Höfel, L., and Cramon, D.Y.V. (2006). Brain correlates of aesthetic judgment of beauty. *Neuroimage* **29**, 276–85.

Johnston, V.S. (2006). Mate choice decisions: the role of facial beauty. *Trends in Cognitive Sciences* **10**, 9–13.

Johnston, V.S. and Franklin, M. (1993). Is beauty in the eye of the beholder? *Ethology and Sociobiology* **14**, 183–99.

Johnston, V.S. and Oliver-Rodriguez, J.C. (1997). Facial beauty and the late positive component of event-related potentials. *Journal of Sex Research* **34**, 188–98.

Jones, D. and Hill, K. (1993). Criteria of facial attractiveness in five populations. *Human Nature* **4**, 271–96.

Kampe, K.K.W., Frith, C.D., Dolan, R.J., et al. (2001). Psychology: reward value of attractiveness and gaze. *Nature* **413**, 589–89.

Kanwisher, N., McDermott, J., and Chun, M.M. (1997). The fusiform face area: a module in human extrastriate cortex specialized for face perception. *Journal of Neuroscience* **17**, 4302–11.

Kawabata, H. and Zeki, S. (2004). Neural correlates of beauty. *Journal of Neurophysiology* **91**, 1699–705.

Kim, H., Adolphs, R., O'Doherty, J.P., et al. (2007). Temporal isolation of neural processes underlying face preference decisions. *Proceedings of the National Academy of Sciences of the USA* **104**, 18253–8.

Klucharev, V., Hytönen, K., Rijpkema, M., et al. (2009). Reinforcement learning signal predicts social conformity. *Neuron* **61**, 140–51.

Knutson, B., Adams, C.M., Fong, G.W., et al. (2001). Anticipation of increasing monetary reward selectively recruits nucleus accumbens. *Journal of Neuroscience* **21**, RC159.

Knutson, B., Rick, S., Wimmer, G.E., et al. (2007). Neural predictors of purchases. *Neuron* **53**, 147–56.

Kober, H. and Wager, T.D. (2010). Meta-analysis of neuroimaging data. *Wiley Interdisciplinary Reviews: Cognitive Science* **1**, 293–300.

Kranz, F. and Ishai, A. (2006). Face perception is modulated by sexual preference. *Current Biology* **16**, 63–8.

Langlois, J.H., Kalakanis, L., Rubenstein, A.J., et al. (2000). Maxims or myths of beauty? A meta-analytic and theoretical review. *Psychological Bulletin* **126**, 390–423.

Langlois, J.H., Ritter, J.M., Roggman, L.A., et al. (1991). Facial diversity and infant preferences for attractive faces. *Developmental Psychology* **27**, 79–84.

Langlois, J.H. and Roggman, L.A. (1990). Attractive faces are only average. *Psychological Science* **1**, 115–21.

Langlois, J.H., Roggman, L.A., Casey, R.J., et al. (1987). Infant preferences for attractive faces: rudiments of a stereotype? *Developmental Psychology* **23**, 363–9.

Le Grand, R., Cooper, P.A., Mondloch, C.J., et al. (2006). What aspects of face processing are impaired in developmental prosopagnosia? *Brain and Cognition* **61**, 139–58.

Liang, X., Zebrowitz, L.A., and Zhang, Y. (2010). Neural activation in the "reward circuit" shows a nonlinear response to facial attractiveness. *Social Neuroscience* **5**, 320–34.

McClure, S.M., Berns, G.S., and Montague, P.R. (2003). Temporal prediction errors in a passive learning task activate human striatum. *Neuron* **38**, 339–46.

Mende-Siedlecki, P., Said, C.P., and Todorov, A. (2012). The social evaluation of faces: a meta-analysis of functional neuroimaging studies. *Social Cognitive and Affective Neuroscience* **8**, 285–99. doi: 10.1093/scan/nsr090.

Møller, A. and Pomiankowski, A. (1993). Fluctuating asymmetry and sexual selection. *Genetica* **89**, 267–79.

Nadal, M., Munar, E., Capó, M.À., et al. (2008). Towards a framework for the study of the neural correlates of aesthetic preference. *Spatial Vision* **21**, 379–96.

Nakamura, K., Kawashima, R., Nagumo, S., et al. (1998). Neuroanatomical correlates of the assessment of facial attractiveness. *Neuroreport* **9**, 753–7.

O'Doherty, J., Dayan, P., Schultz, J., et al. (2004). Dissociable roles of ventral and dorsal striatum in instrumental conditioning. *Science* **304**, 452–4.

O'Doherty, J., Kringelbach, M.L., Rolls, E.T., et al. (2001). Abstract reward and punishment representations in the human orbitofrontal cortex. *Nature Neuroscience* **4**, 95–102.

O'Doherty, J., Winston, J., Critchley, H., et al. (2003). Beauty in a smile: the role of medial orbitofrontal cortex in facial attractiveness. *Neuropsychologia* **41**, 147–55.

Oosterhof, N.N. and Todorov, A. (2008). The functional basis of face evaluation. *Proceedings of the National Academy of Sciences of the USA* **105**, 11087.

Penton-Voak, I.S., Perrett, D.I., Castles, D.L., et al. (1999). Menstrual cycle alters face preference. *Nature* **399**, 741–2.

Perrett, D.I., Burt, D.M., Penton-Voak, I.S., et al. (1999). Symmetry and human facial attractiveness. *Evolution and Human Behavior* **20**, 295–307.

Perrett, D.I., Lee, K.J., Penton-Voak, I., et al. (1998). Effects of sexual dimorphism on facial attractiveness. *Nature* **394**, 884–7.

Plassmann, H., O'Doherty, J., and Rangel, A. (2007). Orbitofrontal cortex encodes willingness to pay in everyday economic transactions. *Journal of Neuroscience* **27**, 9984–8.

Puce, A., Allison, T., Bentin, S., et al. (1998). Temporal cortex activation in humans viewing eye and mouth movements. *Journal of Neuroscience* **18**, 2188–99.

Radua, J. and Mataix-Cols, D. (2012). Meta-analytic methods for neuroimaging data explained. *Biology of Mood Anxiety Disorders* **2**, 6.

Ramachandran, V.S. and Hirstein, W. (1999). The science of art: a neurological theory of aesthetic experience. *Journal of Consciousness Studies* **6**, 15–51.

Rees, G. (2007). Neural correlates of the contents of visual awareness in humans. *Philosophical Transactions of the Royal Society B* **362**, 877–86.

Rellecke, J., Bakirtas, A.M., Sommer, W., et al. (2011). Automaticity in attractive face processing. *NeuroReport* **22**, 706–10.

Rhodes, G. (2006). The evolutionary psychology of facial beauty. *Annual Review of Psychology* **57**, 199–226.

Rhodes, G., Proffitt, F., Grady, J., et al. (1998). Facial symmetry and the perception of beauty. *Psychonomic Bulletin and Review* **5**, 659–69.

Rhodes, G., Sumich, A., and Byatt, G. (1999). Are average facial configurations attractive only because of their symmetry? *Psychological Science* **10**, 52–8.

Rolls, E.T. (2000). Précis of *The Brain and Emotion. Behavioral and Brain Sciences* **23**, 177–234.

Rubenstein, A.J., Kalakanis, L., and Langlois, J.H. (1999). Infant preferences for attractive faces: a cognitive explanation. *Developmental Psychology* **35**, 848–55.

Sadr, J., Duchaine, B.C., and Nakayama, K. (2004). The perception of facial attractiveness in prosopagnosia. *Journal of Vision* **4**, 914.

Said, C.P., Dotsch, R., and Todorov, A. (2010). The amygdala and FFA track both social and non-social face dimensions. *Neuropsychologia* **48**, 3596–605.

Said, C.P., Haxby, J.V., and Todorov, A. (2011). Brain systems for assessing the affective value of faces. *Philosophical Transactions of the Royal Society B* **366**, 1660–70.

Sato, W., Kochiyama, T., Yoshikawa, S., et al. (2001). Emotional expression boosts early visual processing of the face: ERP recording and its decomposition by independent component analysis. *Neuroreport* **12**, 709–14.

Saxe, R. and Kanwisher, N. (2003). People thinking about thinking people: the role of the temporo-parietal junction in "theory of mind." *NeuroImage* **19**, 1835–42.

Senior, C. (2003). Beauty in the brain of the beholder. *Neuron* **38**, 525–8.

Slater, A., Von der Schulenburg, C., Brown, E., et al. (1998). Newborn infants prefer attractive faces. *Infant Behavior and Development* **21**, 345–54.

Smith, D.V., Hayden, B.Y., Truong, T.-K., et al. (2010). Distinct value signals in anterior and posterior ventromedial prefrontal cortex. *Journal of Neuroscience* **30**, 2490–5.

Thornhill, R. and Gangestad, S.W. (1999). Facial attractiveness. *Trends in Cognitive Sciences* **3**, 452–60.

Tobler, P.N., Fiorillo, C.D., and Schultz, W. (2005). Adaptive coding of reward value by dopamine neurons. *Science* **307**, 1642–5.

Tsukiura, T. and Cabeza, R. (2011). Remembering beauty: roles of orbitofrontal and hippocampal regions in successful memory encoding of attractive faces. *NeuroImage* **54**, 653–60.

Van Hooff, J.C., Crawford, H., and Vugt, M. van (2011). The wandering mind of men: ERP evidence for gender differences in attention bias towards attractive opposite sex faces. *Social Cognitive and Affective Neuroscience* **6**, 477–85.

Vartanian, O. and Goel, V. (2004). Neuroanatomical correlates of aesthetic preference for paintings. *Neuroreport* **15**, 893–7.

Vuilleumier, P. and Pourtois, G. (2007). Distributed and interactive brain mechanisms during emotion face perception: evidence from functional neuroimaging. *Neuropsychologia* **45**, 174–94.

Werheid, K., Schacht, A., and Sommer, W. (2007). Facial attractiveness modulates early and late event-related brain potentials. *Biological Psychology* **76**, 100–8.

Whalen, P.J., Shin, L.M., McInerney, S.C., et al. (2001). A functional MRI study of human amygdala responses to facial expressions of fear versus anger. *Emotion* **1**, 70–83.

Winston, J.S., Henson, R.N.A., Fine-Goulden, M.R., et al. (2004). fMRI-adaptation reveals dissociable neural representations of identity and expression in face perception. *Journal of Neurophysio*logy **92**, 1830–9.

Winston, J.S., O'Doherty, J., Kilner, J.M., et al. (2007). Brain systems for assessing facial attractiveness. *Neuropsychologia* **45**, 195–206.

Winston, J.S., Strange, B.A., O'Doherty, J., et al. (2002). Automatic and intentional brain responses during evaluation of trustworthiness of faces. *Nature Neuroscience* **5**, 277–83.

Zaki, J., Schirmer, J., and Mitchell, J.P. (2011). Social influence modulates the neural computation of value. *Psychological Science* **22**, 894–900.

Zebrowitz, L.A. (2004). The origin of first impressions. *Journal of Cultural and Evolutionary Psychology* **2**, 93–108.

Zebrowitz, L.A. and Montepare, J.M. (2008). Social psychological face perception: why appearance matters. *Social and Personality Psychology Compass* **2**, 1497–517.

Zeki, S. (1999). Art and the brain. *Journal of Consciousness Studies* **6**, 76–96

Zhang, Y., Kong, F., Chen, H., et al. (2011). Identifying cognitive preferences for attractive female faces: an event-related potential experiment using a study-test paradigm. *Journal of Neuroscience Research* **89**, 1887–93.

Chapter 7

Indeterminate artworks and the human brain

Robert C. Pepperell and Alumit Ishai

7.1 Visual indeterminacy in art history

This chapter summarizes previously published work by the authors on the topic of visual indeterminacy (Pepperell 2006; Ishai et al. 2007; Fairhall and Ishai 2008; Pepperell 2011). Visual indeterminacy occurs when an image or a scene is structured in such a way as to defy immediate or easy recognition. Artists have long been fascinated by indeterminate images and have frequently exploited their capacity to perplex audiences. In the first major study of unrecognizable images in art history, Gamboni (2002) surveyed the occurrence of visual indeterminacy in works stretching deep into art history. He showed how artists have in various ways deliberately included elements or passages within paintings that confounded the viewers' capacity to identify what they saw. He cites, for example, the work of the symbolist artist Odilon Redon who produced a series of drawings and pastels in the late nineteenth and early twentieth century that are full of richly suggestive forms often lacking clearly definable objects. A major piece, *Roger and Angelica* of 1910 (Museum of Modern Art, New York, Lillie P. Bliss Collection), depicts just recognizable figures in a misty landscape of loosely depicted water and rocks. In his *Confessions of an Artist*, Redon was explicit about his purpose: "The title that occasionally defines my drawings can be superfluous. It is justified only when it is vague, indeterminate, and even equivocal. My drawings inspire and cannot be defined. They do not determine anything. They place us, as music does, in the world of the ambiguous and the indeterminate" (in Dorra 1995, p. 55).

One artist who exploited the phenomenon of visual indeterminacy perhaps more than any other was J.M.W. Turner, the most renowned and influential British painter in modern times, who was widely famed for his atmospheric landscapes and seascapes. Like many artists, Turner's reputation ebbed and flowed throughout his long career. Given how he is regarded now it can be surprising to learn how poorly his works were received by the public and influential critics during his lifetime. Even some of his early works, which we would now judge to be quite clear and precise, were seen by many contemporaries as "foggy," "indistinct," and "unreadable." Yet despite contemporary hostility to his work, Turner's work was to have a profound affect on the future course of art, especially on a generation of young French artists who had taken refuge in London during the Franco–Prussian War in the 1870s. Prominent among them was Claude Monet, who saw many of Turner's paintings during his sojourn (House 1986, p. 113). Monet was to find in Turner the kind of expressive freedom to evoke natural atmospheres that led in 1872 to the painting *Impression, Sunrise* (Musée Marmottan Monet, Paris), the title of which, when exhibited in 1874, gave the Impressionist movement its name. This loose and fluid sketch of the misty sun above a harbour in Le Havre famously provoked a bemused critic to ask: "What does that canvas depict?" (Leroy 1874).

Many artists of Monet's generation were profoundly influenced by contemporary theories of perception, such as those proposed by Hermann von Helmholtz, which stressed the way we infer properties of the world from raw sensory data. In the case of vision, as artists then believed, this raw sensation consisted of patches of color detected by the retina which were then interpreted as objects by the higher brain. In his seminal book *Art and Illusion*, Gombrich (1960, p. 161) referred to this inferential process as the "beholder's share" of the pictorial bargain, this being the part of the meaning generated by the viewer from his or her own cognitive resources. In a famous passage quoted by Gombrich (1960, p. 250) the art critic and champion of Turner, John Ruskin, advised painters to be guided only by what arrives at the "childish" or "innocent eye," and to record only "flat stains of color, merely as such, without consciousness of what they signify—as a blind man would see them if suddenly gifted with sight." This was an approach Monet himself sympathized with, leading him to remark that he "wished he had been born blind and then suddenly gained his sight so that he could have begun to paint in this way without knowing what the objects were that he saw before him" (Nochlin 1966, pp. 35–6).

Thus for Monet, the function of painting was not to obscure but to faithfully depict the appearance of the world. This was to be achieved by recording what he saw, and the way he saw, rather than what he knew from prior knowledge or experience to be in front of him. Take, for example, one of the 30 or so paintings he made of Rouen Cathedral during the 1890s, which capture the appearance of the building's façade in a variety of lighting and weather conditions. These paintings were intended to be less depictions of a particular building and rather the experience of the light reflected by the building (House 1986, p. 221).

Wassily Kandinsky, who is often credited as the originator of abstraction, was profoundly affected by seeing one of Monet's series of paintings depicting haystacks when they were exhibited in Moscow in 1895. Kandinsky was struck by the fact that he could clearly see the painting in front of him but not what it represented. He later recounted: "And suddenly for the first time I saw a picture. That it was a haystack (or rather, a grain stack), the catalog informed me. I did not recognize it . . . And I noticed with surprise and confusion that the picture not only gripped me, but impressed itself ineradicably upon my memory. Painting took on a fairy-tale power and splendor. And, albeit unconsciously, objects were discredited as an essential element within the picture." (Parsons and Gale 1992, p. 255).

Kandinsky later recalled another similar experience which occurred when he returned to his studio at dusk and was astonished to see "an indescribably beautiful picture, pervaded by an inner glow" standing against the wall (Lindsay and Vergo 1982, pp. 369–70). In it he could discern "only forms and colors" and no comprehensible objects. He then realized it was one of his own paintings turned on its side, but he had simply failed to recognize the subject. This experience, compounded by the one he had in front of the Monet, was to lead Kandinsky directly to develop a new form of "non-objective" art. It was form of art that was not purely abstract, as later critics often claimed it to be, but rather indeterminate in the sense that objects were suggested and evoked but never explicitly rendered.

Gerhard Richter is a contemporary German artist working in a number of distinct styles which include both highly rendered photographic-like paintings as well as large-scale expressive pieces produced by spreading and scratching wet paint with huge squeegees. It would be too simplistic to describe either style as straightforwardly "realistic" or "abstract," although these labels are often used. Listening to Richter's own statements it becomes clear that his work is better described as indeterminate. In many of his images Richter's aim is to create sense of mystery or uncertainty about what is being depicted or represented. In the painting *Alps II* (1968, MKM Museum Küppersmühle für Moderne Kunst, Duisburg, Germany, Sylvia and Ulrich Ströher Collection), for example, we are presented with a seemingly abstract mass of gray, formless brush marks. Only through considerable effort can we make out what appears to be a snowy mountain range. Richter says: "Pictures which are interpretable, and which contain a meaning, are bad pictures." A good picture, on the other hand, "demonstrates the endless multiplicity of aspects, it takes away our certainty, because it deprives a thing of its meaning and its name. It shows us the thing in all the manifold significance and infinite variety that preclude the emergence of any single meaning or view" (Elger and Obrist 2009, pp. 32–3).

This brief survey of art and visual indeterminacy demonstrates, first, that artists have understood and tried to exploit the perceptual effects produced by non-specific images. Secondly, it shows that artists, in order to use these effects best, have acted scientifically, exploring the nature of visual perception through their artistic practice. Finally, it shows that visual indeterminacy, far from being a tactic on the part of artists to confound their public audiences, is understood by them to be a potential source of profound aesthetic experience.

Robert Pepperell is a contemporary artist who has a long-standing interest in visual indeterminacy. Over many years he has sought to construct images that are truly indeterminate in that they finely balance recognizability and abstraction. In this way they excite the inquisitiveness of the viewer's visual system while frustrating its capacity for recognition. After a long period of experimentation, with many false starts, Pepperell gradually developed a method of drawing and painting which seemed to produce this effect consistently (Figure 7.1). Many of his paintings use a classical pictorial architecture of the kind frequently found in European paintings made between the 1500s and early 1900s, which incites a strong expectation that the painting will contain a recognizable scene since paintings of this style and period are strongly associated with immediately recognizable content. By using this overall pictorial structure but organizing the content in such a way that no objects could be deciphered as objects, Pepperell was able to achieve a consistently indeterminate pictorial effect (for the artist's personal perspective, see Pepperell 2006 and 2011).

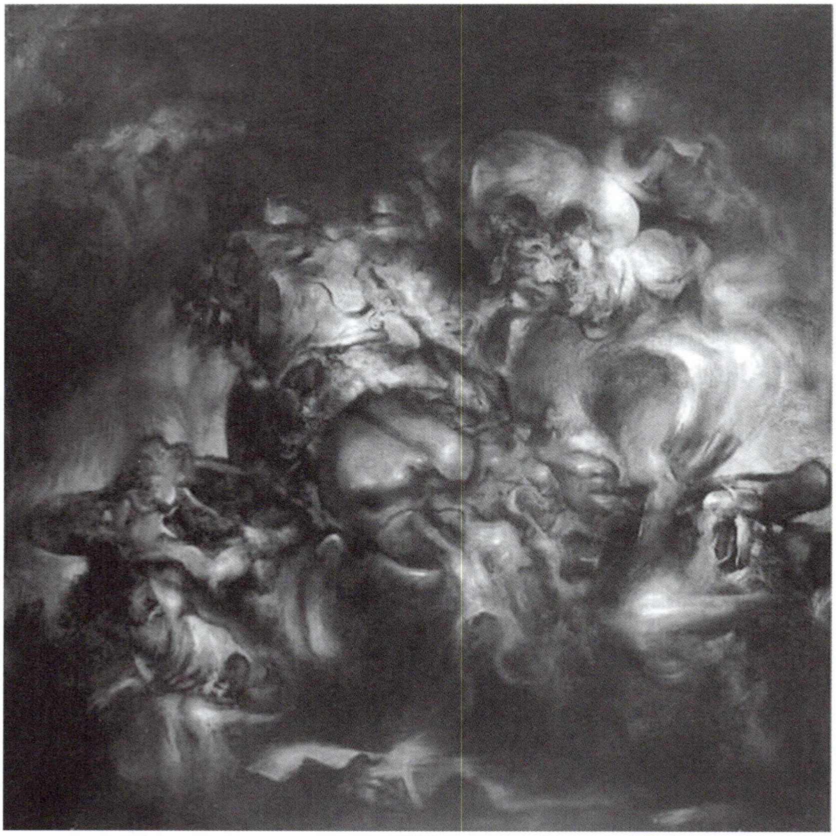

Figure 7.1 *Territory* by Robert Pepperell. In this example of indeterminate painting, natural forms are implied while at the same time resisting easy or immediate identification.

7.2 Scientific background to visual indeterminacy

Visual indeterminacy is related to, but not to be confused with, the phenomenon of ambiguous or reversible images. The Necker cube (a wire-frame cube that can be interpreted in two mutually inconsitent ways), the duck–rabbit illusion (in which a single image is perceived to be either a duck or a rabbit), or the Boring vase image (in which the outline of a vase can be also seen as two faces in profile) are all open to alternating interpretations, and in each case the alternative meanings are quite determinate (Kleinschmidt et al. 1998; Meng and Tong 2004). Similarly, so-called "hidden figures," such as the case of the Dalmatian dog in R.C. James' famous photograph (Gregory 1970; Palmer 1999; Ramachandran and Hirstein 1999), depend on our capacity to find meaningful objects in ambiguous stimuli. However, neither ambiguous nor hidden images are visually indeterminate, in the sense being used here, since what distinguishes indeterminate stimuli is their resistance to final identification of any kind.

An analogy can be drawn between visual indeterminacy and the rare neurological disorder of associative visual agnosia. A widely known case study published by Humphreys and

Riddoch (1987) concerned a patient, John, who had suffered a stroke resulting in a bilateral lesion in the region supplied by the posterior cerebral artery. Much of John's visual system was spared, but his ability to recognize what he saw was greatly impaired. When shown a series of line drawings of everyday objects he was able to identify only a small proportion, and relied on "working out" what was depicted from specific clues within the image, such as the curliness of a pig's tail, rather than by seeing the object "as a whole" (p. 60). Humphreys and Riddoch excluded other possible factors that could have contributed to John's inability to recognize everyday objects when arriving at their diagnosis of associative visual agnosia, including any residual deficit in his stored knowledge or visual sensation. He could still readily identify objects by touch, by describing them from memory, and could copy line drawings of objects, albeit laboriously, "even when he had no idea what the object was" (p. 69).

Humphreys and Riddoch argue that John's case provides evidence to support the idea that visual perception occurs through at least two distinct sets of processes (pp. 101–2). The first processes organize the perceptual "input" data according to position and orientation, and bind multiple visual elements into wholes. A subsequent set of processes match that input data to associations about function and meaning. The authors conclude: "In general terms, (John's) case supports the view that 'perceptual' and 'recognition' processes are separable . . ." (p. 104).

In another important study of visual agnosia, Farah (2004) concurs with Humphreys and Riddoch's distinction between perceptual input and the conceptual associations involved in visual object recognition.

> Visual form agnosia validates the distinction implicit in the labels "early" and "intermediate" vision, on the one hand, and "high-level," "object" vision on the other, by showing the first set of processes can continue to function when the second set is all but obliterated. It shows us a kind of richly elaborated but formless visual "stuff," from which "things" can be derived (p. 156).

The phrase "richly elaborated but formless visual stuff" is an apt description of the appearance of the world at the moment of visual indeterminacy, resulting from the failure to correlate this "visual stuff" to the bank of stored memories and associations. This can occur to us all occasionally, especially when presented with indeterminate images, but for the unfortunate sufferers of visual agnosia it is a permanent condition.

There is some evidence from neuroscientific studies that the perception of indeterminate stimuli brings about measurable changes in brain activity. An EEG study by Supp and colleagues (2005), for example, looked at the changes in cortical networks within the time-window of the event-related potential component N400. Subjects were shown sequences of grayscale pictures that were more or less recognizable although comparable in terms of size, complexity, and structure. During the viewing of unrecognizable pictures, when compared with recognizable ones, there was a marked increase in cooperation in certain parts of the brain and a greater degree of overall coherence between different regions. The researchers concluded that these changes in brain activity reflected the greater demands made on the viewer's perceptual and cognitive resources and consequent "unease" involved in the task of semantically matching the undecipherable stimuli: "the greater number of coherence increases for meaningless object processing suggests enhanced recruitment of more distributed left and right areas during unsuccessful memory search" (p. 1143).

Rainer and colleagues (2004) studied the responses of monkeys to recognizable and unrecognizable images by measuring neural activity in area V4 on exposure to images that were increasingly degraded, from clear to abstract noise. The monkeys learned to recognize familiar images that were degraded compared to novel ones that were treated in the same way. The researchers found that monkeys exposed to indeterminate images showed significantly increased neural activity in both primary and higher cortical areas of the brain than when faced with familiar or recognizable stimuli. Rainer and colleagues concluded that not only are particular loci in the brain recruited in response to indeterminate stimuli, but that the attempt to decipher such stimuli leads to enhanced overall coordination in brain activity. "This suggests that V4 plays a key role in resolving indeterminate visual inputs by coordinated interaction between bottom-up and top-down processing streams" (p. 275).

7.3 Scientific experiments on visual indeterminacy

7.3.1 Behavioral studies

We (Pepperell and Ishai) met at the Max Planck Institute for Biological Cybernetics in Tübingen, and based on mutual interests in art perception, we decided to study together the neural correlates of viewing at indeterminate paintings. To test the effects of visual indeterminacy, we gathered a set of images that included a selection of Robert Pepperell's indeterminate paintings as well as paintings that were similar in visual appearance to Pepperell's but full of readily identifiable objects. These consisted of works by classical artists such as Turner, Tintoretto, Rubens, Michelangelo, and Fuseli (examples can be seen at <http://www.robertpepperell.com/Stimuli/Stimuli.html>). The paintings were divided into two sets, one grayscale and one color, in order to determine whether there was any significant effect between these conditions. Then the subjects were presented with a number of tasks: they were asked first to indicate whether each image contained familiar objects, and secondly they were asked to rate how powerfully the images affected them. On the basis of informal observations made on viewers looking at Pepperell's paintings in the studio and galleries, we hypothesized that compared with paintings depicting meaningful content, object recognition in indeterminate images would take longer. We were also interested in the extent to which aesthetic affect depends on how easily objects in the images could be recognized.

The results showed that it took subjects significantly longer to respond to Pepperell's indeterminate images, and they reported seeing objects in them 24 per cent of the time. It is interesting that subjects were significantly slower in making judgments about indeterminate paintings compared to the determinate ones, and this was true whether or not they recognized familiar objects (Ishai et al. 2007). The findings suggest that when subjects are presented with the problem of resolving indeterminate stimuli a greater degree of cognitive processing is required than would be the case when confronted with immediately recognizable images. Importantly, we found a significant degree of correlation between the amount of time it took subjects to identify objects and the extent to which they rated them as aesthetically affective (Figure 7.2).

Somewhat contrary to expectations, we found that all the images were given very similar aesthetic ratings, whether they were recognizable or not. Previous studies have tended to indicate that subjects with little or no expert art knowledge prefer figurative more than abstract art (Healey and Enns 2002; Vartanian and Goel 2004). These results suggest that in forming their judgment, subjects were less influenced by the semantic content than the overall visual structure of the image. In art historical terms, visual structure would be referred to in terms of the "formal" properties of the work; that is, the shape, color, and texture of an artwork as opposed to its "content," which consists in the meaning, narrative, or objects it depicts. There has been a long-standing debate among art theorists as to which of these properties is the most important in defining the aesthetic value of a work of art (Bell 1914). On the basis of the results we obtained, it would seem that the recognizable content of images is less important in affecting aesthetic evaluation than overall visual appearance.

It is worth noting that we used a somewhat unusual method of measuring aesthetic affect. Non-art specialists often conceive of art in terms of its capacity to evoke feelings of beauty, tranquility, balance, harmony, or order. It is not uncommon to find this assumption held among those in the scientific community wishing to measure the psychological impact of art. Yet art history shows that while many artworks can indeed be described in terms of beauty, harmony, pleasure, and the like, there are many highly important artworks that could be better described as ugly, disturbing, violent, or garish. Take for example Grünewald's *Isenheim Altarpiece* (c.1512, Unterlinden Museum, Colmar, France), which shows an emaciated Christ writhing in agony on the cross, or the grotesque fleshy distortions of Dali's *Soft Construction with Boiled Beans* (1936, Philadelphia Museum of Art,

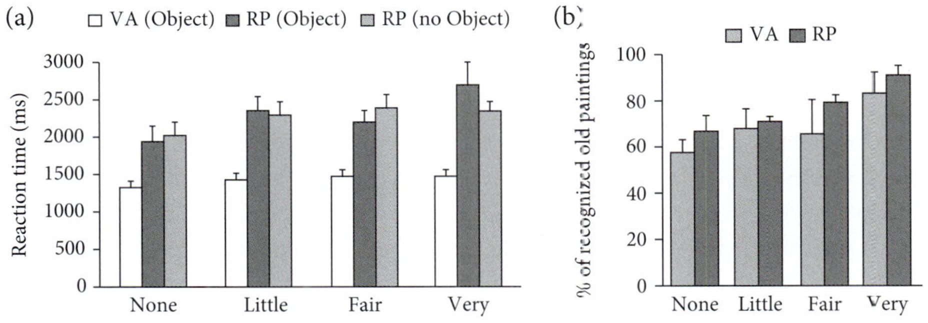

Figure 7.2 Effects of aesthetic judgments on object recognition and memory retrieval. **Left:** Reaction time during the object recognition task predicted subsequent aesthetic affect rating: Longer reaction times in the object recognition task were later associated with stronger affect ratings. **Right:** Aesthetic judgment predicted recognition memory. The stronger the affect rating of an image was, the more likely this painting was to be remembered a week later in a surprise memory test. VA: representational paintings by various artists. RP: indeterminate paintings by Robert Pepperell.

Philadelphia, PA). Such images can have immense power, yet could hardly be described as visually pleasant. We felt, therefore, that measuring the potency of an artwork would be less constraining than merely measuring its beauty, thus we asked subjects to base their aesthetic rating on how powerfully the image affected them on a scale from 1 to 4. The correlation between response latency and heightened power rating suggests that the perceptual struggle to resolve indistinct objects has some aesthetic value.

To test whether memory recall of paintings depends on meaningful content and aesthetic judgment, half the subjects returned for a surprise memory test eight days after participating in the object recognition and aesthetic judgment experiment. The memory test revealed that more representational than indeterminate paintings were remembered, and that affective strength increased the probability of subsequent recall (Figure 7.2), suggesting that meaningful content is critical for incidental memory (Ishai et al. 2007). The lower memory recall for the indeterminate paintings could be due to their lack of meaningful content. An alternative, but not exclusive explanation, is that it was difficult to remember the indeterminate paintings because they had no prior associations to representations stored in memory. Interestingly, we found significant increase in false alarms for indeterminate compared to determinate paintings. Subjects mistook significantly more new indeterminate paintings as old ones, presumably due to their high degree of visual similarity, or because they recognized Pepperell's unique style. Consistently, our studies with portraits, landscapes, and abstract compositions have shown that subjects falsely recognized new paintings as old ones when the new paintings were visually similar to old prototypes that the subjects previously memorized (Yago and Ishai 2006; Wiesmann and Ishai 2008).

Taken collectively, the results of our behavioral studies suggest that perception and memory of art depend on semantic aspects, whereas aesthetic affect depends on formal visual features. The longer latencies associated with indeterminate paintings reflect the underlying cognitive processes that mediate object resolution. Indeterminate artworks therefore comprise a rich set of stimuli with which the neural correlates of visual perception can be investigated (Ishai et al. 2007).

7.3.2 Functional brain imaging studies

We then conducted an fMRI study, in which subjects performed an object recognition task on four classes of paintings: representational, which explicitly depict complex scenes with familiar objects (people, animals, landscapes, still life); indeterminate, in which familiar objects are only suggested; abstract, which do not depict any familiar objects; and scrambled images (Figure 7.3). We hypothesized that subjects would rapidly recognize familiar objects depicted in representational paintings, but would be slower to report the presence or absence of recognizable objects in abstract and indeterminate paintings. Moreover, we predicted a posterior-to-anterior gradient of activation along the ventral visual pathway, such that with increased recognition of familiar content in the paintings, differential activation would be observed in more anterior, higher-tier, object-selective areas. Finally, we postulated that indeterminate paintings would invoke visual imagery-related activation in parietal and prefrontal cortices.

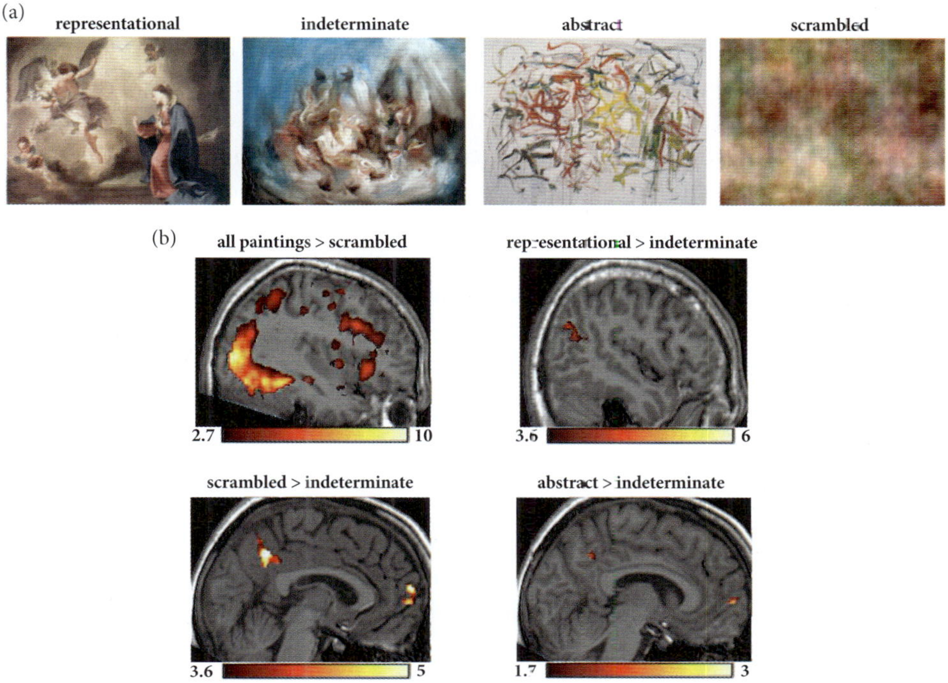

Figure 7.3 Viewing paintings as compared with scrambled pictures elicits activation in a distributed network of visual, limbic, parietal, and frontal regions. Comparing representational with indeterminate paintings revealed activation in the temporoparietal junction, a region that mediates the allocation of spatial attention across visual scenes. Comparing scrambled pictures with indeterminate paintings and abstract paintings with indeterminate ones reveals activation in the precuneus and medial frontal gyrus, regions that mediate the generation and maintenance of mental images from long-term memory.

Replicating our behavioral study, subjects reported seeing familiar objects in 36 per cent of the indeterminate paintings, and in 18 per cent of the abstract paintings, although these images lack object-suggestive content. Moreover, subjects rapidly recognized familiar objects in representational paintings but showed longer reaction times to indeterminate and abstract images (Fairhall and Ishai 2008). These differential response latencies suggest, again, an automatic recognition of objects when they were explicitly depicted, but required more effortful cognitive processes such as visual search, when the objects were ambiguous or suggestive.

All paintings evoked activation within a distributed cortical network that included regions in the visual cortex, as well as parietal, limbic, and prefrontal regions (Figure 7.3). Consistent with our hypotheses, representational paintings with meaningful content evoked stronger activation than abstract and indeterminate paintings in the fusiform gyrus, a region that responds to assorted common objects, including faces, houses, animals, and tools (Chao et al. 1999; Ishai et al. 1999, Ishai et al. 2000b; Haxby et al.

2001). These results are consistent with a study in which enhanced activation in the fusiform gyrus was observed when representational paintings were compared with filtered paintings, namely representational and abstract paintings that were subjected to a median noise filter, which created random distribution of color levels (Vartanian and Goel 2004). The indeterminate paintings, when compared with the representational and the abstract compositions, evoked less activation in the right hippocampus (Fairhall and Ishai 2008). This reduced activation may reflect the poor encoding of the indeterminate paintings, consistent with our behavioral study, in which subjects recalled significantly fewer indeterminate than representational paintings in a surprise memory task (Ishai et al. 2007).

To identify the neural correlates of object indeterminacy further, we compared activation evoked by scrambled paintings with activation evoked by the indeterminate paintings. We found that the scrambled paintings evoked enhanced activation in the precuneus and the medial frontal gyrus, regions of the "imagery network" that mediate the generation and maintenance of mental images from long-term memory (Miyashita and Chang 1988; Fletcher et al. 1995; Mechelli et al. 2004; Ishai et al. 2000a; Ishai et al. 2002). Post-scanning debriefing revealed that most subjects used mental imagery during the perception of the scrambled paintings in order to decide whether the images contained any familiar objects. In contrast, to decide whether the indeterminate paintings, which were rich with suggestive objects, contained any recognizable objects, subjects relied on visual similarity and visual associations.

A direct comparison of representational and indeterminate artworks revealed significant activation in the temporo-parietal junction (TPJ), a region that has been implicated in exerting attentional control over switches from local to global processing (Fink et al. 1996), and the allocation of spatial attention across the visual scene (Corbetta et al. 2005). The enhanced activation within the TPJ for representational paintings reflects the binding of object form and spatial location within these cluttered visual scenes. Thus, the recognition of meaningful, familiar content in artworks is mediated by activation in the TPJ. Incidentally, in another study, the left (but not the right) TPJ was also activated by paintings that subjects rated "beautiful" rather than "neutral" (Kawabata and Zeki 2004).

In sum, the fMRI study shows that perception of art compositions evokes activation in multiple visual regions, the hippocampus, intraparietal sulcus, and inferior frontal gyrus. Content-related modulation in the fusiform gyrus reflects object perception, whereas hippocampal activation reflects memory consolidation. Imagery-related activation was observed for scrambled paintings. Finally, interpreting composite scenes relies on higher order associations in the TPJ, which links the various elements of the visual scene (Fairhall and Ishai 2008). It therefore seems that the human brain is a compulsive object viewer, which automatically segments indeterminate visual input into coherent images. To resolve the visual indeterminacy, higher cognitive functions, such as attention, visual imagery, and memory retrieval, are recruited.

7.4 Object recognition in Cubist paintings

To the naïve observer, Cubist paintings contain geometrical forms in which familiar objects are barely recognizable, even in the presence of a meaningful title. In Cubist artworks, objects are broken up, analysed, and re-assembled to produce abstracted forms, which often depict the same objects from different viewing points. We conducted a new fMRI study to test the extent to which a short training session about Cubism would facilitate object recognition in paintings by Picasso, Braque, and Gris all painted in the period before the First World War. Paintings of this period are characterized as being highly indeterminate; they are directly observed depictions of everyday objects (e.g. tables, fruit, newspapers, glasses) but represented in a fragmented manner that makes immediate identification very difficult. Samples of the stimuli can be seen at: <http://www.robertpepperell.com/Cubism/index.html>. Art experts, and others familiar with the genre, are able to "read" these Cubist scenes and find within them the various forms and objects from which they are constructed. Those without this expertise tend to see only patterns, lines, and textures rather than distinct objects of any kind (Golding 1988).

We hypothesized that subjects who received training would recognize familiar objects faster than control subjects, and would exhibit stronger activation in object-responsive and attention-related regions. Our subjects, students from the University of Zurich, had no formal art education and reported visiting art museums once a year or less. A meaningful title or the word "untitled" appeared before each Cubist painting and subjects had to answer the question "Do you recognize any familiar objects?" by pressing one of two buttons (Yes/No), then answer the question "How many objects did you recognize?" by pressing one of four buttons to indicate "0," "1," "2," or "3," or "more" objects. Thirty minutes before scanning, half the subjects received a short training session, during which they were presented with information about Cubism, viewed examples of Cubist paintings, and practised recognizing familiar objects in these paintings.

Relative to the control group, trained subjects recognized significantly more objects in the paintings and their response latencies were significantly shorter. Moreover, trained subjects took longer to report not recognizing any familiar objects in the paintings. Cubist paintings evoked activation in a distributed cortical network that included extrastriate, parietal, and prefrontal regions. Within the parahippocampal cortex, trained subjects showed a significantly larger spatial extent of activation and a parametric increase in the amplitude of the fMRI signal as a function of the number of objects they recognized. We also found that in trained subjects, the longer response latencies associated with failing to recognize familiar objects were correlated with activation in a fronto-parietal network that mediates spatial attention (Wiesmann and Ishai 2010).

The most surprising and intriguing finding in our study is the enhanced activation in the parahippocampal cortex of trained subjects, the amplitude of which increased as a function of the number of objects recognized. This suggested that the subjects had used broader contextual associations to identify the objects in the paintings rather than the cognitive resources normally linked more specifically to object recognition. The parahippocampal cortex, implicated in the representation and processing of spatial navigation information (Epstein and Kanwisher 1998), episodic memory (Gabrieli et al. 1997), and remote spatial memories (Spiers and Maguire 2007), is a major node in the cortical network for contextual associations (Bar et al. 2008). Associations are formed over time, when repeated patterns and statistical regularities are extracted from the environment and stored in memory. It has been recently suggested that the role of associations is to generate predictions about the immediate future in order to guide behavior (Bar 2007). It is highly likely that due to the short training session, our subjects used contextual associations to perform the tasks. Importantly, our findings provide empirical evidence for Bayesian analysis, which was proposed as a model for object perception (Kersten et al. 2004) and evoked cortical responses (Friston et al. 2003; Friston 2005). According to the Bayes perspective, the short training session enabled our subjects to match the indeterminate visual input with their top-down predictions. It is reasonable to assume that trained subjects were more likely than control subjects to suppress errors and establish a consensus between the actual bottom-up input and the top-down prediction. Thus, minimizing prediction error resulted in faster recognition of more familiar objects in Cubist paintings.

The extent to which titles do or should influence the perception of meaning and the aesthetic impression of art compositions is contentious. In art theoretical terms, critics of a formalist persuasion claim that titles are merely "identification tags" that should not affect the viewer's reading of the work. Others, however, claim titles function as guides to interpretation and provide important contextual cues to engage the attention of the viewer (Gombrich 1960; Fisher 1984). Empirical evidence suggests that titles influence both the understanding and the appreciation of paintings (Leder et al. 2006). In a compelling example of the top-down effects of titles on art perception, viewers' descriptions of the content of paintings varied according to the title (e.g. "Agony" versus "Carnival") they were offered (Franklin et al. 1993). In our experiment, Cubist paintings were preceded by their meaningful title or by the word "Untitled." We found that meaningful titles facilitated object recognition, but only in trained subjects. Thus, relative to control subjects, trained subjects reported recognizing more familiar objects in paintings with meaningful titles. These findings suggest that meaningful titles can provide the top-down solution for ambiguous visual input, but only when prior knowledge or experience exists. Taken together, these observations suggest that recognition of familiar content in artworks is a skill acquired through training. It is also tempting to wonder whether subjects thus trained in recognizing objects in Cubist paintings are also then better at other object recognition tasks.

In sum, our studies suggest that the human brain is not a passive viewer of works of art, but a dynamic interpreter that constantly generates predictions about the content and its meaning based on previous encounters with similar visual input. This is especially evident in the case of indeterminate stimuli that force the subjects to adopt different viewing strategies from more conventionally recognizable images. Understanding the content of modern paintings, which are often difficult to read or recognize, is an acquired, context-dependent skill. It is perhaps not surprising that our emotional response to works of art is also influenced by knowledge and context, as suggested by recent studies of aesthetic judgment of paintings (Kirk et al. 2009). If indeed art evolved as a medium for communication by symbols, it seems that with time, as these non-linguistic symbols became more and more detached from their referents, the human brain had to recruit additional cortical resources to comprehend their meaning.

Acknowledgments

Alumit Ishai is supported by the Swiss National Center for Competence in Research: Neural Plasticity and Repair, and by the Commission on Gender Equality of Zurich University.

References

Bar, M. (2007). The proactive brain: using analogies and associations to generate predictions. *Trends in Cognitive Sciences* **7**, 280–9.

Bar, M., Aminoff, E., and Ishai, A. (2008). Famous faces activate contextual associations in the parahippocampal cortex. *Cerebral Cortex* **18**, 1233–8.

Bell, C. (1914). *Art*. London: Chatto and Windus.

Chao, L.L., Haxby, J.V., and Martin, A. (1999). Attribute-based neural substrates in posterior temporal cortex for perceiving and knowing about objects. *Nature Neuroscience* **2**, 913–9.

Corbetta, M., Kincade, M.J., Lewis, C., et al. (2005). Neural basis and recovery of spatial attention deficits in spatial neglect. *Nature Neuroscience* **8**, 1603–10.

Dorra, H. (1995). *Symbolist Art Theories*. Berkeley, CA: University of California Press.

Elger, D. and Obrist, H.U. (eds) (2009). *Gerhard Richter: Text*. London: Thames and Hudson.

Epstein, R. and Kanwisher, N. (1998). A cortical representation of the local visual environment. *Nature* **392**, 598–601.

Fairhall, S.L. and Ishai, A. (2008). Neural correlates of object indeterminacy in art compositions. *Consciousness and Cognition* **17**, 923–32.

Farah, M. (2004). *Visual Agnosia*. Cambridge, MA: MIT Press.

Fink, G.R., Halligan, P.W., Marshall, J.C., et al. (1996). Where in the brain does visual attention select the forest and the trees? *Nature* **382**, 626–8.

Fisher, J. (1984). Entitling. *Critical Inquiry* **11**, 286–98.

Fletcher, P.C., Frith, C.D., Baker, S.C., et al. (1995). The mind's eye—precuneus activation in memory-related imagery. *NeuroImage* **2**, 195–200.

Franklin, M.B., Becklen, R., and Doyle, C. (1993). The influence of titles on how paintings are seen. *Leonardo* **26**, 103–8.

Friston, K. (2005). A theory of cortical responses. *Philosophical Transactions of the Royal Society B* **360**, 815–36.

Friston, K.J., Harrison, L., and Penny, W. (2003). Dynamic Causal Modeling. *NeuroImage* **19**, 1273–302.

Gabrieli, J.D., Brewer, J.B., Desmond, J.E., et al. (1997). Separate neural bases of two fundamental memory processes in the human medial temporal lobe. *Science* **276**, 264–6.

Gamboni, D. (2002). *Potential Images: Ambiguity and Indeterminacy in Modern Art*. London: Reaktion Books.

Golding, J. (1988). *Cubism: A History and Analysis, 1907-1914*. London: Harvard University Press.

Gombrich, E. (1960). *Art and Illusion: A Study in the Psychology of Pictorial Representation*. London: Phaidon Press.

Gregory, R. (1970). *The Intelligent Eye*. London: Weidenfeld and Nicolson.

Haxby, J.V., Gobbini, M.I., Furey, M.L., et al. (2001). Distributed and overlapping representations of faces and objects in ventral temporal cortex. *Science* **293**, 2425–30.

Healey, C. and Enns, J. (2002). Perception and painting: a search for effective, engaging visualizations. London: IEEE.

House, J. (1986). *Monet: Nature into Art*. London: Yale University Press.

Humphreys, G. and Riddoch, M. (1987). *To See But Not To See: A Case Study of Visual Agnosia*. Hillsdale, NJ: Lawrence Erlbaum Associates.

Ishai, A., Fairhall, S.L., and Pepperell, R. (2007). Perception, memory and aesthetics of indeterminate art. *Brain Research Bulletin* **73**, 314–24.

Ishai, A., Haxby, J.V., and Ungerleider, L.G. (2002). Visual imagery of famous faces: effects of memory and attention revealed by fMRI. *NeuroImage* **17**, 1729–41.

Ishai, A., Ungerleider, L.G., and Haxby, J.V. (2000a). Distributed neural systems for the generation of visual images. *Neuron* **28**, 979–90.

Ishai, A., Ungerleider, L.G., Martin, A., et al. (2000b). The representation of objects in the human occipital and temporal cortex. *Journal of Cognitive Neuroscience* **12**, 35–51.

Ishai, A., Ungerleider, L.G., Martin, A., et al. (1999). Distributed representation of objects in the human ventral visual pathway. *Proceedings of the National Academy of Sciences of the USA* **96**, 9379–84.

Kawabata, H. and Zeki, S. (2004). Neural correlates of beauty. *Journal of Neurophysiology* **91,** 1699–705.

Kersten, D., Mamassian, P., and Yuille, A. (2004). Object perception as Bayesian inference. *Annual Review of Psychology* **55**, 271–304.

Kirk, U., Skov, M., Hulme, O., et al. (2009). Modulation of aesthetic value by semantic context: An fMRI study. *NeuroImage* **44**, 1125–32.

Kleinschmidt, A., Buchel, C., Zeki, S., et al. (1998). Human brain activity during spontaneously reversing perception of ambiguous figures. *Proceedings of the Royal Society B* **265**, 2427–33.

Leder, H., Carbon, C.C., and Ripsas, A.L. (2006). Entitling art: influence of title information on understanding and appreciation of paintings. *Acta Psychologica (Amsterdam)* **121**, 176–98.

Leroy, L. (1874). *The Exhibition of the Impressionists*. Le Charivari, 25 April.

Lindsay, K. and Vergo, P. (1982). *Kandinsky: Complete Writings on Art*. London: Faber and Faber.

Mechelli, A., Price, C.J., Friston, K.J., et al. (2004). Where bottom-up meets top-down: neuronal interactions during perception and imagery. *Cerebral Cortex* **14**, 1256–65.

Meng, M. and Tong, F. (2004). Can attention selectively bias bistable perception? Differences between binocular rivalry and ambiguous figures. *Journal of Vision* **4**, 539–51.

Miyashita, Y. and Chang, H.S. (1988). Neural correlate of pictorial short-term memory in the primate temporal cortex. *Nature* **331**, 68–70.

Nochlin, L. (ed.) (1966). *Impressionism and Post-Impressionism 1874–1904: Sources and Documents*. Englewood Cliffs, NJ: Prentice-Hall.

Palmer, S. (1999). *Vision Science: From Photons to Phenomenology*. Cambridge, MA: MIT Press.

Parsons, T. and Gale, I. (1992). *Postimpressionism, The Rise of Modern Art 1880–1920*. London: Studio editions.

Pepperell, R. (2006). Seeing without objects: visual indeterminacy and art. *Leonardo* **39**, 394–400.

Pepperell, R. (2011). Connecting art and the brain: an artist's perspective on visual indeterminacy. *Frontiers in Human Neuroscience* **5**:84. doi: 10.3389/fnhum.2011.00084.

Rainer, G., Lee, H., and Logothetis, N. (2004). The effect of learning on the function of monkey extrastriate visual cortex. *PLoS Biology* **2**, e44. doi: 10.1371/journal.pbio.0020044.

Ramachandran, V. and Hirstein, W. (1999). The science of art: a neurological theory of aesthetic experience. *Journal of Consciousness Studies* **6**, 15–51.

Spiers, H.J. and Maguire, E.A. (2007). The neuroscience of remote spatial memories: a tale of two cities. *Neuroscience* **149**, 7–27.

Supp, G., Schlogl, A., Fiebach, C., et al. (2005). Semantic memory retrieval: cortical couplings in object recognition in the N400 window. *European Journal of Neuroscience* **21**, 1139–43.

Vartanian, O. and Goel, V. (2004). Neuroanatomical correlates of aesthetic preference for paintings. *Neuroreport* **15**, 893–7.

Wiesmann, M. and Ishai, A. (2008). Recollection- and familiarity-based memory decisions reflect memory strength. *Frontiers in Systems Neuroscience* **2, 1** (1-9). doi: 10.3389/neuro.06.001.2008.

Wiesmann, M. and Ishai, A. (2010). Training facilitates object recognition in cubist paintings. *Frontiers in Human Neuroscience* 4:11. doi: 10.3389/neuro.09.011.2010.

Yago, E. and Ishai, A. (2006). Recognition memory is modulated by visual similarity. *Neuroimage* **31**, 807–17.

Chapter 8

Contextual bias and insulation against bias during aesthetic rating: The roles of VMPFC and DLPFC in neural valuation

Ulrich Kirk and David Freedberg

8.1 Neural aesthetic valuation

In this chapter we concentrate on the respective roles of the ventromedial prefrontal cortex (VMPFC) and the dorsolateral prefrontal cortex (DLPFC) during aesthetic evaluation, particularly the aesthetic evaluation of visual works of art. Activation of these areas has been shown to be modulated by external contextual pressures on viewing (Kirk 2008; Kirk et al. 2009a; Kirk et al. 2011). We also discuss the roles of ACC, hippocampal, and striatal connections in the process of rating and evaluating works of visual art.

One way of considering aesthetic experience is through the examination of the neural correlates of sensory and motor responses works of art and to other visual images. Much attention has already been devoted to the neural substrate of embodied responses to works of art, and to visual images more generally (Freedberg 1989; Freedberg and Gallese 2007; see also Astafiev et al. 2004, and Downing et al. 2006 on responses in extrastriate body area (EBA), and De Gelder 2006 on emotional body language (EBL)). The role of mirror neurons in the felt simulation of action and implied action has played a central role in these discussions (Rizzolatti et al. 1996; Rizzolatti et al. 2001; Freedberg and Gallese 2007; Di Dio et al. 2007). Since such accounts, and all accounts of common coding and shared representation (Prinz 1997; Decety and Sommerville 2003), emphasize not only perception and action coupling but also the links between sensory and motor responses, mirror activations in motor and premotor cortices, as well as in the somatosensory cortices, have been examined at some length (for the visual perception of touch in art, see Keysers et al. 2004; Freedberg and Gallese 2007; for motor responses to art, see also Calvo-Merino 2005; Battaglia et al. 2011). Felt simulatory responses to visual images have become a staple of studies of embodied simulation as well as of empathic responses to images (Gallese et al. 2004; Freedberg 2007, 2009). These have provided a neural account for the kinds of embodied and motor responses to paintings and sculptures described in the work of the French phenomenologist Maurice Merleau Ponty and his predecessors, including the pioneers of empathy theory, such as Robert Vischer and Theodor Lipps, for whom the notion of *Einfühlung* was a fundamental aspect of aesthetic response (e.g. Merleau Ponty 1945; Vischer 1873; Lipps 1903a, 1903b). The connection between movement and emotion as described by Henry James has had a significant revival in the work of Antonio Damasio and others (James 1890; Damasio 1994). The trend towards linking bodily and emotional responses has led to a slew of analyses of the role of amygdalic (LeDoux 1992, 1996; Adolphs et al. 1995; Adolphs 1999) and insular activations in responses to artistic representation (Di Dio et al. 2007; for earlier studies of emotional responses to images, see also the very many studies by Lane and Lang, including Lane 1997; Lang et al. 1993, as well as the wide range of material currently available for the relationship between feelings of disgust and activation of anterior insula). We are now in possession of abundant research on the neural substrates underlying both sensory and affective engagement with works of art (e.g. Freedberg and Gallese 2007). A considerable amount of attention has been paid to the ways in which motor simulation, or a sense of motor simulation of action and movements seen, evokes emotions in the viewer that accord with those intended by the artist (Freedberg 2007). More recently, attention has also been paid to the movements implied by the marks—in other words the traces of the painter's or sculptor's hand—on a painting or sculpture, thus expanding the notion of simulation well beyond figurative art to more abstract forms (and to calligraphy as well) (Umiltà et al. 2012).

Little of this research, however, provides access to the question of aesthetic preference and rating, and even less to understanding the neural correlates of contextual influences on such ratings. While responses in the domain of perception–action coupling provide useful and important evidence of the role of somatic and emotional responses in viewers' engagement with works of art, the degree to which they form the basis of evaluation, ranking, and even judgment is moot. Recent economic and decision-making theory suggests a number of possibilities for the understanding of the neural substrate underlying the (subjective) evaluation of works of art, of the impact of context on hedonic value, of the varieties of insulation against such contextual influences, and of the impact of domain expertise on such insulation (Montague and Berns 2002). Here we set out the role of prefrontal interactions, most notably between VMPFC and DLPFC in the processes of evaluation, and the relationship between expert and non-expert responses.

A number of behavioral studies have shown how information such as titles, labels, texts, and other forms of cognitive supplementation can affect the evaluation of a work of art (e.g. Cupchik et al. 2009; Leder et al. 2004; Russell 2003; Kirk et al. 2008). The neural correlates of aesthetic evaluation began to become clearer with the examination of responses to musical (Blood and Zatorre 2001; Baumgarten 1986/1750) and pictorial stimuli (Cela-Conde et al. 2004). Subjective pleasantness ratings of musical sequences have been shown to correlate with activity in the striatum, amygdala, parahippocampal gyrus, insula, orbitofrontal cortex (OFC), and anterior cingulate cortex (ACC) (Blood and Zatorre 2001; Brown et al. 2004; Menon and Levitin 2005; Koelsch et al. 2006). Similar findings were reported in two studies using paintings as stimuli, where subjective appreciation of works recruited the caudate, ACC, and OFC (Kawabata and Zeki 2004; Vartanian and Goel 2004; Kirk et al. 2008). In a study comparing the networks recruited during aesthetic rating as compared to symmetry judgments, Jacobsen and colleagues found enhanced activity in frontomedian cortex, lateral OFC, inferior frontal gyrus, posterior cingulate, temporal pole, and temporal-parietal junction (Jacobsen et al. 2006). Each of these studies provided significant clues to components of the neural networks underlying aesthetic rating. Kawabata and Zeki's (2004) breakdown of the elements of appreciation into comparatives (determined by a rigorous analysis of behavioral ratings) produced a set of suggestive, if not conclusive, results: when stimuli rated as beautiful versus neutral were compared, activity in medial orbitofrontal cortex (OFC), ACC, and left parietal cortex correlated with such ratings; when stimuli rated as ugly versus beautiful were contrasted, primary motor cortex was activated; when ugly versus neutral stimuli were compared, no activation was observed. In the case of the rating of beautiful versus ugly, medial OFC seemed to be elevated (Kawabata and Zeki 2004). While the sample size here was small and rather vaguely defined (five male and five female students "with no special experience in painting or art theory"), the finding of activity in medial OFC during aesthetic evaluation of beautiful paintings (as rated by subjects) was suggestive enough to use as a base for further experiments that merge with findings from the area of decision making.

Early on, Damasio and colleagues had shown that lesions in medial OFC/VMPFC often lead to decision-making deficits (Bechara et al. 2000; Damasio 1994). The role of VMPFC in emotional appraisal, as in the case of valuation of emotional valence of facial expressions (e.g. Aharon et al. 2001; Winston et al. 2007) also proves to be critical for further understanding of the subjective neural processes involved in decision making—as well, it turns out, for the understanding of the processes underlying the rating of visual works of art and the pressures and biases involved in doing so.

8.2 Ventromedial prefrontal cortex involvement in evaluation

In this section we describe a group of experiments arising out of recent interest in understanding the subjective processes involved in decision making (Montague and Berns 2002). These experiments are unified by the theme of hedonic or affective components in the aesthetic rating of a visual work of art.

Kawabata and Zeki's research showing that the recruitment of medial OFC in the rating of works as beautiful as opposed to ugly (Kawabata and Zeki 2004) suggested the possibility of examining the question of how preference ratings of works of art are modulated by semantic information presented alongside them. Then Kirk and colleagues designed an fMRI study to measure the influence of cognitive information on the modulation of subjective preference for visual works of art (Kirk et al. 2009a). In this study, chromatic reproductions of paintings from the Louisiana Museum of Modern Art in Copenhagen were shown to fourteen subjects with no formal training in any area of art. They were accompanied by labels below each of them, with half identified as from the Louisiana Museum ("gallery" label) and half as computer-generated by the experimenter ("computer" label). Pre-screening ensured that no subject had any familiarity with the paintings shown. Using this experimental design, the aim was to determine whether the contextual information supplied would affect the aesthetic evaluation and the subjective preference associated with it, either in terms of the behavioral ratings or of the neural processing underlying them. It was assumed that the gallery label would induce a higher expectation of reward than the computer label, and that this would be reflected both in the behavioral ratings and in the neural activity engaged in the processing of reward. The question was the degree to which subjective preference might be altered and modulated by framing its status as artistic or as merely computer generated. Whatever the direct sensory appeal of a work, the possibility remained that top-down influences would influence coherent behavioral preferences and recruit a specific neural network that differed from those involved in such appeal. The hypothesis was that one's concept of the status of the art object, rather than its sensory properties, would be the underlying determinant of one's subjective preference for one rather the other.

This, of course, would be consistent with the now long-standing claim in philosophical aesthetics that there are no inherent properties of an object that constitute it as art, and that its very constitution as art is predicated either on its institutional context or its philosophical status (which in many cases, particularly these days, merge) (Danto 1964). Often these are the only criteria, beyond general acclaim, by which works are defined as art or non-art. One might further maintain, consistently with the positions of many contemporary conceptual artists, that it is the labeling of works of art as art that enables them to be experienced as such, precisely as this experiment may seem to suggest. The neural data the experiment provides thus reinforce a central aspect of the new debate around the issue of how art is constituted.

What emerges clearly from this study is the influence of explicit contextual information on neural activity in the VMPFC. Given the role of the VMPFC in value computations and in experienced pleasure during experiential tasks (e.g. Anderson et al. 2003; O'Doherty et al. 2003; Rolls et al. 2003), it is perhaps not so surprising that it should be modulated by contextual information (Plassmann et al. 2008; McClure et al. 2004; de Araujo et al. 2005). Research in the field of neuromarketing and marketing actions further clarifies VMPFC activation during decision making. Moreover, the fact that activity in this area is modulated by contextual information also emerges, for example, in experiments in which subjects evaluate their preference for liquids or odors. One study (de Araujo et al. 2005) showed that subjects rated a test odor as significantly more pleasant when paired with a pleasant verbal descriptor than with an unpleasant one, and that this correlated with activity in the VMPFC. High-level cognitive input such as word labels clearly influence brain activity in the VMPFC, which plays a critical role in experienced pleasantness. Another study (McClure et al. 2004) investigated the neural systems involved in generating preferences produced by two different brands of soft drinks. They found that the rated preferences for unlabeled drinks (i.e. without cognitive influences but based only on sensory information) correlated consistently with VMPFC activation, while brand knowledge produced a strong influence both on expressed ratings and on measured responses in hippocampus (as one might expect), midbrain, and DLPFC. We shall examine the implications of such DLPFC activation later in this chapter. On the other hand, Plassmann and colleagues recently showed, predictably, that wine ratings are higher when the monetary value of a wine is known than when it is unknown, and that there is a neural correlate of this effect in medial OFC/VMPFC (Plassmann et al. 2008).

In their examination of the neural correlates of contextual modulation of subjective preference for paintings, Kirk and colleagues found that aesthetic ratings were significantly higher for pictures labeled "gallery" than those labeled "computer" (Kirk et al. 2009a). This correlated with greater activation of VMPFC under the gallery than under the computer condition. Given the now well-established role of this area in the representation of experienced pleasure (e.g. Anderson et al. 2003; O'Doherty et al. 2003; Rolls et al. 2003), this is hardly surprising. The correlation of VMPFC activation with the higher aesthetic rating assigned to works under the gallery rather than the computer label coincides with current evidence that VMPFC activation represents the subjective preference response involved in the assignment of aesthetic preference to one stimulus over another. The results derived from this experiment also provide evidence that this representation holds even when the attributed preference ratings are modulated by cognitive and semantic input. Taken together, these findings showing modulation of VMPFC by contextual information suggests an interaction between sensory processes and top-down information in this region.

On the other hand, consideration of the neural effects of paintings in the gallery versus computer conditions regardless of aesthetic rating showed increased activation of bilateral parahippocampal gyrus, visual cortex, and bilateral temporal pole. The distinction between these two very different categories of areas of activation calls for comment. Some of the reasons for this difference are clear; others possibly less so. They turn out to be critical for the understanding not just of the cognitive modulation of aesthetic evaluation but also of the role of expertise in suppressing context-induced bias during aesthetic rating.

The study by de Araujo and colleagues of cognitive influences on odor preference showed greater activation in VMPFC when subjects made preference ratings of a set of odors manipulated with a positive word label rather than a negative one. Psychophysical studies had long established the relative inefficiency of olfactory discrimination in humans, to the extent that successful odor identification was known to be highly susceptible to factors such as familiarity and to the semantic connection between an odor and its name (Cain 1979). It is thus hardly surprising that de Araujo and colleagues should have found that verbal or semantic information had a strong influence on both perception and rating of odor attributes. That semantic labels can influence aesthetic and hedonic ratings of works of art is less well documented, however. Russell found that the introduction of title and artist's name upon second ratings of individual abstract and semi-abstract artworks produced an increase in hedonic value (Russell 2003). Kirk and colleagues' research extended this observation by showing that semantic labels (prestigious gallery versus computer generated) influenced subjective preference, even when there was no difference in the stimulus material (Kirk et al. 2009a). Here too they found a neural correlate in VMPFC for the behavioral evidence, just as De Araujo and colleagues had found (de Araujo et al. 2005).

Other studies have shown modulation of VMPC in response to objects of varying reward value. Erk and colleagues found that cultural objects such as sports cars versus small cars modulate reward circuits in male subjects—not only the dopaminergic reward circuitry in ventral striatum, but also in anterior cingulate and in median orbitofrontal cortex (in the usual regions overlapping with VMPFC) (Erk et al. 2002). VMPFC is strongly implicated in signaling basic appetitive aspects of reward. Blood oxygen-level-dependent (BOLD) signal changes in this region correlate with reward value (O'Doherty et al 2003). VMPFC has also been implicated in emotional processing with increased responses to rewarding outcomes (Lane 1997). The likelihood that it is engaged under conditions that require behavioral decision-making is consistent with its involvement in the integration of reward feedback for affective decision-making (Kringelbach et al. 2003; O'Doherty et al. 2003).

This is all very well, but how are these VMPFC results to be interpreted in relation to the question of the neural mechanisms involved in the aesthetic evaluation of visual images, and of art in particular?

McClure and colleagues' study showed that brand knowledge of soft drinks influenced activations in hippocampus, primary visual cortex, and dorsolateral prefrontal cortex, areas that have traditionally been regarded as having strong cognitive rather than flavor-related functionality (McClure et al. 2004). In 2009, Kirk and colleagues demonstrated bilateral activation of the entorhinal cortex in the gallery versus computer condition, irrespective of the actual aesthetic rating (Kirk et al. 2009a). This much stands to reason. The entorhinal cortex adjoins and is interconnected with the hippocampus and is activated in trials in which subjects correctly recollect contextual information compared to trials in which they do not (Cansino et al. 2002). Other findings suggest that midbrain dopaminergic systems involved in reward expectation directly modulate declarative memory formation in the hippocampus. This evidence is consistent with our initial hypothesis that it is the subject's conception of the image, rather than its sensory properties, that primarily/also determine its hedonic value.

This effect might arise from subjects' differing prior expectations of future reward—and thus of hedonic value—as evoked by the stimulus labels. One could thus account for two possible factors in the gallery versus computer-generated experiment. The first would loosely have to do with relative difference in prestige. Despite recent developments in all forms of computer-generated and cyber-art, by and large art galleries remain a more prestigious context both for the display of and as a source of artworks than the Internet. In this case, expectation or prediction of reward arises from a social prior. The more prestigious the gallery, the more competition to display works in it, and a sense of the artist's success may well be correlated with a higher degree of reward expectation. A second, closely related factor might be the greater monetary value attached to showing in a more prestigious than a less prestigious gallery—let alone the monetary value attached to a work that is merely said to be computer-generated, but is in fact a reproduction of a real painting (and not displayed in a more or less prestigious gallery). In many cases—especially outside the rarefied contexts of analysis by students of the history of art and critics—a simple social prior is that the more expensive the artwork, the greater the likelihood that it will have a higher hedonic value.

Cognitive inputs such as semantic word labels clearly have a profound influence on aesthetic rating. The data summarized above suggest that the entorhinal cortex and temporal pole are engaged during recollection of art-related and cultural information that may influence aesthetic preference during gallery conditions, while VMPFC is more involved in the generation of preference computations. These two systems do not appear to function independently of each other, but are interdependently modulated in such a way as to generate aesthetic preferences as induced by semantic context.

8.3 Neural correlates of aesthetic expertise

A related question is that of the degree to which aesthetic expertise modulates aesthetic evaluation. Many recent studies have demonstrated substantial and deep differences between experts and non-experts in their assessment of works of art, as well as in the neural substrates that underlie their preferences and evaluations (Calvo-Merino et al. 2005; Hekkert et al. 1996a, 1996b). In 1966, McWhinnie showed that subjects without art training generally prefer simple and symmetric visual elements, whereas those with training or a background in art tend to favor complex and asymmetric visual elements (McWhinnie 1966). In 1990, Smith and Melara concluded (with equal predictability) that music novices show a greater preference for syntactically prototypical chord progressions than do experts (Smith and Melara 1990). Forms of training (not always adequately specified) and degree of expertise are generally said to contribute to these differential preferences. One could always say—as is often claimed—that training in a particular art, and deeper knowledge, familiarity, and experience, enrich its perceived meaning, and it is usually held that experts incline to respond to aesthetic values (broadly taken) over purely sensory ones. One of the difficulties in the literature so far is that expertise is never clearly defined, nor is it clearly established in exactly what expertise is supposed to consist. Often the rather general supposition is made that experts are more likely than novices to take into consideration concepts of aesthetic value, or the concepts and ideas on which the work is based, or on the supposed (or prevailing) norms of good and bad taste (Smith and Melara 1990; Bourdieu 1979). Preferences expressed by experts are often thought to reflect a disposition for distancing oneself from popular taste (as exemplified by non-expert viewers). Often it is believed that aesthetic experience involves the processing, recognition, and broad understanding of an individual style. Some form of stylistic awareness, analysis, and recognition then becomes a critical marker of aesthetic expertise. As is well-known, "even highly abstract paintings can be constrained by rules, although the underlying principles are not immediately evident to those outside the artist's circle" (Cupchik and Laszlo 1992). There is no question that critics and art historians are frequently more aware of the preferred techniques and rules that distinguish one style from another, however, and that this affects their evaluation of the work under consideration.

An abundance of evidence points to direct expertise effects in the brain, especially in structures related to memory and perception—even on the macro-anatomical scale. In a striking voxel-based morphometric analysis, Maguire and colleagues found that gray-matter volume in the posterior hippocampus of London taxi drivers was greater than in age-matched controls, and seemed to increase correlatively with time spent taxi-driving (Maguire et al. 2000). Increasing numbers of studies have been devoted to the role of expertise effects in responses to music. In their fMRI study of 2006, Bangert and colleagues compared brain activity in groups of musicians and non-musicians as they listened passively to a piano sequence, and found elevated levels of activity in the musicians' brains in regions of the temporal lobe associated with auditory processing, as well as in frontal regions associated with motor control (Bangert et al. 2006). More recently, Jacobsen and others have taken a different approach to comparing aesthetic responses in experts and laypersons in analysing ERP effects in the course of auditory and cognitive processing of chord sequences (Brattico and Jacobsen 2009; cf. also Bigand and Poulin-Charronat 2006).

As already mentioned, a number of fMRI studies have identified cortical areas recruited during the aesthetic evaluation of pictures (Kawabata and Zeki 2004; Jacobsen et al. 2006; Kirk et al. 2008). The results suggest that the computation of preferences largely relies on areas implicated in the processing of reward, especially VMPFC and ACC. To what degree are these reward-related areas modulated by expertise? VMPFC has been found to correlate with subjective preference ratings, and to be involved in coding stimulus value from a variety of sensory modalities (Knutson et al. 2003; O'Doherty et al. 2003; Rolls et al. 2003; McClure et al. 2004; Plassmann et al. 2008). It serves as a critical center both for the tracking of reward value of different stimuli independent of their sensory modality (i.e. taste, smell, etc.), and for relating this value to hedonic experience.

Although it could be argued that the "rules" of art (say of twentieth- and twenty-first-century abstract art) are not effectively equivalent to the technical rules of art or architecture, a further neuroimaging study sought to investigate the extent to which the neural correlates of aesthetic evaluation vary as a function of expertise in architecture (Kirk et al. 2009b). This fMRI study cast considerable light on (1) the question of the impact of expertise on aesthetic preferences, and (2) how differences in assessment strategies between experts and non-experts can be tracked as differences in neural activity. The subjects here consisted of one group professing to have no great interest or expertise in art or architecture, and another group of experts, consisting of graduate students in architecture and professional architects. Each group was asked to rate the aesthetic value of a series of images of architecture (divided between modernist and pre-modernist, public and private) and faces. Aesthetic value was rated on a scale of 1 (very unappealing) to 5 (very appealing) (Kirk et al. 2009b). It was hypothesized that the expert-specific condition (architectural images) would significantly affect both aesthetic ratings and neural activity differentially. Since earlier psychometric studies had found that people in different cultures and of both sexes tend to agree as to which faces were attractive (Langlois et al. 2000), it was predicted that there would be little if any significant differentiation between the two groups' aesthetic ratings of faces and the neural processing underlying such ratings. In other words, it seemed likely that an expertise effect on the neural structures underlying the formation of an aesthetic judgment would be evident when architects rated the buildings, but not when judging faces.

The behavioral responses (i.e. the aesthetic ratings) for both categories of images revealed no significant differences between the two groups. To test whether architectural expertise modulated brain activity underlying aesthetic preference, an analysis was made of those regions where the difference between the responses for the two stimulus conditions varied across the two subject groups. Significant activations were found in VMPFC and bilateral ACC, but were greater in the case of experts' preference ratings of buildings, while remaining essentially balanced across the two groups in the case of face stimuli. The observed effect clearly related to acquired expertise in architecture. Nucleus accumbens (NAcc) showed the same level of response in both groups and to both sets of stimuli, regardless of expertise level. This finding thus replicated previous findings that VPMFC and striatal regions, including the NAcc, play different functional roles in reward processing (Knutson et al. 2001; O'Doherty et al. 2003). A plausible conclusion would be that VMPFC processes evaluation of stimuli, whereas the nucleus accumbens is more involved in prediction of reward and reward expectancy, as has often been shown (Montague et al. 2004).

Nevertheless the fact that the VMPFC is sensitive to the magnitude of aesthetic value accords with recent studies showing that the relative reward value of stimuli is reflected by the amplitude of neural activity in this region, whereby we mean that activity in the VMPFC increases linearly with aesthetic ratings. These studies have also indicated that this region codes the reward-related rather than the sensory aspects of a stimulus (Kringelbach et al. 2003).

One may go a step further. The data from the study of architectural experts versus novices in rating buildings and faces shows that architectural expertise modulates the neural response to buildings even in the absence of any differences in aesthetic rating between experts and non-experts. This further confirms what may seem obvious; namely, that the expertise effect is specific to the domain of expertise.

In conclusion, the results of Kirk and colleagues' study of architectural expertise showed that VMPFC and ACC are differentially engaged as a function of expertise, but that NAcc is equally activated in both experts and non-experts.

8.4 The influence of favors on valuation and decision making

For some time it has been known that decision making can be biased by the offering of favors, or by particular social gestures (Loewenstein et al. 1993). The tendency to be influenced by biases may be rooted in biological mechanisms that subvert cognitive control. In real-life scenarios such biases occur where social gestures from a sender to a receiver encourage some equivalent behavior in return, even if the agreement is not explicit. This feature of response is known as reciprocity. It is clear that in many cases open-ended gestures may be made without any explicit expectation of reciprocity (for recent discussions of the neural mechanisms of reciprocity-eliciting gestures; see Rilling et al. 2002; King-Casas et al. 2005; Li et al. 2009), and it is in this domain that the whole question of how social gestures may manipulate value judgments arises. So too does the still broader question of how "open-loop" favors, in which there is no possibility of reciprocating interactions between sender and receiver, affect decision making and therefore preference judgments. Such cases would seem to be particularly relevant to the case of abstract rewards such as might arise from preference judgments for one painting over another.

The use of monetary favors to influence preference offers particularly striking real-world examples of the manipulation of value judgment by perceived external social value. The media often alleges manipulation of drug choice by monetary favors (cf. also Wazana 2000; Dana and Loewenstein 2003). Harvey and colleagues hypothesized that a monetary favor would affect subjects' preference ratings of works of art (Harvey et al. 2010). It may seem that this sort of favor is a specific and exceptional instance of the external motivation of valuation judgments and rating, but in fact it offers an interesting and critical test case where there is no objectively correct answer, and where the favor is entirely of the "open loop" kind, with no possibility of a reciprocal interaction between sender and receiver. Harvey and colleagues' experiment also makes clear not just that monetary favor biases judgment in domains unrelated to the favor, but also that the neural substrate of the brain modulations induced by monetary favor is not separate from the networks activated by preference judgments more generally.

Furthermore, while monetary favors may not be thought to be particularly relevant to the case of art, in the present art world it very clearly is (and may always have been). The offer of monetary favors stands as exemplary for a wide range of potential biasing factors in judgments about and ratings of works of art. The case seems particularly applicable to the operations of the art market, but in fact could be seen to apply to any other form of social gesture that elicits reciprocity—for example, the use of pictures as forms of marriage solicitations or proposals (a phenomenon that has a long history), or any other such reciprocal arrangement involving supposed beauty or social benefit. Recent experiments have begun to shed light on the neural mechanisms of reciprocity-eliciting gestures (King-Casas et al. 2005; Rilling et al. 2002; Li et al. 2009; van den Bos et al. 2009), but almost nothing is known about the influence of an "open-loop" favor where there is no possibility of reciprocal interactions between the sender and the receiver, or in which one agent makes a gesture or offers a gift without any explicit expectation of reciprocity.

Harvey and colleagues hypothesized that the offer of social favors would bias subjective preferences, even for objects unrelated to the favor itself (as in the case of works of art). They addressed this hypothesis by employing a series of monetary favor and preference experiments, in which a total of 151 subjects participated. Before scanning subjects were told that that they would be sponsored for their participation by one of two companies. At the beginning of the fMRI task they were shown two company logos followed by a screen showing which of the companies would be sponsoring them, as well as the amount of compensation ($30, $100, or $300). During scanning, subjects passively viewed 60 reproductions of paintings, each paired with either a sponsor logo or a non-sponsor logo. After scanning they provided preference ratings for all the pictures they viewed.

In other words, in this experiment a monetary favor was sent from an agent to a subject and the influence of this favor on subjects' preferences for one image over another was tested. This is a domain in which there are no objectively "correct" answers for such preferences, and there is no economic relationship between the favor and the preference judgments. What was striking about this set of experiments is that the mere offer of a favor was able to influence subjective decision making, even without actually paying the favor out. Responses both in VMPFC and posterior cingulate correlated linearly with the behavioral change across the three offer conditions (Harvey et al. 2010).

The behavioral results showed that preference for paintings paired with the sponsoring company's logo was higher than those paired with the non-sponsor logo. The functional imaging results revealed that activity in the VMPFC tracked the value signals of the paintings belonging to the sponsorship category to a higher degree than those that did not thus belong. This really was the first imaging study to indicate that a monetary favor could transfer value to a proxy, and could influence preference judgment of objects that were placed next to the sponsorship logo, with corresponding changes in neural activity. Moreover, the experiment was designed with no reciprocal interaction between the company and the receiver, suggesting that the favor could influence behavior even when the receiver had no explicit expectation of reward in the outcome of the interaction.

What is critical about these results is that they show that the sponsorship effects described do not possess a special and separate network of brain responses but instead modulate responses in neural networks normally activated by a wide range of preference judgments. VMPFC encodes for revealed preference across a whole variety of sensory modalities (Knutson et al. 2003; O'Doherty et al. 2003; Rolls et al. 2003; McClure et al. 2004; Plassman et al. 2008; Kirk et al. 2009a). Activity within VMPFC suggested that the neural networks normally activated by a wide range of preference judgments are also modulated by the strong effect of sponsorship. The effect is so robust that it is not sensitive to changes such as logo size or distance from the stimuli themselves. The results raise the important possibility that monetary favors bias judgments by acting through existing valuation mechanisms, and that individuals may therefore have difficulty detecting the gesture's influence over their subjective preferences—even for objects seemingly unrelated to the favor.

8.5 Mitigation of the effects of favors by domain expertise

In follow-up experiments to those of Harvey and colleagues, the same team examined whether the biasing effects of social gestures on valuation judgments could be mitigated (Kirk et al. 2011). In the case of medical professionals, for example, biases from monetary gifts or other favors from pharmaceutical companies are thought to be mitigated by a variety of mechanisms, including disclosure of potential conflicts of interest, not accepting large gifts or favors, and oversight from institutions regarding biases in prescribing behavior. In addition, the fact that medical professionals have expertise in their domain of decision-making is taken as an argument that they should be more objective in their judgments than laypersons making similar decisions. The obvious question for the field of art is the degree to which expertise and experience in art insulates against judgment bias. To test this directly, a group of participants with expertise in the domain of art were recruited to perform the favor task outlined in the previous section. Expertise criteria included a formal education in a visual art-related area and a minimum of five years of experience working in visual art-related areas. The hypothesis was that training in assessing art would insulate such subjects against the biasing gesture of the sponsoring company, and that if this were indeed the case, there would be neurobiological correlates corresponding to the differences in behavior between art experts and non-experts.

The group of experts was asked to undergo an fMRI scan while viewing images of the artwork, paired either with a sponsor company logo or another non-sponsor company logo. After the fMRI session, each subject rated his or her preference for the art displayed in the scanner. The set-up was identical to the previously discussed experiments, except that student art was used rather than well-known paintings, to eliminate the possibility that differences in results between art experts and non-experts were solely a result of familiarity with the paintings. In the behavioral results, there was no difference between experts' preference for sponsored paintings versus non-sponsored paintings, in contrast to the control subjects who showed a significantly higher preference for sponsored paintings. It was hypothesized that if the experts did not show an effect of sponsorship on preference ratings for paintings, BOLD activity in the VMPFC would also not be sensitive to sponsorship in the way it was in the control subjects (whose behavior reflected the bias towards sponsor-related paintings).

This turned out to be the case: while both the control group and the art experts showed activity in the VMPFC correlating linearly with painting preference for all paintings, the experts did not show differential activity in this region for sponsored compared to non-sponsored paintings. However, if the group of art experts showed no effect of sponsorship either on their behavior or on activity in the VMPFC, what region of the brain was involved in the mitigation of this effect?

To explore this further, activity in the dorsolateral prefrontal cortex (DLPFC) was analysed. DLPFC is known to be involved in executive control (Wagner et al. 2001) and in modulation of valuation (in experiments on goal values and the influence of instruction on reward learning) (Hare et al. 2009; Li et al. 2011). The results showed that relative to the control group, viewing artworks elevated activity in the DLPFC, whereas the control group did not show activity in this region. These results indicate that in the expert group, DLPFC was continuously engaged in the regulation of bias susceptibility. In subsequent analyses, right DLPFC turned out to be functionally connected to the VMPFC, and the coupling of these two regions was stronger during presentation of sponsored paintings than during presentation of non-sponsored paintings.

As recent studies of the neurobiology of self-control have also shown, VMPFC may indeed be modulated by areas of the DLPFC, especially in instances of self-censorship (Hare et al. 2009). It is clear that DLPFC plays a critical role in self-censoring modulations of neural valuation mechanisms such as the VMPFC. It also serves as a more general mechanism by which a person may be insulated from bias and from biasing maneuvers, rather than being specific to individuals with domain expertise, whether in art or in any other field. What does emerge, however, is that in experts or in those with training in art, DLPFC more strongly modulates VMPFC activation in such a way as to insulate them from the kinds of biases that are generated by contextual information and by reward prospects, whether financial or sensory, or by any other of the kinds of elements that attract both non-experts and experts to a work in the first place. The importance of such results for the ways in which experts are insulated from biasing effects (such as those of possible financial gain, or even of prestige effects) cannot be overestimated.

References

Adolphs, R. (1999). The human amygdala and emotion. *Neuroscientist* **5**, 125–37.

Adolphs, R., Tranel, D., Damasio, H., et al. (1995). Fear and the human amygdala. *Journal of Neuroscience* **15**, 5879–91.

Aharon, I., Etcoff, N., Ariely, D., et al. (2001). Beautiful faces have variable reward value: fMRI and behavioural evidence. *Neuron* **32**, 537–51.

Anderson, A.K., Christoff, K., Stappen, I., et al. (2003). Dissociated neural representations of intensity and valence in human olfaction. *Nature Neuroscience* **6**, 196–202.

Astafiev, S.V., Stanley, C.M., Shulman, G.L., et al. (2004). Extrastriate body area in human occipital cortex responds to the performance of motor actions. *Nature Neuroscience* **7**, 542–8.

Bangert, M., Peschel, T., Schlaug, G., et al. (2006). Shared network for auditory and motor processing in professional pianists: evidence from fMRI conjunction. *Neuroimage* **15**, 917–26.

Battaglia, F., Lisanby, S., and Freedberg, D. (2011). Corticomotor excitability during observation and imagination of a work of art. *Frontiers in Neuroscience* **5**, 1–6.

Baumgarten, A.G. (1986/1750). *Aesthetica*. Hildesheim: Olms.

Bechara, A., Damasio, H., and Damasio, A.R. (2000). Descision making and the orbitalfrontal cortex. *Cerebral Cortex* **10**, 295–307.

Bigand, E. and Poulin-Charronat, B. (2006). Are we "experienced listeners"? A review of musical capacities that do not depend on formal musical training. *Cognition* **100**, 100–30.

Blood, A.J. and Zatorre, R.J. (2001). Intensely pleasurable responses to music correlate with activity in brain regions implicated in reward and emotion. *Proceedings of the National Academy of Sciences of the USA* **98**, 11818–23.

Bourdieu, P. (1979). *Distinction: A social critique of the judgment of taste*. Cambridge, MA: Harvard University Press.

Brattico, E. and Jacobsen, T. (2009). Subjective appraisal of music: neuoimaging evidence. *Annals of the New York Academy of Sciences* **1169**, 308–17.

Brown, S., Martinez, M.J., and Parsons, L.M. (2004). Passive music listening spontaneously engages limbic and paralimbic systems. *NeuroReport* **15**, 2033–7.

Cain, W.S. (1979). To know with the nose: keys to odor identification. *Science* **203**, 467–70.

Calvo-Merino, B., Glaser, D.E., Grèzes, J., et al. (2005). Action observation and acquired motor skills: an fMRI study with expert dancers. *Cerebral Cortex* Aug **15**(8), 1243–9.

Cansino, S., Maquet, P., Dolan, R.J., et al. (2002). Brain activity underlying encoding and retrieval of source memory. *Cerebral Cortex* **12**, 1048–56.

Cela-Conde, C.J. Marty, G., Maestú, F., et al. (2004). Activation of the prefrontal cortex in the human visual aesthetic perception. *Proceedings of the National Academy of Sciences of the USA* **101**, 6321–5.

Cupchik, G. and Laszlo, J. (1992). *Emerging visions of the aesthetic process: Psychology, semiology, and philosophy*. New York, NY: Cambridge University Press.

Cupchik, G., Vartanian, O., Crawley, A., et al. (2009). Viewing artworks: contributions of cognitive control and perceptual facilitation to aesthetic experience. *Brain Cognition* **70**, 84–91.

Damasio, A. (1994). *Descartes' Error: Emotion, Reason and the Human Brain*. New York, NY: G.P. Putnam.

Dana, J. and Loewenstein, G. (2003). A social science perspective on gifts to physicians from industry. *JAMA* **290**, 252–5.

Danto, A. (1964). The Artworld. *Journal of Philosophy* **61**, 571–84.

de Araujo, I.E., Rolls, E.T., Velazco, M.I., et al. (2005). Cognitive modulation of olfactory processing. *Neuron* **46**, 671–9.

De Gelder, B. (2006). Toward the neurobiology of emotional body language. *Nature Reviews Neuroscience* **7**, 242–9.

Decety, J. and Sommerville, J. (2003). Shared representations between self and other: a social cognitive neuroscience view. *Trends in Cognitive Sciences* **7**, 527–32.

Di Dio, C., Macaluso, E., and Rizzolatti, G. (2007). The golden beauty: brain response to classical and renaissance sculptures. *PLoS One* **2(11)**, 2–27.

Downing, P.E., Peelen, M.V., Wiggett, A.J., et al. (2006). The role of the extrastriate body area in action perception. *Society for Neuroscience* **1**, 52–62.

Erk, S., Spitzer, M., Wunderlich, A.P., et al. (2002). Cultural objects modulate reward circuitry. *Neuroreport* **13**, 2499–503.

Freedberg, D. (1989). *The Power of Images. Studies in the History and Theory of Response*. Chicago, IL: Chicago University Press.

Freedberg, D. (2007). Empathy, motion and emotion. In K. Herding and A. Kraus-Wahl (eds), *Wie die Gefühle Ausdruck verschaffen*: Emotionen in Nahsicht. Berlin: Driesen, pp. 17–51.

Freedberg, D. (2009). Choirs of Praise: Some Aspects of Action Understanding in Fifteenth Century Painting and Sculpture. In D. Levine and J. Freiberg (eds), *Medieval Renaissance Baroque: A Cat's Cradle for Marilyn Aronberg Lavin*. New York, NY: Italica Press, pp. 65–81.

Freedberg, D. and Gallese, V. (2007). Motion, emotion and empathy in aesthetic experience. *Trends in Cognitive Sciences* **11**, 197–203.

Gallese, V., Keysers, C., and Rizzolatti, G. (2004). A unifying view of the basis of social cognition. *Trends in Cognitive Sciences* **8**, 396–403.

Hare, T.A., Camerer, C.F., and Rangel, A. (2009). Self-control in decision-making involves modulation of the VMPFC valuation system. *Science* **324**, 646–8.

Harvey, A.H., Kirk, U., Denfield, G.H., et al. (2010). Monetary favors and their influence on neural responses and revealed preference. *Journal of Neuroscience* **30**, 9597–602.

Hekkert, P. and van Wieringen, P.C.W. (1996a). The impact of level of expertise on the evaluation of original and altered versions of post-impressionistic paintings. *Acta Psychologia* **94**, 117–31.

Hekkert, P. and van Wieringen, P.C.W. (1996b). Beauty in the eye of expert and nonexpert beholders: a study in the appraisal of art. *American Journal of Psychology* **109**, 389–407.

Jacobsen, T., Schubotz, R.I., Höfel, L., et al. (2006). Brain correlates of aesthetic judgment of beauty. *Neuroimage* **29**, 276–85.

James, W. (1890). *The Principles of Psychology*. New York, NY: H. Holt.

Kawabata, H. and Zeki, S. (2004). Neural correlates of beauty. *Journal of Neurophysiology* **91**, 1699–705.

Keysers, C., Wicker, B., Gazzola, V., et al. (2004). A touching sight: SII/PV activation during the observation and experience of touch. *Neuron* **42**, 335–46.

King-Casas, B., Tomlin, D.A., Anen, C., et al. (2005). Getting to know you: reputation and trust in a two-person economic exchange. *Science* **308**, 78–83.

Kirk, U. (2008). The neural basis of object-context relationships on aesthetic judgment. *PLoS One* **3**, 11.

Kirk, U., Harvey, A.H., and Montague, P.R. (2011). Domain expertise insulates against judgment bias by monetary favors through a modulation of ventromedial prefrontal cortex. *Proceedings of the National Academy of Sciences* **108**, 10332–6.

Kirk, U., Skov, M., Hulme, O., et al. (2009a). Modulation of aesthetic value by semantic context: an fMRI study. *Neuroimage* **44**, 1125–32.

Kirk, U., Skov, M., Nygaard, N., et al. (2009b). Brain correlates of aesthetic expertise: a parametric fMRI study. *Brain and Cognition* **69**, 306–15.

Knutson, B., Fong, G.W., Bennett, S.M., et al. (2003). A region of mesial prefrontal cortex tracks monetarily rewarding outcomes: characterization with rapid event-related fMRI. *Neuroimage* **18**, 263–72.

Koelsch, S., Fritz, T., von Cramon, D.Y., et al. (2006). Investigating emotion with music: an fMRI study. *Human Brain Mapping* **27**, 239–50.

Kringelbach, M.L., O'Doherty, J.P., Rolls, E.T., et al. (2003). Activation of the human orbitofrontal cortex to a liquid food stimulus is correlated with its subjective pleasantness. *Cerebral Cortex* **13**, 1064–71.

Lane, R.D. (1997). Neuroanatomical correlates of pleasant and unpleasant emotion. *Neuropsychologia* **35**, 1437–44.

Lang, P.J., Greenwald, M.K., Bradley, M.M., et al. (1993). Looking at pictures: affective, facial, visceral, and behavioral reactions. *Psychophysiology* **30**, 261–73.

Langlois, J.H., Rubenstein, A.J., Larson, A., et al. (2000). Maxims or myths of beauty? A meta-analytic and theoretical review. *Psychological Bulletin* **126**, 390–423.

LeDoux, J. (1992). Emotion and the Amygdala. In J.P. Aggleton (ed.), *The Amygdala: Neurobiological Aspects of Emotion, Memory and Mental Dysfunction*. New York, NY: Wylie Liss, pp. 339–51.

LeDoux, J.E. (1996). *The Emotional Brain: The Mysterious Underpinnings of Emotional Life*. New York, NY: Simon & Schuster.

Leder, H., Belke, B., Oeberst, A., et al. (2004). A model of aesthetic appreciation and aesthetic judgments. *British Journal of Psychology* **95**, 489–508.

Li, J., Delgado, M.R., and Phelps, E.A. (2011). How instructed knowledge modulates the neural systems of reward learning. *Proceedings of the National Academy of Sciences of the USA* **108**, 55–60.

Li, J., Xiao, E., Houser, D., et al. (2009). Neural responses to sanction threats in two party economic exchange. *Proceedings of the National Academy of Sciences of the USA* 16835–40.

Lipps, T. (1903a). Einfühlung, innere Nachahmung und Organenempfindungen. *Archiv für die Gesamte Psychologie, I* **2–3**, 185–204.

Lipps, T. (1903b). *Aesthetik: Psychologie des Schönen und der Kunst*. Hamburg and Leipzig: L. Voss.

Loewenstein, G., Issacharoff, S., Camerer, C., et al. (1993). Self-serving assessments of fairness and pretrial bargaining. *Journal of Legal Studies* **22**, 135–59.

Maguire, E.A., Gadian, D.G., Johnsrude, I.S., et al. (2000). Navigation-related structural change in the hippocampi of taxi drivers. *Proceedings of the National Academy of Sciences of the USA* **97**, 4414–16.

McClure, S.M., Li, J., Tomlin, D., et al. (2004). Neural correlates of behavioural preference for culturally familiar drinks. *Neuron* **44**, 379–87.

McWhinnie, H.J. (1966). Effects of a learning experience on preference for complexity and asymmetry. *Perceptual and Motor Skills* **23**, 119–22.

Menon, V. and Levitin, D.J. (2005). The rewards of music listening: response and physiological connectivity of the mesolimbic system. *Neuroimage* **28**, 175–84.

Merleau Ponty, M. (1945). *Phénoménologie de la perception*. Paris: Gallimard.

Montague, P.R. and Berns, G.S. (2002). Neural economics and the biological substrates of valuation. *Neuron* **36**, 265–84.

Montague, P.R., Hyman, S., and Cohen, J.D. (2004). Computational roles for dopamine in behavioural control. *Nature* **431**, 760–7.

O'Doherty, J.P., Critchley, H., Deichmann, R., et al. (2003). Dissociating valence of outcome from behavioral control in human orbital and ventral prefrontal cortices. *Journal of Neuroscience* **23**, 7931–39.

Plassmann, H.O., Doherty, J., Shiv, B., et al. (2008). Marketing actions can modulate neural representations of experienced pleasantness. *Proceedings of the National Academy of Sciences* **105**, 1050–54.

Prinz, W. (1997). Perception and action planning. *European Journal of Cognitive Psychology* **9**, 129–54.

Rilling, J.K., Gutman, D.A., Zeh, T.R., et al. (2002). A neural basis for social cooperation. *Neuron* **35**, 395–405.

Rizzolatti, G., Fadiga, L., Gallese, V., et al. (1996). Premotor cortex and the recognition of motor actions. *Cognitive Brain Research* **3**, 131–41.

Rizzolatti, G., Fogassi, L., and Gallese, V. (2001). Neurophysiological mechanisms underlying the understanding and imitation of action. *Nature Reviews Neuroscience* **2**, 661–70.

Rolls, E.T., Kringelbach, M.L., and de Araujo, I.T. (2003). Different representations of pleasant and unpleasant odours in the human brain. *European Journal of Neuroscience* **18**, 695–703.

Russell, P.A. (2003). Effort after meaning and the hedonic value of paintings. *British Journal of Psychology* **94**, 99–110.

Smith, J.D. and Melara, R.J. (1990). Aesthetic preference and syntactic prototypicality in music. *Cognition* **34**, 279–98.

Umiltà, M.A., Berchio, C., Sestito, M., et al. (2012). Cortical motor activation evoked by static images of abstract art: An EEG study. *Frontiers in Human Neuroscience* **6(311)**, 1–9.

van den Bos, W., van Dijk, E., Westenberg, M., et al. (2009). What motivates repayment? Neural correlates of reciprocity in the Trust Game. *Social Cognitive and Affective Neuroscience* **4**, 294–304.

Vartanian, O. and Goel, V. (2004). Neuroanatomical correlates of aesthetic preference for paintings. *NeuroReport* **15**, 893–7.

Vischer, R. (1873). *Ueber das optische Formgefühl; ein Beitrag zur Aesthetik*. Leipzig: H. Credner.

Wagner, A.D., Maril, A., Bjork, R.A., et al. (2001). Prefrontal contributions to executive control: fMRI evidence for functional distinctions within lateral prefrontal cortex. *Neuroimage* **14**, 1337–47.

Wazana, A. (2000). Physicians and the pharmaceutical industry: is a gift ever just a gift? *JAMA* **283**, 373–80.

Winston, J.S., O'Doherty, J., Kilner, J.M., et al. (2007). Brain systems for assessing facial attractiveness. *Neuropsychologia* **45**, 195–206.

Chapter 9

Neuroimaging studies of making aesthetic products

Oshin A. Vartanian

The fields of empirical aesthetics and creativity have largely advanced in parallel. As a result, although we possess fairly sophisticated models that explain how creative production (Gonen-Yaacovi et al. 2013) and aesthetic appreciation (Nadal 2013) are realized in the brain, little is known about how these two processes intersect—psychologically or neurologically (Vartanian 2014). This is a peculiar state of affairs because when artists create a work of art, they are concerned about its eventual appreciation by potential perceivers. Similarly, when perceivers experience artworks, they make inferences about the creative processes that led to their production. Therefore, it would appear that a comprehensive model of the artistic process would necessitate not only models of how creative production and aesthetic appreciation unfold, but also an overarching model that relates the two aspects together.

As the content of this volume makes readily apparent, neuroimaging data have contributed more to our understanding of the neural bases of aesthetic appreciation than they have to our understanding of the neural bases of artistic production. The reasons for this are primarily methodological. Specifically, in the early days of human neuroimaging research few methods existed to enable the collection of ecologically valid data in the scanner in various domains of artistic production. Fortunately, technological advances in recent years have begun to change this landscape (see Tam et al. 2011), and some data are now available on the neural systems that underlie the production of aesthetic products in the scanner.

Because the neurobiological bases of aesthetic and artistic appreciation as well as neuropsychological studies of artistic production will have been reviewed in considerable detail by various contributors to the present volume, my focus here will be on neuroimaging studies of creative production in the domains of music (Limb and Braun 2008), writing (Shah et al. 2013), and drawing (Ellamil et al. 2012). Although there have certainly been studies of the neural bases of other aesthetic behaviors—most notably dance (Brown et al. 2006; see also Brown and Parsons 2008)—the focus of the present chapter will be on studies that zeroed in on the neural bases of the *creative* aspects of generating aesthetic products.

My focus here on the creation of aesthetic products (rather than their appreciation) is motivated by the overarching idea that the creation and appreciation of artworks represent two sides of the same coin, and are intimately linked (see Zeki 1999). Therefore, our neural models of how people respond to artworks will be augmented by a better understanding of neural models of creative output, and vice versa. I will close the chapter by reviewing Tinio's (2013) *Mirror Model of Art*, a very promising theoretical model that provides a testable framework that conjoins the creative and aesthetic aspects of art. It is my hope that the introduction of this model will motivate research geared toward bringing together these two facets of the study of art.

9.1 Creative drawing

One of the most enduring distinctions in the literature on creativity involves its two phases: blind variation (i.e. *generation*) and selective retention (i.e. *evaluation*) of ideas (Campbell 1960). According to the classic interpretation of this distinction, the generative phase of the creative process is not goal-oriented but instead is rather "blind." In other words, the person does a *random* search through the contents of consciousness to initiate the process, including the combination of previously unrelated ideas and concepts and ideas to form new associations (Poincaré 1913). Although many scholars have since questioned the assumption that this process is truly random (Eysenck 1993; Simonton 2011), most agree that the generative phase of the process is where novel combinations of ideas are most likely to occur. In this sense, the generative phase has held much more allure for creativity researchers than the evaluative phase, with relatively less research devoted to the cognitive processes that underlie evaluation (but see Vartanian et al. 2009).

Neuroscientific studies of creative problem solving have also attempted to isolate the neural bases of generation versus evaluation. For example, Goel and Vartanian (2005) demonstrated that the right ventral lateral prefrontal cortex (PFC) and left dorsal lateral PFC were activated to a relatively greater extent in the generation versus evaluation of matchstick problems, a well-established paradigm for studying divergent thinking (Guilford 1967). Furthermore, additional analyses demonstrated that the activation of right ventral lateral PFC could be attributed to its involvement in "set shifting" or "lateral transformation" processes (i.e. breaking of conceptual or perceptual constraints imposed on the task). This interpretation is supported by a large body of neuropsychological and neuroimaging data implicating the right PFC in the generative phase of creative problem-solving tasks (e.g. Miller and Tippett 1996), ill-structured design tasks (e.g. Goel and Grafman 2000), and well-structured problem solving tasks (e.g. Newman et al. 2003). Thus, although the neural basis of creativity is known to be distributed in the brain (Gonen-Yaacovi et al. 2013), the lateral aspects of right PFC appear to play an important role in the generative phase of creative problem solving, especially in cases where this involves breaking the constraints imposed on the problem.

Ellamil and colleagues (2012) implemented an ingenious experiment in a functional magnetic resonance imaging (fMRI) scanner to dissociate the generative and evaluative phases of creative drawing. This was made possible by their use of an MRI-compatible tablet (Tam et al. 2011), enabling the participants to draw on the digital device. Importantly, participants were undergraduates from the Emily Carr University of Art and Design. In other words, they had training and experience in the visual arts. The experiment consisted of two key blocks (as well as additional baseline blocks). The "generate" block began with the visual presentation (using words) of book descriptions involving abstract concepts (e.g. war, immigration, religion), and participants were instructed to draw (and write) their ideas for a corresponding book cover using the MRI-compatible tablet. Then, in the "evaluate" block, they were instructed to draw (and write) their evaluations of their book cover ideas, again using the MRI-compatible tablet. Following the termination of a specific number of generate–evaluate cycles the participants were prompted to talk about their assessments of how well they were able to engage in generation and evaluation of these book covers.

Here I will focus on a subset of the findings most relevant for our purposes. The generate–evaluate contrast (what is unique about generation over and above evaluation) revealed greater activation extensively in the medial temporal lobes (MTL) bilaterally, including the hippocampus and the parahippocampus. In addition, activations were observed bilaterally in the inferior and superior parietal lobule, fusiform gyrus, middle temporal gyrus, and the premotor area, as well as left inferior frontal gyrus and cerebellum. In contrast, evaluate–generate contrast (what is unique about evalutaion over and above generation) revealed greater activation in a large distributed network including regions within the executive network such as dorsal anterior cingulate cortex (ACC) and bilateral dorsal lateral PFC, regions within the default network including medial PFC, posterior cingulate cortex (bordering on precuneus), and right temporoparietal junction, and bilateral rostrolateral PFC, bilateral cerebellum, bilateral temporopolar cortex, left anterior insula, supplementary motor area, bilateral inferior frontal gyrus, bilateral superior parietal lobule, bilateral middle temporal gyrus, bilateral lingual gyrus, bilateral middle occipital gyrus, and bilateral cuneus. In addition, based on their subjective reports, for each phase the investigators assessed the degree of correlation between the activated regions and success in the respective task. The analyses revealed that in the generate condition, activation in MTL regions including the hippocampus and parahippocampus was correlated with perceived success, whereas in the evaluate condition, activation in executive function and default network regions was correlated with perceived success.

The results revealed that MTL regions—specifically the hippocampus and parahippocampus—contribute to the generation of ideas during creative drawing. The involvement of MTL can be attributed to at least two sources. First, MTL is known to play an important role in the formation and retrieval of semantic and episodic associations (Henke et al. 1999), and its involvement here is consistent with the role of associative processes in creative cognition (Mednick 1962). However, another intriguing possibility is suggested by the composition of the sample. Remember that the participants recruited in this study were not new to drawing. In fact, they were university students majoring in the visual arts. As such, it might also be that the activation observed in MTL during idea generation might be due to the mnemonic recall of relevant information stored in long-term memory—also a well-established function of MTL. Importantly, these two possibilities are not mutually exclusive, and point to the important role of MTL in the generation phase of creative drawing, at least in trained and experienced participants.

Compared to the generative phase of creative drawing, the evaluative phase was associated with activations in executive function and default network regions, among others. The involvement of the executive function region is not surprising, given that cognitive control in the service of analytic processing would appear to play an important role in judging the quality of generated products (Dosenbach et al. 2008). By contrast, at first glance the involvement of the default network region is surprising. However, Ellamil and colleagues (2012) argued that its involvement in the evaluative phase is less surprising when one considers the broader role of this network in the processing of internally generated, affective information (Buckner et al. 2008). When viewed as such, the default network's role could be to "perform inductive inferences based on internal affective information to draw conclusions that guide behavior" (p. 1790)—in this case opting to select or discard a generated idea (i.e. drawing or writing).

In conclusion, the results of Ellamil and her team (2012) demonstrate that the generative and evaluative aspects of creative drawing involve dissociable neural systems. This is consistent with the idea that these two phases involve different sets of mental processes. Importantly, by elucidating the involvement of the default network in the evaluative phase, the results of this study illustrate the value of neuroimaging data in improving our understanding of the creative process underlying the generation of aesthetic products. Specifically, the results point to a possible role for affective processes in evaluating the quality of aesthetic products—a process not emphasized by any of the main generate–evaluate models of the creative process (see Campbell 1960; Eysenck 1993; Simonton 2011).

9.2 Creative writing

Few endeavors are as quintessentially creative as writing. In this sense, the fMRI study conducted by Shah and co-workers (2013) is a welcome addition to the family of studies on the neural bases of creative production. Unlike the research undertaken by Ellamil and colleagues (2012), who were motivated to dissociate the neural systems underlying generation and evaluation in creative drawing, Shah's team's experiment was designed to test Flower and Hayes' (1981) model of writing. According to this model, the writing process can be broken down into three stages: planning, translating, and reviewing. The planning stage involves generating ideas based on the contents of long-term memory. The intermediate stage of translating involves transcribing those ideas into text based on linguistic rules. Finally, the reviewing stage consists of the evaluation of the written text, possibly involving revision. Clearly, Flower and Hayes' (1981) three-stage model shares much in common with Campbell's (1960) two-stage model, where the only (explicitly) missing part is translation. I suspect this is because Campbell's model was not meant to be domain-specific, and was therefore made to account for all types of creativity each of which would (obviously) engender their own domain-specific translation rules. Nevertheless, despite their superficial differences, it is clear that the results of Shah and her colleagues' and Ellamil and her team's work can be compared to examine whether any aspect of the neural bases of the creative process is shared in spite of differences in the specific domain under consideration.

Shah and colleagues (2013) recruited 28 participants, none of whom had no special training in creative writing. Ellamil's team (2012) had employed an MRI-compatible tablet to record drawings and writings inside the scanner, whereas Shah's team (2013) used a design where the arm of the participant was extended outside the scanner to enable writing on paper, and was positioned to minimize movement artifacts. The design consisted of four phases: (a) covert reading (of a text), and (b) copying part of the read text (both of which functioned as control conditions). The theoretically interesting phases were (c) brainstorming, and (d) creative writing. In the brainstorming phase participants were instructed to think how they might continue the passage without actually writing any of it down. Finally, in the creative writing component they were instructed to write an original but appropriate continuation of the given text. They were allowed to think beyond ideas generated during "'brainstorming'" if they wished (p. 1090). These instructions were used to mimic the phases of creative writing as closely as possible, a process where modifications to the product (writing) are commonly made (Flower and Hayes 1981).

Here I will focus on brain activations during the phases of brainstorming and creative writing. The brainstorming phase was associated with a distributed network (more strongly activated in the left hemisphere) including the left inferior frontal gyrus (BA 44, 45, and 47), the left dorsal lateral PFC (BA 9/46), and dorsal ACC, and sizable temporal-parietal cluster of regions (including Wernicke's area, BA 39/40, BA 22, superior temporal sulcus) extending to anterior temporal regions (BA 38). The left inferior frontal gyrus (BA 44, 45, and 47) is known to play a key role in language processing (production) and verbal fluency (Price 2010). Its involvement here is not surprising, given that the research explicitly instructed the participants to engage in verbal (linguistic) creativity. In addition, left dorsal lateral PFC (BA 9/46) and dorsal ACC are key regions in the brain's executive function network and are critical cogs for planning and cognitive control. Their activation here suggests that during brainstorming, the participants were engaged in some level of selective retention and pruning of the fruits of the generative process, based on the self-imposed criteria of the quality of writing that they sought. Finally, the temporal-parietal cluster of regions (extending to anterior temporal regions) is another well-established region in language processing, including sentence comprehension and word association (Price 2010), and its activation here suggests that "functions of these regions were related to the flexibility of thinking and involved in imagination and fantasy" (Shah et al. 2013, p. 1096). In other words, the involvement of this region in brainstorming can be attributed to its role in the formation of potentially distant (linguistically derived) conceptual associations.

In turn, the creative writing phase was associated with activation in motor areas including the primary motor cortex, bilateral secondary motor areas, and corresponding somatosensory areas. In addition, frontal activations were observed in the bilateral inferior frontal gyrus (BA 44, 45, 47) and left dorsal lateral PFC (BA 46), as well as activations in the temporal lobes including the posterior part of the superior temporal gyrus (BA 22, superior temporal sulcus), among others. The involvement of the motor areas can be attributed to the realization of motor movement in the service of the mechanics of writing. As was the case with brainstorming, the involvement of the left dorsal lateral PFC (BA 46) points to the role of cognitive control processes for selecting the products of the brainstorming phase for actual writing. In addition, the involvement of the left inferior frontal gyrus and the specific regions in the temporal lobes can be attributed to their role in linguistic processing, and their contribution to creative writing. Summarizing their results, Shah and colleagues (2013) noted that creative writing can be understood as "translating the conceptual and verbal ideas into a handwriting process after selection, semantic integration, and motor coordination" (p. 1097).

Do these results shed light on Flower and Hayes' (1981) model of writing? Remember that according to that model, the writing process can be broken down into the three stages of planning, translating, and reviewing. By and large, the results support this view. Remember too that according to that model, planning involves generating ideas based on the contents of long-term memory. The main effect of brainstorming would appear to support this notion, given that it activated a core set of structures involved in language and memory. In addition, the involvement of the left dorsal lateral PFC (BA 9/46) and dorsal ACC in brainstorming can be attributed to their contribution to planning and cognitive control. In turn, translating was hypothesized to involve transcribing those ideas into text based on linguistic rules, which appears to be reflected by the observed activations in motor areas during creative writing. Reviewing may be reflected in frontal activations observed during both brainstorming and creative writing conditions.

Furthermore, Shah and colleagues (2013) argued that their study also sheds light on the role of domain specificity in the creative writing process. Specifically, they argued that their results revealed:

> . . . the importance of domain-relevant knowledge for creative writing, including language processing, working memory functions, and predominantly, the memory system: episodic and semantic memory retrieval and the semantic integration in an original and coherent story concept. Furthermore, free association and spontaneous cognition are typical characteristics for actual story writing. (p. 1097)

Thus, as has been suggested elsewhere (Vartanian 2012), although there are certain cognitive functions (and corresponding neural structures) that can be expected to be operational in most instances of creativity (e.g. working memory), there are likely to be other regions whose involvement will be determined by the specific content and/or context under consideration, such as linguistic regions that operate on verbal material or draw on linguistic knowledge.

9.3 Jazz improvisation

Neuroscientists have long been interested in the brain bases of musical perception and appreciation, as evidenced by studies conducted on the experience of "musical chills" in response to hearing particularly evocative pieces of music (Blood and Zatorre 2001).

Much less is known about the neurobiology of musical production, specifically its creative components. Toward this end, Limb and Braun (2008) conducted a study to examine the neural bases of the creative component of musical performance in the context of jazz improvisation. Improvisation is a form of spontaneous musical performance—defined as "the immediate, on-line improvisation of novel melodic, harmonic, and rhythmic musical elements within a relevant musical context" (Limb and Braun 2008, p. 1). A major aim of their study was to attain a high degree of ecological validity, recreating an environment during data collection that mimics critical aspects of jazz improvisation in real-world settings as closely as possible.

For this purpose, they recruited six skilled professional jazz musicians, all of whom were highly proficient at playing jazz piano. They underwent fMRI scanning while improvising, using an MRI-compatible piano keyboard. Data were collected under two conditions: one that was relatively high in musical complexity, and one that was relatively low. In the control condition of the low complexity condition, participants repeatedly played a one-octave C major scale in quarter notes. In turn, in the corresponding improvisation condition, they improvised a melody, but they were restricted to the use of C major-scale quarter notes within the same octave. For the high complexity condition (i.e. jazz paradigm), participants were asked to memorize an original jazz composition prior to the scan date. In the corresponding control condition they played the composition with the auditory accompaniment of a pre-recorded jazz quartet. In contrast, in the corresponding improvisation condition they were given freedom to improvise, using the chord structure of the composition and the same auditory accompaniment as the basis for improvisation.

Here the focus will be on a subset of the findings. First and foremost, improvisation in both paradigms (i.e. high and low complexity) was associated with widespread deactivation in PFC, specifically in dorsal lateral PFC and lateral orbital PFC. In turn, this was combined with activation in medial PFC, specifically in BA10. In addition, activations were also observed in sensorimotor areas, as well as deactivations of limbic activity in the amygdala and the hippocampus. This pattern of brain activity tells a rich tale about the cognitive and affective processes that underlie jazz improvisation. As discussed earlier in this chapter, the lateral aspects of PFC—specifically dorsal lateral PFC—are heavily implicated in planned behavior and reflect the involvement of executive function and cognitive control. The fact that they were deactivated during improvisation suggests that this activity is associated with a reduction of the top-down control mechanisms that normally govern planned behavior. In contrast, activations were observed in BA10, an area strongly associated with autobiographical narrative, the self, and internally generated cognitions (Buckner et al. 2008; Christoff and Gabrieli 2000). According to Limb and Braun (2008), this pattern suggests that:

> . . . creative intuition may operate when an attenuated DLPFC no longer regulates the contents of consciousness, allowing unfiltered, unconscious, or random thoughts and sensations to emerge. Therefore, rather than operating in accordance with conscious strategies and expectations, musical improvisation may be associated with behaviors that conform to rules implemented by the medial PFC outside of conscious awareness (p. 4).

The important message here is that improvisation is facilitated by the *dynamic interplay* between systems that place opposing constraints on cognition. The emerging picture is also consistent with the idea that jazz improvisation is borne out of the realization of a personal voice, in this case neurally mediated by medial PFC.

That strong activations were observed in sensorimotor, premotor, and motor areas, as well as selective activation during improvisation in supplementary motor areas and portions of the primary motor cortex is consistent with the idea that the mechanics of jazz improvisation are accompanied by the engagement of the brain's motor circuitry and point to domain specificity in the neural correlates of creative production. Finally, the modulation of activity in the limbic system is also informative, although future studies are needed to determine the contributions of reduced activation in the amygdala and the hippocampus to jazz improvisation. In summary, the results of Limb and Braun (2008) paint a rich picture of the neural bases of creative musical performance among highly skilled performers, and point to the interactive effect of various neural systems to this activity.

9.4 Artistic creativity

Thus far I have reviewed the results of three neuroimaging studies on the brain bases of creative production in the domains of drawing, writing, and musical performance. As mentioned earlier, there is a great degree of domain specificity in the observed patterns of activation. For example, creative generation in the domain of drawing was reflected in activation in the MTL (Ellamil et al. 2012), whereas creative writing was associated with widespread activation in regions that underlie language (Shah et al. 2013), and creative jazz improvisation activated the motor and sensorimotor regions particularly strongly (Limb and Braun 2008). These activations can be broadly understood as context effects (i.e. reflecting the specific content or activities associated with the target task under consideration). However, it is important to note that in each case the observed activations might also reflect other processes. For example, in relation to the activity observed in the motor regions during jazz improvisation, Limb and Braun (2008) noted that "rather than reflecting an increase in motor activity per se, these activations may be associated with encoding and implementation of novel motor programs that characterize spontaneous improvisation". In other words, these domain-specific activities might contribute to the creative aspects of performance over and above their supplementary, supportive contributions.

Equally importantly, certain regions appear to contribute to creative production across modalities. Chief among those are the dorsal and lateral aspects of PFC. It is well established that these regions play a key role in planned behavior, including executive function, inhibition, and cognitive control. Their involvement in various types of creativity—whether in the form of activations or deactivations—reflects their domain-general contributions to creative production (Vartanian 2011). The available evidence suggests that creative production in various domains is borne out of the interaction between domain-general systems such as dorsal and lateral PFC and domain-specific systems that are critical for the target type of activity under consideration (e.g. drawing).

9.5 Mirror model of art

This chapter began with a call toward the development of an overarching model that can bridge the two sides of the coin: artistic production and aesthetic appreciation. This bi-directional model can be used to study the communication of intended ideas from the creator to the audience via the medium of the work. Recently, a significant advance in this direction was taken by Tinio (2013) in the form of his *Mirror Model of Art*. The explicit aim of this model is to provide a verifiable theoretical framework that conjoins the creation and perception aspects of art. The basic premise of the model is that the temporal sequence of art perception mirrors the temporal sequence of art creation. As a result, the psychological processes the perceiver engages in during the early stages of art perception mimic the psychological processes the creator is engaged in during the late stages of art creation, and vice versa.

The evidence supporting this model comes from two sources: direct observation of artists at work, and the retrospective studies of archival material from artists (Locher 2010). Both of these approaches have converged to show that the creative process in the (visual) arts occurs in three stages. In the first stage (*initialization*) the artist explores the viability of various ideas. At this stage, sketching can play a key role because it enables the artist to develop emerging ideas, leading to a basic structural configuration of the painting. This basic structure is reflected in the under-drawing of the painting. During the initialization stage, ideational flexibility is key. In the second stage (*expansion and adaptation*), the artist fine-tunes and reworks the under-drawing. At this stage, the characteristics of the objects and the subjects in the painting undergo numerous changes, reflected in modifications that are made to specific elements of the work. In the third and last stage (*finalizing*), work on major structural elements of the artwork stops, and time is devoted to enhancing the surface layer of the work in preparation for viewing. Here, focus is shifted to color and texture—elements that are likely to grab the viewer's attention first when encountering the artwork.

Critically, the artist communicates ideas to the perceiver via various *layers* of materials that will be encountered in *reverse* order by the perceiver. In this sense the medium (i.e. layering) functions as a communicative tool (that mirrors the stages in reverse order). The concept (i.e. the initial idea) that triggered the piece is encapsulated in the work conducted during initialization, and represents the final destination in the viewer's journey toward discovering the essential meaning of the artwork. This is because when a perceiver first encounters an artwork, he or she is likely to engage in early automatic processing of low-level features such as color and contrast that are driven by the surface characteristics of the artwork (Chatterjee 2003; Leder et al. 2004), features that were added during the finalizing stage of creation. The next stage of perception involves intermediate memory-based processing of content and style, driven by elements such as objects that were modified during the intermediate expansion and adaptation stage of creation. Finally, during the last stage that corresponds to meaning making and the generation of aesthetic judgments and emotions, the focus is on deciphering the intentions of the artist for creating the artwork. This is a higher-order cognitive state that draws from the structural organization (i.e. under-drawing) of the painting.

9.6 Aesthetic processing in the brain

Tinio's (2013) model suggests that aesthetic appreciation is a process that begins with an analysis of the surface characteristics of (visual) artworks such as color and contrast, eventually moving via memory-based processing of content and style to the deepest level of analysis involving the underlying meaning of the work—both personal and cultural. To date, no neuroimaging study has investigated the temporal dynamics of aesthetic appreciation to test this progression. However, it is true that typically experimenters have only provided subjects with mere seconds to view each individual stimulus (i.e. painting) during data collection. Assuming that the three stages described by Tinio (2013) unfold largely in sequence (rather than in parallel), one can conclude that most neuroimaging studies of aesthetic appreciation to date have only isolated brain structures that are most likely to contribute to the early and intermediate stages of exposure to artworks, rather than the later stages that are most likely require more time than mere seconds for meaning extraction. Consistent with this possibility, a recent quantitative meta-analysis of 15 fMRI studies conducted by Vartanian and Skov (2014) has demonstrated that the act of viewing paintings was correlated with activation in a distributed system including the occipital lobes, temporal lobe structures in the ventral stream involved in object (fusiform gyrus) and scene (parahippocampal gyrus) perception, and the anterior insula—a key structure in experience of emotion and reward. In addition, activation was also observed in the posterior cingulate cortex bilaterally—part of the brain's default network. Primarily, the isolated regions underlie visual representation and object recognition, emotion and reward processing, and representations of internalized cognitions. The picture that emerges supports the idea that activated regions are more likely to contribute to early and intermediate rather than late processing in painting viewing. As noted by Nadal (2013), determining how the neural network that underlies the experience of art is impacted by personal and cultural meaning constitutes an important challenge to the field. In turn, this will facilitate the development of a framework for connecting the neural foundations of the experience of art to the neural foundations of art creation.

References

Blood, A.J. and Zatorre, R.J. (2001). Intensely pleasurable responses to music correlate with activity in brain regions implicated in reward and emotion. *Proceedings of the National Academy of Sciences USA* **98**, 11818–23.

Brown, S., Martinez, M.J., and Parsons, L.M. (2006). The neural basis of human dance. *Cerebral Cortex* **16**, 1157–67.

Brown, S. and Parsons, L.M. (2008). The neuroscience of dance. *Scientific American* **299**, 32–7.

Buckner, R.L., Andrews-Hanna, J.R., and Schacter, D.L. (2008). The brain's default network. *Annals of the New York Academy of Sciences* **1124**, 1–38.

Campbell, D.T. (1960). Blind variation and selective retention in creative thought as in other knowledge processes. *Psychological Review* **67**, 380–400.

Chatterjee, A. (2003). Prospects for a cognitive neuroscience of visual aesthetics. *Bulletin of Psychology and the Arts* **4**, 55–60.

Christoff, K. and Gabrieli, J.D.E. (2000). The frontopolar cortex and human cognition: evidence for a rostrocaudal hierarchical organization within the human prefrontal cortex. *Psychobiology* **28**, 168–86.

Dosenbach, N.U.F., Fair, D.A., Cohen, A.L., et al. (2008). A dual-networks architecture of top-down control. *Trends in Cognitive Sciences* **12**, 99–105.

Ellamil, M., Dobson, C., Beeman, M., et al. (2012). Evaluative and generative modes of thought during the creative process. *Neuroimage* **59**, 1783–94.

Eysenck, H.J. (1993). Creativity and personality: Suggestions for a theory. *Psychological Inquiry* **4**, 147–78.

Flower, L.S. and Hayes, J.R. (1981). A cognitive process theory of writing. *College Composition and Communication* **32**, 365–87.

Goel, V. and Grafman, J. (2000). Role of the right prefrontal cortex in ill-structured planning. *Cognitive Neuropsychology* **17**, 415–36.

Goel, V. and Vartanian, O. (2005). Dissociating the roles of right ventral lateral and dorsal lateral prefrontal cortex in generation and maintenance of hypotheses in set-shift problems. *Cerebral Cortex* **15**, 1170–7.

Gonen-Yaacovi, G., de Souza, L.C., Levy, R., et al. (2013). Rostral and caudal prefrontal contribution to creativity: A meta-analysis of functional imaging data. *Frontiers in Human Neuroscience* 7, Article number 00465.

Guilford, J.P. (1967). *The nature of human intelligence*. New York, NY: McGraw-Hill.

Henke, K., Weber, B., Kneifel, S., et al. (1999). Human hippocampus associates information in memory. *Proceedings of the National Academy of Sciences USA* **96**, 5884–9.

Leder, H., Belke, B., Oeberst, A., et al. (2004). A model of aesthetic appreciation and aesthetic judgments. *British Journal of Psychology* **95**, 489–508.

Limb, C.J. and Braun, A.R. (2008). Neural substrates of spontaneous musical performance: an FMRI study of jazz improvisation. *PLoS One* **3**, e1679.

Locher, P.J. (2010). How does a visual artist create an artwork? In J. Kaufman and R. Sternberg (eds), *Cambridge Handbook of Creativity*. Cambridge: Cambridge University Press, pp. 131–44.

Mednick, S.A. (1962). The associative basis of the creative process. *Psychological Review* **69**, 220–32.

Miller, L.A. and Tippett, L.J. (1996). Effects of focal brain lesions on visual problem-solving. *Neuropsychologia* **34**, 387–98.

Nadal, M. (2013). The experience of art: insights from neuroimaging. *Progress in Brain Research* **204**, 135–58.

Newman, S.D., Carpenter, P.A., Varma, S., et al. (2003). Frontal and parietal participation in problem solving in the tower of London: fMRI and computational modeling of planning and high-level perception. *Neuropsychologia* **41**, 1668–82.

Poincaré, H. (1913). *The foundations of science*. Lancaster, PA: Science Press.

Price, C.J. (2010). The anatomy of language: a review of 100 fMRI studies published in 2009. *Annals of the New York Academy of Sciences* **1191**, 62–88.

Shah, C., Erhard, K., Ortheil, H.J., et al. (2013). Neural correlates of creative writing: an fMRI Study. *Human Brain Mapping* **34**, 1088–101.

Simonton, D.K. (2011). Creativity and discovery as blind variation and selective retention: Multiple-variant definition and blind-sighted integration. *Psychology of Aesthetics, Creativity, and the Arts* **5**, 222–8.

Tam, F., Churchill, N.W., Strother, S.C., et al. (2011). A new tablet for writing and drawing during functional MRI. *Human Brain Mapping* 32, 240–8.

Tinio, P.P.L. (2013). From artistic creation to aesthetic reception: the mirror model of art. *Psychology of Aesthetics, Creativity, and the Arts* **7**, 265–75.

Vartanian, O. (2011). Brain and neuropsychology. In M. Runco and S. Pritzker (eds), *Encyclopedia of creativity* (2nd edn). London: Academic Press, pp. 164–9.

Vartanian, O. (2012). Dissociable neural systems for analogy and metaphor: Implications for the neuroscience of creativity. *British Journal of Psychology* **103**, 302–16.

Vartanian, O. (2014). Empirical aesthetics: hindsight and foresight. In J.K. Smith and P. Tinio (eds), *Cambridge Handbook on the Psychology of Aesthetics, Creativity, and the Arts*. Cambridge: Cambridge University Press, pp. 6–34.

Vartanian, O., Martindale, C., and Matthews, J. (2009). Divergent thinking is related to faster relatedness judgments. *Psychology of Aesthetics, Creativity, and the Arts* **3**, 99–103.

Vartanian, O. and Skov, M. (2014). Neural correlates of viewing paintings: Evidence from a quantitative meta-analysis of functional magnetic resonance imaging data. *Brain and Cognition* **87**, 52–6.

Zeki, S. (1999). *Inner vision*. New York, NY: Oxford University Press.

Section 3

Cognitive neuroscience of dance

Chapter 10

Beautiful embodiment: The shaping of aesthetic preference by personal experience

Emily S. Cross

Dance is the only art of which we ourselves are the stuff of which it is made.
—*Ted Shawn, American Modern Dance Pioneer*

10.1 Introduction

Imagine the following situation: you arrive at the train station bright and early on a Friday morning to start your usual commute into the office. As you cross the main concourse, the train platform departure announcements fade out and are suddenly replaced by the opening lyrics to the Isley Brothers 1959 classic song "Shout." At the same time, a lone figure in the middle of the concourse bursts into energetic dance. A moment later, a woman behind him casts aside her shopping bags and joins in, just as two middle-aged men in dark suits with briefcases kick up their heels and join as well. Within seconds, it appears the entire concourse has come alive with everyday commuters dancing with incredible energy, precision, and joy as they spin, kick, twist, and jazz-hand their way through a mash-up of decades' worth of popular songs. You and the seemingly dwindling crowd of those whose bodies are unfamiliar with this captivating choreography watch in amazement, as the faces of all the "ordinary" commuters light up with delight to be surrounded by this spontaneous injection of song and dance into a public place.

For those of us who were not present at London's Liverpool Street Station on 15 January 2009, this flashmob, filmed as an advertisement for T-Mobile, can be witnessed on YouTube (see <https://www.youtube.com/watch?v=VQ3d3KigPQM>). For those who were there, as well as the rest of us who watch online, witnessing a dancing flashmob is something that most people find thrilling. A flashmob is defined as "a spontaneous public performance, usually a dance, that appears seemingly out of nowhere with just one person, grows to hundreds of people, then quickly disappears" (<http://www.flashmobamerica.com>). The Liverpool Street Station example is one of many; from dance performances of "Do Re Mi" in the Antwerp train station to "Gangnam Style" in a Jakarta fountain to

"Loveshack" on Bondi Beach, the flashmob is gaining momentum as a much-loved, often-reproduced way to bring excitement, levity, and even a little bit of art into people's lives where they might least expect it. As I explore in this chapter, one possible explanation for the flashmob's success and appeal is that it transforms seemingly normal pedestrians to extraordinary performers through dance.

Throughout history, dance has maintained a critical presence across human cultures, defying barriers of class, race, and status (Dils and Albright 2001). The co-evolution of dance with the human race has fuelled a rich debate on the function of art and aesthetic experience, engaging numerous artists, philosophers, and scientists. While dance shares features with other art forms, one unique attribute is that it is expressed (only) through movement of the human body. Because of this, social scientists and neuroscientists are turning to dance to help answer questions of how the brain coordinates the body to perform complex, precise, and beautiful movements. For example, when we watch a ballet dancer leap through the air in a perfect split or a breakdancer perform five consecutive head spins, how might our own dance abilities (or lack thereof) influence how we perceive performers who excite and impress us? In this chapter, I explore how research with dancers and dance-learning paradigms can help illuminate issues related to artistic creation and aesthetic preferences, and also consider more fundamental questions concerning the relationship between action and perception. I begin with a section that briefly summarizes the development of a multidisciplinary field whose aim is to explore how action production informs action perception, with a special emphasis on how studies with dancers have advanced this line of inquiry. Next, I focus on a specific conceptualization of the inverse of this relationship; namely, how simple observation of actions can profoundly influence an observers' motor repertoire. Here again, pioneering work using dance underscores how powerful visual experience can be in shaping our motor abilities. While the first two sections mainly use dance as a means for exploring fundamental links between action and perception, the final two sections explore the utility of dance as an art form to inform our understanding of an observer's aesthetic experience of watching another body in motion. After a cursory introduction of issues of concern when studying the aesthetics of action in the penultimate section, the final section explores the relationship between observers' experience and the dancers they behold, in an attempt to establish connections between the two (and in so doing, perhaps help us to understand why we might derive so much pleasure from being caught up in a flashmob in person or watching one on the Internet).

10.2 The shaping of perception by action

While we are inclined to take it for granted, our ability to use perceptual information about other people's movements to inform our subsequent actions is essential for successful interactions with the environment, and thus, for survival. Consequently, it is no surprise that inquiry into how the human brain negotiates the path between action execution and perception has intrigued venerable thinkers throughout the millennia, including

Aristotle, Descartes, and William James. Until the mid-nineteenth century, thinking about this relationship was dominated by the Cartesian sensorimotor view, which advanced the idea of independence between action perception and action production (Descartes writing in 1664, as discussed by Prinz 1997). As knowledge about the human brain continued to accrue, pioneering scholars began to explore the idea that perceptual processes might indeed converge or overlap to some extent with motor performance (James 1890). Empirical investigation into the brain's ability to use perceptual information to shape movement began in the mid-twentieth century, when information processing explanations proposing complex transformations from perception to the organization and execution of action gathered momentum (Massaro and Friedman 1990; Mountcastle et al. 1975; Mountcastle 1975; Sanders 1967, 1983; Welford 1968).

In the mid-1970s, Mountcastle and colleagues began the first investigations into putative neurophysiological mechanisms linking action and perception (Mountcastle et al. 1975; Mountcastle 1975; Yin and Mountcastle 1977). In these seminal studies, Mountcastle and colleagues used single-unit recording techniques to track the activity of neurons within the superior parietal lobule and posterior parietal areas of the macaque cerebral cortex while the animals performed simple behavioral acts in response to sensory stimuli. In one such study, Mountcastle and colleagues investigated the specific properties of neurons in the posterior parietal association cortex, and determined that these neurons respond to visual stimulation to encode the meaning of an action, such as the goal of the perceived movement (Mountcastle et al. 1975). These researchers concluded that this region of the parietal cortex is a likely candidate for cross-modal convergence of perception and action, and established the foundation for the next several decades of further inquiry into how these modalities interact (Andersen et al. 2004).

In the 1990s, research into the intersection of action perception and production experienced an extraordinary surge in interest that was due in large part to the discovery of so-called mirror neurons in the ventral premotor cortex of the monkey. These neurons were found to discharge in a similar manner both when a monkey performed an action as well as when it observed another monkey or human perform the same action (di Pellegrino et al. 1992; Gallese et al. 1996; Rizzolatti et al. 1996). As such, these specialized neurons have prompted these researchers and others to propose that action perception and production processes form a bi-directional, interactive loop within the primate brain, and that action understanding might be explained by the observer's brain simulating the observed movements of another individual (Fadiga et al. 1995; Fadiga et al. 1999; Grafton et al. 1996; Rizzolatti et al. 2001).

This hypothesis has sparked hundreds of studies and an ongoing lively debate among researchers regarding the specific parameters, scope, and limitations of such an action simulation system within the human brain (Gallese et al. 2011; Gallese and Sinigaglia 2011). While an exhaustive review of the past decade and a half of research undertaken in this domain is beyond the scope of this chapter (the interested reader is directed to several outstanding overviews of this work, such as those by Grosbras et al. 2012; Molenberghs et al. 2012; Rizzolatti and Sinigaglia 2010), what a burgeoning corpus of research demonstrates is that neural tissue found within parietal and premotor cortices of the human brain is engaged when actions are performed or when they are observed. This work has given rise to the notion of a human mirror system, which comprises multiple cortical regions and shows evidence for behaving in a similar manner to individual mirror neurons identified within non-human primate brains (Molenberghs et al. 2012). In our laboratory, however, our research focus is on what we term the "action observation network" (AON; Cross et al. 2009; Grafton 2009). The AON comprises those brain regions implicated in the human mirror system (and shown to contain mirror neurons in monkeys), as well as broader swathes of premotor and occipitotemporal cortices (Figure 10.1). The interest I have in this network and its role in motor learning and performance is perhaps best summed up by Scott T. Grafton, who said "[w]hat is striking about this network is that a perceptual stimulus can lead to such a large-scale recruitment of the brain with a complete disregard of textbook divisions between sensory and motor portions of the cortex" (Grafton 2009, p. 100).

The bulk of research conducted on the neural correspondence between action and perception, across human and non-human primates, has focused on small, constrained movements of the hands and fingers (cf. Grosbras et al. 2012; Rizzolatti and Sinigaglia 2010). However, my colleagues and I have devoted a considerable amount of attention to investigating the neural underpinnings of dance performance and perception which at first might appear to be rather far removed from simple hand and finger actions. However, we use dance to study the AON for three main reasons. First, for most of us, daily life requires the efficient coordination of the entire body, not just the fingers and hands. Thus, dance provides a variety of ideal paradigms for studying whole-body coordination. Secondly, dance is generally not object-directed, and thus enables the study of "movement for movement's sake," where the goal of an action is the action itself (rather than opening a bottle, tying a shoelace, etc.). Finally, by studying a motor behavior that is much more complex and requires the coordination of multiple effectors, we are better placed for understanding a broader range of motor behaviors, from the simple to complex. Such an approach should, in turn, foster greater understanding of failures of the sensory and motor systems, such as those caused by neurological injury.

Thus, we turn to dance as a useful means of exploring the impact of action experience on perception. The first study to use dance to explore the relationship between action and perception was the seminal work by Beatriz Calvo-Merino and colleagues. These authors investigated how highly specialized physical training in one dance style (classical ballet or capoeira) influenced perception of the familiar dance style compared to the unfamiliar dance style (Calvo-Merino et al. 2005; see also Chapter 11, this volume). The authors reported that when trained ballet dancers watched ballet, or when trained capoeira dancers watched capoeira, greater activity was seen throughout the AON, compared to when ballet dancers watched capoeira or capoeira dancers watched ballet. The authors interpreted this as evidence that years of physical practice shape action resonance processes that occur within sensorimotor cortices active during action perception and action production, such that they are finer-tuned to respond to movement that is familiar due to extensive physical practice.

As this study by Calvo-Merino and colleagues (2005) was underway, my colleagues and I were addressing a similar question, with additional interest in whether it is possible to track the emergence of motor expertise in something closer to real time with expert dancers (Cross et al. 2006). In our study, we followed an ensemble of expert contemporary dancers as they learned Laura Dean's 25-minute piece entitled *Skylight* (Dean 1982) across an eight-week rehearsal period. We invited the dancers into the laboratory for six consecutive weeks to scan their brains with functional magnetic resonance imaging (fMRI) as they watched and imagined themselves performing short segments of choreography from *Skylight*, as well as kinematically similar dance movements that were never seen or rehearsed in the studio. Across the rehearsal period, dancers unsurprisingly became more adept at physically performing the choreography they rehearsed in the studio each day. However, the most interesting finding to emerge from this longitudinal study is that as dancers became more adept performers, activity within two core regions of the AON, left inferior parietal lobule and left ventral premotor cortex, showed increased activity during observation the better a dancer was at performing a particular sequence. We took this as evidence that the better one is at performing an action, the more one embodies it, as demonstrated by increased parietal/premotor activity tracking with increased performance ability (Cross et al. 2006; although see also Cross et al. (2012) for an updated account of the relationship between the magnitude of AON response and motor familiarity).

Since these two early studies investigating the brains of expert dancers, a number of research teams have combined dancers with brain imaging techniques (Brown et al. 2006; Orgs et al. 2008) and behavioral paradigms (Calvo-Merino et al. 2009; Jola et al. 2011; Stevens et al. 2010) in innovative ways to illuminate further how highly skilled movement repertoires influence perception of others (for a comprehensive overview, the interested reader is referred to Bläsing et al. 2012). However, from the earliest conceptualizations of action–perception coupling, it has been asserted that not only does action influence perception (as we have seen with the expert dance studies in this section), but perception also influences action (Prinz 1990). The following section examines how dance paradigms help illustrate the impact of perception on action.

10.3 The shaping of action by perception

While it is an indisputable fact that one's prior motor experience influences how one sees the world (Prinz 2006), how we see others moving around us also shapes our own interactions with the environment. In particular, we rely heavily upon our visual system to help us learn new actions. One of the hallmark ways in which we learn new skills in life, from baking a cake to dancing the tango, is by watching others (Flynn and Whiten 2010). Behavioral research on action learning suggests that simultaneously observing and reproducing a pattern of movements results in the quickest and most accurate learning (e.g. Badets et al. 2006; Bandura 1977, 1986; Blandin et al. 1999; Blandin and Proteau 2000; Schmidt 1975; Sheffield 1961). Nevertheless, the ability to learn a new skill or improve an existing one by observation alone, without concurrent practice, is also a powerful capacity in humans (Mattar and Gribble 2005; Torriero et al. 2007).

Early behavioral investigations of observational learning by Sheffield (1961) led to the proposal that observation of a motor sequence improved learning by providing a "perceptual blueprint," or standard of reference for how the task should be performed. Behavioral studies comparing observational and physical learning support the value of a perceptual blueprint (Blandin and Proteau 2000; Carroll and Bandura 1990; Doody et al. 1985; Lee et al. 1990; Zelaznik and Spring 1976). While the bulk of observational learning research has focused on learning from an expert human model (see Hodges et al. 2007 for a review), the use of a human actor performing the target behavior is not a requirement for forming a perceptual blueprint. A more inclusive conceptualization of observational learning encompasses encoding any instruction, whether physical or symbolic, that can provide a sufficient model of the actions to be performed (e.g. Cisek and Kalaska 2004). The key distinction of what defines observational learning is the subject not performing concurrent physical practice at the time instructions are provided.

In a study using psychophysical and electromyographic (EMG) measures with human participants, Mattar and Gribble (2005) demonstrated that participants' learning of a novel, complex motor task was facilitated if they previously observed another individual learning to perform that same task, compared to watching another individual perform the task without learning or learning to perform a different task. This established the specificity of a perceptual blueprint formed during the observational period on subsequent physical practice. What emerges from these studies and others (Barzouka et al. 2007; Bouquet et al. 2007; Ferrari 1996; McCullagh et al. 1989; Schmidt and Lee 2005) is the hypothesis that observation of movement sequences and physical rehearsal share common cognitive mechanisms, and training of this circuit by either means improves novel motor-skill learning.

Evidence from non-human primate neurophysiology as well as psychophysical and EMG work with humans suggest that physical and observational learning might share common neural substrates, at least in terms of mental rehearsal (Cisek and Kalaska 2004; Frey and Gerry 2006; Mattar and Gribble 2005; Wiggett et al. 2012). In one such study, Cisek and Kalaska (2004) recorded extracellular spike activity from single neurons within the dorsal premotor cortex (PMd) of monkeys while they observed the sensory events associated with accurate performance of a saccadic motor task. They reported that when the monkeys passively observed a symbolic representation of accurate task performance, PMd neurons responded in a similar manner as when the monkeys physically performed the task. These authors were thus the first to measure mental rehearsal of a to-be-learned task occurring at the single-neuron level.

The studies discussed previously provide a foundation for exploring areas of overlap and divergence between observational and physical learning in the human brain. However, important questions remain concerning how physical practice compared to observational practice affects brain and behavior. To begin addressing such questions, my colleagues and I again turned to dance to determine how brain and behavior are influenced by only watching a new dance sequence, compared to physically practising it (Cross et al. 2009). In this study, we incorporated the dance video-game "StepMania" (<http://www.stepmania.com>;

a freeware programme similar to the popular video game "Dance Dance Revolution™" (Konami Digital Entertainment, Redwood City, CA)) into our paradigm. To play this game, participants must attend to scrolling arrow cues on a computer screen that correspond to particular foot positions on a dance mat attached to the computer. When the steps are performed in quick succession with music, this approximates dancing (to some). Importantly, however, the video-game context is fun, engaging, and ideal for studying motor learning, as participants receive real-time feedback about the accuracy of each movement via points scored. In this study, we recruited 17 young adults with no formal dance training and asked them to practise six different dance sequences for one week, to observe six different dance sequences, and a further six sequences were untrained in any way. We collected scans of participants' brains with fMRI as they watched and listened to all 18 sequences before and after a one-week training period. Following all training and scanning procedures, participants returned to the laboratory to attempt to perform all 18 dance sequences.

Based on the wealth of behavioral data showing evidence for learning by observation, we hypothesized that participants' ability to perform the sequences from the observation condition would be better than their performance for the untrained sequences. This is indeed what our post-training follow-up dance test revealed (Cross et al. 2009). Of greater interest was how the brain was affected by physical compared to observational experience with the previously novel dance sequences. Importantly, in this study, participants were not instructed to try and learn during the observation blocks of the training period: instead, they were told to just rest and watch several music videos of dance sequences they had not trained on while their heart rates recovered. Thus, they were surprised when they were asked to perform the sequences they had observed throughout the training week (as well as the untrained sequences, which were experienced only during scanning). The fMRI data revealed that physical and observational learning appeared to affect brain regions in a similar manner. Specifically, responses within the left inferior parietal lobule and right premotor cortex were quantifiably more similar when observing danced and watched sequences compared to untrained sequences. Thus, observational training, even when one might not explicitly be trying to learn, shapes responses within the AON in a similar manner to physical practice. This was among the first brain-based evidence to suggest that mere observation of an action has the potential to build a similar neural representation of that action as physical practice (Cross et al. 2009), which added to prior behavioral support (e.g. Badets et al. 2006; Bandura 1986; Hodges et al. 2007) for a common neurocognitive representation of performed and perceived actions.

Of course, this is not to say that observational learning is so powerful that physical practice is unnecessary for learning new motor skills. However, it does suggest that the parts of the human brain that are active during physical training might be active in a similar manner when observing others in action, and this activity appears to be related to the efficacy of learning by observation. As with the studies discussed in the first section, when examining how observation shapes action, studies using dance (as well as other highly skilled full-body actions, such as gymnastics and contortion) have advanced our understanding of how observation affects motor regions of the brain (e.g. Calvo-Merino et al. 2006; Cross et al. 2010; Cross et al. 2013; Jola et al. 2012). While some evidence suggests that visual experience alone is not enough to affect motor regions within the brain (Calvo-Merino et al. 2006), other research suggests that even when an individual observes movements that he or she could not possibly physically perform, extensive visual experience with these actions is associated with the emergence of neural representations within sensorimotor regions (Cross et al. 2013; Jola et al. 2012). Thus, while the influence of perception on action might not be as strong as the inverse, evidence continues to accumulate to support the notion that how we see others move influences how we ourselves move. In the next section, the focus on action perception is broadened to include how we *feel* when we watch others move; a natural segue way into the aesthetics of action.

10.4 **Aesthetics of action**

When watching an attractive stranger dancing in a club, a solo ballerina pirouetting with ease on stage, or a street performer's breathtaking breakdancing moves, motor resonance processes are engaged. Another major factor that guides and holds our attention on the mover is how much we *enjoy* watching them move. When considering the aesthetics of action at either the brain or behavioral level, two key issues warrant discussion. The first is what, precisely, is meant by aesthetics, and the second is what kinds of entities or experiences should be "eligible" for aesthetic evaluation. In a forward-thinking piece on the future and utility of neuroaesthetics as a field, Brown and Dissanayake (2009) present a strong case for broadening the focus of aesthetics to include any perceptual phenomenon that "promotes selective attention and positive emotional responses" (p. 44; see also Orians 2001). Consequently, they challenge the nascent field of neuroaesthetics to develop and evaluate a broader, more general conception of aesthetic responses in the brain, rather than focusing exclusively (and narrowly) on Eurocentric conceptions of "art," which has typically comprised works created for a wealthy elite in the classical/fine art traditions (Brown and Dissanayake 2009; see also Brown et al. 2011; Ramachandran and Hirstein 1999).

One of the most prominent theoretical conceptualizations of neuroaesthetics was put forth by Freedberg and Gallese (2007). These authors argue that an important factor in shaping an observer's affective (or aesthetic) experience is the simulation of actions, emotions, and bodily sensations visible or implied in a work of art. In this theory, Freedberg and Gallese (2007) focus almost exclusively on static works of visual art in the form of paintings and sculpture, and suggest that an observer can experience "embodied resonance" when viewing a piece of art based on the content of the work itself or via the visible traces of the artistic medium. While a compelling case can be built to support these ideas when beholding paintings or sculpture, in many ways it seems just as intuitive and valuable to extend this concept beyond classical conceptions of visual art. Without a broader evidence base for considering aesthetic experiences, we risk under-appreciating the breadth and depth of aesthetic appreciation of all manner of experiences in the real world. Thus, by accepting Brown and Dissanayake's challenge that the study of (neuro) aesthetics should include a wide range of phenomena and experiences beyond classical paintings and sculpture, we are free to explore affective responses to all manner of objects and experiences around us, from street dance and folk music to product design and culinary creations.

Along these lines, it is instructive to consider an elegant study by Hayes and colleagues that looked at aesthetic appraisals of performing or observing simple actions directed toward common, everyday objects (Hayes et al. 2008). The study of such simple actions can provide a useful template to build toward understanding affective appraisals of more complex actions, such as dance. In a set of three experiments, Hayes and colleagues examined how perceptual or motor fluency in terms of how an object is grasped influences participants' liking of that object. In one experiment, participants were asked to perform a fluid reaching action to grasp an object (such as a salt-shaker), or they were asked to grab the same salt-shaker, but the only possible way to grab it was by performing a more awkward grasp so as to avoid an obstacle. Hayes and colleagues report that participants rated the objects with which they interacted fluidly as more likeable. In a follow-up experiment, Hayes and colleagues showed participants video clips of an actor interacting fluently or awkwardly with a number of different everyday objects. They asked participants to rate how likeable they found the objects based on viewing these interactions. Here again, simply the perception of a fluent grasp toward an object significantly increased participants' liking ratings.

The authors conclude that fluent interactions with objects, defined in this situation by smooth, efficient movement, engender positive affect toward those objects (Hayes et al. 2008). Moreover, it does not matter whether fluent movements are performed or simply observed for the effect to manifest. Hayes and colleagues suggest that these findings are consistent with models of hedonic fluency, which posit that when a stimulus is easily processed, it is liked more (Winkielman and Cacioppo 2001; see also Berlyne's (1974) account of perceptual fluency). What Hayes and colleagues (2008) added to earlier models was evidence that movement itself adheres to these same principals. In other words, objects participants attempted to grasp or watched an actor grasp with an awkward or fluid movement are susceptible to being liked more or less depending on the movement directed toward them.

This evidence reported by Hayes and colleagues (2008) fits well with Freedberg and Gallese's (2007) embodied simulation notion of aesthetic processing. Perhaps unsurprisingly, since Freedberg and Gallese's (2007) theoretical proposal and the behavioral work performed with simple grasping actions, a small number of research labs have begun the empirical evaluation of aesthetic responses in the human brain when watching dance (Calvo-Merino et al. 2008; Cross et al. 2011; Cross and Ticini 2012; Kirsch et al. in press). In particular, my colleagues and I have been interested in exploring the relationship between the observer's own experience and how much enjoyment they derive from watching a dancer perform.

10.5 How action experience shapes preferences in dance spectators

The motivation for the first empirical study my colleagues and I performed that investigated neuroaesthetic responses to watching dance came from a number of sources. The first was the "Skylight" study that showed greater activity within the AON when (dance-experienced) participants watched movements they were highly skilled at performing (Cross et al. 2006). The second was Friedberg and Gallese's (2007) embodied simulation account of neuroaesthetics, suggesting that we use our sensorimotor cortices when making aesthetic evaluations of artworks. The third and most significant motivation behind our investigation was the first study to investigate aesthetic responses of spectators when watching dance (Calvo-Merino et al. 2008). In the study by Calvo-Merino and colleagues (2008), six dance-naïve spectators watched 24 movements performed by an expert ballet dancer or an expert capoeira dancer whilst undergoing fMRI. Approximately one year later, the investigators invited these six participants back to the laboratory to watch the same 24 movements. For this testing session, participants simply watched the videos on a computer screen and responded to simple questions designed to query their aesthetic experience (after Berlyne 1974). The authors then used participants' group averaged ratings to interrogate brain regions that showed neural responses of higher magnitude when watching movements that were generally liked. They reported two important findings that were relevant for the follow-up study performed by my team: first, that bilateral medial visual cortices as well as right premotor cortex showed greater responses when viewing liked movements; and second, that participants as a group tended to like watching whole-body movements and those that involved significant displacement of the whole body across space (such as a large horizontal jump) compared to smaller or less explosive movements (Calvo-Merino et al. 2008).

The study by Calvo-Merino and colleagues (2008) offered an intriguing first glimpse into the neural underpinnings of an aesthetic response when watching dance. It also sparked many new questions concerning the relationship between sensorimotor brain regions and the pleasure derived from watching dance that were ripe for exploration. One main question concerned how individuals' online ratings of aesthetic value relate to brain function. In Calvo-Merino and colleagues' (2008) paradigm, all six participants' aesthetic ratings were averaged into one value for each dance movement, and aesthetic ratings were collated in a later testing session outside the fMRI scanner. Thus, this paradigm did not allow the authors to address individual differences in aesthetic preferences the first time the movements were watched. In our study, we wished to address these specific issues of individual differences in aesthetic preferences, as well as how an observer's physical ability to perform a movement interacts with their enjoyment of watching a movement (Cross et al. 2011).

In our experiment, participants' main task was to rate 64 3-second video clips of dance movements performed by either a male or female professional ballet dancer. When we were filming the two dancers to construct the stimuli, we asked them to perform a range of movements from both classical ballet and contemporary dance vocabularies, and to vary the size, intensity, and complexity of the movements, so that we ended up with the widest possible range of different movements. We then piloted 100 video clips on several dozen participants (who did not take part in the main experiment), whom we asked to rate each movement on how difficult they thought it would be to reproduce, as well as how much they liked watching it. From these pilot ratings, we selected the 64 dance clips that elicited the broadest range of ratings in terms of both liking and reproducibility.

For the main experiment, we recruited a study sample of 22 dance-naïve participants who underwent fMRI scanning while watching the dance video stimuli and answering one question after each video clip. The question was either "how much did you enjoying watching the previous video?" or "how well do you think you could reproduce the previous movement?". Participants responded to each question according to a 1 to 4 response scale, with 1 corresponding to "liked it a lot" or "could perfectly reproduce the movement" and 4 corresponding to "did not like it at all" or "could not reproduce the movement at all." To understand the relationship between embodiment and aesthetics within our study design, we assessed what the behavioral data revealed in terms of the relationship between liking and reproducibility ratings, as well as what the brain imaging data revealed about how the action observation network responded to movements that could be physically reproduced compared to those that were pleasing to watch. If our data were to support dominant theories of the AON, which state these brain regions respond most to movements that a participant can perform or are present in their action repertoire (e.g. Calvo-Merino et al. 2005; Cross et al. 2006), then we would predict most AON activation when watching the movements that participants rated as easily reproducible, independent of aesthetic value. However, if the embodied simulation theory of aesthetics more accurately explains how our brains respond to aesthetically pleasing stimuli (Freedberg and Gallese 2007), then we would predict strongest AON activity when watching movements that participants ranked as most pleasing, independent of how possible it is to reproduce them.

The results yielded several interesting and surprising findings. First, we found a strong negative relationship between enjoyment and perceived ability to physically reproduce a movement (Figure 10.2a). This analysis demonstrates that in our study, liking and physical ability ratings were not independent, as participants showed a tendency to prefer those movements that they rated as extremely difficult to reproduce. As far as the fMRI data were concerned, we chose to apply a series of parametric analyses to our data, which enabled us to look at participants' individual liking and reproducibility ratings for each stimulus, and query how brain responses change as movements are rated as increasingly (or decreasingly) likeable or increasingly (or decreasingly) easy to reproduce. We found that several regions within the AON, specifically occipitotemporal cortex and the inferior parietal lobule, show increasingly strong activation the more a movement is liked. Curiously, when we asked which brain regions show greater activity the more easily a movement is reproduced, we failed to replicate the findings from our previous study with expert dancers (Cross et al. 2006). Our 2006 study demonstrated greater parietal and premotor engagement when expert dancers rated movements as increasingly possible to reproduce. In our 2011 study with non-dancers, no brain regions emerged from a contrast evaluating brain areas showing increased activity the better an observer is at performing an observed action. When we evaluated the inverse contrast, designed to reveal brain regions showing stronger activity the more difficult a movement appears, activity emerged within bilateral visual cortices. Figure 10.2b illustrates the overlap between these two analyses within the occipital cortices of the brain.

The most interesting analysis in our view, however, directly compared the interaction between liking and perceived physical ability. Put another way, this analysis asked which brain regions respond more strongly when participants watch a movement they really like, but which they also judge to be extremely difficult to reproduce. As with the simple parametric analysis of liking only, here we saw robust activation of bilateral occipitotemporal cortices, as well as the right inferior parietal lobule. Thus, it appears that watching exciting (or enjoyable) movements that would be impossible or extremely difficult to perform engages certain portions of the AON that are implicated in action embodiment (cf. Calvo-Merino et al. 2005; Cross et al. 2006).

When the behavioral and imaging findings are considered together, they appear somewhat contradictory to the findings reported by Hayes and colleagues (2008) and what might be predicted by the perceptual or hedonic fluency accounts of stimulus processing (Berlyne 1974; Winkielman and Cacioppo 2001). One reading of what these accounts would predict is that participants should find less pleasing those movements that would be extremely difficult for them to perform. However, it seems that the perception trumps action ability when dance-naïve participants watch dance movements for the first time, and seeing a dancer perform a highly complex movement with apparent ease and grace is extremely enjoyable, even if (or even especially because?) a participant could not come close to achieving the same thing with his or her own body. This explanation finds much anecdotal support, of course, when considering that people pay large sums to money to see physical feats that most of us could never perform ourselves, such as the incredible precision of dancers in the Forsythe Company, the acrobats in Cirque du Soleil, the slam-dunking professional basketball players in the NBA, or the action hero who narrowly escapes death by leaping across rooftops of burning buildings and jumping to safety on a passing helicopter.

What we discuss in our paper is that watching an aesthetically pleasing movement might drive activity within certain parts of the AON because interest in the movement drives sensorimotor regions to "work harder" to try and embody the movement (Cross et al. 2011). As such, we argue our findings lend tentative support to Freedberg and Gallese's (2007) proposal that a perceiver's use of his or her own body to simulate what is seen in an artwork (or any stimulus) is related to his or her aesthetic experience of that work. Of course, this relationship between lack of physical ability, favorable aesthetic evaluation, and AON activity reflecting an increased drive to embody a movement remains speculative at this stage, and it will require careful empirical evaluation. Moreover, my team's 2011 study is not without shortcomings. One major limitation to this study is that we relied on participants' self report of physical ability to reproduce movements, instead of actually asking them to attempt to perform each movement they watched.[1]

As a next step toward understanding how physical experience is related to aesthetic evaluation, work in my laboratory is underway to test how learning to perform a dance sequence affects participants' aesthetic appraisal of it. In one experiment, we implemented a between-subjects training design with 60 dance-naïve participants. Our question was whether aesthetic ratings of a dance sequence increase or decrease as a participant learns to perform that sequence. Because actual physical ability was assessed, we again chose to use a video-game paradigm that evaluates dance performance in an unbiased manner. However, we have taken a step forward in the complexity of movement scoring by using the Xbox Kinect™ system in conjunction with the "Dance Central 2" video game. This system uses sophisticated whole-body motion tracking and relays real-time feedback to participants as they dance along with an avatar on the monitor. In this behavioral study, participants were split into three groups of 20. They came into the laboratory across five consecutive days for training, and took part in a pre-training rating assessment before the first training session and a post-training rating assessment after the last training session. Among the three training groups, one group physically practised several routines of hip-hop/dance-club movements choreographed to a popular song, a second group just watched the avatar dance to same sequences while also listening to the music, and a third group only listened to the music from the dance sequences in each training session.

Contrary to what we found when dance naïve participants watched professional ballet dancers perform short dance sequences, the data from this between-subjects behavioral training study suggest that participants who physically practise a dance sequence across a week of training show significant increases in their aesthetic ratings of the dance movements on several dimensions (Kirsch et al. 2013). Notably, just watching the sequences and listening to the corresponding music, or just listening to the music alone did not correspond with an increase in enjoyment of any dimensions of the dance sequences, so we can be fairly confident that it is not a mere exposure or familiarity effect in the audio or visual domains. Rather, it appears physical performance of these particular dance sequences contributes to positive aesthetic appraisals. Data from our most recent training study, which implemented the dance training procedures described for Kirsch and colleagues (2013) and added pre- and post-training fMRI measures, replicate the behavioral effects described by Kirsch and colleagues (2013). Moreover, these latest neuroimaging data provide evidence for a shift in neural engagement from subcortical regions associated with dopaminergic reward processing when an observed movements are novel to cortical regions associated with multisensory integration and higher-level emotional processing once dance movements have been rehearsed (Kirsch et al., in press).

In terms of what might explain the contradictory findings between our fMRI investigation with novice dancers watching professional ballet dancers (Cross et al. 2011) and our latest training studies with novice dancers learning to perform hip-hop/street dance moves in a video game context (Kirsch et al. 2013; in press), we suggest that there are likely to be multiple ways in which embodiment and aesthetic preferences interact. One possibility is that a U-shaped function, similar to the one we proposed in a related study on action familiarity (Cross et al. 2012), might partially explain why we derive intense pleasure from watching movement that is familiar, practised, and well within our physical capabilities, as well as movements that are so spectacular and novel and far beyond our ability to reproduce them. Returning to the flashmob example discussed at the start of the chapter, I propose that the reason dancing flashmobs are so effective in engaging and delighting onlookers is because they satisfy both possibilities, thus resulting in maximum enjoyment. First, flashmobs feature dancers that look like all other pedestrians in the crowded public space in which they materialize, and onlookers can easily imagine dropping their own shopping bags and joining in (if only they knew the choreography). Second, dancing flash mobs elevate the common pedestrian to the exalted status of a dancer who is extremely adept at performing a sequence of movements that are often highly complex and have required hours of practice.[2] Thus, the appeal of flashmobs might lie in the fact that they are made up of individuals who look much like the onlookers, and yet move in a complex, coordinated way that invites others to dance along, even if they do not know the moves.

10.6 Next steps in using dance to explore embodiment and aesthetics

The overarching aim of this chapter has been to highlight two areas where the convergence of dance and neuroscience has the potential to further our understanding of the human brain. The first is that action perception and production share common neural substrates, and dance provides a useful means for exploring this relationship. The second is that notions of embodiment and an observer's prior sensorimotor experience might help to illuminate the parameters guiding why and when we derive pleasure from watching others move. As has now been suggested by a number of researchers with expertise that spans the dance and neuroscience domains, ongoing collaboration between the dance and scientific communities stands to benefit both disciplines (Bläsing et al. 2010, 2012; Cross 2010; Cross and Ticini 2012; Hagendoorn 2004, 2010; Jola 2010; Orgs et al. in press). While most of the work described in this chapter has used dancers as participants or dance learning/perception as a paradigm for studying full-body movement without much concern for the artistic properties of dance per se, the field of neuroaesthetics is one where science has the potential to benefit the dance world (depending, of course, on the right questions being asked). Along these lines, one platform that is encouraging collaborations between scientists and dancers from the Forsythe Company is the Dance

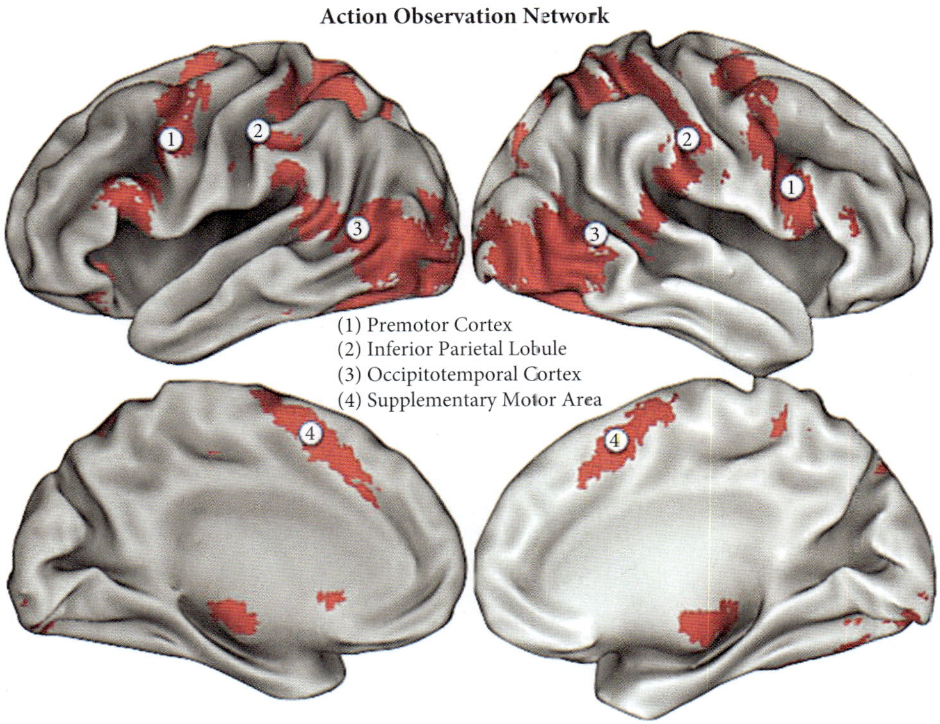

Figure 10.1 Schematic of the Action Observation Network.
The action observation network is distributed bilaterally and comprises brain regions active when watching other people's bodies in motion. The core regions of this network are: (1) the ventral and dorsal premotor cortices; (2) the inferior parietal lobule, often extending posteriorly along the intraparietal sulcus; (3) more anterior portions of the occipitotemporal cortex, including posterior superior temporal sulcus as well as the lateral occipital complex; and (4) the supplementary and pre-supplementary cortices located on the medial surface of each hemisphere. The particular set of brain regions illustrated here came from a contrast that compared watching dancers in motion compared to dancers standing still.
Figure as originally published in Emily S. Cross, Louise Kirsch, Luca F. Ticini, and Simone Schütz-Bosbach (2011), The impact of aesthetic evaluation and physical ability on dance perception, *Frontiers in Human Neuroscience*, 5, figure 2 doi: 10.3389/fnhum.2011.00102.

Engaging Science interdisciplinary research workshop series (<http://motionbank.org/en/research-2/>). As a consequence of this series, a number of exciting new collaborations focused on how the neuroscience of action and perception relates to movement aesthetics are being forged with equal input from dancers and scientists (e.g. Waterhouse et al. 2014). While such interdisciplinary work necessarily faces myriad challenges of accommodating different vocabularies, priorities, and goals, it also sets us on a path worth exploring, and will hopefully further elucidate the links between the embodiment and enjoyment when watching or performing dance.

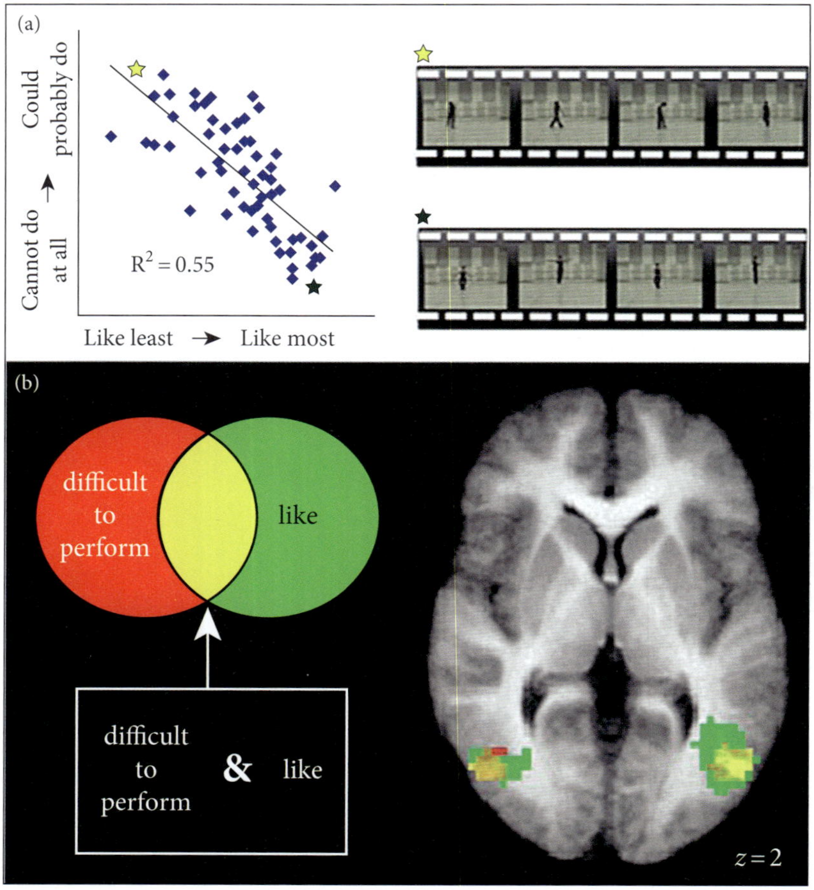

Figure 10.2 Summary of behavioral and neuroimaging findings from Cross et al. 2011.
a: Panel A illustrates the negative correlation between liking ratings and perceived physical ability ratings. The scatter plot shows how participants liked least those movements they thought they could easily reproduce, such as a dancer walking simply in a circle. The light-colored star at the top left of the plot corresponds to the movement detailed in the top video clip on the right. Participants liked most those movements that they rated as extremely difficult to reproduce. The dark-colored star at the bottom right of the scatter plot corresponds to the video clip on the bottom right of panel A. In this sequence, the dancer performs an *entrechat six*, followed by a *demi-plié* as a preparation for a double *tour en l'air* (put simply, this is a jump where the dancer performs six crossings of his pointed feet, followed by a jump with straight legs where he performs two consecutive rotations). **b:** Panel B illustrates some of the occipitoteomporal brain regions that emerge from the conjunction analysis interrogating regions that increase in activation the more difficult participants rate the movements to be (in red), brain regions that become increasingly active the more participants like a movement (in green), and brain regions that increase in activity both as a movement is rated as more enjoyable to watch *and* more difficult to perform (in yellow). Figure as originally published in Emily S. Cross, Louise Kirsch, Luca F. Ticini, and Simone Schütz-Bosbach (2011), The impact of aesthetic evaluation and physical ability on dance perception, *Frontiers in Human Neuroscience*, 5, figure 2 doi: 10.3389/fnhum.2011.00102.

Acknowledgments

Thanks are due to Richard Ramsey and Alison Wiggett for helpful comments on earlier versions of this chapter. The author's work presented in this chapter was performed primarily during doctoral training at Dartmouth College (USA) and postdoctoral training at the Max Planck Institute for Cognitive and Brain Sciences (Germany), and forms the foundation of a larger series of studies underway within the Social Brain in Action Laboratory at Bangor University. Financial support from the National Institute of Neurological Disorders and Stroke, the Max Planck Society, the Alexander von Humboldt Foundation, the Volkswagen Foundation, the Economic and Social Research Council, and the Dutch Science Foundation is gratefully acknowledged.

Notes

1 While this is a shortcoming of the study by Cross and colleagues (2011), it must be stated that we used the same physical ability ratings scale that we used in the earlier study by Cross and team (2006). We found a close correspondence between self-report ratings of physical ability and actual performance ability, but in the 2006 study, participants were expert dancers who, it could be argued, are much more sensitive to their own physical abilities and better at assessing what movements they could or could not reproduce.

2 Of course, there are undoubtedly elements of awe and respect felt by the spectators as the dancers bare their souls through movement in very close proximity to other people in public—unfortunately, addressing larger issues of performance context, audience expectations, and how they relate to the enjoyment of art is beyond the scope of this chapter.

References

Andersen, R.A., Meeker, D., Pesaran, B., et al. (2004). Sensorimotor transformations in the posterior parietal cortex. In M. Gazzaniga (ed.), *The Cognitive Neurosciences III*. Cambridge, MA: The MIT Press, pp. 463–74.

Badets, A., Blandin, Y., and Shea, C.H. (2006). Intention in motor learning through observation. *Quarterly Journal of Experimental Psychology (Colchester)* **59**(2), 377–86.

Bandura, A. (1977). *Social learning theory*. Englewood Cliffs, NJ: Prentice-Hall.

Bandura, A. (1986). *Social foundations of thought and action: A social cognitive theory*. Englewood Cliffs, NJ: Prentice-Hall.

Barzouka, K., Bergeles, N., and Hatziharistos, D. (2007). Effect of simultaneous model observation and self-modeling of volleyball skill acquisition. *Perceptual and Motor Skills* **104**(1), 32–42.

Berlyne, D.E. (1974). *Studies in the new experimental aesthetics: Steps toward an objective psychology of aesthetic appreciation*. Washington DC: Hemisphere Co.

Blandin, Y., Lhuisset, L., and Proteau, L. (1999). Cognitive processes underlying observational learning of motor skills. *Quarterly Journal of Experimental Psychology: Human Experimental Psychology* **52A**, 957–79.

Blandin, Y. and Proteau, L. (2000). On the cognitive basis of observational learning: development of mechanisms for the detection and correction of errors. *Quarterly Journal of Experimental Psychology* **53**(3), 846–67.

Bläsing, B., Calvo-Merino, B., Cross, E.S., et al. (2012). Neurocognitive control in dance perception and performance. *Acta Psychologica* **139**(2), 300–8.

Bläsing, B., Puttke, M., and Schack, T. (eds) (2010). *The Neurocognition of Dance*. Hove: Psychology Press.

Bouquet, C.A., Gaurier, V., Shipley, T., et al. (2007). Influence of the perception of biological or non-biological motion on movement execution. *Journal of Sports Science* **25**(5), 519–30.

Brown, S., Gao, X., Tisdelle, L., et al. (2011). Naturalizing aesthetics: brain areas for aesthetic appraisal across sensory modalities. *NeuroImage* **58**(1), 250–8.

Brown, S. and Dissanayake, E. (2009). The arts are more than aesthetics: Neuroaesthetics as narrow aesthetics. In M. Skov and O. Vartanian (eds), *Neuroaesthetics*. Amityville, NY: Baywood, pp. 43–57.

Brown, S., Martinez, M.J., and Parsons, L.M. (2006). The neural basis of human dance. *Cerebral Cortex* **16**(8), 1157–67.

Calvo-Merino, B., Glaser, D.E., Grezes, J., et al. (2005). Action observation and acquired motor skills: an FMRI study with expert dancers. *Cerebral Cortex* **15**(8), 1243–9.

Calvo-Merino, B., Grezes, J., Glaser, D.E., et al. (2006). Seeing or doing? Influence of visual and motor familiarity in action observation. *Current Biology* **16**(19), 1905–10.

Calvo-Merino, B., Jola, C., Glaser, D.E., et al. (2008). Towards a sensorimotor aesthetics of performing art. *Conscious Cognition* **17**(3), 911–22.

Calvo-Merino, B., Ehrenberg, S., Leung, D., et al. (2010). Experts see it all: configural effects in action observation. *Psychology Research* **74**(4), 400–6.

Carroll, W.R. and Bandura, A. (1990). Representational guidance of action production in observational learning: a causal analysis. *Journal of Motor Behavior* **22**(1), 85–97.

Cisek, P. and Kalaska, J.F. (2004). Neural correlates of mental rehearsal in dorsal premotor cortex. *Nature* **431**(7011), 993–6.

Cross, E.S., Kraemer, D., Hamilton, A.F.deC., et al. (2009). Sensitivity of the action observation network to physical and observational learning. *Cerebral Cortex* **19**(2), 315–26.

Cross, E.S. (2010). Building a dance in the human brain: insights from expert and novice dancers. In B. Bläsing, M. Puttke, and T. Schack (eds), *The Neurocognition of Dance*. London: Psychology Press.

Cross, E.S., Mackie, E., Wolford, G., et al. (2010). Contorted and ordinary body postures in the human brain. *Experimental Brain Research* **204**(3), 397–407.

Cross, E. S., Kirsch, L., Ticini, L., et al. (2011). The impact of aesthetic evaluation and physical ability on dance perception. *Frontiers in Human Neuroscience* **5**, 102.

Cross, E.S., Liepelt, R., Hamilton, A.F., et al. (2012). Robotic movement preferentially engages the action observation network. *Human Brain Mapping* **33**(9), 2238–54.

Cross, E.S., Stadler, W., Parkinson, J., et al. (2013). The influence of visual training on predicting complex action sequences. *Human Brain Mapping* **34**(2), 467–86.

Cross, E.S., Hamilton, A.F., and Grafton, S.T. (2006). Building a motor simulation de novo: Observation of dance by dancers. *Neuroimage* **31**(3), 1257–67.

Cross, E.S. and Ticini, L.F. (2012). Neuroaesthetics and beyond: new horizons in applying the science of the brain to the art of dance. *Phenomenology and the Cognitive Sciences* **11**(1), 5–16.

Dean, L. (1982). "Skylight," (Charleston: Spoleto Festival, USA).

Dils, A. and Albright, A.C. (2001). *Moving History/Dancing Cultures: A Dance History Reader*. Middletown, CT: Wesleyan University Press.

di Pellegrino, G., Fadiga, L., Fogassi, L., et al. (1992). Understanding motor events: a neurophysiological study. *Experimental Brain Research* **91**, 176–80.

Doody, S.G., Bird, A.M., and Ross, D. (1985). The effect of auditory and visual models on acquisition of a timing task. *Human Movement Science* **4**, 271–81.

Fadiga, L., Fogassi, L., Pavesi, G., et al. (1995). Motor facilitation during action observation: a magnetic stimulation study. *Journal of Neurophysiology* **73**(6), 2608–11.

Fadiga, L., Buccino, G., Craighero, L., et al. (1999). Corticospinal excitability is specifically modulated by motor imagery: a magnetic stimulation study. *Neuropsychologia* **37**(2), 147–58.

Ferrari, M. (1996). Observing the observer: self-regulation in the observational learning of motor skills. *Developmental Review* **16**, 203–40.

Flynn, E. and Whiten, A. (2010). Studying children's social learning experimentally 'in the wild'. *Learning and Behavior* **38**(3), 284–96.

Freedberg, D. and Gallese, V. (2007). Motion, emotion and empathy in esthetic experience. *Trends in Cognitive Sciences* **11**(5), 197–203.

Frey, S.H. and Gerry, V.E. (2006). Modulation of neural activity during observational learning of actions and their sequential orders. *Journal of Neuroscience* **26**(51), 13194–201.

Gallese, V., Fadiga, L., Fogassi, L., et al. (1996). Action recognition in the premotor cortex. *Brain* **119**(Pt 2), 593–609.

Gallese, V., Gernsbacher, M.A., Heyes, C., et al. (2011). Mirror Neuron Forum. *Perspectives on Psychological Science* **6**, 369–407.

Gallese, V. and Sinigaglia, C. (2011). What is so special about embodied simulation? *Trends in Cognitive Sciences* **15**(11), 512–9.

Grafton, S.T., Arbib, M.A., Fadiga, L., et al. (1996). Localization of grasp representations in humans by positron emission tomography. 2. Observation compared with imagination. *Experimental Brain Research* **112**(1), 103–11.

Grafton, S.T. (2009). Embodied cognition and the simulation of action to understand others. *Annals of the New York Academy of Sciences* **1156**, 97–117.

Grosbras, M.H., Beaton, S., and Eickhoff, S.B. (2012). Brain regions involved in human movement perception: a quantitative voxel-based meta-analysis. *Human Brain Mapping* **33**(2), 431–54.

Hagendoorn, I.G. (2004). Towards a neurocritique of dance. *BalletTanz Yearbook* 62–7.

Hagendoorn, I.G. (2010). Dance, choreography, and the brain. In D. Melcher and F. Bacci (eds), *Art and the Senses*. Oxford: Oxford University Press, pp. 499–514.

Hayes, A.E., Paul, M.A., Beuger, B., et al. (2008). Self produced and observed actions influence emotion: the roles of action fluency and eye gaze. *Psychological Research* **72**(4), 461–72.

Hodges, N.J., Williams, A.M., Hayes, S.J., et al. (2007). What is modelled during observational learning? *Journal of Sports Science* **25**(5), 531–45.

James, W. (1890). *Principles of psychology*. New York, NY: Holt.

Jola, C. (2010). Research and choreography: merging dance and cognitive neuroscience. In B. Bläsing, M. Puttke, and T. Schack (eds), *The Neurocognition of Dance*. Hove: Psychology Press, pp. 203–34.

Jola, C., Abedian-Amiri, A., Kuppuswamy, A., et al. (2012). Motor simulation without motor expertise: enhanced corticospinal excitability in visually experienced dance spectators. *PloS One* **7**(3), e33343.

Jola, C., Davis, A., and Haggard, P. (2011). Proprioceptive integration and body representation: insights into dancers' expertise. *Experimental Brain Research* **213**(2–3), 257–65.

Kirsch, L., Dawson, K., and Cross, E.S. (in press). Dance experience sculpts aesthetic perception and related brain circuits. *Annals of the New York Academy of Sciences*.

Kirsch, L., Drommelschmidt, K., and Cross, E.S. (2013). The impact of sensorimotor experience on affective evaluation of dance. *Frontiers in Human Neuroscience* **7**, 521.

Lee, T.D., White, M.A., and Carnahan, H. (1990). On the role of knowledge of results in motor learning: exploring the guidance hypothesis. *Journal of Motor Behavior* **22**(2), 191–208.

Massaro, D.W. and Friedman, D. (1990). Models of integration given multiple sources of information. *Psychological Review* **97**(2), 225–52.

Mattar, A.A. and Gribble, P.L. (2005). Motor learning by observing. *Neuron* **46**(1), 153–60.

McCullagh, P., Weiss, M.R., and Ross, D. (1989). Modeling considerations in motor skill acquisition and performance: an integrated approach. *Exercise and Sports Science Reviews* **17**, 475–513.

Molenberghs, P., Cunnington, R., and Mattingley, J.B. (2012). Brain regions with mirror properties: a meta-analysis of 125 human fMRI studies. *Neuroscience and Biobehavioral Reviews* **36**(1), 341–9.

Mountcastle, V.B. (1975). The view from within: pathways to the study of perception. *Johns Hopkins Medical Journal* **136**(3), 109–31.

Mountcastle, V.B., Lynch, J.C., Georgopoulos, A., et al. (1975). Posterior parietal association cortex of the monkey: command functions for operations within extrapersonal space. *Journal of Neurophysiology* **38**(4), 871–908.

Orgs, G., Caspersen, D., and Haggard, P. (in press). You move, I watch, it matters: aesthetic communication in dance. To appear in S. Obhi and E. Cross (eds), *Shared Representations: Sensorimotor Foundations of Understanding Others*. Cambridge: Cambridge University Press.

Orgs, G., Dombrowski, J.H., Heil, M., et al. (2008). Expertise in dance modulates alpha/beta event-related desynchronization during action observation. *European Journal of Neuroscience* **27**(12), 3380–4.

Orians, G.H. (2001). An evolutionary perspective on aesthetics. *Bulletin of Psychology and the Arts* **2**(1), 25–9.

Prinz, W. (1990). A common coding approach to perception and action. In O. Neumann and W. Prinz (eds), *Relationships between perception and action: Current approaches*. Berlin: Springer-Verlag, pp. 167–201.

Prinz, W. (1997). Perception and action planning. *European Journal of Cognitive Psychology* **9**, 129–54.

Prinz, W. (2006). What re-enactment earns us. *Cortex* **42**(4), 515–7.

Ramachandran, V.S. and Hirstein, W. (1999). The science of art. *Journal of Consciousness Studies* **6**(6–7), 15–51.

Rizzolatti, G., Fadiga, L., Gallese, V., et al. (1996). Premotor cortex and the recognition of motor actions. *Brain Research and Cognitive Brain Research* **3**(2), 131–41.

Rizzolatti, G., Fogassi, L., and Gallese, V. (2001). Neurophysiological mechanisms underlying the understanding and imitation of action. *Nature Reviews Neuroscience* **2**(9), 661–70.

Rizzolatti, G. and Sinigaglia, C. (2010). The functional role of the parieto-frontal mirror circuit: interpretations and misinterpretations. *Nature Reviews Neuroscience* **11**(4), 264–74.

Sanders, A.F. (1967). Some aspects of reaction processes. *Acta Psychologica (Amsterdam)* **27**, 115–30.

Sanders, A.F. (1983). Towards a model of stress and human performance. *Acta Psychologica (Amsterdam)* **53**(1), 61–97.

Schmidt, R.A. (1975). A schema theory of discrete motor skill learning. *Psychological Review* **82**, 225–60.

Schmidt, R.A. and Lee, T.D. (2005). *Motor control and learning: A behavioral emphasis* (4th edn). Champaign, IN: Human Kinetics.

Sheffield, F.D. (1961). Theoretical consideration in the learning of complex sequential task from demonstration and practice. In A.A. Lumsdaine (ed.), *Student response in programmed instruction*. Washington, DC: National Academy of Sciences—National Research Council.

Stevens, C., Winskel, H., Howell, C., et al. (2010). Perceiving dance: schematic expectations guide experts' scanning of a contemporary dance film. *Journal of Dance Medicine and Science* **14**(1), 19–25.

Torriero, S., Oliveri, M., Koch, G., et al. (2007). The what and how of observational learning. *Journal of Cognitive Neuroscience* **19**(10), 1656–63.

Waterhouse, E., Watts, R., and Bläsing, B. (2014). Doing *Duo* – a case of entrainment in William Forsythe's choreography "*Duo*". *Frontiers in Human Neuroscience* **15**, 1008.

Welford, A.T. (1968). *Fundamentals of skill*. London: Methuen.

Wiggett, A.J., Hudson, M., Clifford, A., et al. (2012). Doing, seeing, or both: effects of learning condition on subsequent action perception. *Social Neuroscience* **7**(6), 606–21.

Winkielman, P. and Cacioppo, J.T. (2001). Mind at ease puts a smile on the face: psychophysiological evidence that processing facilitation increases positive affect. *Journal of Personality and Social Psychology* **81**, 989–1000.

Yin, T.C. and Mountcastle, V.B. (1977). Visual input to the visuomotor mechanisms of the monkey's parietal lobe. *Science* **197**(4311), 1381–3.

Zelaznik, H. and Spring, J. (1976). Feedback in response recognition and production. *Journal of Motor Behavior* **8**(4), 309–12.

Chapter 11

Sensorimotor aesthetics: Neural correlates of aesthetic perception of dance

Beatriz Calvo-Merino

11.1 Do we need a neuroaesthetics of dance?

The study of neuroaesthetics has expanded in the last decade, and with the use of cognitive neuroscience techniques, such as functional magnetic resonance imaging (fMRI) or magnetoencephalography (MEG), this recent work has aimed at elucidating the neural underpinnings of aesthetic experience. Many of these studies have focused on the aesthetic perception of beautiful compared to less beautiful artworks (classified according to individual preference), revealing some neural mechanisms for the visual perception of pleasant stimuli, or for aesthetic judgment itself (Cela-Conde et al. 2004; Jacobson et al. 2006; Kawabata and Zeki 2004). Alongside studying the reception of paintings and sculptures as artworks (di Dio et al. 2007), a smaller proportion of research has studied the aesthetic perception and neural underpinnings of music (Blood and Zatorre 2001; Petrini et al. 2011). By and large, this work is based on our in-depth knowledge about visual perception, auditory processing, or decision making (for an extended review on neuroaesthetics of artwork and music studies, see other chapters in this volume; di Dio and Gallese 2009; Nadal et al. 2008).

Another large body of work in the cognitive neurosciences investigates the neural bases of movement processing, with a growing number of studies investigating the role of human actions and the human body (Jeannerod 1997) for social interactions (Johnson and Shiffrar 2013). This research has set the necessary neurocognitive grounds to start exploring the processing of a different form of art: the performing arts.

The idea of a neuroaesthetics of performing arts (and dance in particular) has won a dedicated following in the recent neuroaesthetic literature. To date, a few studies have focused on understanding the neural processes associated with watching and enjoying dance. One may wonder if the study of the neural correlates of the aesthetic experience of performing arts, or dance in particular, is any different from the neuroaesthetic studies of any other type of art, such as painting, sculpture, or music, and therefore whether the neuroaesthetics of dance demands a special role in this type of research. This chapter, and Chapters 10 and 12, provide a series of arguments as to why it should. Here, we will revise the current knowledge about the neural mechanisms associated with the aesthetic perception of dance, and we evaluate avenues for future research. We will describe three reasons that justify the consolidation of the neuroaesthetics of dance as a special discipline within the broader umbrella of neuroaesthetics. The two following chapters will focus on two very important issues that relate to the aesthetic appreciation of dance: (1) the specific effect of dance experience and familiarity during aesthetic perception of dance (Chapter 10), and (2) the evolution of dance research towards a more ecological approach to investigate aesthetic experience in performing arts (Chapter 13).

11.2 The body as a dynamic human canvas

The first, simplest, and yet fundamental differences between dance and other kinds of art is that in dance, both the artist and the artwork are contained within the human body. The body itself (or to be more precise, the dancer's body) acts as a kind of dynamic canvas from which dance movements flow. The spectator of a dance performance is visually exposed to intertwined bodies and movements. From a neuroscientific perspective, the mere observation of such a scene engages specific brain mechanisms that process information about bodily movements. The underlying brain regions processing this information are significantly different from those associated with static visual stimuli (such as objects) or auditory information (music). Understanding the specific neural mechanisms for the processing of bodily movements therefore provides a starting point for many of the experimental studies that will be presented here and in the following chapters. We start with a brief introduction about what is currently known about the processing of bodily movements.

11.2.1 Specific brain mechanism for observing bodily movements

Neuroscience has shown that biologically relevant stimuli such as faces and bodies are processed in a different way to other non-biological stimuli (i.e. objects). At the neural level, one specific brain region, the fusiform face area (FFA), has been found to respond specifically to the perception of human faces (Kanwisher et al. 1997). It is posited that a similarly specialized brain region participates during processing of human bodies. Experiments using fMRI have identified two body-specific brain regions. The first region was found in the lateral occipito-temporal part of the human brain. This area has been labeled the extrastriate body area (EBA) (Downing et al. 2001) and it responds to the observation of human bodies or body parts (arms, legs, etc.), as well as to the observation of bodies depicted in photographs, drawings, or movies (Peelen and Downing 2005). A second body-selective area was recently located in the fusiform gyrus, very close to where the FFA has been located. This region is known as the fusiform body area (FBA) (Peelen and Downing 2005), and it responds to the observation of whole bodies and body parts. These two body areas seem to have separate functions during body processing. For example, while the EBA responds to simple presentation of bodies and body parts, the FFA is largely activated during the presentation of complex body configurations (rather than a simple body part, i.e. an arm) (Taylor et al. 2007).

This neural specificity for faces and human bodies is also accompanied by a particular perceptual way of seeing. Two types of visual mechanisms have been described in perception: configural and analytical processing. While the perceptual process of objects involves an analytical perception of the different parts composing the object, the perception of faces and bodies triggers a more configural way of seeing (Valentine 1988; Reed et al. 2003). In the latter, the stimulus is processed as a whole, providing a holistic perception of the observed stimuli from a very early visual stage (Maurer et al. 2002). Configural processing has been interpreted as a more efficient and fluent machinery for understanding external events. Research has concluded that this special processing scheme may have appeared during evolution due to the importance of faces and bodies in life, and the high level of visual exposure that we have acquired over our evolutionary history.

11.2.2 Specific brain mechanisms to watch the dynamic human body

Only through movement can humans interact with the world. It is through movement that we are able to communicate with others, express our emotions, and interact with external objects. The performing arts, and dance in particular, have made this particular human feature their essential tool. Dance is dynamic and involves the displacement of a human body in time and space. As a consequence, dance observation may recruit a particular neural mechanism involved in movement processing that is qualitatively different from processes involved in merely watching static objects.

Traditionally, movement research has been concerned with the neural foundations of movement execution and motor control (Jeannerod 1997). Each acquired movement is internally represented in our own neural and cognitive system in the form of motor representations. Recent research has focused on the internal components of such representations rather than the overt movement. These representations play an important role during action execution, but interestingly for our discussion, also during action observation, movement understanding, and as we will see later, aesthetic perception of movement.

To summarize some of this work, let us look into Giacomo Rizzolatti's laboratory in Parma University in the early 1990s. In a series of studies using cell recoding in primates, his team discovered a group of neurons in the ventral premotor cortex (vPMc) of a primate that spiked in a similar fashion when a monkey was making a movement, and when the animal was merely observing the experimenter performing the same action (di Pellegrino et al. 1992; Rizzolatti et al. 1996). These neurophysiological studies marked a turning point in the field because they described for the first time a common coding mechanism for executed and observed actions. Similar results were found in other areas in the primate brain, such as the parietal cortex (Gallese et al. 1996; see Rizzolatti and Craighero 2004 for a review). Collectively, this work suggests that the motor system is recruited in a similar manner during both motor execution and the observation of movements. Movement representations thus appear to store motor properties and information associated with acquired movements (movements that we have learnt to perform), and provide an entrée to internal motor knowledge just by action observation. In other words, we can see other people's movement with our own movement repertoire. Neurons with these visual and motor properties were called *mirror neurons*, as they seemed to reflect, like a mirror, observed actions as much as actions from our own motor repertoire.

The first evidence for coexisting action observation and action execution systems in the human brain came from a transcranial magnetic stimulation (TMS) study performed by Fadiga and colleagues (1995). These authors reported similar muscle excitability patterns during execution and observation of the same action. This is a fascinating result that illustrates how the muscles from the physical effectors involved in an action (i.e. arm and hand during a grasping movement) are also activated in a similar manner when we observe someone else performing that same action (i.e. when we observe someone else grasping). This study and subsequent work (Strafella and Paus 2000; Baldissera et al. 2001) have provided increasing evidence for internal motor resonance in the human brain.

Consistent with the original work in non-human primates, human fMRI studies have sought to identify brain regions that participate in both action execution and action observation. Brain activation during action observation has consistently been reported in ventral and dorsal premotor cortex (vPM, dPM), areas in the parietal cortices, such as the inferior parietal lobe (IPL) and superior parietal lobe (SPL) and superior temporal sulcus (STS) (Decety et al. 1997; Grafton et al. 1996; Iacoboni et al. 2001; Rizzolatti et al. 2001). From these regions, the premotor and parietal cortex have well-defined functions during action execution, and may represent the core of the so-called human mirror neurons system that links action execution and action observation. The STS responds to action observation, but this area does not assume any motor functions and therefore, strictly speaking, does not fall into the category of "mirror region." However, the STS may be part of a general action observation system (that goes beyond mirror neuron regions), because it responds to visual (and non-visual) presentation of other type of social stimuli (Allison et al. 2000; Puce and Perret 2003). Finally, some other areas that are classically identified during action execution, such as the supplementary motor area and primary motor cortex, are normally not activated during action observation, unless there is a motor preparation element involved; for example, if the task includes a subsequent motor imitation (Grèzes and Decety 2001). For a more detailed overview of the human mirror neuron system see Rizzolatti and Sinigaglia (2010).

Finally, cognitive psychology has also explored whether movement processing engages a series of specific mechanisms similar to the ones described for faces and human bodies. Several studies have shown that recognizing the performing agent, his/her gender, or identifying the action may require a kind of global or configural process of the movement (Mather and Murdoch 1994; Troje 2002). This configural process is particularly engaged when observing familiar actions (Calvo-Merino et al. 2010). Recognizing other action features such as the emotion represented in a movement can use more analytical or visual analysis (Pollick et al. 2001), although other studies have shown that a similar task of reading emotion can be performed using only configural information (Dittrich et al. 1996).

Overall, this series of studies sheds light on several neurocognitive processes for body and movement observation that should be taken into account when investigating the aesthetic appreciation of dance.

11.2.3 The dancer's body: a human gem

A second though no less important issue to consider when investigating the neuroaesthetics of dance is the special quality of a dancer's body. The physical training entailed in learning this art makes dancers unique by affording them a set of movements that far exceeds the type of movements our body articulations normally allow. Dancers' flexibility and smoothness during movement execution is not comparable to the motor ability of any of the rest of us. Dancers' bodies are special, and this is reflected not only in their physical bodies and how they move, but also in some of their cognitive abilities (Blasing et al. 2012), ranging from basic motor control mechanism, including equilibrium, posture, and complex movement execution, to the sense of timing, motor memory, motor imagery, or spatial transformations. Here, we will briefly describe how dancers' acquired motor skills influence the way action and movements are processed.

The first imaging study describing an expertise effect during action observation compared brain activity of different types of motor experts (i.e. ballet dancers and capoeristas) when observing familiar and non-familiar movements (Calvo-Merino et al. 2005). Both groups observed a number of three-second video clips of standard ballet dance and capoeira movements. These movements were performed by professional dancers dressed in black garments in front of a blue background. The dancers' faces were blurred. (These last observations are important as the study aimed to isolate brain activity related to the observation of movements, whilst controlling for other elements often associated to a dance performance such as costumes, choreography, music, or facial expression.) This study demonstrated different brain activations when dancers were observing dance movements that were familiar to them (ballet dancers watching ballet movements, and capoeristas watching capoeira movements; see Chapter 10) compared to watching unfamiliar movements (ballet dancers watching capoeira movements, and capoeristas watching ballet movements). Observation of familiar movements elicited stronger brain responses in the bilateral premotor (ventral and dorsal regions), parietal cortex (bilateral intraparietal sulcus, and right superior parietal cortex) and left posterior superior temporal sulcus (pSTS).

These results are interesting because premotor and parietal regions have been included as part of a mirror neuron system. This suggests that if we observe an action we are familiar with, and therefore an action that is part of our motor repertoire, we automatically enhance the response of the corresponding motor representations when we merely watch that action. In general, the information content in these motor representations may be used during observation of that action and may help understand the meaning of this action, predict its output, and plan a complementary response. Interestingly, other areas, such as the pSTS, also responded more strongly when observing familiar movements, suggesting a complementary response associated with expertise outside the mirror mechanism, but also in areas associated with visual body and human processing (Calvo-Merino et al. 2005).

This latter result triggered new experiments that sought to clarify how much of this effect during observation of familiar movements is due to motor familiarity (having acquired the observed movement in our motor repertoire) or whether visual experience alone can explain these results (i.e. strong or prolonged visual exposure to the observed action, without having physically trained them). To disentangle this question, we developed an fMRI study that allowed isolation of motor response from visual responses (Calvo-Merino et al. 2006). With the help of choreographer Tom Sapsford, four sets of stimuli were created. The first two involved a male and a female dancer performing gender-specific ballet movements, whilst the other two involved a male and female dancer performing common ballet movements. It is important to note that classical ballet movements are well categorized. For example, some movements are gender specific and are therefore only performed by either male or female dancers. However, because both males and females train together, they both have extended visual familiarity with all the dance movements. In the experiment, male and female dancers therefore performed together a series of common movements with whom all dancers had extensive visual and motor training. The comparison between females and males watching a gender-specific movement (and using the common movements to control for gender differences while observing) then allowed for the identification of the specific brain responses associated only with motor familiarity. This comparison revealed stronger brain activations in premotor and parietal cortices, the classical mirror neurons regions. This study thus provided some of the first evidence for internal motor resonance during action observation.

The studies cited above made use of professional dancers in order to understand a question initiated in the cognitive neuroscience domain. They also showed that dancers' brain responses in some mirror neuron regions are stronger than the brain responses in non-dancers. These results demonstrate that the physical abilities of professional dancers provide them with a unique repertoire of movements that influence how they can observe and see actions, including actions that untrained individuals would find difficult to perform or to understand. Other studies using dancers have shown similar results (Cross et al. 2006; Orgs et al. 2008). Finally, it is important to note that the key comparison here is about the differences in brain activation between dancers and non-dancers. This marked difference does not imply that non-dancers do not have the ability for action observation, but it does suggest that the "resonance" is larger when we are familiar with the observed movement.

Beside differential neural responses during observation of dance movements, dancers' motor skills have also been shown to influence responses at the behavioral level. For example, in a similar expertise paradigm to the one described earlier, dancers showed better visual discrimination of familiar dance movements than non-dancers (Calvo-Merino et al. 2010). This suggests that the expertise effect that surrounds dancers (due to their exquisite motor skills) during action observation is observable at the neural and at the behavioral level. A further description of expertise modulations and how they are related to the aesthetic processing of dance is described in the preceding chapter.

11.2.4 The spectator's body: mirror, mirror

Despite the extraordinary motor abilities of dancers, their body-configural features remain similar to those of non-trained bodies. Human bodies of dancers and spectators (non-dancers) share the same structure (i.e. body parts, head, and limbs, organized in a specific configuration). This similarity evokes a unique relationship between the observer and the dancer (who is himself the artwork) at the sensorimotor or embodiment level.

Who has not closed their eyes or felt goosebumps when watching the beginning of Dali's film "Perro andaluz," or a dramatic painting representing a human in pain? When these physical human body features are depicted in art, they are often experienced viscerally (Freedberg and Gallese 2007). The previous sections have described the neural bases of experiencing our own and other people's body/movements in the context of action perception. This self–other matching mechanism (or mirror neuron mechanism) has been generalized to other aspects of social perception, such understanding of other's emotions, feeling empathy (Gallese 2005; de Vignemont and Singer 2006), and, potentially, aesthetic perception.

11.3 Neural correlates of aesthetics experience of dance

As a type of performing art, dance is intrinsically linked to the concept of the aesthetic. However, how is the human brain aware that dance may warrant an aesthetic response? Is there an ongoing aesthetic response every time we observe a dance performance? Or is there only a specific response when we tune in our aesthetic senses during the dance observation? These and other questions arise when we study any behavior associated with aesthetics in general. Here we aim to provide some answers, but it is worth noting that the field of the neuroaesthetics of dance is quite new, and the complex relationships between the observer/spectator, the dancer, the movement, and associated aesthetic processes offer a rich seam for the curious researcher.

11.3.1 Implicit neuroaesthetics of dance

The first study to focus on the neural correlates of implicit appreciation of dance sought to define brain responses that correlated with the level of implicit liking towards the observed dance movement (Calvo-Merino et al. 2008). The experiment was divided in two parts. The first session measured brain responses associated with the visual process of watching a series of dance movements comprising a variety of kinetic properties (speed, dance style, amount of displacement). In a second session, individual preferences of those movements were obtained. This enabled exploring how neural responses to observed movements related to the preference or liking of these movements. More specifically, during the fMRI session, brain activations were measured from participants without any dance experience whilst watching dance movements from two different disciplines (classical ballet and capoeira). The participants were asked to look at the dance movements and indicate how tiring they judged each of the movements to be. Importantly, during this session, there was no question or reference to the potential aesthetic quality of the movements. In a second session outside the scanner, participants were asked to explicitly rate all dance movements across several dimensions related to aesthetics experience. Five dichotomous scales were used: liking–disliking, simple–complex, interesting–dull, tense–relaxed, weak–powerful (Berlyne 1974). From this set of ratings, only the "liking–disliking" dimension correlated with brain activity changes. To explore the meaning of these activations, further analyses were performed. Dance movements were divided into two sets: one set contained the movements with the highest scores across the group of participants, and the other those with the lowest scores. This permitted identifying brain areas that were sensitive to this categorization. This analysis showed two regions that were sensitive to generally appealing and well-liked movements. These aesthetics-sensitive areas were localized in the early visual cortex (in the medial region—precuneus), and in the premotor cortex of the right hemisphere. The opposite comparison showed no significant results (i.e. when looking for brain regions sensitive to viewing less liked rather than liked movements). This finding indicates the existence of an automatic sensorimotor brain mechanism concerned with the observation of positive appreciation of dance movements. In particular, sensorimotor responses may be stronger when we see dance movements that we implicitly like (Calvo-Merino et al. 2008).

11.3.2 Explicit neuroaesthetics of dance

Recent work has extended this finding and investigated the brain activity changes associated with making an explicit aesthetic judgment about a dance movement (Cross et al. 2011). In this study, participants' brain responses were measured while they observed a series of dance movements. After each observation, participants were instructed to rate how much they liked each individual dance movement, or how well they could reproduce it (we will discuss this part of task later). The authors focused on brain regions that are known to respond to movement as compared to other features such as the dancers' body or costume. Their results showed that activation in the right parietal cortex (inferior parietal lobe, IPL) and occipitotemporal cortices (namely, motion area V5, MT, and body regions in inferior temporal gyrus, ITG, and middle temporal gyrus, MTG) correlated positively with the degree of liking. In other words, the higher the liking ratings for any particular movement, the stronger were the neural responses in these brain regions. For a detailed description of this study, please see Chapter 10, this volume.

The causal or correlational role of classical movement (premotor) and body areas (EBA) was investigated in a TMS study. Calvo-Merino and colleagues (2010) presented pairs of dance postures to participants with no dance experience who were asked to indicate which of two dance postures they liked more. Scrambled versions of the same body posture were used as control stimuli. Results demonstrated that repetitive TMS (rTMS) over the EBA region blunted aesthetic judgments about body postures relative to ventral PMC rTMS, thus disrupting the pattern of aesthetic preference observed for each participant in a rating session without stimulation.

This shows that both EBA and the premotor cortex contribute in a complementary manner to body aesthetic evaluation. Interestingly, we interpreted these results within the framework of the "dual-route model" of visual body perception (Urgesi et al. 2007), suggesting that disruption of the local system, housed in the EBA region, blunted the aesthetic sensitivity. To the opposite, the disruption of the ventral premotor cortex, involved in the configural processing of whole-body postures, heightened aesthetic sensitivity.

11.3.3 Sensorimotor aesthetics of dance

The available studies on neuroaesthetics of dance show that a combination of implicit and explicit aesthetic perception of dance movements activates a set of regions that involve the premotor cortex (in the dorsal region), parietal cortex (IPL), and occipitotemporal region (V5, MT, ITG, MTG, precuneus) (Calvo-Merino et al. 2008; Cross et al. 2011). These regions have one function in common: they all are activated during action observation. The premotor and parietal regions are part of the mirror neuron system and thus of a shared mechanism for both executing and seeing movements that could represent the neural links between the observed agent and the observer. Occipitotemporal regions have also been frequently observed during action observation and body perception. Interestingly, these areas do not have visuomotor properties (as is the case of premotor and parietal cortices). They are, however, part of a network of regions involved in body processing that seems to have been biologically tuned to respond to particular components of the human figure (Urgesi et al. 2007).

A broader neuroaesthetics perspective describes how the organization of artworks may reflect the properties of the human visual system (Zeki 1999). In the specific case of performing arts, the few studies that have explored the topic suggest a strong relationship between the sensory process involved in perceiving the dance movement and the associated positive aesthetic response. It is important to notice that both the studies by the teams led by Calvo-Merino and Cross found a number of brain regions in which activity was related to the degree of liking of observed actions. In other words, higher liking ratings correlate with stronger brain response in the corresponding sensorimotor regions. However, both studies failed to find brain regions in which activity related to the degree of disliking. This suggests that while we know that some brain regions are sensitive to the level of positive liking during movement perception, more research is needed to clarify the brain regions tuned when we observed dance movements that we dislike (or more accurately, have received lower ratings in the liking scale).

11.4 Neural correlates of dance's physical properties

Most studies conducted up until now have followed an individual or subjective approach that aimed to understand the brain responses correlating with preference behavior, regardless of what has been the output of that preference. This research focuses on the individual's brain not on his choices. However, in the neuroaesthetic of dance, there is a growing keenness to understand the relation between the dance itself (the movement) and the associated aesthetic response.

In one of our previous studies (Calvo-Merino et al. 2008), a first attempt to identify kinematic properties intrinsically linked to neural processes which are linked to aesthetic experience, movements were selected on the basis of four kinematic properties: speed, body part used, direction of movement, and vertical and horizontal displacement. In the analysis, we eliminated differences between individuals when generating a group average or consensus of liked versus less-liked movements. This permitted the identification of stimulus quantities that specifically modulated activity related to those aesthetically sensitive brain areas of that group of participants. In this way, we identified which specific dance movements were responsible for maximal and minimal activation for the two aesthetically sensitive brain regions previously described (dorsal premotor and occipital cortices). The results suggested that on average, these areas "preferred" whole-body movements, such as jumping in place, or with a significant displacement of the entire body in space (e.g. horizontal jumps). At the behavioral level, a similar type of analysis suggests that the kinematic properties of the movements that received the highest and lowest consensus on the liking score in the subjective rating show clear correspondence with the movements that target the brain areas revealed as aesthetically relevant in the functional imaging analysis. Overall these results suggest a high level of correspondence between some dynamic properties of the dance movements (speed, high vertical and horizontal displacement—jumps), a positive liking behavior, and the maximal activations in the above-mentioned aesthetic-sensitive regions.

Another movement parameter that has often been associated with the idea of dance as a domain for motor experts is the level of difficulty associated with performing the movement. Dancers have the extraordinary ability to perform extremely challenging movements in an effortless manner, and the non-trained observer often uses his or her judgment about the difficulty level during their aesthetic judgment. Perceived motor difficulty can also be described as the observer's evaluation on how well he or she could possibly perform the observed movements. Following this idea, Cross and colleagues (2011) asked about aesthetic experience during the fMRI session and participants rated how well they thought they could reproduced the observed movement. While no significant activations were found in the action observation network, the behavioral data indicated that liking and "perceived" physical or motor ability ratings were not entirely independent. In other words, those dance movements that they rated as difficult to perform were liked by participants more.

The idea behind this research opens a line of correspondence between neuroaesthetic research and dance. This can help to explore still further the relationship between aesthetic sensitivity (or in this case, neuroaesthetic sensitivity) and movement kinematics. Despite the different methodologies used by Calvo-Merino and colleagues (2008) and Cross and her team (2011), and the lack of opportunity for the direct comparison between studies, we could speculate that the physical movement descriptions of our study (i.e. jumps, large displacement) may also correspond to those that participants would have rated as more difficult to perform in the study by Cross and colleagues (2011). Having said that, it is interesting to note that while our study revealed aesthetically sensitive areas that correlated with the liking dimension, the study did not reveal significant brain activity changes related to other aesthetic components or dimensions described in Berlyne (1974), such as complexity or symmetry. Liking was the only dimension that evoked a significant aesthetic response. However, the complexity of the stimuli, or the perceived complexity, can be closely interpreted by observers as a feature to guide their interpretation of how well they think they could perform such movement. Further studies are needed to explore these matters and the relation between physical and aesthetic dimensions and the kinematic features associated with them.

11.5 Future directions

This chapter sought to provide a brief introduction to the neural mechanisms that participate during the aesthetic experience of dance perception. The direct correspondence between the observer's body and the performing dancers permits the consideration of the neuroaesthetics of dance from a sensorimotor or bodily perspective. This suggests that there is an ongoing internal representation of the dancer's body and movements in the spectator's mind while he or she watches a dance performance. This simulation may be passive but at a certain level it influences the final aesthetic judgment made of the observed dance piece. The judgment is also modulated by a large number of factors that have not been described here. Chapter 10 and Chapter 12 focus on understanding (a) how experience and familiarity participate in the aesthetic experience of dance, and (b) how research can be carried out on dance aesthetics beyond the simple body and movement embodiment approach, including essential features for an ecological dance experience.

Acknowledgment

Beatriz Calvo-Merino was supported by the Pump Priming Scheme (City University London), and the research grants funded by the Spanish Ministerio de Economía y Competitividad (RYC-2008-03090 and PSI2012-34558/PSIC). Special thanks to dancers and choreographers (Tom Sapsford and MavinKhoo) for their help and assistance during some of the projects described throughout this chapter.

References

Allison, T., Puce, A., and McCarthy, G. (2000). Social perception from visual cues: role of the STS region. *Trends in Cognitive Science* **4**, 267–78.

Baldissera, F., Cavallari, P., Craighero, L., et al. (2001). Modulation of spinal excitability during observation of hand actions in humans. *European Journal of Neuroscience* **13**, 190–4.

Berlyne, D.E. (1974). The new experimental aesthetics. In D.E. Berlyne (ed.), *Studies in the new experimental aesthetics: Steps toward an objective psychology of aesthetic appreciation*. Washington, DC: Hemisphere Publishing Corporation.

Bläsing, B., Calvo-Merino, B., Cross, E.S., et al. (2012). Neurocognitive control in dance perception and performance. *Acta Psychologica* **139**(2), 300–8.

Blood, A.J. and Zatorre, R.J. (2001). Intensely pleasurable responses to music correlate with activity in brain regions implicated in reward and emotion. *Proceedings of the National Academy of Sciences of the USA* **98**(20), 11818–23.

Calvo-Merino, B., Glaser, D.E., Grèzes, J., et al. (2005). Action observation and acquired motor skills: an fMRI study with expert dancers. *Cerebral Cortex* **15**(8), 1243–9.

Calvo-Merino, B., Grèzes, J., Glaser, D.E., et al. (2006). Seeing or doing? Influence of visual and motor familiarity in action observation. *Current Biology* **16**(22), 2277–2277.

Calvo-Merino, B., Jola, C., Glaser, D.E., et al. (2008). Towards a sensorimotor aesthetics of performing art. *Consciousness and Cognition* **17**(3), 911–22.

Calvo-Merino, B., Urgesi, C., Orgs, G., et al. (2010). Extrastriate body area underlies aesthetic evaluation of body stimuli. *Experimental Brain Research* **74**, 400–6.

Cela-Conde, C.J., Marty, G., Maestú, F., et al. (2004). Activation of the prefrontal cortex in the human visual aesthetic perception. *Proceedings of the National Academy of Sciences of the USA* **101**, 6321–25.

Cela-Conde, C.J., Ayala, F.J., Munar, E., et al. (2009). Sex-related similarities and differences in the neural correlates of beauty. *Proceedings of the National Academy of Sciences of the USA* **106**(10), 3847–52.

Cross, E.S., Hamilton, A.F.d.C., and Grafton, S.T. (2006). Building a motor simulation de novo: observation of dance by dancers. *Neuroimage* **31**(3), 1257–67.

Cross, E.S., Kirsch, L., Ticini, L.F., et al. (2011). The impact of aesthetic evaluation and physical ability on dance perception. *Frontiers in Human Neuroscience* **5**, 102.

Decety, J., Grèzes, J., Costes, N., et al. (1997). Brain activity during observation of actions. Influence of action content and subject's strategy. *Brain* **120**, 1763–77.

de Vignemont, F. and Singer, T. (2006). The empathic brain: how, when and why? *Trends in Cognitive Sciences* **10**(10), 435–41.

Di Dio, C. and Gallese, V. (2009). Neuroaesthetics: a review. *Current Opinion in Neurobiology* **19**(6), 682–7.

Di Dio, C., Macaluso, E., and Rizzolatti, G. (2007). The golden beauty: brain response to classical and renaissance sculptures. *PLoS One* **2**(11), e1201.

di Pellegrino, G., Fadiga, L., Fogassi, L., et al. (1992). Understanding motor events: a neurophysiological study. *Experimental Brain Research* **91**(1), 176–80.

Dittrich, W.H., Troscianko, T., Lea, S.E.G., et al. (1996). Perception of emotion from dynamic point-light displays represented in dance. *Perception* **25**(6), 727–38.

Downing, P.E., Jiang, Y., Shuman, M., et al. (2001). A cortical area selective for visual processing of the human body. *Science* **293**, 2470–3.

Fadiga, L., Fogassi, L., Pavesi, G., et al. (1995). Motor facilitation during action observation: a magnetic stimulation study. *Journal of Neurophysiology* **73**, 2608–11.

Freedberg, D. and Gallese, V. (2007). Motion, emotion and empathy in esthetic experience. *Trends in Cognitive Science* **11**, 197–203.

Gallese, V. (2005). Embodied simulation: from neurons to phenomenal experience. *Phenomenology and the Cognitive Sciences* **4**, 23–48.

Gallese, V., Fadiga, L., Fogassi, L., et al. (1996). Action recognition in the premotor cortex. *Brain* **119**, 593–609.

Grafton, S.T., Arbib, M.A., Fadiga, L., et al. (1996). Localization of grasp representations in humans by positron emission tomography. 2. Observation compared with imagination. *Experimental Brain Research* **112**, 103–11.

Grezes, J. and Decety, J. (2001). Functional anatomy of execution, mental simulation, observation, and verb generation of actions: a meta-analysis. *Human Brain Mapping* **12**, 1–9.

Iacoboni, M., Koski, L.M., Brass, M., et al. (2001). Reafferent copies of imitated actions in the right superior temporal cortex. *Proceedings of the National Academy of Sciences of the USA* **98**, 13995–9.

Jacobsen, T., Schubotz, R.I., Hofel, L., et al. (2006). Brain correlates of aesthetic judgment of beauty. *Neuroimage* **29**, 276–85.

Jeannerod, M. (1997). *The Cognitive Neuroscience of Action*. Oxford: Wiley-Blackwell.

Johnson, K.L. and Shiffrar, M. (2013). *People Watching: Social, Perceptual, and Neurophysiological Studies of Body Perception*. New York, NY: Oxford University Press.

Kanwisher, N., McDermott, J., and Chun, M.M. (1997). The fusiform face area: a module in human extrastriate cortex specialized for face perception. *Journal of Neuroscience* **17**(11), 4302–11.

Kawabata, H. and Zeki, S. (2004). Neural correlates of beauty. *Journal of Neurophysiology* **91**(4), 1699–705.

Mather, G. and Murdoch, L. (1994). Gender discrimination in biological motion displays based on dynamic cues. *Proceedings of the Royal Society B* **258**, 273–9.

Maurer, D., Le Grand, R., and Mondloch, C.J. (2002). The many faces of configural processing. *Trends in Cognitive Science* **6**, 255–60.

Nadal, M., Munar E., Capó, M.A., et al. (2008). Towards a framework for the study of the neural correlates of aesthetic preference. *Spatial Vision* **21**(3–5), 379–96.

Orgs, G., Dombrowski, J.H., Heil, M., et al. (2008). Expertise in dance modulates alpha/beta event-related desynchronization during action observation. *European Journal of Neuroscience* **27**(12), 3380–4.

Peelen, M.V. and Downing, P.E. (2005). Selectivity for the human body in the fusiform gyrus. *Journal of Neurophysiology* **93**, 603–8.

Petrini, K., Crabbe, F., Sheridan, C., et al. (2011). The music of your emotions: Neural substrates involved in detection of emotional correspondence between auditory and visual music actions. *PLoS One* **6**(4), 10. doi: e1916510.1371/journal.pone.0019165.

Pollick, F.E., Paterson, H.M., Bruderlin, A., et al. (2001). Perceiving affect from arm movement. *Cognition* **82**(2), B51–61.

Puce, A. and Perret, D. (2003). Electrophysiology and brain imaging of biological motion. *Philosophical Transactions of the Royal Society B* **358**(1431), 435–45.

Reed, C.L., Stone, V.E., Bozova, S., et al. (2003). The body inversion effect. *Psychological Science* **14**, 302–8.

Rizzolatti, G. and Craighero, L. (2004). The mirror-neuron system. *Annual Review of Neuroscience* **27**, 169–92.

Rizzolatti, G., Fadiga, L., Matelli, M., et al. (1996). Localization of grasp representations in humans by PET: 1. Observation versus execution. *Experimental Brain Research* **111**, 246–52.

Rizzolatti, G., Fogassi, L., and Gallese, V. (2001). Neurophysiological mechanisms underlying the understanding and imitation of action. *Nature Review Neuroscience* **2**, 661–70.

Rizzolatti, G. and Sinigaglia, C. (2010). The functional role of the parieto-frontal mirror circuit: Interpretations and misinterpretations. *Nature Reviews Neuroscience* **11**, 264–74.

Strafella, A.P. and Paus, T. (2000). Modulation of cortical excitability during action observation a transcranial magnetic stimulation study. *Neuroreport* **11**, 2289–92.

Taylor, J.C., Wiggett, A.J., and Downing, P.E. (2007). Functional MRI analysis of body and body part representations in the extrastriate and fusiform body areas. *Journal of Neurophysiology* **98**, 1626–33.

Troje, N.F. (2002). Decomposing biological motion: A framework for analysis and synthesis of human gait patterns. *Journal of Vision* **2**, 371–87.

Urgesi, C., Calvo-Merino, B., Haggard, P., et al. (2007). Transcranial magnetic stimulation reveals two cortical pathways for visual body processing. *Journal of Neuroscience* **27**, 8023–30.

Valentine, T. (1988). Upside-down faces—a review of the effect of inversion upon face recognition. *British Journal of Psychology* **79**, 471–91.

Zeki, S. (1999). *Inner Vision: An Exploration of Art and the Brain*. Oxford: Oxford University Press.

Chapter 12

Moving towards ecological validity in empirical aesthetics of dance

Julia F. Christensen and Corinne Jola

Both authors contributed to the chapter equally.

12.1 Far from ecological validity

12.1.1 Positioning dance in experimental neuroaesthetics

In order to conduct experimental research on the effects aesthetic objects can have on its recipients, empirical psychologists accept the reality of aesthetic experiences. An aesthetic experience can and has been described as a mental state (e.g. Bullot and Reber 2013; Shusterman and Tomlin 2008) or a composition of psychological processes (e.g. Leder et al. 2004) evoked within a spectator through his or her perception of an object located in the external world. Often, but not exclusively, such an object can be a work of art. Notably, in psychology, the aesthetic responses to *visual* art have been studied for over a century. This approach has its roots in psychophysics, although in recent years aesthetic appreciation has also been approached from the perspective of cognitive neurosciences, giving rise to the field of neuroaesthetics (Zeki 1999). Across all these domains (neuroaesthetics, experimental aesthetics, philosophy), the study of aesthetic experience in response to watching *dance* is still limited by a focus on the features derived from visual artifacts as the determinants of aesthetic value.

This chapter seeks to pave the way to a more thorough discourse of aesthetic experience in the field of dance. In order to achieve this, we aimed to acknowledge three dimensions that have been found to be particularly relevant in the concept of aesthetic experiences (Shusterman 1997): aesthetic experiences are linked to *evaluative* processes (inherently valuable and in some ways pleasurable), to *phenomenological* occurrences (distinctively felt and identified amongst other sensations), and to instances of *semantic* qualities (meaningful). These dimensions—as well as the concept of aesthetic experiences per se—are not without criticism. With regards to evaluative processes, for example, the concept of beauty is central in most studies on aesthetic appreciation (e.g. Chatterjee 2010). However, it is important to bear in mind that aesthetic responses to art and other classes of objects is much more varied.

Although dance is predominantly received in the visual domain, it is important to acknowledge its other realms: dance involves not just the eyes, but the entire body. Dance has thus been described as a multisensory, ephemeral, kinaesthetic art, something we will discuss later in the chapter and is expressed in Table 12.1. Indeed, aesthetic experiences are not proposed to occur exclusively in response to the visual arts. Aesthetic experiences studied in response to sensory modalities other than vision are the tactile sense (Hintz and Nelson 1971; Silverthorne et al. 1972), the auditory sense (Konecni et al. 2007; Kornysheva et al. 2009), and more recently through increased interest in dance, the kinaesthetic sense (Calvo-Merino et al. 2008; Cross et al. 2011; Jola et al. 2012b; Nieminen et al. 2011). Clearly, we know through experience that a visual work of art can also provoke a kinaesthetic experience, and a staged dance performance can also be visually pleasing (Reason and Reynolds 2010). Here we do not dissect different sensory modes of aesthetic responses. On the contrary, as it is particularly evident in the case of dance, we base our chapter on the assumption that aesthetic experiences are *a combination of sensory responses*, and that the potential of aesthetic experiences and therefore related aesthetic judgments are present across sensory modes at all times. Moreover, what is different between aesthetic experiences evoked through watching dance (or listening to music) and the sensations provoked by looking at an object of art is the *continuation* of sensory stimulation: the notion of time is a critical factor when attempting to capture aesthetic experiences in dance.

In this chapter, we thus focus on the challenges that the predisposition to engage in aesthetic experiences in the form of a continuing multisensory, immediate, meaningful, and evaluative process poses on empirical research, as is the case when studying aesthetic processes in response to watching dance. We emphasize the relevance of employing dance stimuli in their original form (i.e. close to real-life experience) and back our argument up with recent evidence. We argue that every so often, and most likely due to attempts of experimental control and scientific reductionism and partly due to a lack of knowledge of the subject matter, art in general (Silvia 2012) and dance in particular has been employed in such a manner that it was no longer recognizable as either art or dance (see Jola 2010; Jola et al. 2011). In extreme cases, arguments made for using dance as a particular form of art can actually be lost by the way dance is employed. Enhanced ecological validity is a necessary step towards a better understanding of the psychological and neural processes involved in the aesthetic experience of dance.

12.1.2 Dissecting dance: scientific reductionism

One common assumption grounding many scientific experiments is that complex systems can be reduced to their individual elements without changing the relevant conceptual properties of the elements as they were originally present in the complex system (reductionism). The process of reducing and abstracting complexity to individual conceptual properties has been defended by the scientific community as a necessity in order to achieve reliable results in controlled environments.

Consistent with more recent criticism of this approach in the study of visual perception (Felsen and Dan 2005), as well as in other fields that have an applied focus (e.g. Nairn 2014), we question the extent to which reductionism is applicable to the observation of dance movements (see also Jola et al. 2011) and encourage the use of genuine dance material. A reductionist approach can be considered a risk, given that it requires the scientist to know what the relevant conceptual dimensions are. Sometimes these are not necessarily known as in the case of aesthetic experiences (Silvia 2012). Neuroaesthetics, as well as the study of spectators' responses to watching dance, is a novel field which has much work to do on understanding the relevant dimensional variables (but see Graham and Field 2007 or Chatterjee et al. 2010). Moreover, conceptual dimensions change over time, as has been apparent in general concepts such as the example given by Nairn (2014) of the changing concept of gender, the changing interest and use of the concept of perspective in the history of art (e.g. Kubovy 1986), or the changing choreographic principles in dance (Butterwoth and Wildschutt 2009). In its extreme form, the manipulation of conceptual dimensions is itself the subject of a choreographer's artistic practice (Klein 2007). Not only does choreography define the work, it also demands cognitive engagement of the spectators in order to unfold such conceptual dimensions, as in conceptual dance (Jola 2010; see also section 12.2.1). Hence, complex systems in art and dance may be based on different relevant conceptual dimensions for each and every work.

In addition, a reductionist approach focuses on explaining complex systems through the study of low-level causation. However, the relationship of the whole and its units is more complex and entails the possibility of top-down causation as has been shown even at the neuronal level (i.e. through adaptation). As outlined in more detail later in this chapter, there is significant evidence that aesthetic appreciation of an object relies on its context, its complexity, and spectators' expertise, all existing beyond the object's own features (Leder et al. 2004; Locher et al. 2005). Research thus far in empirical aesthetics of dance has provided a preliminary understanding of visual features that make an individual derive pleasure from watching dance. Nevertheless, we propose that it is possible to deepen our understanding through the use of more ecologically valid stimuli. Alongside the problem of identifying the conceptual dimensions relevant to this area, and the emphasis on low-level causation in reductionist approaches, there is a third concern which we must raise here: the frequently low aesthetic validity of dance stimuli in empirical aesthetics, which is based on the field's history (see also Cross et al. 2011).

The number of experimental studies that have employed dance stimuli to study the human brain and behavior has been increasing since Calvo-Merino, Glaser, and colleagues' (2005a) seminal brain imaging work with dancers. Dance has since proved to be useful in studying how training in a particular domain of movement—such as dance—modifies the cognitive and neural structure of the brain (Bläsing et al. 2009; Hänggi et al. 2010), or how our brain "tangoes" with or without music (Brown et al. 2006). Predominantly, however, dance has been employed to study how movement expertise affects action observation processes in the passive spectator (Calvo-Merino et al. 2005a; Calvo-Merino et al. 2005b; Calvo-Merino et al. 2006; Cross et al. 2006; Orgs et al. 2008). In general, these studies showed that neural processes of motor resonance in the so-called mirror neuron network are enhanced when the observed action is part of the passive observers' motor repertoire (for further information on mirror-neuron theories and dance, see review in Sevdalis and Keller 2011).

With the growth of empirical aesthetics and neuroaesthetics (Chatterjee 2003, 2011), dance has re-entered the realm of science as a genuine *art* form (Cross and Ticini 2011;

Jola et al. 2011; Christensen and Calvo-Merino 2013). Some of the earliest evidence on the neural underpinnings of the aesthetic experience derived from watching a dance was published in *Consciousness and Cognition* in 2008 (Calvo-Merino et al. 2008). The authors showed how enhanced processing in visual and sensorimotor regions of the brain correlated positively with the subjective aesthetic judgment dance novices made about the movements watched outside a scanner when they were recalled for an additional study about a year later. A number of subsequent studies found further evidence for how specific aspects of a dance-movement modulate the aesthetic experience (Cross and Ticini 2011).

In this chapter we are specifically interested in "how" dance has been employed in these studies. Notably, the first experimental stimuli used in empirical aesthetics of dance were similar to the stimuli used in studies of cognitive neuroscience of action observation. Hence, because they were not informed by dance theory, they lack ecological validity. From a dancer's point of view, these stimuli are have limited artistic quality and thus in their potential to trigger genuine aesthetic experiences in the observers. While one could argue that aesthetic aspects may have been less relevant in the exploration of brain processes involved in action observation, it is obvious that the aesthetic appearance of stimuli is crucial when investigating aesthetic appreciation. However, aesthetic processes are potentially automatic processes, and thus affect the manner in which stimuli are processed even when the experiment does not target aesthetic explorations.

Although researchers are aware of the intricacy of dance and understand that dance is more than just a type of movement pattern, assumptions about "dance" are sometimes made that do not correspond to contemporary understanding within the domain. Hence, many researchers still use somewhat impoverished dance stimuli. We will be addressing what potential effects this can have. However, if one were to employ dance in its original form, many variables would remain uncontrolled. This chapter attempts to address in particular why a more ecological approach to studying the human brain and behavior is necessary, using the example of employing dance stimuli.

12.1.3 Is it still dance?

As visible in Table 12.1, studies that refer to having employed dance in their research vary notably in their basic assumptions about what dance may reveal (column 2), what dance is (column 3), how it looks (column 4), and which task is best employed to measure it (column 5). For example, empirical psychological investigation usually starts with a rationale for the objectives of the study and the sensory stimulation leading to a clear description with the exact parameters of the experimental stimuli in order to allow reproducibility. This applies also for more complex, natural stimuli, that need to be clearly documented. In our case, the style of dance that is used, for instance, should be linked to the objectives of the study and clearly defined. Notably, however, the stimuli are rarely available for full inspection. Further, several published papers that have used dance stimuli have defined dance in an unmethodical manner. We argue that the definition of dance provided by the authors is critical because it indicates the key features on which the selection and creation of the stimuli were based and it further highlights which features of the stimuli were relevant for the research. Hence, it explains the purpose of using *dance* (rather than some other type of movement), and also offers insight into what level of understanding of dance was

Table 12.1 Published empirical studies (2005–2013) from the fields of cognitive neuroscience and empirical aesthetics that use dance movements as stimuli

Paper	Research question	Definition dance and/or movement	Stimuli	Task
Calvo-Merino et al. (2005a)	How acquired motor skills influence the neural processing of familiar and non-familiar actions (fMRI)	*Dance is a kind of acquired motor skill that "involves arbitrary, intransitive movements of the whole body. . . . dance movements need not involve either external objects or spatial target location."* *Ballet and capoeira ". . . involve a well-established, distinctive set of movements" of which "many male ballet and capoeira moves are kinematically comparable" with "quite similar components at the muscle level. . . ." (pp. 1244–6)*	- Ballet, capoeira (kinematically matched) - 3-s-long video clips (color) - Two male dancers, one ballet and one capoeira - Faces blurred - Training clothes (black) - Background: neutral (blue) - Movements "involved the whole body, with large, proximal arm and leg movements"	- Aesthetics task: None - Other tasks: Indicate how tiring the movement was on one of three button presses (randomly assigned across trials) during the scanning.
Calvo-Merino et al. (2006)	How visual and motor experience modulate brain activity when observing familiar and non-familiar movements (fMRI)	*None* *"The dance movements studied here did not involve external objects." (p. 1907)*	- Gender-specific ballet moves (kinematically matched) - 3-s-long video clips (color) - Two professional ballet dancers, one female, one male - Faces blurred - Training clothes (black) - Background: Doorframe, white/gray	- Aesthetics task: None - Other task during the scanner: Indicate "how symmetric" each video was on one of three button presses (randomly assigned across trials) - Other task outside the scanner: Questionnaire on ballet terms of the movement seen; "How often do you see this move" (visual familiarity) and "How often do you do this move" (motor familiarity) between 0 (never) and 10 (very often).

Table 12.1 (continued) Published empirical studies (2005–2013) from the fields of cognitive neuroscience and empirical aesthetics that use dance movements as stimuli

Paper	Research question	Definition dance and/or movement	Stimuli	Task
Cross et al. (2006)	Cognitive representation of movements/neural bases of action simulation and how neural activity increases with motor experience (fMRI)	*None*	- Modern dance - 5-s-long video clips - One dancer - Training clothes (black) - Dance studio with bar in the background	- Aesthetics task: None - Other tasks: "'How well could you dance this movement right now?.'" Four possible responses: from 1 (extremely well), 2 (well), 3 (adequate), to 4 (not well/need to see again).
Urgesi et al. (2007)	How premotor, visual, and parietal areas are causally related in the configural processing of human bodies (TMS)	*None*	- Ballet, Indian, mock dance positions, i.e. mixed ballet and Indian dance positions - Static black and white photographs - One male dancer - Training clothes - Neutral (white) background	- Aesthetics task: None - Other tasks: Participants performed a match-to-sample task on pairs of stimuli
Calvo-Merino et al. (2008)	"aesthetic appreciation" (fMRI)	*Dance is a dynamic visual form of artistic expression that has accompanied mankind since ancient times. (p. 1)*	As in Calvo-Merino et al. (2005a), see above	- Aesthetics task: Post-scan session: five 5-point Likert scales with the labels: (1) simple–complex; (2) dull–interesting; (3) tense–relaxed; (4) weak–powerful and (5) like–dislike - Other tasks: During the scan: dummy task: "how tiring is the movement?"

Table 12.1 (continued) Published empirical studies (2005–2013) from the fields of cognitive neuroscience and empirical aesthetics that use dance movements as stimuli

Paper	Research question	Definition dance and/or movement	Stimuli	Task
Orgs et al. (2008)	How desynchronization in alpha and beta frequency bands is modulated by expertise in dance when watching dance movements (EEG)	*No definition of dance present. The authors refer to the style of dance movements contrasted to everyday movements such as hand waving (no other example is given) as "Contemporary" taken from exercises taught in daily classes of this style at the Folkwang University in Essen.* *Notably, this is dependent of the time of the study and the particular teachers employed. No other description of the dance movements are given other than that they included the arms, hands, and torso while the head and lower limbs were eliminated from the display "in order to minimize distraction, eye movements and cognitive processes related to face identification." (p. 3381)*	- Contemporary - 6 to 12-s-long video clips (head and lower limbs outside the picture) - One male dancer - Training clothes (black) - Neutral background (white)	- Aesthetics task: None - Other tasks: At the end of each video, participants indicated by button press whether the movement they had just seen was dance or an everyday movement.
Daprati et al. (2009)	How audience preferences map the biomechanical changes in ballet positions over time (behavioral)	*Ballet is a dance form, which uses a single, codified set of positions of the human body to express emotions and other mental states. (p. 2)*	- Ballet - Static ballet positions (photos) - Performance setting - Live pictures and stick figures made from the pictures - Different female dancers	- Aesthetics task: Stimuli were presented in pairs of all possible combinations of stimuli and participants were asked to indicate which picture/stick figure of each pair they preferred.

Table 12.1 (continued) Published empirical studies (2005–2013) from the fields of cognitive neuroscience and empirical aesthetics that use dance movements as stimuli

Paper	Research question	Definition dance and/or movement	Stimuli	Task
Calvo-Merino et al. (2010b)	How ventral premotor cortex and extrastriate body area in the superior temporal sulcus of the human cortex are causally related to the aesthetic sensitivity for body stimuli (TMS)	*In the case of performing arts such as dance, the observer's aesthetic experience is presumably grounded in the responses of their neural sensory, motor and affective circuits to the expressive actions of the dancer's body. (p. 448)*	- Ballet, Indian, mock dance positions, mixture of ballet and Indian dance - Static black and white photographs - One male dancer - Training clothes (black) - Neutral background (white)	- Aesthetics task: Participants indicated on a visual analogue scale how much they liked each movement, ranging from 0 "I do not like it at all" to 100 "I like it very much"
Miura et al. (2010)	How motion smoothness of expressive body actions relate to neural activity and the subjective experience of smoothness in dance movements (fMRI)	[. . .] *dance consists of a sequence of expressive body actions assigned by choreography and tied to music. (p. 42)*	- Modern dance, "smooth" and "awkward" movements - 30-s-long video clips - Small biped humanoid robot named QRIO, white body on blue background - One female dancer, gray clothes	- Aesthetics task: None - Other tasks: After fMRI scanning, participants evaluated the motion smoothness of the performance in each clip by means of a scale from 0 to 10 (0 = not smooth; 10 = very smooth)
Calvo-Merino et al. (2010a)	How visual and motor experience modulates perceptual mechanisms for configural processing of familiar and non-familiar actions.	*Dance = a kind of human motor action. (~ p. 405)*	- Ballet - 3-s-long video clips - Three female dancers - Point-light displays, white dots on black background	- Aesthetics task: None - Other tasks: Stimuli were presented in pairs and participants were asked to indicate whether the two were the same of different

Table 12.1 (continued) Published empirical studies (2005–2013) from the fields of cognitive neuroscience and empirical aesthetics that use dance movements as stimuli

Paper	Research question	Definition dance and/or movement	Stimuli	Task
Jola, Pollick et al. (2011)	How specific events, numbers of dancers, and the time-course of a 2.5-h performance modulates audiences' neurophysiological (motor corticospinal excitability) and experiential (interview, drawing) responses [Study highlights the importance of time as a relevant factor in dance performance] (TMS)	*Watching dance is a multisensory experience. In dance, movements are intertwined with music and/or sound, costumes, the narrative, light and stage design (in summary)* *Watching a dance performance is a multisensory experience formed by the perception of a stream of audiovisual events. People are prepared to pay for this experience, quite often without knowing for certain whether they will like it or not. Dance can also be described as a complex socio-cultural practice that connects to a diverse range of motives, impulses and pleasures. Different types of theater dances have emerged and made watching dance, too, a cultural practice with a diverse range of objectives, including exploration of movement, narrative interpretation and spectacle (Introduction)*	- Ballet (Sleeping Beauty) - 2.5-h live performance with live orchestra (dress rehearsal)	- Aesthetics task: None - Other tasks: Accessing subjective experience: interview, chart drawings on memorized events and engagement in the performance
Jang and Pollick (2011)	How the neural networks underlying action understanding are influenced by visual and/or motor experience (fMRI)	*Dance is generally a social activity and so is watching dance, and although we might only show an isolated time segment of an isolated dancer in our experiment, we can expect the mechanisms of social perception and cognition to be what are shaped by dance experience (p. 367)*	- Ballet - 2-s-long video clips; point light displays - One dancer - Static postures from the point-lights, stick figures - White dots on black background	- Aesthetics task: None - Other tasks: Participants were asked to (a) merely observe, or (b) observe and mentally imitate the movement/position they saw.

Table 12.1 (continued) Published empirical studies (2005–2013) from the fields of cognitive neuroscience and empirical aesthetics that use dance movements as stimuli

Paper	Research question	Definition dance and/or movement	Stimuli	Task
Cross et al. (2011)	How the liking of a movement is influenced by the subjectively evaluated ability of the observer to perform the movement herself (fMRI)	*None. Dancers from the Leipziger Ballet performed a "range of movements" based on a classical and contemporary vocabulary. The authors selected a number of movements. No indication of the representation of complexity, speed, difficulty, and size is in the selected sample is given (p. 3).*	- Ballet, contemporary dance - 3-s-long video clips - One male one female dancer - Training clothes (black) - Dance studio with bar and windows in the background	- Aesthetics task: "How much did you like it?" and performative task: "How well could you reproduce it?"
Jola, Abedian-Amiri et al. (2012)	How corticospinal excitability during watching dance can be modulated by visual experience and fantasy proneness (TMS)	*Dance is multicultural, multilayered, and multisensory, but several dance styles are defined by their specific kinematic motion patterns, types of music, costumes, lights, and stage settings. . . . (p. 2)*	- Ballet, Indian dance (Bharatanatyam) - 5-min life performances with music, participants were seated in the actual dance studio - Two dancers, one ballet, one Indian dancer - Stage clothes, tutu and Indian costume - Dance studio with bar and windows in the background	- Aesthetics task: Aesthetic questions in the interview. Post-hoc questionnaire: "How much did you like it?" On a scale from 1 to 10 (0 = not at all; 10 = a lot) - Other tasks: During the performance, participants were asked to simply enjoy the live performances they were watching

Table 12.1 (continued) Published empirical studies (2005–2013) from the fields of cognitive neuroscience and empirical aesthetics that use dance movements as stimuli

Paper	Research question	Definition dance and/or movement	Stimuli	Task
Grosbras et al. (2012)	How cortical stimulation over emotion processing brain regions affects liking (fMRI and TMS)	*None.*	- Contemporary dance - Three 3:38-min-long video clips, played three times, with three different musical accompaniments - Two dancers, one female, one male - Background: dimly lit stage, black, performance clothes	- Aesthetics task: Participants continuously rated their positive or negative emotional response to the dances on a linear analogue scale, using a slider
Jola and Grosbras (2013)	Does the co-presence of the performer and the spectator in a live situation enhance corticospinal excitability in novices compared to a video presentation on a computer monitor. (TMS)	*None*	- See Jola et al. 2012a - Presentation live in a rehearsal space versus on a computer screen	- Aesthetic task: Participants rated their enjoyment in a dedicated questionnaire after the testing.
Jola et al. (2013b)	Does the multisensory presentation of dance and music enhance neuronal correlation across spectators? (fMRI)	*None*	- Bharatanatyam dance performance - Costumes - With and without music - Wooden flooring, neutral background - 6 min 26s	- Aesthetic task: Free viewing. Participants were asked to view and enjoy the performances

Table 12.1 (continued) Published empirical studies (2005–2013) from the fields of cognitive neuroscience and empirical aesthetics that use dance movements as stimuli

Paper	Research question	Definition dance and/or movement	Stimuli	Task
Orgs et al. (2013).	To use "apparent biological motion to investigate the aesthetic effects of three levels of movement representation: body postures, movement transitions and choreographic structure"	*"Dance composition involves a language-like grammar that includes movement specific and general compositional levels"*	- 8 body postures taken from filmed choreographed dance movements - Postures had no meaning in any dance style and have no corresponding label - The postures were used to create 64 apparent movement sequences - The sequences were either symmetrical or asymmetrical, and exhibited good continuity or did not	- Aesthetic task: Rating liking of the sequences - Aesthetic task: Rating liking of the component postures
Torrents et al. (2013)	To "identify some of the kinematic parameters of expert dancers' movements that influence the subjective aesthetic perception of observers in relation to specific skills of contemporary dance"	*"Dance is the artistic expression of human movement, the art form that exists simultaneously in time and space"*	- 48 animated avatars created by drawing lines joining markers attached to dancers' body depicting the head, thorax, upper arm, lower arm, hand, pelvis, thigh, leg, foot. - Stimuli included recordings of four dancers performing four contemporary dance motor skills in three different movement qualities, according to Laban's criteria	- Aesthetic task: Provide ratings on a 7-point semantic differential anchored by ugly and beautiful

the basis for the study. This, in turn, can reveal the artistic quality of the stimulus material, which again is particularly important for empirical aesthetics.

Finally, it is worth noting that the subjects' task (Table 12.1), last column has a large impact in neuroaesthetics on how stimuli should be contextualized (Leder et al. 2005). For example, asking how "tiring" a movement appears may relate more to a different subjective and neural signature than when asked how much a person "likes" what he or she sees. We will describe this in more detail at a later point in the chapter. In Figure 12.1, we also provide a visual representation of the evolution from simple to more complex dance stimuli as notable in Table 12.1 over the last ten years.

Reviewing the definitions of dance in Table 12.1, there appears to be a tendency in many instances to oversimplify dance and define it merely as a "type of movement pattern." In addition, when considering some of the stimulus materials used to date one may come to the conclusion that scientists often wrongly assume that "everybody knows when a movement is a dance." While this is probably true for informal social dances, it is certainly not the case for certain types of classic, formal, and post-modern dance movements particularly when taken out of their normal contexts.

In addition, a significant portion of the studies provides no definition of dance, and, only a few studies described the specifics of the particular dance style employed. In fact, the dance styles used in cognitive neuroscience and empirical aesthetics research to date are not only very different from each other in the types of movements or their repertoires, but also different in their function, social meaning, and origin. Hence, dance's meaning for participants in a study is not necessarily shared by a different study participant. In order to design a study employing dance as a means to study the human brain and behavior, it is important to understand what exactly makes a movement a "dance," and to acknowledge that *dance* is something unique—and thus very different—from other kinds of movement patterns. It is therefore going to be very helpful for us to start with a thorough definition of dance.

12.2 **Defining dance**

12.2.1 **Dance as a social phenomena**

According to the online *Oxford English Dictionary's* definition of dance (<http://oxforddictionaries.com>), a dance is composed of a series of steps and movements that match the speed and rhythm of a piece of music. That dance steps and movements can be considered either as an activity (which can be a social gathering) or as an art form is also noted. The clarification of the role of the movement is important because, although the movement is arguably the most prominent characteristic of dance, definitions such as "a person moving rhythmically to music," or "a moving body in time and space," are not sufficient to describe dance. Moreover, movement-focused definitions are not applicable to many contemporary dance forms. For instance, conceptual dance works may not contain much movement at all. In Ivana Müller's "*While we are holding it together*" the dancers stay still for almost the entire performance (Jola 2010). Movement itself, the often claimed predominant characteristic of dance that gives it a "fluid" appearance, leads to the adjective frequently used

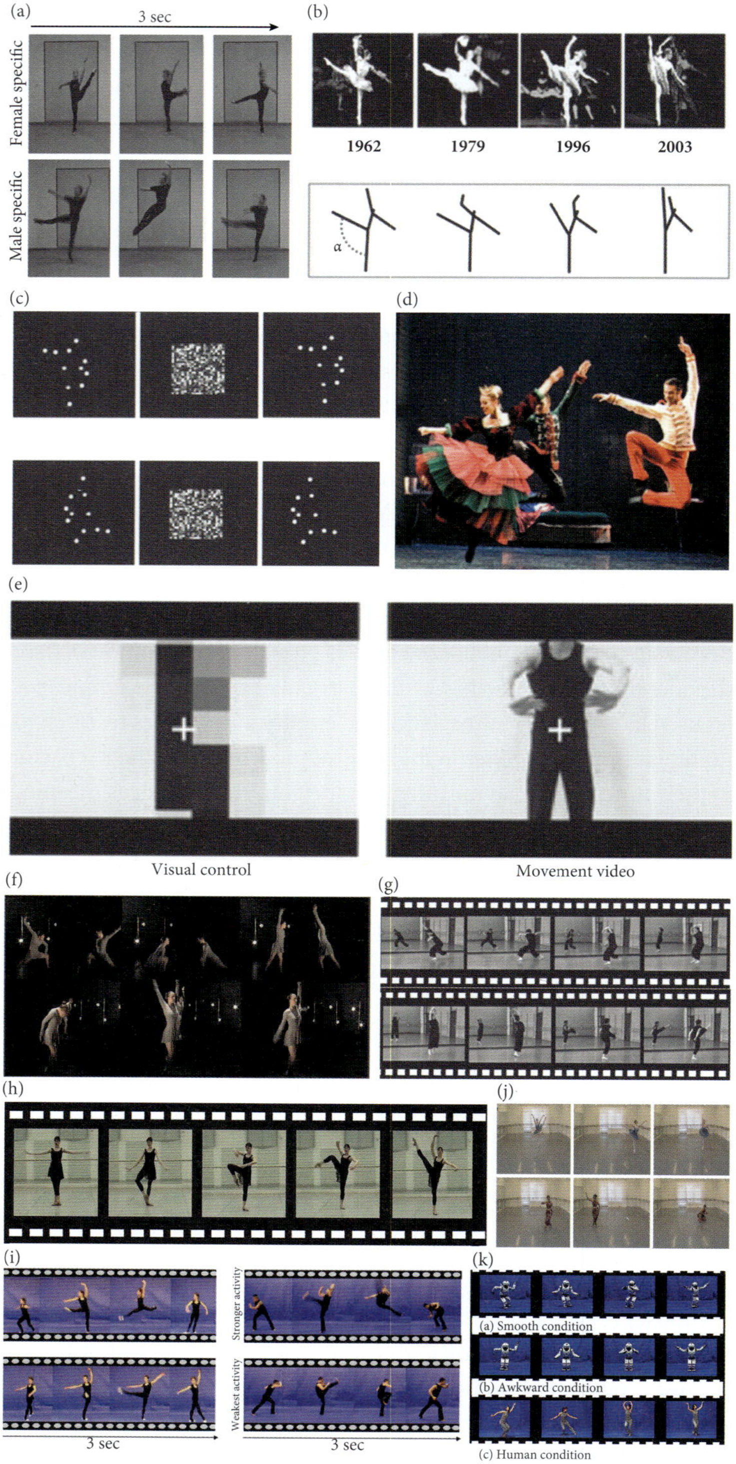

Figure 12.1 Selection of stimulus materials with dance movements. **(a)** Video clips of academic male and female-specific ballet movements, 3 seconds. Reproduced from *Current Biology*, 16 (19), Beatriz Calvo-Merino, Julie Grèzes, Daniel E. Glaser, Richard E. Passingham, and Patrick

in connection with dance—"ephemeral" (Conroy 2012). Dance is thus difficult to capture, literally as well as metaphorically: dance *is* fluid, always redefining itself.

Dance—in all its various forms and appearances—is not without purpose. We support definitions that elevate dance from a mere kind of movement pattern. A useful definition

Figure 12.1 *(Continued)* Haggard, Seeing or Doing? Influence of Visual and Motor Familiarity in Action Observation, pp. 1905–1910, Copyright (2006), 2006 Elsevier Ltd. Published by Elsevier Inc. User rights governed by an Open Access license. **(b)** Photographs from live performances of classical ballets. Reproduced from Elena Daprati, Marco Iosa, Patrick Haggard, A dance to the music of time: aesthetically-relevant changes in body posture in performing art, *PLoS ONE*, 4 (3), p. 4, figure 2b, DOI: 10.1371/journal.pone.0005023 © 2009, Elena Daprati, Marco Iosa, Patrick Haggard. User rights governed by an Open Access license. **(c)** Point-light video clips of academic ballet movements, 3 seconds. Reproduced from *Psychological Research Psychologische Forschung*, 74 (4), 2009, p. 402, Experts see it all: configural effects in action observation, Beatriz Calvo-Merino. © 2009, Springer Science and Business Media. With kind permission from Springer Science and Business Media. **(d)** Still from the live ballet performance "The Sleeping Beauty" by Ashley Page with the Scottish Ballet at the Theatre Royal Glasgow, Photograph by Andrew Ross. © Andrew Ross, 2011. Reproduced from *Phenomenology and the Cognitive Sciences*, 11(1), pp. 17–37, The experience of watching dance: phenomenological–neuroscience duets, Corinne Jola, Shantel Ehrenberg, and Dee Reynolds, doi: 10.1007/s11097-010-9191-x, © 2011, Springer Science and Business Media. With kind permission from Springer Science and Business Media. **(e)** Video clips of contemporary dance movements, 8–12 seconds. Reproduced from Guido Orgs, Jan-Henryk Dombrowski, Martin Heil, and Petra Jansen-Osmann, Expertise in dance modulates alpha/beta event-related desynchronization during action observation, *European Journal of Neuroscience*, 27 (12), pp. 3380–3384 DOI: 10.1111/j.1460-9568.2008.06271.x © The Authors (2008). Journal Compilation © Federation of European Neuroscience Societies and Blackwell Publishing Ltd, with permission. **(f)** Video clips of contemporary dance, ~ 4 minutes, screenshots of stimuli used by Grosbras et al. (2012). Available online at <http://paco.psy.gla.ac.uk/watchingdance>. Reprinted from *Brain Stimulation*, 5 (2), Marie-Hélène Grosbras, Haodan Tan, and Frank Pollick, Dance and emotion in posterior parietal cortex: A low-frequency rTMS study, pp. 130–136, Copyright (2012), with permission from Elsevier. **(g)** Video clips of contemporary dance, 5 seconds. Reprinted from *NeuroImage*, 31 (3), Emily S. Cross, Antonia F. de C. Hamilton, and Scott T. Grafton, Building a motor simulation de novo: Observation of dance by dancers, pp. 1257–1267, Copyright (2006), with permission from Elsevier. **(h)** Video clips of academic ballet movements, 3 seconds. Reprinted from Emily S. Cross, Louise Kirsch, Luca F. Ticini, and Simone Schütz-Bosbach, The impact of aesthetic evaluation and physical ability on dance perception, *Frontiers in Human Neuroscience*, 5, figure 2, doi: 10.3389/fnhum.2011.00102 © 2011, Frontiers Media S.A. **(i)** Video clips of academic ballet movements and capoeira movements, 3 seconds, used by Calvo-Merino et al. (2008). Reprinted from *Consciousness and Cognition*, 17 (3), B. Calvo-Merino, C. Jola, D.E. Glaser, and P. Haggard, Towards a sensorimotor aesthetics of performing art, pp. 911–922, Copyright (2008), with permission from Elsevier. **(j)** Video clips and live performances of ballet and Indian dance solos. Available online at <http://paco.psy.gla.ac.uk/watchingdance>. Data from *PLoS ONE*, 7 (3), Corinne Jola, Ali Abedian-Amiri, Annapoorna Kuppuswamy, Frank E. Pollick, and Marie-Hélène Grosbras, DOI: 10.1371/journal.pone.0033343. User rights governed by an Open Access license. **(k)** Video clips of contemporary dance by a robot and by a human, 30 seconds. Reproduced from Effect of motion smoothness on brain activity while observing a dance: An fMRI study using a humanoid robot, Naoki Miura, Motoaki Sugiura, Makoto Takahashi, et al., *Social Neuroscience*, 5 (1), pp. 40–58 © 2010, Taylor & Francis, reprinted by permission of the publisher (Taylor & Francis Ltd, <http://www.tandf.co.uk/journals>).

of dance should thus include the *functions* of the specific type of dance used (e.g. emotional expression, creative engagement, rituals). Furthermore, it should take into account how these are achieved. This includes the *complex interplay* of the movement (the dance) with other parameters, the so-called strands of dance (e.g. performer, sound, space, costumes, lights) as well as the *context* (e.g. location or historical, social, and political position, the sociocultural background and experience of the spectator), through which spectators and participants experience the meaning of a dance. Even when dance movements are abstract in form with no obvious everyday propositional or referential meaning (such as fist shaking, movements of longing, gestures like putting a hand on the heart, etc.), they still convey messages and emotions to the spectators (whether intended or not by the dancer). In other words, when compared to other skillful movements such as those prevalent in sport, dance is unique because of these three key elements that unfold together: purpose, compositional strands (of which movement is one), and context.

As with other art forms, the main purpose of the choreographer is not necessarily to produce a beautiful dance piece consisting of aesthetically pleasing movements. Clearly, the mastery of movements is the main goal in the early years of dance training. Over the last century, one could indeed observe a general tendency towards increased physical virtuosity, from new dance styles performed mostly by young people (e.g. hip-hop, breakdance) to established dance forms (e.g. classical ballet), which are supposedly perceived as more beautiful (Daprati et al. 2009). For contemporary dancers, risk taking is an important aspect of their practice as well as the drive to acquire skillful athletic movements (Dumont 2009). Dance is, however, more than a technical exercise in mastering particularly difficult movements—unlike rhythmic and artistic gymnastics, for example, where the skillful execution of a range of increasingly difficult types of movements is the ultimate goal.

In early neuroscientific studies using dance movements, dance was simply used to study, for instance, mirror neuron mechanisms for intransitive movements (movements that are not directed towards an object) compared to transitive movements (towards an object, goal-oriented actions). It is worth nothing that dance movements can be intransitive but still meaningful. In fact, from the dance experts' point of view, it is clear that the conditions under which a movement is without meaning are rare.

Furthermore, intransitive movements can be propositional or non-propositional. In speech production, for example, non-propositional speech refers to utterances of overused expressions/phrases. In movement terms, however, propositional movements are those that have a culturally recognizable communicative intention (such as waving, nodding, or shaking the head, stretching out the arms in an expression of longing, etc.), while non-propositional movements are abstract. The latter may be meaningful as well, but they achieve this without having a clear cultural meaning in the form of a gesture. In dance, we can find both propositional as well as non-propositional movements. While some dance forms do include the use of propositional movements, such as certain classical ballets (e.g. *Les Sylphides*), other dances predominantly consist of non-propositional movements, such as numerous contemporary dance performances (e.g. creations by random dance of

Wayne McGregor; see May et al. 2011). Hence, dance employs different forms of actions (transitive–intransitive; propositional–non-propositional, gestural–physical).

Finally, movement types are closely linked with the music and the other strands of the dance. The music, scenery, costumes, lighting, and other special effects are as previously indicated strands of the dance, which add to the experience of the spectator (Preston-Dunlop and Sanchez-Colberg 2010). As early research was very much dealing with movement; it is only recently that the strands that crucially influence the form of dance have been brought into focus. It is likely that spectators' aesthetic experience, including their affective, aesthetic, and kinaesthetic responses, are specific to different types of dance forms, and therefore careful choices need to be made when using such stimuli, whether in the study of action observation processes or in neuroaesthetics.

In addition, it is important to consider that dance happens in a specific context. Elements of the context may be (but are not limited to) its location and its historical, cultural, and socio-political position. The context may be the key to truly understanding a dance, especially as far as contemporary dances are concerned—unlike acrobatics which is astonishing to watch, no matter where, when, and in what context. In acrobatics and some twenty-first-century forms of dance, context is employed to increase physical challenges; however, it does not add any other meaning to the movement beyond demonstrating the capability of the performer.

Thus, dance differs from other movement because (i) *it has a communicative intention*, (ii) this intention is brought to the observer by different constructs of motion accompanied with *other strands*, (iii) it happens in a specific *context* whose elements may be the key for the understanding and cognitive (and emotional) mastering of the dance. This combination makes dance unique. How a dance and its meaning is received and processed in the mind of a spectator under these premises is the objective of the next section.

12.2.2 Modeling the aesthetic experience of dance

In dance, aesthetic attributes are generally assigned to the dancer's movements. Nevertheless, as indicated above, many other elements contribute to the aesthetics of dance. These include, but are not limited to, part of the stage setting (lighting, clothing, objects on stage, and their arrangement, and so on) and the synchronicity between dancers or parts of the dancer's body. Even when movement is the most relevant feature of dance, aesthetic decisions about all strands are necessarily to be made when staging a dance performance To a dance experts' eye it is evident when beautifully crafted movements are situated in a performance context that lacks coherent artistic choices. In other words, a spectators' aesthetic experience is the outcome of a holistic impression of all elements of a given performance.

Importantly, the aesthetics of the dance itself, the choreography (i.e. the movements and their composition), the staging of the performer (i.e. the performers' presence), as well as the combination of movement and music, are dynamic elements. Thus, dance's aesthetic appearance unfolds and decays with time. Despite its fleeting character, dance can evoke resonating aesthetic experiences of emotional, affective, kinaesthetic, and visual-aesthetic forms in spectators (Reason and Reynolds 2010). But how to begin to account for the aesthetic contributions of so many elements at these different levels? A good starting point might be the psychological model of aesthetic appreciation proposed by Leder and his team (2004).

They (Leder et al. 2004) sought to understand the processes involved in an aesthetic experience in response to exposure to an artwork and how they trigger aesthetic pleasure as well as the formation of aesthetic judgments. Since the model was specifically designed to explain the aesthetic experience derived from *modern* visual art, it adopts a very broad view when taking into account aspects of the context, the perceiver, and the overall situation in which an artwork is encountered. By describing the interrelated levels of visual and cognitive processes of looking at art, it indirectly illustrates the appreciation of the *multilayered* nature of an artwork.

Applied to dance, Leder and colleagues' (2004) model requires further development. As already indicated, one very distinctive feature of dance, as compared to the visual arts, is its *temporality*: dance unfolds over time and so does the spectators' aesthetic experience with it. While the processing stages of the original model (perceptual analysis, implicit memory integration, explicit classification, cognitive mastering, evaluation, affective processing, and aesthetic judgment/emotions) are adept in the reception of dance works, the temporal dimension needs to be included. The interrelated processing stages are thus not only influenced by variables such as the *previous experience* of the individual (including specific knowledge about the particular artwork, general declarative knowledge about style and content, personal interest and taste), the *social discourse* and *context* in which the aesthetic experience takes place (museum, gallery, experiment, etc.), but also intertwined by the *continuing flow of motion*. At a given moment in time, viewers may be in the process of classifying a particular movement they see according to previous knowledge ("*explicit classification*"), while they could also be at the stage of *cognitive mastering* regarding previous dance movement sequences whose deeper meaning they are in the process of grasping, and they may also be undergoing a comparative process of the predicted movement with the movement seen. All these processes are constantly updated, receiving new inputs as a dance unfolds. Judgments and emotions are thus modulated by the continuous updating by experience. Below we outline the specific findings of the continuous influence of context, movement, and experience undergone when watching dance, according to the six processing steps of Leder and colleagues' (2004) model.

12.2.2.1 Input: pre-classification and context

The historical and social context of a dance work is highly relevant to its appreciation (and understanding). For example, when focusing on theater dance, as its name suggest, the theater setting is of relevance. Furthermore, while the particular movements of a dance recording may match those of the live dance as it has taken place in the theater, spectators' experiences for these two viewing formats are not the same. This has been supported by recent empirical research by Jola and colleagues (2012a, 2013a), who measured spectators' internal motor resonance when watching dance live and on video. The authors assessed spectators' motor resonance through the indirect measure of motor corticospinal excitability whereby the amplitude of motor-evoked potentials in the finger and the arm muscle groups induced by transcranial magnetic stimulation (TMS) indicate the level of corticospinal excitability. In the first study (Jola et al. 2012a), visual expertise in ballet was found to enhance muscle-specific motor resonance when watching a classical ballet solo live. This finding expands previous research on dancers' physical expertise. Furthermore, it partly contradicts previous findings from other fields than dance (i.e. basketball), where visual expertise was reported to be insufficient for muscle-specific motor resonance (Aglioti et al. 2008). In a second study, Jola and Grosbras (2013) compared two groups of novices watching a ballet solo, a classical Indian dance solo (Bharatanatyam), and an acting control condition either live in a rehearsal studio or on video in the laboratory. Not only did the novice live-spectator group show enhanced motor corticospinal excitability for the gestural-prone Indian dance, they also reported to have liked watching all performances more than the video group. In addition, the liking ratings differed significantly between the individual performance types when watched on video. This is of huge relevance, as it indicates that in a video setting, affective components such as enjoyment could modify the general affective state and potentially act as a confounding variable, leading to different aesthetic experiences as well as impacting upon neuronal processes. Indeed, a number of recent studies support a general effect of "liveness" on spectators' responses to the observation of everyday actions (Järveläinen et al. 2001; Shimada and Hiraki 2006).

Overall, it seems that at least for novice spectators, the context of a live performance changes their responses to watching dance. It is thus important to look further at how effects of expertise as have been reported in earlier studies may have been affected by the artificial setting of dance when taken out of its theater context. It is worth noting that most studies recorded dance movements in a rehearsal space, where the gap of familiarity between spectators with physical expertise (who were very likely to have danced in similar spaces) and novices (who may have never been in such spaces) can be expected to be large. Hence, studying live face-to-face interaction may be the big challenge of future research in cognitive neuroscience, as it may require further consolidation as a necessary condition for producing empirically valid and reliable results (Redcay et al. 2010). Additional aspects of spectators' experiences could also include the specificity of an event, the sharing of the experience with other spectators, related expenditures (e.g. ticket costs, traveling expenses), and so forth.

Hence, the social context stands out as particularly relevant in the case of dance. The theater setting in which dance is commonly "consumed" provides a context of vivid social interaction and discourse, in the intervals or during the performance where the spectator perceives other spectators' reactions around him/her. Besides, the presence of both the performer and the spectator is specific in a live setting. A good performer uses the feedback from spectators; he plays with the interaction. Thus, he will adjust the performance to the spectator's needs. Optimally this means that each spectator will be presented with the performance expression that at his or her own level of intensity. This possibility has, however, not yet been empirically verified.

12.2.2.2 Evaluation processes: I perceptual analyses

Perceptual variables that have been found to affect aesthetic responses include contrast, visual complexity, intensity, brightness, saturation, size, color, symmetry, and grouping. According to Leder and colleagues' (2004) model, preferences will vary according to changes in these perceptual dimensions. For instance, curves and symmetric shapes have been found to be preferred over lines and edgy shapes in visual art (Tinio and Leder 2009; Leder et al. 2011). Clearly, some of these dimensions are also prevalent in the context of dance and their relevance has been discussed; for example symmetry (Hagendoorn 2005; Aronoff 2006) or the edginess of shapes (Daprati et al. 2009). Using dance photographs, Daprati and colleagues (2009) found evidence that more vertical dance positions (i.e. positions than involved a higher extension of the legs) were liked more than less vertical ones. The authors related their finding to a general preference of certain shapes over others, independent of the particular dance position.

Dance further makes use of the lighting as a stylistic element, colors, costume textures and the like to highlight specific visual aspects of the movements and to capture spectators' attention. Undoubtedly, different levels of luminance influence subsequent cognitive processes. Sometimes, different light settings are so extreme that video recordings from an original stimulus (i.e. a live performance) cannot be used without modification for empirical research (e.g. too dark for visibility). Thus, for a truly ecological approach to dance, further studies are needed in order to determine how different perceptual stimuli properties can influence the appreciative process of the individual when contemplating dance. A better understanding of perceptual parameters will enhance targeted control in the service of experimental rigor, while still allowing a genuine aesthetic experience of dance to unfold.

12.2.2.3 Evaluation processes: II implicit memory integration

In the aesthetic appreciation of visual artworks (Tinio and Leder 2009) as well as that experienced when watching dance (Calvo-Merino et al. 2008; Jang and Pollick 2011), novelty and familiarity with the stimuli were found to influence motor resonance significantly as well as influencing aesthetic appreciation (Cross and Ticini 2011). Hence, a careful control of participants' previous dance experience is important. Notably, a large number of potential experimental participants have had formal dance training in their youth. While they may not be presently practising any dance, their prior exposure to—normally—a particular kind of systematized dance training constitutes *at least* implicit and/or embodied knowledge of a given dance style (if not also extensive *explicit* knowledge). An example of how implicit knowledge influences low-level perception of a dance piece was offered by Stevens and colleagues (2010). The authors demonstrated that expertise with a dance piece led to anticipatory eye movements when watching a section of a dancing duo on video (Stevens et al. 2010).

Another important consideration is the spectator's cultural knowledge and the environment in which the study is carried out. For example, traditional Indian dances are displays of popular fairytales and movements from them are bound to elicit related memories in an individual with an Indian cultural background. Therefore, an experimental participant from an Indian cultural background will "perceive" more or more refined associations in an Indian dance movement or position than a Scandinavian participant with no experience in or knowledge of Indian culture. Thus, even if the research question is not primarily about the content of a dance, specific dance stimuli will elicit specific cognitive and affective processes in some individuals but not in others. Some studies have investigated the effects of culture on action observation processes (Désy and Théoret 2007; Molnar-Szakacs et al. 2007) with, however, contradicting results. Further studies are needed to improve our understanding of when and how top-down processes are shadowing the aesthetic experience.

12.2.2.4 Evaluation processes: III explicit classification

Explicit knowledge of paintings was found to affect naïve spectators' subjective preferences (Augustin and Leder 2006; Belke et al. 2006). Specifically, participants' aesthetic judgments have been found to vary according to (i) whether the paintings' titles were presented or not (Leder et al. 2006; Kirk et al. 2009); (ii) whether information was given about the paintings' style (Lengger et al. 2007; Augustin et al. 2008); and (iii) whether participants were made aware of the artist (Leder 2001). Hence, explicit knowledge influences the evaluation of an artwork.

Leder and colleagues (2004) proposed that experts classify an artwork based on prototypes. No research has yet investigated prototypical classification in dance. It is worth noting that the early study of motor expertise in certain dance styles (e.g. Calvo-Merino et al. 2005a, 2005b) has compared two dominant styles (ballet and capoeira), but has not investigated prototype perception. However, we suggest that dance professionals might perceive dance according to their individually formed prototypes (subject to training and personal taste). Hence, a movement may represent a perfect prototype, and experts may evaluate variations of that movement on the basis of their prototype representation. According to Leder and colleagues (2004), non-experts' aesthetic experience is more heavily based on "stereotypes" than experts' aesthetic experience. Research into the aesthetic appreciation of dance would thus benefit greatly from knowing about people's stereotypes about dance and different dance forms.

We argue that the dance stimuli used in experiments should be regarded by experts as of high artistic quality, adhering to style and content conventions, independently of the research question. Only when a movement or a dance posture is "correct" (i.e. of high quality according to the style used) will expert spectators be able to engage fully in aesthetic contemplation (observation of poorer quality movements or postures may lead to a fast and frugal aesthetic evaluation or a response of disgust/disagreement).

12.2.2.5 Evaluation processes: IV cognitive mastering and evaluation

According to Leder and colleagues (2004), the successful perception and processing of an artwork will ultimately lead to a *cognitive mastering* which is the level at which an individual *understands* the artwork. Understanding is supposedly based on art-specific and/or self-related interpretation and can lead to either a positive experience of liking or a negative one—the individual can come to the conclusion that she doesn't like what she sees.

In dance, the cognitive mastering based on art-specific interpretation is likely to be influenced by elements of the context that help relate the dance to its historical and socio-political framework (Bullot and Reber 2013). The temporal evolution during a dance performance is also particularly relevant here as different processing stages are constantly influencing each other (bottom-up as well as top-down). Furthermore, dance often has a clear *narrative* (i.e. the storyline) and *discourse* (the conceptual metaphoric, expressive intention of the artist). A spectator can only comprehend those aspects as the dance unfolds over time. Thus, accessing the discourse of a dance is likely to be possible only if a full performance is watched. In previous research that used short dance stimuli (approximately 3 seconds), this level of cognitive mastering may not have been reached. We thus suggest adding to the art-related and self-related interpretation a within-work related interpretation.

12.2.2.6 Affective and emotional processing

Many studies in the field of empirical aesthetics of paintings focus on the affective responses participants derive from looking at them (Vartanian and Goel 2004; Kuchinke et al. 2009; Vartanian 2009). For dance, such empirical findings are still elusive despite the fact that choreographers frequently seek to craft works that convey emotional valence to the observer.

Some suggestions have been made that both novelty and, contrarily, familiarity of actions lead to emotional positive affective responses in the observers (e.g. Hagendoorn 2010). Further, watching different types of non-dance everyday movements has been found reliably to modify physiological as well as subjective emotional experiences (De Meijer 1989; Atkinson et al. 2004; de Gelder 2006; Peelen et al. 2007; Atkinson et al. 2012). Some of this research has also harnessed dance movements (Dittrich et al. 1996; Hejmadi et al. 2000; Pollick et al. 2002; Camurri et al. 2003). This indicates that emotion is indeed reliably induced when watching dance. However, so far no emotion-related brain activity has been found in neuroaesthetics experiments with dance movements (with the exception of Grosbras et al. 2012).[1] It is possible that the reason for the lack of evidence for changes in the affective (i.e. subjective, physiological, and neural) responses resides in the stimulus material used in research thus far.

12.2.2.7 Recapitulation of the model

The evaluative stage is obviously not only attained at the very end of a performance but forms part of the processing stages a spectator is constantly going through and updating in response to the unfolding dance on stage. The cognitive mastering of one sequence of movement may be successful at one point of time while another one is not, *except* if the whole piece is watched. Insight about the meaning of a particular sequence of movements may reveal itself at a much later point in time of the performance than in the very moment of display. The interacting and constantly updating nature of the whole aesthetic intercourse between the spectator, the dance, its different strands, and the other elements of the context is represented in the figure with the often bi-directional arrows. Full-length and complete ecological stimulus materials certainly provide the best grounds for such a genuine aesthetic experience to unfold.

12.3 Dance employed as ecologically valid stimuli

> Aesthetic episodes, whether in a museum or laboratory, have basic features in common. They formally begin when a viewer comes into contact with a particular art work and end when the viewer ceases to explore it. Viewers in both settings usually encounter a series of artworks. The experimenter, like a museum curator, selects the set of stimuli in accordance with a particular theme or goal. Indeed, the boundary between laboratory and museum is blurred when visitors are formally asked to rate paintings at an exhibition (Temme 1983). The structure of an aesthetic episode as an encounter between a viewer and a stimulating object makes it amenable to empirical study.
>
> (Cupchik and Gebotys 1988, pp. 1–2)

Cupchik and Gebotys (1988) provide an interesting analogy between aesthetic episodes evoked in art institutions and those experienced in laboratory settings. As the authors state, the scientist could be considered a kind of curator, selecting the theme displayed as well as the individual art works. However, in many instances the scientist does not display the original material. Moreover, the works are often modified in order to test particular hypotheses. The interference of the scientist with the artwork itself could cause considerable alteration of the original, and sometimes its destruction. In the case of empirical aesthetics of dance, the stimuli used in research have rarely been related to original choreographed works. One could thus argue that there is no ecological validity of such stimuli as they are not only out of context but they have no context at all.

There are three aspects to the validity of a measurement: *internal validity* designates the degree to which causal relationships between independent and dependent variables exist. For high internal validity, research conditions such as stimuli or the context must be controlled, and any potentially confounding variables accounted for completely. *External validity* is confined to the generalizability of the results (i.e. whether or not the same results would be obtained in a different context, time, and with other participants than those included in the study). *Ecological validity* refers to whether or not the results obtained under laboratory conditions are applicable also to the natural context where the objective phenomenon of the study usually unfolds. Overall, when we request more ecological stimuli for empirical aesthetics of dance, we refer to stimuli that more reliably capture dance as a social artistic phenomenon and that are capable of eliciting a genuine aesthetic experience in the observer.

Undoubtedly, studying the neural processes of observation or appreciation of "original" dance is more challenging than studying the same processes when looking at a painting or listening to music. Paintings have the benefit of being static over time (while looking at a particular painting), and music has infinite degrees of freedom (in the sense that any note can follow another). Dance works, however, change over time (while watching) and have numerous but biologically limited degrees of freedom (not every posture can follow another without physiological transition). The scientist faces these challenges when attempting to use ecological valid stimuli in dance. Other associated issues are organizational (i.e. accessing theaters and coinciding with their schedules), financial (filming, lighting, and space is expensive), interference (of non-controlled environments), unpredictability (at any one performance, unforeseen things can happen and they may not be repeated), and analysis issues (the analysis of complex data of long duration is not possible using conventional approaches). Nevertheless, some pioneering studies showed that it is possible to use ecological valid dance stimuli when studying spectators' motor resonance or aesthetic appreciation, with varying degrees of success (e.g. Jola et al. 2011; Grosbras et al. 2012; Noble et al. 2014; Jola et al. 2013a; 2013b). While some of these studies showed less clear and potentially unreliable results, they provided a valuable testing ground for our understanding of dealing with complex natural stimuli. We hope that researchers are encouraged to explore such approaches further. For example, Jola and colleagues (2009) measured motor corticospinal excitability in dance spectators when watching a live technical rehearsal (including light, costumes, orchestra) of the *Sleeping Beauty* in the Theatre Royal in Glasgow. This study did not show the expected results (i.e. enhanced motor simulation for physically challenging movement phrases compared to theatrical moments without dance) but instead showed a tendency for a decrease in motor simulation over time. The data had, however, several outliers and a more controlled study would be required to test the possibility of reduced motor simulation over time. We thus support the use of video recordings of genuine dance performances as a valuable alternative to life performances. This allows improving the quality of the data through a better control of the complex environment while employing stimuli of long duration.

Short video clips have been employed most frequently to study action observation of dance movements (e.g. ~2 seconds, cf. Jang and Pollick 2011; ~3 seconds, cf. Calvo-Merino et al. 2005a; and ~5 seconds, cf. Cross et al. 2006). Stimulus materials in such studies are commonly very thoroughly controlled regarding all information which could represent a distraction from the movement, such as color, music, stage props, and the face of the dancer. All of these are commonly removed. This enhances internal validity as potential responses to the dance that could otherwise be modulated by the addition of these strands of expressive information. At the neural level, a face consistently triggers the activation of the fusiform gyrus, i.e. the fusiform face area (Kanwisher et al. 1997). Color helps object recognition and its processing requires distinctive and additional neural pathways when compared to black and white processing (superior parietal lobule and precuneus bilaterally, the right hippocampus, and the right fusifom gyrus (V4); Bramao et al. 2010). Finally, research on music has shown that it clearly has its own potential to induce emotion (Blood and Zatorre 2001; Salimpoor et al. 2009).

The use of such short stimulus durations is not exclusive to the field of action observation. In neuroaesthetics, dance stimuli of short durations have also predominantly been used. As the characteristics of dance movements can change considerably over time, short sequences do indeed allow for a more uniform selection of movements, and participants' reactions can be measured in response to specific movements (Calvo-Merino et al. 2008; Calvo-Merino et al. 2010; Cross et al. 2011).

A conceptual concern about short stimuli is that the artistic quality is constrained. We argue that for a genuine aesthetic experience to unfold, more than 3 seconds (s) of continuous dance movements are necessary. This allows the spectator to follow a narrative as well as the discourse level of the choreography. However, short stimulus durations are methodologically better suited to neuroimaging as well as behavioral studies. Nevertheless, with novel analyses methods that focus on the time components, longer stimuli become feasible (e.g. Orgs et al. 2008).

Therefore, if the methodology of a chosen study requires short stimulus durations, then a careful selection of the stimuli should include a dance professional in order to select stimuli that are artistically as expressive as possible, in spite of being only of a couple of seconds long. For this, it is especially important to respect the sequencing of the dance movements. Non-dancers are likely to be less accurate than dancers in sequencing dance movements correctly. The elicitation of an aesthetic experience from dance movements depends crucially on the choice of particular movements and their being started and completed adequately as specified by the audiences' perception (Noble et al. 2014).

Within the field of neuroaesthetics, long dance video clips have been chosen more frequently than short dance clips as stimuli (e.g. ~3 minutes, cf. Grosbras et al. 2012; ~5 minutes, cf. Jola et al. 2012a; ~6 minutes, cf. Jola et al. 2013a; 2013b; and even 2.5 hours, cf. Jola et al. 2011b). Clearly, longer stimuli allow a more holistic approach to the actual unfolding aesthetic experience of dance as described earlier (i.e. narrative and discourse). Pitfalls of such an approach using long complex stimuli presentation are that they do not allow high levels of control and they require novel types of analyses. Nevertheless, what is an "adequate" level of experimental control in the stimulus materials for neuroaesthetics studies is critically dependent on the objectives of a study. If the research question is about the genuine aesthetic experience of dance, then longer, more ecologically valid stimuli should be presented and later combined with psychophysiological (e.g. respiration, skin conductance; Galvanic skin response (GSR); heart rate, electrocardiogram (ECG)) and subjective measures (e.g. visual analogical scales with an online slider option, or verbal reports) that accompany the experiences.

Similar qualms as those raised with regard to the long video clips have been voiced about the use of live dance performances as a suitable location for neuroaesthetics of dance research. The clear advantage of this approach is that the performance is fully ecological and also includes the very relevant *social* and communicative aspect of "an evening at the ballet," in addition to measures related to the co-presence of the performer and the spectator. In other words, the experience is genuine and only subtly altered due to the apparatus of the experiment (Stevens et al. 2009; Jola et al. 2011a; Jola and Grosbras 2013). Live performances will be experienced differently by each participant. While this may at first sight represent an issue, there is an important aspect that deserves a second look. Dancers in live performances adapt their dance to the time of the performance (time of the day, historical point, light, etc.), and—critically—in response to the atmosphere created by a particular audience. It may be a very daring suggestion, but consider the following as a tentative working hypothesis. When dancers actually "tailor" their performance to the particular audience they have at a given time, they will automatically modulate different aspects of their dance according to the needs of that particular audience. This means that in the end, a number of performances of the same piece of dance will have been performed with different expressive and artistic qualities depending on the feedback the dancers received during the performance. Some performances will have been "stronger" because the audience needed more stimulation, while other performances may have felt more suppressed. Thus, in the end, the same piece performed at different times and in front of different audiences will have elicited a similar level of aesthetic experience in the observers because these observers were receiving the performance tailored to their individual stimulation requirements. Hence, tentatively, we could say that responses obtained by means of such a set-up are actually *more* internally valid than a those elicited by watching a video, because this set-up ensures that each audience member is being "moved" according to his or her own needs (Jola and Grosbras 2013; Jola et al. 2011a). Such an experimental should be approached with caution though, and only under the supervision of dance professionals who can ensure that the experience is genuine. Besides, adjustments are likely to depend on the number of people in the audience, on the importance of an audience–dancer interaction of the particular dance style, and on whether the dancer is sensitive and skilled enough to actually tailor the performance to the particular needs of an audience.

Finally, dance photographs have been used as stimuli. These have varied from very controlled static academic dance positions in front of a neutral background, with neutral clothes and the dancers' face blurred out (Calvo-Merino et al. 2010), to photographs from genuine performance settings, reproduced in color, including the dancers' faces and costumes (Daprati et al. 2009; Jola et al. 2012b). Concerns about this kind of stimuli are mainly conceptual: first, *photographs* are not strictly speaking *dance*; they are a type of visual art. Second, especially if static academic positions are used (i.e. positions that are from the basic syllabus of movements and positions of a dance style), this is not technically speaking *dance*. Dance positions are merely the "tools of the trade." Accordingly, they are of questionable artistic quality when presented in isolation. However, photographs can be valuable experimental materials. By controlling the temporal dimension of the dance to a minimum (a dance photograph is only a snapshot of a distinctive moment in time of an unfolding dance), it allows studying spectators' reactions to particular, well-isolated features of a dance movement. Thus, in a recent kinaesthetic empathy study it was shown how an embodied response of movement could be elicited through ecological dance photographs when particular parameters such as movement direction (horizontal versus vertical) were carefully controlled in the stimulus materials.

12.4 Challenges—past and present

We have shown that dance stimuli can and have been employed in experimental research in and beyond neuroaesthetics. Other scientific investigations that have already benefited greatly from using dance as a movement stimulus did not aim at making assumptions about dance itself or the specific dance styles used. Dance in these cases is just a tool to study the human brain and human behavior. We argue, however, that since aesthetic appreciation processes are inherent, aesthetic experiences are always attendant when exposed to watching dance. Hence, careful creation and selection of stimuli creation is crucially independent of whether dance is the main subject of the study or not. In order to meet inherent aesthetic expectations, we argued that high ecological validity is needed when dance is employed in empirical research. As outlined, this naturally poses several challenges. In this final part, we suggest how to overcome these challenges when choosing an ecological approach to dance as an experimental stimulus.

In many cases, scientists have been very scrupulous and have carefully controlled for interfering variables by employing a reductionist experimental approach. For dance stimuli, however, our arguments go in two seemingly opposing directions. One the one hand, dance stimuli need to be better controlled; for example, for motion parameters. On the other hand, dance stimuli need to be presented closer to how they are in real life, something which inherently introduces an increase in potentially confounding factors. Clearly, the two arguments are related: i.e. the higher the ecologically validity, the more uncontrolled variables are included, versus the higher the control over confounding variables, the less likely the stimuli resemble what we perceive as dance. Furthermore, we have not yet discussed that beyond the choice of the stimulus content and presentation, aspects of the experimental conduction are crucial. In other words, the instruction given to participants as well as the participants' tasks can modify aesthetic appreciation. Finally, as most studies showed differences between experts and novices, it is hard to dispute the fact that the participants' demographics (e.g. experience) requires attention. For example, the aesthetic appearance is especially relevant for experts. Also, as dance is not acknowledged to be on the same artistic level as visual art or music, it is relevant to consider possible prejudices when testing novices.

The particular parameters that need to be better controlled for relate to the multiple interrelated strands of dance (i.e. movements, music, costume, presence, lighting, narrative as introduced in section 12.2). The combination of strands is particularly relevant for ecologically valid stimuli as dance is often performed with music, costumes, etc. The control for all of these parameters, however, is challenging and as yet we know little about how the individual strands in combination with movement affect neuronal processes when watching dance. Scientific focus on this artform oonly began about ten years ago (e.g. Calvo-Merino et al. 2005; Jola and Mast 2005). Hence, we have only recently begun to disentangle how relevant parameters such as features related to movement, music and/or sound, level of performance, cultural background of performer and spectator, determine neuronal processes when watching dance. If scientists were to create dance stimuli based on knowledge about perception and cognition from basic science (for analogies see Hagendoorn 2005, 2010; Jola 2010), we would end up with stimuli that are not recognizable as dance.

It is not only the movement dynamics that have affected motion perception (Pollick et al. 2001; Lagerlöf and Djerf 2002), but also the pure quantity of motion (Noble et al. 2014), the predictability of motion (Cross et al. 2011), the duration of motion (Jola et al. 2011b), and whether the dance contains gestures or abstract actions (Jola et al. 2012a; Jola and Grosbras

2013). All have an influence on action observation processes. However, further studies are needed to understand their role in aesthetic appreciation. For instance, the movement direction seems to have a great impact on the aesthetic appreciation of dance (Calvo-Merino et al. 2008). A recent study by Jola et al. (2012b) further supported this finding. Participants responded to experience a stronger kinaesthetic sensation in vertical compared to horizontal implied dance moves. Hence, not only does one have to consider that processes related to movement perception may be influenced by the quantity of motion, but also may be influenced by the direction of movements. Hence, one should control for such potential confounding factors if they are not already implemented as experimental factors.

We have also shown that the background of the dance stimuli has varied extensively across existing studies. A dance recording taken in a ballet studio often shows barres and mirrors. The barre is a tool used during classical warm-up exercises in ballet. The dancers hold on to the barre with one hand to support their balance, integrate it as a reference point for posture alignment, and use it as a tool for stretching exercises. The barre is thus a very familiar element of training in particular for ballet dancers. Hence, a barre in the background may be pleasurably nostalgic to some people (with memories of idling or stretching), eliciting unintended representations and a mindset of rehearsal and daily training, but may be entirely meaningless to others. One could argue that it has the potential to enhance motor simulation processes due to its association with a training context. Therefore, the visibility of barre or mirror acts as a reference to a training context and does not qualify as a natural setting for watching a professional dance performance. We hope that we have illustrated with this example that the context in which an aesthetic experience usually unfolds needs careful consideration and preparation. A number of very recent studies investigating spectators' responses to watching dance performances have indeed used recordings from the stage (Grosbras et al. 2012; Reason et al. in review).

Regarding participants' instructions, it is important how the dance movements are brought to the participant as well as how the participant is brought to the dance. Instructions that emphasize the artistic nature of the experiment are one way to ensure that the participant understands that aesthetic expectations are welcome, especially if the experiment takes place in a laboratory and is artificially controlled with the risk of a deprivation of artistic features. As shown in a neuroimaging study using paintings (Cupchik et al. 2009), neural activity in response to visual artworks depends on the participants' task, When participants were asked to engaged in an irrelevant object identification task, activity in the fusiform gyrus (right) was found (in accordance with the object identification task). However, if participants were in the mindset of aesthetic appreciation, enhanced bilateral activity in the insula was found, a region crucially involved in affective processing. Interestingly, brain imaging studies of action observation and dance appreciation show similar cortical activity in visual, parietal, temporal, and frontal areas (for a review see Christensen and Calvo-Merino 2013), whereas evidence for limbic structures' involvement (i.e. structures important for affective processing) in aesthetic experience has been shown frequently. Brown and colleagues (2011), for example, showed in their meta-analysis that the insula appears to be key as it was found to show increased activation in so many studies on aesthetic appreciation.

Forgas (1995) proposed the affect infusion model (AIM) which suggests that *current mood* and *affective priming* has an effect on judgments. According to this view, judgments to non-familiar stimuli are particularly likely to be influenced by "affect infusion." For empirical aesthetics this means that for artworks one has little knowledge or no knowledge about, the current affective state will influence one's judgment more than for artworks with which one is familiar. This was recently experimentally assessed by means of a priming paradigm where happy and disgusted facial emotional expressions were presented with very brief stimuli onset asynchronies (SOA) just before instances of unknown paintings by Pollock. The happy primes increased the aesthetic experience, while disgust primes reduced it (assessed by means of subjective ratings of "liking" on a Likert scale) (Flexas et al. 2014). This shows that aesthetic judgment crucially depends on the spectator's affective state, be it triggered by the artwork itself or by some stimulus which is independent and may even be *irrelevant* to the art work itself (as were the faces in the case of the mentioned study). Thus, the "affective state" prior to the exposure to an artwork potentially forms part of the whole "mindset" an individual is in when he looks at an artwork.

Moreover, regardless of whether the mood or affective state prior to the exposure to an artwork is positive or negative, it is crucial that there is affective engagement with the artful object. Higgins (2006) argued that in general terms, the experience of value of an object can be influenced by the *strength* of the (attentional and affective) *engagement* with that object (for empirical aesthetics this *value* can be assessed by measuring the subjects' subjective responses: "liking," "aesthetic quality," etc.; or their psychophysiological responses: GSR, ECG, etc.). This is because *engagement strength* is said to be accompanied by the feeling of a motivational force which will be either of attraction or of repulsion from the object. According to this view, pleasure experienced about something involves a component of motivation, which is a classical hedonic principle. This is supposedly due to the fact that people are not "just" reacting to a stimulus; they also assign meaning and significance to it as a function of their own current needs, desires, goals, and motives. Higgins has related the concept of engagement strength to the experience of "flow" as it was proposed by (Csikszentmihalyi 1990). A high level of engagement strength is a prerequisite to enter that state of flow. This state has been suggested as a state of mind a spectator of an artwork enters when confronted with an artwork and successfully masters all aspects of the appreciative process and, eventually, is aesthetically moved by it (Marković 2012). It is a type of transcendent understanding of the expressive qualities of an artwork and is related to the ultimate enjoyment of it. For this to happen, there must be an affective predisposition to let the artwork genuinely get "under your skin" and to engage willingly with it, otherwise this state of flow is not likely to be attained, irrespectively of how long a person looks at an artwork.

The importance of carefully designing an experiment's instructions so that they facilitate an affective predisposition and strong engagement with the artful object (here: an art stimulus) holds for all art forms. What makes dance special in this respect is that real dance is a multisensory experience. Thus, experimenters must be skillful at transmitting the special nature of attending a dance performance. Besides, whilst the vast majority—if not all—people have looked at paintings in an art gallery, or in poster shops, comparatively few people have experienced dance performances first-hand. Thus trying to instil this particular mindset in an experimental participant can be particularly challenging with dance stimuli, especially, if the researcher has no own dance or dance viewing experience.

12.5 Requirements for the future

For a fruitful future of dance in science to occur, a number of points require careful consideration. When investigating cerebral processes in response to watching dance, known parameters are motion, culture, presence, and experimental instruction. We mentioned that dance needs to be considered in its complete form. How do we achieve all of this? For the design of experiments using ecological valid stimuli, there are at least three requirements to follow that support adequate consideration of these points: (1) the experience of the participant; (2) the knowledge of the experimenter; (3) the multisensory nature of dance.

Requirement 1: Consideration of participants' sensory experiences. We have discussed the fact that the aesthetic appreciation of an experimental subject is never independent of context. We provided a detailed example of how the background on video recordings can change the context of a dance performance. Other contexts are related to time (historical context, see Gombrich 2006), location (cultural context, see Jola et al. 2012a), presentation form (i.e. duration, see Jola et al. 2011b), and presentation size (Wildschut 2010). Hence, the context of the stimuli needs careful consideration when creating the stimuli and should be described in the experimental outline. Furthermore, the participants' reported experiences when watching the stimuli offers relevant insight into the quality and appropriateness of the stimuli. Hence, we propose that subjective experiences should be collected and reported as part of any study using complex aesthetic stimuli such as dance. In other words, it is important to recognize that the aesthetic appreciation is a form of experience and thus the stimuli should be created in such a way that they first allow such an aesthetic experience to evolve, and secondly, that this aesthetic experience can be measured and reported in some way. Finally, since the context affects subjects' experience dependent on their expertise of the matter, it is highly relevant that the subjects' expertise is clearly described.

Requirement 2: State the background and knowledge of the researcher. It is largely those scientists who have at least a marginal interest in dance that employ dance in their studies, yet the level of expertise varies hugely. Expertise in dance is necessary to understand to at least some degree what we know about our subject of study (see e.g. Jola et al. 2011b). Moreover, as Gombrich makes very clear, knowledge affects what one sees (Gombrich 2006). We thus argue that not only is it relevant for the scientist to spend time investigating the human brain, but also to spend time investigating the stimuli used, a practice described as embodied neuroscience (Jola 2010). In addition, we argue that it would be beneficial if the scientists' approach to dance were made transparent by citing a clear definition of terms, especially that of dance, at the very least.

Requirement 3: Investigating the multisensory nature of dance. Aesthetic experiences based on visual perception (as in visual form of art) can evoke various forms of sensory experiences (kinaesthetic tactile, etc.). As we have outlined in this chapter, the aesthetic appreciation of dance is a sensory experience that touches different sensory modalities, not just the visual or kinaesthetic sense. It is thus crucial to improve our understanding of what forms of sensory experiences dance can evoke, particularly as different forms of dance supposedly evoke different forms of sensory experiences. Furthermore, we believe that these different sensory experiences can be very subtle and are only evident if the dance is presented in its natural, multisensory form (see e.g. Jola et al. 2014): dance is often performed to music, with audible footsteps and breathing, seeing dancers' sweat, etc. Hence, this is why ecological valid forms of dance stimuli are crucial: dance needs to speak to us to be appreciated otherwise it is a condensation of purely physical motion. Therefore, a definition of dance underlying the selection of the dance stimuli of a study may be helpful.

12.6 Conclusion

In many instances, the scientific employment of dance does not live up to the natural phenomenon of dance. As outlined earlier, a number of studies have shown evidence that the way one presents dance affects perceptual, cognitive, and evidently neuronal processes significantly. Therefore, we argued for a protocol of *necessities* when using dance as a performative artform in studying the human brain and human behavior.

We consider it crucial to use "*genuine* dance," dance that can be seen and enjoyed watching on stage or in other aesthetic contexts. Hence, it could be presented in the form of static displays (photographs), video clips, or live performances. Due to the multifaceted nature of dance, its complexity should be recognized and acknowledged in order to reach ecological validity. In other words, the dance should be seen as dance by experienced spectators. For this, qualitative data should be collected and reported not only from participants, but optimally also from dance experts specifically consulted for the study. These reports should include the overall experience of the experiment, appreciation of the stimuli. and a description of what participants think they saw in the experiment. All stimuli used should further be made available to readers (online, or by request). Parameters of the dance strands should be analysed and/or described, and these are motion, gestural components, and a description of the dance form used.

We want to encourage researchers in empirical aesthetics as well as in general (i) to define more thoroughly what they mean by "dance" when using dance stimuli, (ii) to design their stimuli according to a theory of what dance is that also bears the scrutiny of dance experts (only in this way can one be certain to test the responses to the "the real thing"), and finally (iii) to embed their studies in the general framework of empirical aesthetics. Using dance as a stimulus, in our view, involves a commitment, thus conclusions drawn from results obtained with not carelessly designed dance stimuli are based on shaky grounds. The difficulty for the scientist resides especially in understanding what exactly makes dance *dance*, and not just a motor pattern. What if the enhanced motor response found so often in experts when they watch (impoverished) dance movements of their area of expertise is not due to the much-suggested "enhanced motor simulation," but rather due to the dancers' compensatory cognitive and neural mechanisms trying to "make up" for the pitfalls of the movements they see?

To fulfil these criteria, we propose cross-disciplinary collaborations. The existing understanding of dance by dance practitioners, choreographers, scenographers, as well as dance scholars (i.e. humanities, philosophy, phenomenology, and languages) is useful in the study of the aesthetic experiences of watching dance. Furthermore, collaborations with other scientists working in specific areas of perception, cognition, physiology, and computation are hugely beneficial in order to control for parameters of the strands of dance.

Empirical aesthetics began with the aesthetic experiences in response to visual art. The frameworks were based on models of visual perception. These are only partially applicable to the study of aesthetic experiences when watching dance. Although dance is perceived visually, the sensory experiences clearly extend beyond the visual sense. Kinaesthetic responses are important and so are auditory experiences (as dance is often accompanied by music or physical sounds). One could even argue that dance evokes tactile sensory aesthetic experiences through the observation of the touch and interaction between dancers on stage. Dance has come late onto the map of experimental cognitive psychology and neuroscience. This is why the models provided by these fields may not yet be conclusive or encompassing for dance but as shown here, provide a good starting point.

Acknowledgments

Julia F. Christensen was supported by a FPU Ph.D. scholarship from the Spanish Ministry of Education (AP2009–2889), and by the research project FFI2010–20 759 funded by the Spanish Ministry of Science and Innovation. Corinne Jola was supported by the University of Surrey, UK and INSERM-CEA NeuroSpin Center, France through l'Agence Nationale de la Recherche France (ANR) and the laboratoire d'excellence des Arts et Médiations humaines (Labex Arts-H2H), Université Paris 8.

Correspondence concerning this chapter should be addressed to Julia F. Christensen, University of the Balearic Islands, University Campus, Department of Psychology, Guillem Cifre de Colonya building, 07 122 Palma, Spain, julia.christensen@uib.es, or to Corinne Jola, Division of Psychology, School of Social & Health Sciences, Abertay University, Bell Street, Kydd Building, Level 5, Dundee, DD1 1HG, UK, c.jola@abertay.ac.uk.

References

Aglioti, S.M., Cesari, P., Romani, M., et al. (2008). Action anticipation and motor resonance in elite basketball players. *Nature Neuroscience* **11**, 1109–16.

Aronoff, J. (2006). How we recognize angry and happy emotion in people, places and things. *Cross-Cultural Research* **40**, 83–105.

Atkinson, A.P., Dittrich, W.H., Gemmell, A.J., et al. (2004). Emotion perception from dynamic and static body expressions in point-light and full-light displays. *Perception* **33**, 717–46.

Atkinson, A.P., Vuong, Q.C., and Smithson, H.E. (2012). Modulation of the face- and body-selective visual regions by the motion and emotion of point-light face and body stimuli. *NeuroImage* **59**, 1700–12.

Augustin, D. and Leder, H. (2006). Art expertise: a study of concepts and conceptual spaces. *Psychological Science* **48**, 135–57.

Augustin, D.M., Leder, H., Hutzler, F., and Carbon, C.C. (2008). Style follows content. On the microgenesis of art perception. *Acta Psychologica* **128**, 127–38.

Belke, B., Leder, H., and Augustin, D. (2006). Mastering style. Effects of explicit style-related information, art knowledge and affective state on appreciation of abstract paintings. *Psychology Science* **48**, 115–35.

Bläsing, B., Tenenbaum, G., and Schack, T. (2009). The cognitive structure of movements in classical dance. *Psychology of Sport and Exercise* **10**, 350–60.

Blood, A.J. and Zatorre, R.J. (2001). Intensely pleasurable responses to music correlate with activity in brain regions implicated in reward and emotion. *Proceedings of the National Academy of Sciences of the USA* **98**, 11818–23.

Bramao, I., Faísca, L., Forkstam, C., Reis, A., and Petersson, K.M. (2010). Cortical brain regions associated with color processing: an FMRI study. *The Open Neuroimaging Journal* **4**, 164–73.

Brown, S., Martinez, M.J., and Parsons, L.M. (2006). The neural basis of human dance. *Cerebral Cortex* **16**, 1157–67.

Brown, S., Gao, X., Tisdelle, L., et al. (2008). Naturalizing aesthtics: brain areas for aesthetic appraisal across sensory modalities. *Neuroimage* **58**, 250–58.

Bullot, N.J. and Reber, R. (2013). The artful mind meets art history: toward a psycho-historical framework for the science of art appreciation. *Behavioral and Brain Sciences* **36**, 123–37.

Butterworth, J. and Wildschutt, L. (2009). *Contemporary Choreography. A Critical Reader* (Section 5: Changing Aesthetics). New York, NY: Routledge.

Calvo-Merino, B., Ehrenberg, S., Leung, D., et al. (2010a). Experts see it all: configural effects in action observation. *Psychological Research* **74**, 400–6.

Calvo-Merino, B., Glaser, D.E., Grèzes, J., et al. (2005a). Action observation and acquired motor skills: an fMRI study with expert dancers. *Cerebral Cortex* **15**, 1243–9.

Calvo-Merino, B., Grèzes, J., Glaser, D.E., et al. (2005b). The influence of visual and motor familiarity during action observation: An fMRI study using expertise. *Journal of Cognitive Neuroscience* **Supplement S**, 115–115.

Calvo-Merino, B., Grèzes, J., Glaser, D.E., et al. (2006). Seeing or doing? Influence of visual and motor familiarity in action observation. *Current Biology* **16**, 1905–10.

Calvo-Merino, B., Jola, C., Glaser, D.E., et al. (2008). Towards a sensorimotor aesthetics of performing art. *Consciousness and Cognition* **17**, 911–22.

Calvo-Merino, B., Urgesi, C., Orgs, G., et al. (2010b). Extrastriate body area underlies aesthetic evaluation of body stimuli. *Experimental Brain Research* **204**, 447–56.

Camurri, A., Lagerlof, I., and Volpe, G. (2003). Recognizing emotion from dance movement: comparison of spectator recognition and automated techniques. *International Journal of Human-Computer Studies* **59**, 213–25.

Chatterjee, A. (2003). Prospects for a cognitive neuroscience of visual aesthetics. *Bulletin of Psychology and the Arts* **4**, 55–60.

Chatterjee, A. (2011). Neuroaesthetics: a coming of age story. *Journal of Cognitive Neuroscience* **23**, 53–62.

Chatterjee, A., Widick, P., Sternschein, R., et al. (2010). The assessment of art attributes. *Empirical Studies of the Arts* **28**, 207–22.

Christensen, J.F. and Calvo-Merino, B. (2013). Dance as a subject for empirical aesthetics. *Psychology of Aesthetics, Creativity, and the Arts* **7**, 76–88.

Conroy, R.M. (2012). *Dance.* In A.C. Ribeiro (ed.), *The Continuum Companion to Aesthetics.* London: Continuum International Publishing Group, pp. 156–70.

Cross, E.S., Hamilton, A.F.d.C., and Grafton, S.T. (2006). Building a motor simulation de novo: observation of dance by dancers. *NeuroImage* **31**, 1257–67.

Cross, E.S., Kirsch, L., Ticini, L.F., et al. (2011). The impact of aesthetic evaluation and physical ability on dance perception. *Frontiers in Human Neuroscience* **5**, doi: 10.3389/fnhum.2011.00102.

Cross, E.S. and Ticini, L.F. (2011). Neuroaesthetics and beyond: new horizons in applying the science of the brain to the art of dance. *Phenomenology and the Cognitive Sciences* **11**, 5–16.

Csikszentmihalyi, M. (1990). *FLOW: The Psychology of Optimal Experience.* New York, Harper and Row.

Cupchik, G.C. and Gebotys, R. (1988). The experience of time, pleasure, and interest during aesthetic episodes. *Empirical Studies of the Arts* **6**, 1–12.

Cupchik, G.C., Vartanian, O., Crawley, A., et al. (2009). Viewing artworks: contributions of cognitive control and perceptual facilitation to aesthetic experience. *Brain and Cognition* **70**, 84–91.

Daprati, E., Iosa, M., and Haggard, P. (2009). A dance to the music of time: aesthetically-relevant changes in body posture in performing art. *PloS* One **4**(3), 1–11.

de Gelder, B. (2006). Towards the neurobiology of emotional body language. *Nature Reviews Neuroscience* **7**, 242–9.

De Meijer, M. (1989). The contribution of general features of body movement to the attribution of emotions. *Journal of Nonverbal Behavior* **13**, 247–68.

Désy, M.-C. and Théoret, H. (2007). Modulation of motor cortex excitability by physical similarity with an observed hand action. *PLoS* One **2**, e971.

Dittrich, W.H., Troscianko, T., Lea, S.E., et al. (1996). Perception of emotion from dynamic point-light displays represented in dance. *Perception* **25**, 727–38.

Dumont, A. (2009). Une perte de controle, une perte de confort: Louise Lecavalier. La Question du Risque dans un parcours d'Interprete. *Repères, Cahier de Danse* **28**, 15–7 doi: <http://revue-interrogations.org/Danseurs-contemporains-virtuosite>.

Felsen, G. and Dan, Y. (2005). A natural approach to studying vision. *Nature Neuroscience* **8**, 1643–6.

Flexas, A., Rossello-Mir, J., de Miguel, P., et al. (2014). Cognitive control and unusual decisions about beauty: an fMRI study. *Frontiers in Human Neuroscience* **5**, doi: 10.3389/fnhum.2014.00520.

Forgas, J.P. (1995). Mood and judgment: the affect infusion model (AIM). *Psychological Bulletin* **117**, 39–66.

Gombrich, E.H. (2006). *The Story of Art.* Pocket Edition, London: Phaidon Press Ltd.

Graham, D.J. and Field, D.J. (2007). Statistical regularities of art images and natural scenes: spectra, sparseness and nonlinearities. *Spatial Vision* **21**, 149–64.

Grosbras, M.-H., Tan, H., and Pollick, F. (2012). Dance and emotion in posterior parietal cortex: a low-frequency rTMS study. *Brain Stimulation* **5**, 130–6.

Hagendoorn, I. (2005). Dance perception and the brain. In R. Grove, C. Stevens, and S. McKechnie (eds), *Thinking in Four Dimensions. Creativity and Cognition in Contemporary Dance.* Melbourne: Melbourne University Press, pp. 137–48.

Hagendoorn, I. (2010). Dance, choreography and the brain. In D. Melcher and F. Bacci (eds), *Art and the Senses.* Oxford: Oxford University Press, pp. 499–514.

Hänggi, J., Koeneke, S., Bezzola, L., et al. (2010). Structural neuroplasticity in the sensorimotor network of professional female ballet dancers. *Human Brain Mapping* **31**, 1196–206.

Hejmadi, A., Davidson, R.J., and Rozin, P. (2000). Exploring Hindu Indian emotion expressions: evidence for accurate recognition by Americans and Indians. *Psychological Science* **11**, 183–7.

Higgins, E.T. (2006). Value from hedonic experience and engagement. *Psychological Review* **113**, 439–60.

Hintz, J.M. and Nelson, T.M. (1971). Haptic aesthetic value of the golden section. *British Journal of Psychology* **62**, 217–23.

Jang, S.H. and Pollick, F.E. (2011). Experience influences brain mechanisms of watching dance. *Dance Research Journal* **29**, 352–77.

Järveläinen, J., Schürmann, M., Avikainen, S., et al. (2001). Stronger reactivity of the human primary motor cortex during observation of live rather than video motor acts. *NeuroReport* **12**, 3493–5.

Jola, C. (2010). Research and choreography. Merging dance and neuroscience. In B. Bläsing, M. Puttke, and T. Schacke (eds), *The Neurocognition of Dance. Mind, Movement, and Motor Skill*. Hove: Psychology Press, pp. 203–34.

Jola, C., Abedian-Amiri, A., Kuppuswamy, A., et al. (2012a). Motor simulation without motor expertise: enhanced corticospinal excitability in visually experienced dance spectators. *PLoS One* 7(3), e33343. doi:10.1371/journal.pone.0033343.

Jola, C., Clements, L., and Christensen, J.F. (2012b). Moved by stills: kinesthetic sensory experiences in viewing dance photographs. *Seeing and Perceiving* **25** (supplement), 80–1.

Jola, C., Ehrenberg, S., and Reynolds, D. (2011a). The experience of watching dance: phenomenological-neuroscience duets. *Phenomenological Cognitive Sciences* **11**, 17–37.

Jola, C. and Grosbras, M.H. (2013). In the here and now: enhanced motor corticospinal excitability in novices when watching live compared to video recorded dance. *Cognitive Neuroscience* **4**, 90–8.

Jola, C. and Mast, F.W. (2005). Mental object rotation and egocentric body transformation: two dissociable processes? *Spatial Cognition and Computation* **5**, 217–37.

Jola, C., McAleer, P., Grosbras, M.-H., et al. (2013). Uni- and multisensory brain areas are synchronized across spectators when watching unedited dance recordings. *i-Perception* **4**, 1–20.

Jola, C., Pollick, F.E., and Calvo-Merino, B. (2014) "Some like it hot": spectators who score high on the personality trait openness enjoy the excitement of hearing dancers breathing without music. *Frontiers in Human Neuroscience* **8**, 718. doi: 10.3389/fnhum.2014.00718.

Jola, C., Pollick, F.E., and Grosbras, M.-H. (2011b). Arousal decrease in Sleeping Beauty: audiences' neurophysiological correlates to watching a narrative dance performance of 2.5 hrs. *Dance Research Journal* **29**, 378–403.

Kanwisher, N., McDermott, J., and Chun, M.M. (1997). The fusiform face area: a module in human extrastriate cortex specialized for face perception. *Journal of Neuroscience* **17**, 4302–11.

Kirk, U., Skov, M., Hulme, O., Christensen, M.S., et al. (2009). Modulation of aesthetic value by semantic context: An fMRI study. *NeuroImage* **44**, 1125–32.

Klein, M. (2007). Choreography: a pattern language. *Kybernetes* **36**, 1081–8.

Konecni, V.J., Wanic, R.A., and Brown, A. (2007). Emotional and aesthetic antecedents and consequences of music-induced thrills. *American Journal of Psychology* **120**, 619–43.

Kornysheva, K., von Cramon, D.Y., Jacobsen, T., et al. (2009). Tuning-in to the beat: Aesthetic appreciation of musical rhythms correlates with a premotor activity boost. *Human Brain Mapping* **31**, 48–64.

Kubovy, M. (1986). *The Psychology of Perspective and Renaissance Art*. Cambridge MA: Cambridge University Press.

Kuchinke, L., Trapp, S., Jacobs, A.M., et al. (2009). Pupillary responses in art appreciation: effects of aesthetic emotions. *Psychology of Aesthetics, Creativity and the Arts* **3**, 156–63.

Lagerlöf, I. and Djerf, M. (2002). Children's understanding of emotion in dance. *European Journal of Developmental Psychology* **6**, 409–31.

Leder, H. (2001). Determinants of preference. When do we like what we know? *Empirical Studies of the Arts* **19**, 201–11.

Leder, H., Augustin, D., and Belke, B. (2005). Art and cognition! Consequences for experimental aesthetics. *Bulletin of Psychology and the Arts* **2**, 11–20.

Leder, H., Belke, B., Oeberst, A., et al. (2004). A model of aesthetic appreciation and aesthetic judgments. *British Journal of Psychology* **95**, 489–508.

Leder, H., Carbon, C.C., and Ripsas, A.L. (2006). Entitling art: influence of title information on understanding and appreciation of paintings. *Acta Psychologica* **121**, 176–98.

Leder, H., Tinio, P.P.L., and Bar, M. (2011). Emotional valence modulates the preference for curved objects. *Perception* **40**, 649–55.

Lengger, P.G., Fischmeister, F.P., Leder, H., et al. (2007). Functional neuroanatomy of the perception of modern art. A DC-EEG-Study on the influence of stylistic information on aesthetic experience. *Brain Research* **1158**, 93–102.

Locher, P.J., Overbeeke, K., and Stappers, P.J. (2005). Spatial balance of color triads in the abstract art of Piet Mondrian. *Perception* **34**, 169–89.

Marković, S. (2012). Components of aesthetic experience: aesthetic fascination, aesthetic appraisal, and aesthetic emotion. *i-Perception* **3**, 1–17.

May, J., Calvo-Merino, B., deLahunta, S., et al. (2011). Points in mental space: an interdisciplinary study of imagery in movement creation. *Dance Research Journal* **29**, 404–32.

Miura, N., Sugiura, M., Takahashi, M., et al. (2010). Effect on motion smoothness on brain activity while observing a dance: an fMRI study using a humanoid robot. *Social Neuroscience* **5**, 40–58.

Molnar-Szakacs, I., Wu, A.D., Robles, F.J., et al. (2007). Do you see what I mean? Corticospinal excitability during observation of culture-specific gestures. *PLoS One* **2**, e626.

Nairn, S. (2014). Nursing and the new biology: towards a realist, anti-reductionist approach to nursing knowledge. *Nursing Philosophy* **15**, 261–73.

Nieminen, S., Istók, E., Brattico, E., et al. (2011). The development of aesthetic responses to music and their underlying neural and psychological mechanisms. *Cortex* **47**, 1138–46.

Noble, K., Glowinski, D., Murphy, H., et al. (2014). Event segmentation and biological motion perception in watching dance. *Art & Perception* **2**, 59–74.

Orgs, G., Dombrowski, J.H., Heil, M., et al. (2008). Expertise in dance modulates alpha/beta event-related desynchronization during action observation. *European Journal of Neuroscience* **27**, 3380–4.

Orgs, G., Hagura, N., and Haggard, P. (2013). Learning to like it: aesthetic perception of bodies, movements and choreographic structure. *Consciousness and Cognition* **22**, 603–12.

Peelen, M.V., Atkinson, A.P., Andersson, F., et al. (2007). Emotional modulation of body-selective visual areas. *Social Cognitive and Affective Neuroscience* **2**, 274–83.

Pollick, F.E., Lestou, V., Ryu, J., et al. (2002). Estimating the efficiency of recognizing gender and affect from biological motion. *Vision Research* **42**, 2345–55.

Pollick, F.E., Paterson, H.M., Bruderlin, A., et al. (2001). Perceiving affect from arm movement. *Cognition* **82**, B51–B61.

Preston-Dunlop, V. and Sanchez-Colberg, A. (2010). *Dance and the Performative, a choreological perspective—Laban and beyond*. Alton: Dance Books Ltd.

Reason, M., Jola, C., Kay, R., et al. (under review). Spectators' aesthetic experience of sound and movement in dance performance: a transdisciplinary investigation.

Reason, M. and Reynolds, D. (2010). Kinesthesia, empathy and related pleasures: an inquiry into audience experiences of watching dance. *Dance Research Journal* **42**, 49–75.

Redcay, E., Dodell-Feder, D., Pearrow, M.J., et al. (2010). Live face-to-face interaction during fMRI: a new tool for social cognitive neuroscience. *NeuroImage* **50**, 1639–47.

Salimpoor, V.N., Benovoy, M., Longo, G., et al. (2009). The rewarding aspects of music listening are related to degree of emotional arousal. *PLoS ONE* **4**, e7487. doi:10.1371/journal.pone.0007487.

Sevdalis, V. and Keller, P.E. (2011). Captured by motion: dance, action understanding, and social cognition. *Brain and Cognition* **77**, 231–6.

Shimada, S. and Hiraki, K. (2006). Infant's brain responses to live and televised action. *NeuroImage* **32**, 930–9.

Shusterman, R. (1997). The end of aesthetic experience. *The Journal of Aesthetics and Art Criticism* **55**, 29–41.

Shusterman, R. and Tomlin, A. (2008). *Aesthetic Experience*. NY: Routledge.

Silverthorne, C., Noreen, C., Hunt, T., et al. (1972). The effects of tactile stimulation on visual experience. *Journal of Social Psychology* **88**, 153–4.

Silvia, P.J. (2012). Human emotions and aesthetic experience. An overview of empirical aesthetics. In A.P. Shimamura and S.E. Palmer (eds), *Aesthetic Science: Connecting Minds, Brain, and Experience*. New York: Oxford University Press, pp. 250–75.

Stevens, C.J., Schubert, E., Morris, R.H., et al. (2009). Cognition and the temporal arts: investigating audience response to dance using PDAs that record continuous data during live performance. *International Journal of Human-Computer Studies* **67**, 800–13.

Stevens, C., Winskel, H., Howell, C., et al. (2010). Perceiving dance: schematic expectations guide experts' scanning of a contemporary dance film. *Journal of Dance Medicine & Science* **14**, 19–25.

Temme, J.E.V. (1983). *Accounting for Tastes*. Ultrecht, Dnikkerij Elinkwijk.

Tinio, P.P.L. and Leder, H. (2009). Just how stable are stable aesthetic features? Symmetry, complexity, and the jaws of massive familiarization. *Acta Psychologica* **130**, 241–50.

Torrents, C., Castañer, M., Jofre, T., et al. (2013). Kinematic parameters that influence the aesthetic perception of beauty in contemporary dance. *Perception* **42**, 447–58.

Urgesi, C., Calvo-Merino, B., Haggard, P., et al. (2007). Transcranial magnetic stimulation reveals two cortical pathways for visual body processing. *Journal of Neuroscience* **27**, 8023–30.

Vartanian, O. (2009). Conscious experience of pleasure in art. In M. Skov and O. Vartanian (eds), *Neuroaesthetics*. Amityville, NY: Baywood, pp. 261–54.

Vartanian, O. and Goel, V. (2004). Neuroanatomical correlates of aesthetic preference for paintings. *NeuroReport* **15**, 893–7.

Wildschut, E.M.M. (2010). Becoming one body. Manifestations of kinesthetic empathy while watching dancers close or further away. In R. Naber, B. Nieuwboer, and E.M.M. Wildschut (eds), *Danswetenschap in Nederland*. Amsterdam: Vereniging voor Dansonderzoek, pp. 75–81.

Zeki, S. (1999). *Inner Vision. An Exploration of Art and the Brain*. Oxford: Oxford University Press.

Section 4

Cognitive neuroscience of music

Chapter 13

Liking music: Genres, contextual factors, and individual differences

Kathleen A. Corrigall and E. Glenn Schellenberg

13.1 Introduction

The amount of time and money that people spend on music provides ample evidence that listening to it is important for most of us. We choose to listen to music at concerts, on the radio, and on portable media players (e.g. MP3 players), computers, and CD players. We also hear music that others choose for us in television shows and movies, in shopping centers and restaurants, and at significant events such as weddings and funerals. In 2012, the International Federation of the Phonographic Industry (2013) reported that global revenues from digital sales alone (i.e. not including CDs or vinyl) reached approximately US$5.6 billion, which confirms that music is valued worldwide.

The amount of resources devoted to music consumption establishes unequivocally that at a fundamental level, people *enjoy* listening to music. However, different individuals like different kinds of music, and the same individual may exhibit greater or lesser appreciation for a particular genre or piece of music depending on the circumstances in which it is heard. The goal of this chapter is to provide an overview of research that has examined preferences for genres and liking individual pieces of music.

Over the past few decades, research interest in music perception and cognition has grown (Jones et al. 2010), as has interest in emotional responses to music (Juslin and Sloboda 2010). Nevertheless, research on liking or disliking music—the most basic emotional response—has received relatively little attention. For example, Juslin and Västfjäll (2008) proposed several mechanisms to explain why listeners respond emotionally to music but they did not consider basic liking. Instead they focused on how music evokes emotional responses such as happiness or sadness, ignoring the fact that responding emotionally to music occurs on two levels (Hunter and Schellenberg 2010). One is *evaluative*—whether the listener likes the music or considers it to be good. The other level is the specific *feeling*, such as happiness, that is expressed and/or evoked. In the present chapter, our focus is on the evaluative level. What makes someone like a particular song or genre of music? How are individual differences related to music preferences? What factors influence how liking music develops or changes? Our goal is to provide an overview of what is known and unknown in response to these questions, focusing primarily on studies published since 2000.

13.2 Liking different genres of music

Most research on music preferences has focused on characterizing the kinds of people who like different genres of music (e.g. pop, rock, country, rap/hip-hop, classical, jazz, etc.), asking whether individual differences in demographic, cognitive, and personality variables are correlated with how much a person likes particular genres. Although the concept of genre may seem poorly defined, genres can be immediately obvious to the average listener. In one instance, college freshmen registered in a psychology course performed at above-chance levels on a genre-identification task after hearing excerpts of music that were only 250 ms in duration (Gjerdingen and Perrott 2008).

Because recognizing and classifying genres can be virtually instantaneous, genres are likely to play an important role in evaluative responses to music. Early research tended to focus on documenting positive associations between preferences for genres, particularly music that sounded aggressive or negative, and personality characteristics such as rebelliousness or sensation seeking (Arnett 1992; Carpentier et al. 2003; Cattell and Saunders 1954; Christenson and Peterson 1988; Dollinger 1993; Glasgow et al. 1985; Hakanen and Wells 1993; Hansen and Hansen 1991; Litle and Zuckerman 1986; McNamara and Ballard 1999; Payne 1980; Rawlings and Ciancarelli 1997; Rawlings et al. 1995, 2000; Roe 1985; Schwartz and Fouts 2003; Weaver 1991; Wheeler 1985; see Rentfrow and McDonald 2010 for a review). More recently, researchers have begun to study associations between genre preferences and personality more comprehensively.

13.2.1 The structure of music preferences

Rentfrow and Gosling's (2003) research was the first to examine systematically the underlying structure of music preferences and whether these preferences are associated with contemporary personality constructs—the "Big Five" (McCrae and Costa 1987). The Big Five considers individual differences in personality to vary according to five major dimensions: *openness-to-experience*, *conscientiousness*, *extroversion*, *agreeableness*, and *neuroticism* (forming the acronym OCEAN). In a series of studies, Rentfrow and Gosling examined links between genre preferences and the Big Five as well as other personality, demographic, and attitudinal variables. Rather than using judgments made by the researchers to select which music genres to include, a preliminary study established participants' familiarity with various genres and subgenres. The authors used these data to create the Short Test of Music Preferences (STOMP), a list of 14 genres that were familiar to most college students in the United States. The STOMP requires participants to provide a rating from 1 (*dislike strongly*) to 7 (*like strongly*) for each genre. In different samples that comprised thousands of participants from different geographical regions in the USA, Rentfrow and Gosling measured music preferences with the STOMP, as well as by analysing participants' online music libraries. These analyses allowed them to determine (1) whether fans of a particular genre, such as rock, also tend to be fans of other genres, such as alternative or heavy metal; (2) how many truly independent dimensions of preferences there are; and (3) whether liking particular genres varies systematically with other non-musical differences among individuals.

Principal components analysis revealed four main factors (i.e. latent variables) of music preferences (Rentfrow and Gosling 2003). The first, which the authors designated *Reflective and Complex*, represented preferences for classical, jazz, folk, and blues, genres that were also considered to be more musically complex than other kinds of music even though the blues in particular usually has a very simple harmonic structure. In any event, liking these genres was associated positively with openness-to-experience and liberal political views. It was also associated positively with self-reports of intelligence, but negatively with self-reports of athletic ability. The second factor, *Intense and Rebellious*, described preferences for rock, alternative, and heavy-metal music. These preferences were also correlated positively with openness-to-experience and with self-reports of intelligence *and* athletic ability. Fans of *Upbeat and Conventional* genres, which included pop, soundtracks, religious, and country music, tended to be extroverted, agreeable, conscientious, and politically conservative. They also considered themselves to be physically attractive and athletic but they scored low in openness-to-experience. The fourth and final factor, *Energetic and Rhythmic*, reflected preferences for rap/hip-hop, soul/funk, and electronica/dance music. Liking for these genres was correlated positively with extroversion, agreeableness, and political liberalism, and with self-reports of attractiveness and athletic ability.

In a series of follow-up studies, Rentfrow and his colleagues showed that people easily form impressions about fans of different music genres that are stereotypical but are nonetheless often quite accurate (Rentfrow and Gosling 2006, 2007; Rentfrow et al. 2009). These findings are consistent with the idea that music preferences are an important part of an individual's identity, and that such preferences can communicate important information about a person's personality, values, and lifestyle (e.g. North and Hargreaves 1999). Other research suggests that similarities in music preferences between individuals are considered to reflect a common value system, which in turn provides a basis for social attraction (Boer et al. 2011).

In some instances, studies conducted in countries other than the Unites States have found results similar to those of Rentfrow and Gosling (2003) concerning the underlying dimensional structure of music preferences. For example, Langmeyer and colleagues (2012) replicated and extended Rentfrow and Gosling's results with a sample of young German adults. The results validated the four-factor structure of music preferences using real audio samples of popular songs that exemplified the relevant genres instead of a paper-and-pencil questionnaire. Nevertheless, results from many other studies differed from the original findings. For example, in another German sample with a wider age range, a six-factor structure was obtained (Schäfer and Sedlmeier 2009). In a large sample of Dutch adolescents, moreover, a four-factor structure of music preferences emerged but some genres were attributed to different character traits than in the American samples, specifically those that were either more (techno/trance) or less (gospel) popular in the Netherlands than in the United States (Delsing et al. 2008).

In another sample of Dutch adults between 22 and 60 years of age, Dunn and colleagues (2011) found that a four-factor structure explained underlying music preferences, but attribution of the individual genres onto the attributes meant that the factors needed to be defined differently than in the original study (Rentfrow and Gosling 2003). One factor represented traditional musical genres (classical, folk, blues, country, religious), another represented harder rock genres (rock, alternative, heavy metal), a third represented genres in which rhythm plays a major role (rap, dance, soul, jazz), and the fourth represented softer rock genres (pop, soundtracks). The fact that a preference for classical and for country music was seen to belong to the same factor is particularly noteworthy, and as one might expect, the diverse nature of jazz music (e.g. Diana Krall and Ornette Coleman are both jazz artists but they sound very different) meant that liking jazz was associated with more than one factor (i.e. also with traditional genres and soft rock).

Colley (2008) found that the structure of British undergraduates' music preferences was similar to that of Rentfrow and Gosling's (2003) American samples, but with some additional gender differences. Female students exhibited greater liking than male students for Top 40 pop music, but males showed greater liking for hard rock. In a Canadian study, an eight-factor structure emerged when participants provided liking ratings for 30 different genres instead of the 14 from the STOMP (George et al. 2007). The sample comprised many undergraduates from a Christian college as well as adults from the surrounding area who also tended to be quite religious. As a consequence, participants provided ratings for four different genres of Christian music, and two of these loaded onto one factor (called *Christian Contemporary*) and two loaded onto a different factor (*Traditional Christian*). The clearest non-musical individual differences were evident for the factors labeled *Rebellious* (alternative, classic rock, grunge, heavy metal, punk), *Rhythmic and Intense* (hip-hop/rap, pop, reggae, rhythm and blues), and *Classical* (choral, classical instrumental, Disney/Broadway tunes, opera/ballet, piano). Greater levels of liking for genres that belonging to the Rebellious or Rhythmic and Intense factors were evident among participants who were younger, under-employed, less educated, and less spiritually minded. These individuals also tended to have poorer social skills, higher levels of neuroticism and hostility, and lower levels of self-esteem, conscientiousness, and agreeableness. People who liked genres that loaded onto the Classical factor showed essentially the opposite pattern.

Brown (2012) asked university students in Japan to provide liking ratings for 12 different genres, six of which were taken from the STOMP. Six others were included because they were relatively well known to Japanese listeners. A four-factor structure was revealed that closely resembled Rentfrow and Gosling's (2003) original findings, and participants with high scores on openness-to-experience tended to exhibit greater liking for so-called Reflective and Complex music. In another study conducted in Japan, however, the original STOMP led to a four-factor structure but several genres loaded onto factors different from those in the original study (Oshio 2012).

Even Rentfrow and his colleagues have found that the dimensional structure of music preferences differs from the solution they reported initially. In a series of follow-up studies, they asked listeners to provide liking ratings for many (i.e. 25 to 94 depending on the particular study) 15-second excerpts of music taken from unfamiliar recordings that represented 26 different genres (Rentfrow et al. 2011). As in the study undertaken by Langmeyer and colleagues (2012), instead of making liking ratings for genre labels, participants made liking ratings for actual music. A separate group of participants made 14 ratings about each excerpt, seven based on musical descriptors (dense, distorted, electric, fast, instrumental, loud, percussive), and seven based on psychological descriptors (aggressive, complex, inspiring, intelligent, relaxing, romantic, sad). Principal components analysis revealed that liking ratings varied along five distinct dimensions rather than four, which were labeled so that they formed the acronym MUSIC (Mellow, Unpretentious, Sophisticated, Intense, Contemporary).

The authors explained loadings onto the particular dimensions as a function of the different genres and the average ratings for the 14 descriptors (Rentfrow et al. 2011). *Mellow* music was judged to sound slow, soft, romantic, relaxing, and non-aggressive, and the excerpts belonged primarily to soft-rock, R&B/soul, and adult-contemporary genres. *Unprententious* music was vocal, non-distorted, soft, non-aggressive, and simple, and the excerpts came from country and bluegrass genres. *Sophisticated* music was non-electric, non-percussive, non-distorted, intelligent, and inspiring, and the excerpts were from recordings of classical and marching-band music. *Intense* music was distorted, loud, electric, aggressive, and non-relaxing, with excerpts coming from rock, punk, and metal genres. Finally, *Contemporary* music sounded electric, percussive, and happy, and the excerpts were primarily from rap, electronica, and Latin genres.

Although the results of the study by Rentfrow and his colleagues (2011) revealed that music preferences varied reliably as a function of musical attributes (as measured by the descriptors) and genre membership, genres predicted liking ratings slightly better than attributes. In a follow-up study, however, the same five-factor (MUSIC) solution emerged when *all* of the stimuli were excerpts from jazz music, or all were excerpts from rock music (Rentfrow et al. 2012). In these instances, the underlying structure of music preferences was a sole consequence of musical attributes (i.e. differences in how the pieces sounded) rather than genre differences.

Thus, music preferences involve liking for particular musical attributes as well as liking for genres, although music attributes may actually play a larger role. The MUSIC model provides a new foundation for future research that seeks to examine what makes different individuals like different pieces of music. Considered as a whole, though, the available literature suggests that the structure of underlying musical preferences differs across cultures, ages, and historical periods, because different genres are more or less popular with different groups of people. The factors (latent variables) that emerge in principal components analyses also vary depending on which manifest variables (genres) participants are asked to rate, and whether participants rate their liking for genre labels or actual excerpts from recordings. It is also clear that the researchers' own musical preferences influence how the factors that emerge are labeled, such that any dimension associated with classical music may be labeled "complex" or "sophisticated" even though the blues, marching-band, or country music belong to the same factor.

13.2.2 Genre preferences and individual differences

Findings from studies of associations between music preferences and personality dimensions are more consistent, particularly for openness-to-experience and extroversion. In a number of studies, openness-to-experience was associated with preferences for elite or highbrow genres such as classical and jazz, as well as with intense and rebellious genres such as rock (e.g. Brown 2012; Delsing et al. 2008; Dunn et al. 2011; George et al. 2007; Langmeyer et al. 2012; Rentfrow and Gosling 2003; Zweigenhaft 2008). Several studies have also reported links between extroversion and liking genres of music that are considered upbeat and conventional or energetic and rhythmic, including pop, rap/hip-hop, and dance (e.g. Delsing et al. 2008; Langmeyer et al. 2012; Rentfrow and Gosling 2003; Zweigenhaft 2008). By contrast, associations between music preferences and the other three dimensions of personality that comprise the Big Five (i.e. agreeableness, conscientiousness, and neuroticism) have been much less consistent (e.g. Brown 2012; Delsing et al. 2008; Dunn et al. 2011; George et al. 2007; Langmeyer et al. 2012; Rentfrow and Gosling 2003; Zweigenhaft 2008).

A few recent studies also examined associations between more specific personality facets (rather than broad dimensions) and music preferences (Dunn et al. 2011; Zweigenhaft 2008). For example, the dimension openness-to-experience comprises six facets: *imagination* (being prone to fantasy), *artistic interests* (appreciating poetry, art, music, and beauty in nature), *emotionality* (awareness of one's own feelings), *adventurousness* (eagerness to try new things), *intellect* (receptiveness to new ideas), and *liberalism* (readiness to challenge authority, convention, and tradition). In principle, future research could delineate personality differences between fans of classical music and fans of rock music, who both have relatively high levels of openness-to-experience. For example, rock fans may be more liberal but classical fans more emotional. More generally, some facets may be better than other facets as predictors of music preferences, and better than broad dimensions as well.

Other scholars have examined associations between music preferences and individual-difference variables other than the Big Five personality dimensions. Brown (2012) found that liking classical, jazz, and gospel music was associated with honesty–humility, a trait that includes sincerity, fairness, greed avoidance, and modesty, which some scholars consider to be the sixth broad dimension of personality (Ashton and Lee 2008; Lee and Ashton 2008). Others have reported that men and women prefer harder and softer styles of popular music, respectively (Colley 2008), and that dichotomous thinking (i.e. black-and-white or all-or-none thinking) is associated positively with liking many genres of popular music but negatively with liking classical music, at least among Japanese undergraduates (Oshio 2012).

In a more comprehensive approach with a large sample of individuals living in the UK, North and Hargreaves (2007a, 2007b, 2007c) reported associations between genre-based music preferences and a wide variety of lifestyle variables including income, education, living arrangements, employment, health, leisure interests, and beliefs. Listeners who liked "problem" music (e.g. hip-hop/rap, DJ music, dance/house) tended to hold somewhat anti-social views; although they had liberal social attitudes, they did not have particularly liberal political views in terms of taxation, healthcare, supporting recycling or use of alternative energy sources (North and Hargreaves 2007a). In general, fans of opera and classical music were more likely than other individuals to live a lifestyle associated with the upper-middle and upper classes (with regard to travel, finances, education, employment; North and Hargreaves 2007c). They were also more likely to have sophisticated preferences in non-musical domains (e.g. TV and radio, reading, leisure activities; North and Hargreaves 2007b). Unfortunately, many links between music preferences and individual-difference variables other than the Big Five dimensions of personality have yet to be replicated, and some of the observed associations are bound to be culture- and/or cohort-specific.

One finding that has been replicated in several studies is a positive association between musicianship and liking classical and jazz music, whether musicianship is defined as music training or the ability to play an instrument (Colley 2008; George et al. 2007; Getz et al. 2014). An interesting avenue for future research would be to examine whether associations between music preferences and music training—or any individual differences other than the Big Five personality dimensions—are nevertheless explained by personality. For example, because musically trained individuals also tend to be high in openness-to-experience (Corrigall et al. 2013), openness could explain, in part or in full, the association between music training and liking for classical and jazz music.

In sum, since 2000, research has revealed much about the kinds of people who like different genres of music. One issue that has been widely acknowledged in the field is the difficulty in choosing which genres to include in empirical research and which to exclude. For example, country music is popular in America but it has significantly fewer fans in other parts of the world. As noted, the genres that are included in any study appear to influence the underlying dimensional structure of music preferences, such that solutions from principal components analysis fail to be capable of more broad generalization. In unpublished research conducted in our own lab, we administered a version of the STOMP to approximately 2000 students registered in introductory psychology for several years in a row. The underlying factor structure differed from year to year and was never identical to that reported by Rentfrow and Gosling (2003).

Although researchers must ensure that the genres included in measures of music preference comprise only those that are well known in the particular culture and age group being studied, in one instance, this approach did not accurately reflect the music preferences of almost one-third of a sample of Finnish university students (Ferrer et al. 2013). In other words, measures at the level of genre may not be precise enough to capture the subtleties of many listeners' preferences. Another problem is the lack of specificity in defining musical genres (e.g. what distinguishes a rock song from a pop song?), and the fact that many genres are likely to be conceptualized differently by different individuals. This issue is further complicated by the fact that genres evolve over time. Supposedly objective measures of genre such as those provided by iTunes are similarly inconsistent. For example, original studio albums by the group Radiohead are classified as "alternative," yet their greatest hits collection—comprising recordings from the same albums—is classified as "rock."

In a documentary film (Dunn et al. 2005) and a follow-up TV series (aptly named *Metal Evolution*; Banger Films, Inc. 2011), anthropologist Sam Dunn mapped a "Heavy Metal Family Tree" that identified the origins of heavy metal and the progression of its subgenres over 40 years. Dunn's analysis of a single genre shows why it is difficult to define broad and evolving categories of music that contain distinctive subgenres that change over the years. It also shows how even within a particular genre, people often have strong opinions about their liked and disliked subgenres. The non-musical social and cultural connotations associated with different genres, which also change over time, also influence music preferences. For example, rock music was associated with deviant behavior and delinquency many years ago, but it is now among the most mainstream genres in many places around the world. Finally, even within a preferred genre, some individual songs are liked better than others. It is also common to like a song that comes from a genre that one usually dislikes. In the next section, we examine what factors influence whether or not a listener will like a particular piece of music, and how such liking is associated with contextual factors and individual differences.

13.3 Liking for individual musical pieces

The study of liking individual pieces of music can include excerpts taken from a wide variety of genres or, alternatively, many different excerpts from a single genre. In either case, researchers can examine what factors influence liking responses while (1) minimizing the influence of extra-musical social and cultural connotations that are associated with different genres, and (2) reducing random variation that arises from individual differences in listeners' understanding of how genres differ according to their labels. The use of *unfamiliar* music excerpts rather than verbal labels or descriptions minimizes the effects of episodic memories that may be associated with familiar pieces.

13.3.1 Liking and exposure

Anecdotally, listeners often report that they do not like a piece of music initially, even though they may grow to like it after repeated listening. Indeed, frequency of exposure, or familiarity with a stimulus, has been known to influence liking ever since Zajonc's (1968) seminal research documenting the *mere exposure effect*. This refers to the fact that originally neutral stimuli (e.g. line drawings, Chinese characters) are liked better when the perceiver has encountered them previously, even when the perceiver cannot remember them explicitly. Early research provided evidence that the effect extended to excerpts of unfamiliar music, which are liked better if the perceiver has heard them before (Brickman and D'Amato 1975; Heingartner and Hall 1974; Krugman 1943; Meyer 1903; Mull 1957; Obermiller 1985; Peretz et al. 1998; Wilson 1979).

Nevertheless, people often complain that a particular song has been "overplayed." In other words, listeners can come to dislike a song that they have heard too frequently, especially over a short period of time. Consistent with these anecdotal observations, some theorists have proposed that liking varies as an inverted-U shaped function of novelty (Berlyne 1970; Stang 1974). Completely unfamiliar stimuli are potential threats and therefore somewhat disliked. With initial exposure that has no negative consequences, liking increases because the stimulus is more predictable and less of a threat. As exposure increases further, however, the stimulus can become too predictable (i.e. boring), such that liking decreases. Early research provided evidence consistent with the proposal that overexposure to music decreases liking (Bartlett 1973; Brentar et al. 1994; Getz 1966; Hargreaves 1984; Heyduk 1975; Verveer et al. 1933).

In the real world, music listening occurs in a context. In a typical experimental setting, the listener is asked to attend closely to the music stimuli, usually while providing some kind of rating or judgment. Although everyday music listening can be equally focused, such as during a concert, it frequently occurs under more passive or incidental listening conditions. For example, music is often present in the background while people drive, shop, do housework, cook, surf the Internet, watch television, and exercise. In these contexts, attention is divided between music listening and the task at hand, and familiarity with the music may develop more slowly. As such, one important question is whether the effect of exposure on liking varies under different listening conditions.

To answer this question, Szpunar and colleagues (2004, Experiment 2) presented listeners with excerpts from concerti (i.e. orchestral music with a featured lead instrument) twice, eight times, or 32 times under one of two conditions. In the "focused" listening condition, participants actively attended to the excerpts because they were required to identify the lead instrument in each presentation. In the "incidental" listening condition, listeners listened closely to a narrated story presented in their right ear while the music excerpts were presented at a low volume in their left ear. The results revealed an inverted-U shaped function of liking as a consequence of exposure for focused listeners, with liking increasing from baseline (no previous exposure) up to eight presentations, and then decreasing down to baseline levels with 32 exposures. By contrast, in the incidental listening condition there was a linear increase in liking as a function of exposure. These findings reveal that familiarity with a piece of music increases liking for it, up to the point at which it becomes too familiar. Such overfamiliarity occurs only, or at least more quickly, when listeners are attending closely to the music rather than overhearing it playing in the background.

Frequency of exposure may also affect liking differently across individuals, as a consequence of differences in personality or other variables. In one study, Hunter and Schellenberg (2011a) re-analysed the data from the focused listening conditions of two previously reported experiments (Schellenberg et al. 2008; Szpunar et al. 2004, Experiment 2). As a group, participants exhibited an inverted-U shaped pattern of liking as a function of exposure frequency, as reported in the original published articles. When participants were examined individually, however, less than half of them actually exhibited this pattern. The other participants exhibited monotonic increases, monotonic decreases, or an upright-U pattern of liking with increasing exposure.

To examine whether personality can explain such individual differences, Hunter and Schellenberg (2011a) replicated the findings from the focused-listening condition from Szpunar and colleagues (2004, Experiment 2), but they also included a measure of the Big Five personality dimensions. Compared to individuals who were low in openness-to-experience, individuals who scored high in this dimension liked novel musical excerpts more and excerpts they had heard repeatedly less, showing particularly rapid satiation as a function of exposure. Moreover, although the inverted-U shaped pattern for liking was common among listeners who scored low in openness, the most common pattern for listeners who scored high in openness was a monotonic decrease in liking with increasing exposure. This finding was expected because a hallmark of openness-to-experience is a preference for variety over routine. In sum, the effect of exposure on liking interacts with a variety of other factors including initial liking, the listening context, and personality.

Other factors that are likely to influence how quickly an individual will tire of hearing a particular song can be difficult to test in a laboratory setting. For example, we do not typically hear the same musical excerpt multiple times in one day, let alone in one hour, which can be the case in an experiment. One would expect it to take longer for listeners to dislike an overplayed song if the exposures were spread out over days or weeks compared to when they occur repeatedly in a short time frame.

Another factor that interacts with the effect of exposure on liking is how much the excerpt of music is liked initially. Repeatedly playing a song that a listener dislikes strongly is unlikely to shift such displeasure in a positive direction. In one study, listeners heard 12 excerpts of unfamiliar orchestral music six times each (Witvliet and Vrana 2007). They rated how much they liked each piece after they heard it the first time, and again after they heard it the sixth time. The 12 pieces were pre-selected so that half sounded pleasant (i.e. with positive valence) and the other half sounded unpleasant (negative valence). As in many other studies (e.g. Hunter et al. 2008, 2010, 2011a, 2011b; Husain et al. 2002; Khalfa et al. 2008; Ladinig and Schellenberg 2012; Schellenberg et al. 2008; Thompson et al. 2001; Vieillard et al. 2008), these adult listeners initially preferred positively valenced over negatively valenced music. After repeated exposure to both types of music, however, responses became more polarized. Positively valenced excerpts were liked even more, but negatively valenced excerpts were liked even less. It is unknown whether liking for the unpleasant-sounding music increased initially (i.e. between the first and sixth exposures) before it began to decrease.

13.3.2 Liking and complexity

The idea that music liking peaks after a *moderate* number of exposures raises the possibility that moderation in general may be fundamental to liking. Berlyne (1970) also proposed that liking should exhibit an inverted-U shaped function according to complexity. In the case of music, it will be liked most when its melodic, rhythmic, and metrical structures are not too simple, predictable, and uniform, nor too surprising, unpredictable, and erratic. Using piano pieces composed specifically for experimental purposes, Heyduk (1975) found evidence consistent with this view. His participants preferred pieces that exhibited moderate levels of complexity (i.e. those that made use of a moderate number of different chords and a moderate amount of rhythmic syncopation) over pieces that were either too simple or too complex. These results have also been replicated with more ecologically valid stimuli. For example, when listeners were asked to rate unfamiliar excerpts of new age and house music in terms of liking and subjective complexity, they showed a similar preference for excerpts with moderate complexity (North and Hargreaves 1995).

Nevertheless, the effect of complexity on liking may not be uniform across all genres of music. For example, Orr and Ohlsson (2001) presented listeners with jazz and bluegrass improvisations that were created by professional musicians to vary in complexity. For bluegrass excerpts, listeners exhibited a clear preference for a moderate level of complexity. For jazz excerpts, however, they preferred simplicity. It is possible that the inclusion of even simpler jazz improvisations may have revealed increases followed by decreases in liking as complexity increased. It is also possible that the effect of complexity on liking differs across genres, or that other factors such as initial liking or general familiarity with a genre interact with the effect of complexity on liking.

Moreover, complexity may not influence liking for all music listeners. Orr and Ohlsson (2005) found that the effect of musical complexity on liking was weaker or nonexistent among professional and amateur musicians compared to individuals with no music training. The authors suggested that musically trained individuals use a different set of criteria in their aesthetic judgments compared to other people, placing less importance on complexity. There is also some evidence that preferences for simple or complex music are affected by situational factors. For example, in one study, listeners preferred to hear simple music in the background while they performed a difficult sensorimotor task, but they preferred complex music while they performed a simple task (Arkes et al. 1986). Thus, the preferred level of musical complexity may vary with an individual's attentional capacity to process the music at a given point in time.

13.3.3 Liking and feelings

One of the most common reasons that people give for listening to music is because of the way that it makes them feel, or because it can change how they are currently feeling (Juslin and Laukka 2004; Lonsdale and North 2011). In other words, emotional responding on the evaluative level can be influenced by responding on the feeling level. When Juslin and Laukka (2004) analysed listeners' free responses to the question "why do you listen to music?" the most common response (47 per cent) was "to express, release, and influence emotions", followed by other emotion and liking-related reasons such as "to relax and settle down", "because it makes me feel good", "because I like/love music", and "to get energized". Other research confirms that people particularly enjoy music that evokes intense feelings (Ladinig and Schellenberg 2012; Schellenberg et al. 2012).

One especially strong positive response to music is the physical sensation of *chills* (or *thrills*; Goldstein 1980). Chills are described as shivers down the spine, or a tingling sensation that is usually felt at the back of the neck and often accompanied by goosebumps. In the first study to describe these kinds of physical responses to music, approximately half the participants reported experiencing music-induced chills (Goldstein 1980). Subsequent research revealed that individuals who are high in openness-to-experience tend to experience chills more frequently than other individuals (McCrae 2007; Nusbaum and Silvia 2011; Silvia and Nusbaum 2011). Although chills can also occur in response to other stimuli, including visual art, tactile stimulation (think of a feather tickling the back of your neck), and even fantasies or memories, chills elicited by music tend to be experienced as particularly pleasant (Goldstein 1980; Grewe et al. 2011; Panksepp 1995). There is also some evidence that individuals tend to experience chills in response to unexpected or surprising musical events, such as an unexpected harmony or a sudden change in volume or orchestral texture (e.g. Grewe et al. 2007; Guhn et al. 2007; Panksepp 1995; Sloboda 1991).

Because chills are indicative of peak emotional experiences, pieces that elicit chills tend to be accompanied by higher ratings of the intensity of the listener's emotional response, faster heart rate, and a more intense skin conductance response (Grewe et al. 2009). Chills also tend to occur more frequently in response to emotionally expressive music compared to relaxing or arousing music that is less emotionally expressive (Rickard 2004). Chills most often occur in response to familiar and well-liked music (Grewe et al. 2009), possibly because familiar music is more likely than unfamiliar music to induce powerful emotions.

Consistent with the view that chills represent particularly emotional experiences, several studies have shown that that music-induced chills are associated with physiological arousal (e.g. Craig 2005; Grewe et al. 2011, 2009; Guhn et al. 2007; Rickard 2004; Salimpoor et al. 2011, 2009) as well as with activation in brain areas that are involved in reward and emotion processing (e.g. Blood and Zatorre 2001; Salimpoor et al. 2011). In fact, participants' subjective ratings of the intensity of the chill response are correlated with the amount of dopamine released in the nucleus accumbens, also known as the brain's pleasure center (Salimpoor et al. 2011). In short, results from behavioral, physiological, and neuroimaging studies confirm that chills are particularly strong, positive emotional experiences to well-liked music, and, conversely, that well-liked music often tends to be music that influences how the listener feels.

13.3.4 Liking and specific feelings

Listeners tend to experience positive emotions more frequently than negative emotions in response to music (e.g. Gabrielsson 2001; Juslin and Laukka 2004; Juslin et al. 2008), and they generally prefer music that expresses positive emotions such as happiness over music that expresses negative emotions such as sadness (e.g. Hunter et al. 2008, 2010, 2011a, b; Husain et al. 2002; Khalfa et al. 2008; Ladinig and Schellenberg 2012; Schellenberg et al. 2008; Thompson et al. 2001; Vieillard et al. 2008). This preference for music that is positive in valence is evident even among individuals who have experienced damage to the emotion-processing centers of the brain, including the amygdala (Gosselin et al. 2005). Thus, the elicitation of positive emotions appears to be central to the universal appeal of music. At the same time, we know that people often listen to sad-sounding music. But why would people choose to listen to sad-sounding music? And when in development does the preference for happy-sounding music emerge?

In one developmental study, adults and 5-, 8-, and 11-year-old children were asked to rate how much they liked short excerpts of music (Hunter et al. 2011b). The excerpts were chosen to sound unequivocally happy, sad, scary, or peaceful. These particular emotions vary on two dimensions that are relevant to emotion research: arousal and valence (Russell 1980). Arousal refers to activation or energy level, whereas valence refers to pleasantness. Happiness has high arousal and positive valence, sadness has low arousal and negative valence, fear has high arousal and negative valence, and peacefulness has low arousal and positive valence. Adults showed a preference for excerpts with positive valence (happiness and peacefulness), but arousal did not influence their liking ratings. By contrast, children of all ages had elevated levels of liking for excerpts that expressed high-arousal emotions (happiness and fear), but valence did not matter. Because musical tempo (fast or slow) is associated with arousal, whereas mode (major or minor) is associated with valence (Husain et al. 2002), these results point to a basic preference for faster music in childhood, which turns into a preference for major-key music in adulthood. Other findings from the same study revealed a preference for music with positive valence among female listeners, which is likely to be related to the female preference for Top 40 music discussed earlier (Colley 2008).

The children and adults in the study by Hunter and colleagues (2011b) were also asked to identify which of the four emotions each excerpt conveyed. Young girls were better than young boys at this task, but both genders performed as well as adults by the time they reached 11 years of age. Among children of all ages, individual differences in identification accuracy were associated with liking. Greater accuracy predicted higher levels of liking for positively valenced music among the 5- and 8-year-olds, but lower levels for the 11-year-olds. In other words, as 11-year-olds became more accurate at identifying music as happy-, sad-, scary-, or peaceful-sounding, they also tended to dislike music that sounds happy or peaceful; that is, music their parents (particularly their mothers) like.

The phenomenon of liking sad-sounding music, which has been documented empirically (Garrido and Schubert 2011; Kawakami et al. 2013; Kreutz et al. 2008; Vuoskoski and Eerola 2012; Vuoskoski et al. 2012; Zentner et al. 2008), is particularly interesting for psychologists because it demonstrates that music that expresses a negative feeling can nevertheless be evaluated positively. In fact, although sad-sounding music may be perceived to sound tragic, it can simultaneously decrease tragic feelings but increase romantic and blithe feelings (Kawakami et al. 2013). But when, where, and why do individuals enjoy listening to music that expresses a negative emotion?

One factor that matters is the emotional character of previously heard songs. Schellenberg and colleagues (2012) collected emotion intensity and liking ratings of excerpts of classical piano music that expressed happiness or sadness unambiguously. Happy-sounding excerpts were in a major key with a fast tempo. Sad-sounding excerpts were minor and slow. For each participant, one of these emotions was designated as the "background" emotion and the other as the "contrasting" emotion. Excerpts that expressed the contrasting emotion were presented after a varying number of different excerpts that expressed the background emotion. For example, some of the listeners heard a happy-sounding excerpt after hearing one, two, four, or eight different excerpts that sounded sad; others heard a sad-sounding excerpt after hearing one, two, four, or eight different excerpts that sounded happy. As the presentation frequency of the background emotion increased, emotion intensity and liking ratings decreased, such that there was greater appreciation and more intense emotional responding for excerpts that conveyed the contrasting emotion. In other words, sad-sounding music is appreciated more after listeners have been inundated with happy-sounding music (and vice versa) because it represents a change from emotional monotony.

Other research suggests that sad-sounding music may be liked more after listeners are fatigued or in a negative mood. Schellenberg and colleagues (2008) collected liking ratings of musical excerpts that expressed happiness or sadness after they had been heard 0, 2, 8, or 32 times previously. Half the participants heard the music in a focused-listening condition, which required them to identify whether each excerpt expressed happiness or sadness. The other half—incidental listeners—completed the same task used by Szpunar and colleagues (2004) while the music was presented softly in the background. Specifically, they heard a narrated story in their right ear and were required to press one button every time they heard the word "and", another button every time they heard the word "the", and to count the number of times they heard the word "but". In their left ear, the happy- and sad-sounding excerpts were presented audibly but very quietly. Exposure frequency had the usual effect on liking. For focused listeners, liking first increased but then decreased as the number of exposures increased. For incidental listeners, liking increased linearly for excerpts heard more frequently. But there was also an interaction between listening condition and emotion that was independent of number of exposures. Focused listeners exhibited the typical bias, liking happy-sounding more than sad-sounding excerpts. For incidental listeners, however, sad-sounding excerpts were liked just as much as happy-sounding excerpts. Importantly, this difference was driven by greater liking for sad-sounding pieces in the incidental compared to the focused listening condition.

Why was sad-sounding music liked more for some listeners? Schellenberg and colleagues (2008) suggested that the difficult and attention-demanding task for incidental listeners may have affected their emotional state negatively. After completing the task, participants may have been fatigued or even slightly agitated, such that they found the sad music to be calming. Alternatively, participants may have displayed a mood-congruency effect, preferring negatively valenced music because it matched their negative mood. To test the second hypothesis, Hunter and colleagues (2011a) induced a sad, neutral, or happy mood in participants by asking them to rate and describe how different pictures (e.g. a birthday party for happy, a building for neutral, an injured animal for sad) made them feel. Participants were subsequently presented with either a happy- or sad-sounding excerpt (Hunter et al. 2011a, Experiment 1) and asked to judge how much they liked it. The results revealed that although the typical preference for happy-sounding music was present after a happy or neutral mood was induced, it disappeared when listeners were in a sad mood. Sad listeners liked happy- and sad-sounding excerpts equally. In a second experiment, sad moods were found to increase the perception of sadness in neutral-sounding music. These findings extended the results of the study by Schellenberg and his team (2008) by showing that mood congruency influences what music people like and their perception of emotion in music. In short, the expression *misery loves company* extends to music listening.

But why are people more inclined to appreciate sad music when they are in a sad mood? Is it because they prefer to listen to music that expresses their current feelings, or is it because they feel that listening to mood-incongruent (i.e. happy) music would be inappropriate or ineffective at lifting their mood? Friedman and colleagues (2012, Experiment 3) explored participants' motivations for listening to music that expressed different emotions after inducing a happy, neutral, or sad mood. Participants first generated a short list of their favorite happy and sad songs. After watching a film clip for the purposes of mood induction, they were then asked to rate: (1) their desire to listen to happy- or sad-sounding songs, (2) how listening to happy or sad songs would make them feel, and (3) how appropriate it would be to listen to happy or sad songs. The results revealed that participants who were in a happy mood were averse to listening to sad-sounding music, and participants who were in a sad mood were averse to listening to happy music (the neutral group was not averse to either). Participants in a happy mood believed that listening to sad songs would worsen their mood, whereas those in a sad mood reported that listening to happy-sounding music would feel inappropriate or wrong, which would, presumably, counteract any emotion-contagion effects of happy music. In some instances, then, an appreciation for sad-sounding music while in a sad mood is better described as an aversion to happy-sounding music.

Other studies have examined whether individual differences based on personality or musical expertise help to explain why some people like to listen to sad-sounding music. In one study, Ladinig and Schellenberg (2012) presented listeners with unfamiliar excerpts of music taken from a wide variety of genres. The excerpts varied in tempo (fast or slow) and mode (major or minor), such that half were obviously happy- or sad-sounding (fast-major or slow-minor, respectively), and half contained mixed cues to happiness and sadness (fast-minor or slow-major). Participants first completed a background questionnaire that assessed their scores on the Big Five personality traits and how much music training they had received. They then listened to each excerpt and rated how much they liked it, how intense their emotional response was, and which emotion(s) they felt. Liking for music that conveyed happiness did not differ from individual to individual. By contrast, liking for music that conveyed sadness or mixed happiness and sadness was influenced by personality and music training. Individuals who had a stronger appreciation for sad-sounding music tended to score low on extroversion (or high on introversion) and high on openness-to-experience, whereas individuals who enjoyed music with mixed cues to emotion tended to have more music training.

Future research could examine whether introverts like sad-sounding music because of its slow tempo. We know that tempo is associated with the arousal dimension of emotion (Husain et al. 2002), and that introverts tend to avoid excitement and activity (McCrae and Costa 1987), possibly because their optimal level of arousal is relatively low (Eysenck 1981). Introverts also tend to experience positive emotions less frequently than extroverts (Matthews et al. 1990), which may explain their greater appreciation for sad-sounding music. As for individuals who are high in openness-to-experience, their aesthetic sensitivity and preference for variety (Costa and McCrae 1992) may explain why they like sad-sounding music. More specifically, these individuals may be more likely to perceive the beauty of music that expresses sadness and emotions other than happiness. Finally, Ladinig and Schellenberg (2012) suggested that musically trained individuals have a propensity for enjoying music with mixed cues to emotion because they are more sophisticated listeners. In other words, musically trained individuals may be more likely to perceive nuanced and subtle emotions in music with mixed cues, leading them to appreciate it for its emotional complexity, even though they may not care much about musical complexity (Orr and Ohlsson 2005). Other reports substantiate the association between openness-to-experience and liking sad-sounding music (Vuoskoski and Eerola 2011; Vuoskoski et al. 2012), and show additionally that empathy is associated positively with experiencing intense emotions in response to music and with enjoying sad-sounding pieces (Vuoskoski et al. 2012). Formal hypotheses about *why* particular groups of individuals are more likely than others to appreciate sad-sounding music have yet to be developed or tested empirically.

At a more macro level, there is evidence that among the American music-buying population, liking sad-sounding and emotionally ambiguous music has become more common over the years. When Hunter and colleagues (2008) were choosing their stimulus excerpts, they were surprised to discover that it was relatively difficult to find recent recordings that were clearly happy sounding (fast tempo and major mode). To examine whether there has been a shift away from unambiguously happy-sounding music over the past five decades, Schellenberg and von Scheve (2012) analysed several musical characteristics (e.g. tempo, mode, duration) from over 1000 songs that reached the Top 40 between 1965 and 2009. In line with their hypothesis, they documented that the use of the minor mode doubled during that time, and that popular recordings became slower in tempo and longer in duration. Moreover, mean tempo decreased over time more for songs that were composed in major keys compared to songs in minor keys, pointing to a more frequent use of relatively slow tempi in music composed in major keys, but relatively fast tempi in music composed in minor keys. In short, popular music has moved away from the primarily happy-sounding songs of the 1960s, with increasing use of musical cues to sadness (slow tempo or minor mode) and emotional ambiguity (slow tempo and major mode, fast tempo and minor mode). Even when looking beyond the level of the individual or listening context, liking for clearly happy-sounding music has decreased over the past five decades in American society as a whole.

13.4 **Future directions**

Research on music preferences is in its infancy and many questions remain unanswered. Rentfrow and McDonald (2010) provided a thorough review of issues in the field that have yet to be resolved, and they proposed new avenues for future research. One major limitation of the current body of work is that it focuses almost exclusively on older adolescents and young adults. Although there is some evidence that music preferences crystallize between 20 and 30 years of age (Holbrook and Schindler 1989), it is important to study how music preferences change, if at all, in middle and late adulthood, and to examine how preferences develop in the first place during infancy, childhood, and adolescence. A related issue concerns the stability of music preferences over the life span. In the present review, only one study examined liking music among child participants, using a cross-sectional approach (Hunter et al. 2011b). Although a few studies have examined this issue longitudinally over a few years (e.g. approximately 2–3 years; Delsing et al. 2008; Mulder et al. 2010), studies of longer duration could reveal more about whether preferences change or stay the same, especially at important transitional periods in the lifespan (e.g. from childhood to adolescence and from adolescence to adulthood). Studying music liking from a developmental perspective could provide additional insight into why music is such an important part of people's lives.

References

Arkes, H.R., Rettig, L.E., and Scougale, J.D. Jr. (1986). The effect of concurrent task complexity and music experience on preference for simple and complex music. *Psychomusicology* **6**, 51–60.

Arnett, J. (1992). The soundtrack of recklessness: musical preferences and reckless behavior among adolescents. *Journal of Adolescent Research* **7**, 313–31.

Ashton, M.C. and Lee, K. (2008). The prediction of honesty–humility-related criteria by the HEXACO and Five-Factor models of personality. *Journal of Research in Personality* **42**, 1216–28.

Banger Films, Inc. (2011). *Metal Evolution* [Television series]. Canada: Tricon Films and Television.

Bartlett, D.L. (1973). Effect of repeated listenings on structural discrimination and affective response. *Journal of Research in Music Education* **21**, 302–17.

Berlyne, D.E. (1970). Novelty, complexity, and hedonic value. *Perception and Psychophysics* **8**, 279–86.

Blood, A.J. and Zatorre, R.J. (2001). Intensely pleasurable responses to music correlate with activity in brain regions implicated in reward and emotion. *Proceedings of the National Academy of Sciences of the USA* **98**, 11818–23.

Boer, D., Fischer, R., Strack, M., et al. (2011). How shared preferences in music create bonds between people: values as the missing link. *Personality and Social Psychology Bulletin* **37**, 1159–71.

Brentar, J.E., Neuendorf, K.A., and Armstrong, G.B. (1994). Exposure effects and affective responses to music. *Communication Monographs* **61**, 161–81.

Brickman, P. and D'Amato, B. (1975). Exposure effects in a free choice situation. *Journal of Personality and Social Psychology* **32**, 415–20.

Brown, R.A. (2012). Music preferences and personality among Japanese university students. *International Journal of Psychology* **47**, 259–68.

Carpentier, F.D., Knobloch, S., and Zillmann, D. (2003). Rock, rap, and rebellion: comparisons of traits predicting selective exposure to defiant music. *Personality and Individual Differences* **35**, 1643–55.

Cattell, R.B. and Saunders, D.R. (1954). Musical preferences and personality diagnosis: a factorization of one hundred and twenty themes. *Journal of Social Psychology* **39**, 3–24.

Christenson, P.G. and Peterson, J.B. (1988). Genre and gender in the structure of music preferences. *Communication Research* **15**, 282–301.

Colley, A. (2008). Young people's musical taste: relationship with gender and gender-related traits. *Journal of Applied Social Psychology* **38**, 2039–55.

Corrigall, K.A., Schellenberg, E.G., and Misura, N.M. (2013). Music training, cognition, and personality. *Frontiers in Psychology* **4**:222. doi: 10.3389/fpsyg.2013.00222.

Costa, P.T. Jr. and McCrae, R.R. (1992). *Revised NEO Personality Inventory (NEO-PI-R) and NEO Five-Factor Inventory (NEO-FFI) Professional Manual.* Odessa, FL: Psychological Assessment Resources.

Craig, D.G. (2005). An exploratory study of physiological changes during "chills" induced by music. *Musicae Scientiae* **9**, 273–87.

Delsing, M.J.M.H., ter Bogt, T.F.M., Engels, R.C.M.E., et al. (2008). Adolescents' music preferences and personality characteristics. *European Journal of Personality* **22**, 109–30.

Dollinger, S.J. (1993). Research note: personality and music preference: extraversion and excitement seeking or openness to experience? *Psychology of Music* **21**, 73–7.

Dunn, P.G., de Ruyter, B., and Bouwhuis, D.G. (2011). Toward a better understanding of the relation between music preference, listening behavior, and personality. *Psychology of Music* **40**, 411–28.

Dunn, S. (Producer and Director), McFayden, S. (Producer and Director), and Wise, J.J. (Director). (2005). *Metal: A Headbanger's Journey* [Documentary]. Canada: Seville Pictures.

Eysenck, M.W. (1981). Learning, memory and personality. In H.J. Eysenck (ed.), *A Model for Personality*. Heidelberg: Springer, pp. 169–209.

Ferrer, R., Eerola, T., and Vuoskoski, J.K. (2013). Enhancing genre-based measures of music preference by user-defined liking and social tags. *Psychology of Music* **41**, 499–518. doi:10.1177/0305735612440611.

Friedman, R.S., Gordis, E., and Förster, J. (2012). Re-exploring the influence of sad mood on music preference. *Media Psychology* **15**, 249–66.

Gabrielsson, A. (2001). Emotions in strong experiences with music. In P.N. Juslin and J.A. Sloboda (eds), *Music and Emotion: Theory and Research*. New York, NY: Oxford University Press, pp. 431–49.

Garrido, S. and Schubert, E. (2011). Individual differences in the enjoyment of negative emotion in music: a literature review and experiment. *Music Perception* **28**, 279–96.

George, D., Stickle, K., Rachid, F., et al. (2007). The association between types of music enjoyed and cognitive, behavioral, and personality factors of those who listen. *Psychomusicology* **19**, 32–56.

Getz, R.P. (1966). The effects of repetition on listening response. *Journal of Research in Music Education* **14**, 178–92.

Getz, L.M., Marks, S., and Roy, M. (2014). The influence of stress, optimism, and music training on music uses and preferences. *Psychology of Music* **42**, 71–85.

Gjerdingen, R.O. and Perrott, D. (2008). Scanning the dial: the rapid recognition of music genres. *Journal of New Music Research* **37**, 93–100.

Glasgow, M.R., Cartier, A.M., and Wilson, G.D. (1985). Conservatism, sensation seeking and music preferences. *Personality and Individual Differences* **6**, 395–6.

Goldstein, A. (1980). Thrills in response to music and other stimuli. *Physiological Psychology* **8**, 126–9.

Gosselin, N., Peretz, I., and Noulhiane, M., et al. (2005). Impaired recognition of scary music following unilateral temporal lobe excision. *Brain* **128**, 628–40.

Grewe, O., Katzur, B., Kopiez, R., et al. (2011). Chills in different sensory domains: frisson elicited by acoustical, visual, tactile and gustatory stimuli. *Psychology of Music* **39**, 220–39.

Grewe, O., Kopiez, R., and Altenmüller, E. (2009). The chill parameter: goose bumps and shivers as promising measures in emotion research. *Music Perception* **27**, 61–74.

Grewe, O., Nagel, F., Kopiez, R., et al. (2007). Listening to music as a re-creative process: physiological, psychological, and psychoacoustical correlates of chills and strong emotions. *Music Perception* **24**, 297–314.

Guhn, M., Hamm, A., and Zentner, M. (2007). Physiological and musico-acoustic correlates of the chill response. *Music Perception* **24**, 473–83.

Hakanen, E.A. and Wells, A. (1993). Music preference and taste cultures among adolescents. *Popular Music and Society* **17**, 55–69.

Hansen, C.H. and Hansen, R.D. (1991). Schematic information processing of heavy metal lyrics. *Communication Research* **18**, 373–411.

Hargreaves, D.J. (1984). The effects of repetition on liking for music. *Journal of Research in Music Education* **32**, 35–47.

Heingartner, A. and Hall, J.V. (1974). Affective consequences in adults and children of repeated exposure to auditory stimuli. *Journal of Personality and Social Psychology* **29**, 719–23.

Heyduk, R.G. (1975). Rated preference for musical compositions as it relates to complexity and exposure frequency. *Perception and Psychophysics* **17**, 84–91.

Holbrook, M.B. and Schindler, R.M. (1989). Some exploratory findings on the development of musical tastes. *Journal of Consumer Research* **16**, 119–24.

Hunter, P.G. and Schellenberg, E.G. (2010). Music and emotion. In M.R. Jones, R.R. Fay, and A.N. Popper (eds), *Music Perception*. New York, NY: Springer, pp. 129–64.

Hunter, P.G. and Schellenberg, E.G. (2011). Interactive effects of personality and frequency of exposure on liking for music. *Personality and Individual Differences* **50**, 175–9.

Hunter, P.G., Schellenberg, E.G., and Griffith, A.T. (2011a). Misery loves company: Mood-congruent emotional responding to music. *Emotion* **11**, 1068–72.

Hunter, P.G., Schellenberg, E.G., and Schimmack, U. (2008). Mixed affective responses to music with conflicting cues. *Cognition and Emotion* **22**, 327–52.

Hunter, P.G., Schellenberg, E.G., and Schimmack, U. (2010). Feelings and perceptions of happiness and sadness induced by music: Similarities, differences, and mixed emotions. *Psychology of Aesthetics, Creativity, and the Arts* **4**, 47–56.

Hunter, P.G., Schellenberg, E.G., and Stalinski, S.M. (2011b). Liking and identifying emotionally expressive music: age and gender differences. *Journal of Experimental Child Psychology* **110**, 80–93.

Husain, G., Thompson, W.F., and Schellenberg, E.G. (2002). Effects of musical tempo and mode on arousal, mood, and spatial abilities. *Music Perception* **20**, 149–69.

International Federation of the Phonographic Industry (2013). IFPI digital music report 2013: engine of a digital world. Available at: <http://www.ifpi.org/content/library/dmr2013.pdf>.

Jones, M.R., Fay, R.R., and Popper, A.N. (eds) (2010). *Music Perception*. New York, NY: Springer.

Juslin, P.N. and Laukka, P. (2004). Expression, perception, and induction of musical emotions: a review and a questionnaire study of everyday listening. *Journal of New Music Research* **33**, 217–38.

Juslin, P.N., Liljeström, S., Västfjäll, D., et al. (2008). An experience sampling study of emotional reactions to music: listener, music, and situation. *Emotion* **5**, 668–83.

Juslin, P.N. and Sloboda, J.A. (eds) (2010). *Handbook of Music and Emotion: Theory, Research, Applications*. New York, NY: Oxford University Press.

Juslin, P.N. and Västfjäll, D. (2008). Emotional responses to music: the need to consider underlying mechanisms. *Behavioral and Brain Sciences* **31**, 559–621.

Kawakami, A., Furukawa, K., Katahira, K., et al. (2013). Sad music induces pleasant emotion. *Frontiers in Psychology* **4**: 311. doi: 10.3389/fpsyg. 2013.00311.

Khalfa, S., Roy, M., Rainville, P., et al. (2008). Role of tempo entrainment in psychophysiological differentiation of happy and sad music? *International Journal Psychophysiology* **68**, 17–26.

Kreutz, G., Ott, U., Teichmann, D., et al. (2008). Using music to induce emotions: influences of musical preference and absorption. *Psychology of Music* **36**, 101–26.

Krugman, H.F. (1943). Affective responses to music as a function of familiarity. *Journal of Abnormal and Social Psychology* **38**, 388–92.

Ladinig, O. and Schellenberg, E.G. (2012). Liking unfamiliar music: effects of felt emotion and individual differences. *Psychology of Aesthetics, Creativity, and the Arts* **6**, 146–54.

Langmeyer, A., Guglhör-Rudan, A., and Tarnai, C. (2012). What do music preferences reveal about personality? A cross-cultural replication using self-ratings and ratings of music samples. *Journal of Individual Differences* **33**, 119–30.

Lee, K. and Ashton, M.C. (2008). The HEXACO personality factors in the indigenous personality lexicons of English and 11 other languages. *Journal of Personality* **76**, 1001–53.

Litle, P. and Zuckerman, M. (1986). Sensation seeking and music preferences. *Personality and Individual Differences* 7, 575–7.

Lonsdale, A.J. and North, A.C. (2011). Why do we listen to music? A uses and gratifications analysis. *British Journal of Psychology* **102**, 108–34.

Matthews, G., Jones, D.M., and Chamberlain, A. (1990). Refining the measurement of mood: the UWIST Mood Adjective Checklist. *British Journal of Psychology* **81**, 17–42.

McCrae, R.R. (2007). Aesthetic chills as a universal marker of openness to experience. *Motivation and Emotion* **31**, 5–11.

McCrae, R.R. and Costa, P.T. Jr. (1987). Validation of the five-factor model of personality across instruments and observers. *Journal of Personality and Social Psychology* **52**, 81–90.

McNamara, L. and Ballard, M.E. (1999). Resting arousal, sensation seeking, and music preference. *Genetic, Social, and General Psychology Monographs* **125**, 229–50.

Meyer, M. (1903). Experimental studies in the psychology of music. *American Journal of Psychology* **14**, 192–214.

Mulder, J., Ter Bogt, T.F.M., Raaijmakers, Q.A.W., and Sikkema, P. (2010). From death metal to R&B? Consistency of music preferences among Dutch adolescents and young adults. *Psychology of Music* **38**, 67–83.

Mull, H.K. (1957). The effect of repetition upon the enjoyment of modern music. *Journal of Psychology* **43**, 155–62.

North, A.C. and Hargreaves, D.J. (1995). Subjective complexity, familiarity, and liking for popular music. *Psychomusicology* **14**, 77–93.

North, A.C. and Hargreaves, D.J. (1999). Music and adolescent identity. *Music Education Research* **1**, 75–92.

North, A.C. and Hargreaves, D.J. (2007a). Lifestyle correlates of musical preference: 1. relationships, living arrangements, beliefs, and crime. *Psychology of Music* **35**, 58–87.

North, A.C. and Hargreaves, D.J. (2007b). Lifestyle correlates of musical preference: 2. media, leisure time and music. *Psychology of Music* **35**, 179–200.

North, A.C. and Hargreaves, D.J. (2007c). Lifestyle correlates of musical preference: 3. travel, money, education, employment and health. *Psychology of Music* **35**, 473–97.

Nusbaum, E.C. and Silvia, P.J. (2011). Shivers and timbres: personality and the experience of chills from music. *Social Psychological and Personality Science* **2**, 199–204.

Obermiller, C. (1985). Varieties of mere exposure: the effects of processing style and repetition on affective responses. *Journal of Consumer Research* **12**, 17–30.

Orr, M.G. and Ohlsson, S. (2001). The relationship between musical complexity and liking in jazz and bluegrass. *Psychology of Music* **29**, 108–27.

Orr, M.G. and Ohlsson, S. (2005). Relationship between complexity and liking as a function of expertise. *Music Perception* **22**, 583–611.

Oshio, A. (2012). The relationship between dichotomous thinking and music preferences among Japanese undergraduates. *Social Behavior and Personality* **40**, 567–74.

Panksepp, J. (1995). The emotional sources of "chills" induced by music. *Music Perception* **13**, 171–207.

Payne, E. (1980). Towards an understanding of music appreciation. *Psychology of Music* **8**, 31–41.

Peretz, I., Gaudreau, D., and Bonnel, A.-M. (1998). Exposure effects on music preference and recognition. *Memory and Cognition* **26**, 884–902.

Rawlings, D., Barrantes i Vidal, N., and Furnham, A. (2000). Personality and aesthetic preference in Spain and England: two studies relating sensation seeking and openness to experience to liking for paintings and music. *European Journal of Personality* **14**, 553–76.

Rawlings, D. and Ciancarelli, V. (1997). Music preference and the five-factor model of the NEO Personality Inventory. *Psychology of Music* **25**, 120–32.

Rawlings, D., Hodge, M., Sherr, D., et al. (1995). Toughmindedness and preference for musical excerpts, categories and triads. *Psychology of Music* **23**, 63–80.

Rentfrow, P.J., Goldberg, L.R., and Levitin, D.J. (2011). The structure of musical preferences: a five-factor model. *Journal of Personality and Social Psychology* **100**, 1139–57.

Rentfrow, P.J., Goldberg, L.R., Stillwell, D.J., et al. (2012). The song remains the same: replication and extension of the MUSIC model. *Music Perception* **30**, 161–85.

Rentfrow, P.J. and Gosling, S.D. (2003). The do re mi's of everyday life: the structure and personality correlates of music preferences. *Journal of Personality and Social Psychology* **84**, 1236–56.

Rentfrow, P.J. and Gosling, S.D. (2006). Message in a ballad: the role of music preferences in interpersonal perception. *Psychological Science* **17**, 236–42.

Rentfrow, P.J. and Gosling, S.D. (2007). The content and validity of music-genre stereotypes among college students. *Psychology of Music* **35**, 306–26.

Rentfrow, P.J. and McDonald, J.A. (2010). Preference, personality, and emotion. In P.N. Juslin and J.A. Sloboda (eds), *Handbook of Music and Emotion: Theory, Research, Applications*. Oxford: Oxford University Press, pp. 669–95.

Rentfrow, P.J., McDonald, J.A., and Oldmeadow, J.A. (2009). You are what you listen to: young people's stereotypes about music fans. *Group Processes and Intergroup Relations* **12**, 329–44.

Rickard, N.S. (2004). Intense emotional response to music: a test of the physiological arousal hypothesis. *Psychology of Music* **32**, 371–88.

Roe, K. (1985). Swedish youth and music: listening patterns and motivations. *Communication Research* **12**, 353–62.

Russell, J.A. (1980). A circumplex model of affect. *Journal of Personality and Social Psychology* **39**, 1161–78.

Salimpoor, V.N., Benovoy, M., Larcher, K., et al. (2011). Anatomically distinct dopamine release during anticipation and experience of peak emotion to music. *Nature Neuroscience* **14**, 257–62.

Salimpoor, V.N., Benovoy, M., Longo, G., et al. (2009). The rewarding aspects of music listening are related to degree of emotional arousal. *PLoS One* **4**(10), e7487. doi:10.1371/journal.pone.0007487.

Schäfer, T. and Sedlmeier, P. (2009). From the functions of music to music preference. *Psychology of Music* **37**, 279–300.

Schellenberg, E.G., Corrigall, K.A., Ladinig, O., et al. (2012). Changing the tune: listeners like music that expresses a contrasting emotion. *Frontiers in Psychology* **3**: 574. doi: 10.3389/fpsyg.2012.00574.

Schellenberg, E.G., Peretz, I., and Vieillard, S. (2008). Liking for happy and sad sounding music: effects of exposure. *Cognition and Emotion* **22**, 218–37.

Schellenberg, E.G. and von Scheve, C. (2012). Emotional cues in American popular music: five decades of the Top 40. *Psychology of Aesthetics, Creativity, and the Arts* **6**, 196–203.

Schwartz, K.D. and Fouts, G.T. (2003). Music preferences, personality style, and developmental issues of adolescents. *Journal of Youth and Adolescence* **32**, 205–13.

Silvia, P.J. and Nusbaum, E.C. (2011). On personality and piloerection: individual differences in aesthetic chills and other unusual aesthetic experiences. *Psychology of Aesthetics, Creativity, and the Arts* **5**, 208–14.

Sloboda, J.A. (1991). Music structure and emotional response: some empirical findings. *Psychology of Music* **19**, 110–20.

Stang, D.J. (1974). Methodological factors in mere exposure research. *Psychological Bulletin* **81**, 1014–25.

Szpunar, K.K., Schellenberg, E.G., and Pliner, P. (2004). Liking and memory for musical stimuli as a function of exposure. *Journal of Experimental Psychology: Learning, Memory, and Cognition* **30**, 370–81.

Thompson, W.F., Schellenberg, E.G., and Husain, G. (2001). Arousal, mood, and the Mozart effect. *Psychological Science* **12**, 248–51.

Verveer, E.M., Barry, H., and Bousfield, W.A. (1933). Change in affectivity with repetition. *American Journal of Psychology* **45**, 130–4.

Vieillard, S., Peretz, I., Gosselin, N., et al. (2008). Happy, sad, scary and peaceful musical excerpts for research on emotions. *Cognition and Emotion* **22**, 720–52.

Vuoskoski, J.K. and Eerola, T. (2011). The role of mood and personality in the perception of emotions represented by music. *Cortex* **47**, 1099–06.

Vuoskoski, J.K. and Eerola, T. (2012). Can sad music really make you sad? Indirect measures of affective states induced by music and autobiographical memories. *Psychology of Aesthetics, Creativity, and the Arts* **6**, 204–13.

Vuoskoski, J.K., Thompson, W.F., McIlwain, D., et al. (2012). Who enjoys listening to sad music and why? *Music Perception* **29**, 311–7.

Weaver, J.B. (1991). Exploring the links between personality and media preferences. *Personality and Individual Differences* **12**, 1293–9.

Wheeler, B.L. (1985). Relationship of personal characteristics to mood and enjoyment after hearing live and recorded music and to musical taste. *Psychology of Music* **13**, 81–92.

Wilson, W.R. (1979). Feeling more than we can know: exposure effects without learning. *Journal of Personality and Social Psychology* **37**, 811–21.

Witvliet, C.V.O. and Vrana, S.R. (2007). Play it again Sam: repeated exposure to emotionally evocative music polarizes liking and smiling responses, and influences other affective reports, facial EMG, and heart rate. *Cognition and Emotion* **21**, 3–25.

Zajonc, R.B. (1968). Attitudinal effects of mere exposure. *Journal of Personality and Social Psychology Monographs* **9**(2, Pt. 2), 1–27.

Zentner, M.R., Grandjean, D., and Scherer, K.R. (2008). Emotions evoked by the sound of music: characterization, classification, and measurement. *Emotion* **8**, 494–521.

Zweigenhaft, R.L. (2008). A do re mi encore: a closer look at the personality correlates of music preferences. *Journal of Individual Differences* **29**, 45–55.

Chapter 14

Tension–resolution patterns as a key element of aesthetic experience: Psychological principles and underlying brain mechanisms

Moritz Lehne and Stefan Koelsch

14.1 Introduction

Art forms like music, film, or literature play a major role in our cultural lives, but what exactly motivates humans to spend large amounts of their time listening to music, watching movies, or reading novels? One major incentive for "consuming" art is without a doubt its power to elicit strong emotions. A piece of music can move us to tears,[1] a painting can induce feelings of awe and fascination, a poem can evoke feelings of melancholy, elation, or joy, and a suspenseful movie can make our hearts race in excitement. However, the emotional power of art leads to another conundrum: how can a piece of art evoke any emotion at all? Assuming that emotions are elicited by events that have some intrinsic significance (either favorable or harmful) to our individual concerns, or—in biological terms—some kind of survival value, the aesthetic emotions evoked by works of art remain largely mysterious. Unlike the imminent threat posed by the hungry grizzly bear one might accidentally run into on a hiking trip, or the feelings of excitement after having hit the lottery jackpot, the imaginary events of *Romeo and Juliet*, the brush strokes of Munch's *The Scream*, or the musical events of Mozart's *Requiem* seem to lack any relevance to our lives as biological organisms. So how can aesthetic emotions be integrated into a theoretical framework of human affect? What are their underlying psychological principles? And how do they relate to neuronal processes in the brain? In this chapter, we propose that studying *patterns of tension and resolution* in works of art can help to answer these questions and advance the understanding of aesthetic experience in art (and of emotions in general).

Tension–resolution patterns refer to the opposite poles of tension—a state of instability or dissonance associated with feelings of hopeful (or fearful) expectation—and its counterpart, resolution, relaxation, release, or repose. In music, for example, a sequence of dissonant chords may gradually build up tension that is finally resolved by a consonant and stable chord at the end of sequence. In literature, tension—and its close relative, suspense—can be created on the level of the plot (e.g. the protagonist of a novel being threatened by impending peril) as well as on more micro-structural levels of the text (e.g. a single sentence creating tension via its elaborate syntactic composition). Although tension–resolution patterns are most apparent in art forms that unfold in time (e.g. music, literature, film, or dance), they can also be observed in non-temporal visual arts such as painting, sculpture, or architecture, because even in "static" pieces of art, the process of aesthetic experience and appreciation requires time to unfold (Fitch et al. 2009). An abstract painting, for example, can create tension by triggering a process of interpretation—a drive for understanding—that is resolved when a satisfying interpretation is found. This omnipresence of tension–resolution patterns in a wide variety of different art forms suggests that they play a central role in the aesthetic experience of art, motivating a closer investigation of underlying psychological principles and possible brain mechanisms.

Although the opposite poles of tension and resolution were already proposed by Wilhelm Wundt as a key component of affective experience in his tri-dimensional theory of emotion (Wundt 1896, 1911), psychological research has largely neglected the role of tension and resolution in emotion (for rather recent considerations see Schimmack and Reisenzein 2002; Thayer 1989; Watson et al. 1999). One of our aims in this chapter is to reawaken the interest in tension and resolution phenomena as a valuable tool that may help to unravel the psychological and neuronal mechanisms underlying aesthetic emotions, and emotions in general. To some extent, many pieces of art can be regarded as forms of "applied psychology," and Zeki (1999) even suggests considering artists as "neurologists," who creatively explore the processing characteristics of the human brain. Accordingly, tension and resolution patterns constitute an important artistic device by which artists take advantage of general principles of human cognition to evoke emotional responses. Thus, studying tension–resolution patterns in works of art—apart from promising new insights into the psychological mechanisms of aesthetic experience—can also inform psychological and neuroscientific research on emotions evoked in non-artistic contexts. In particular, the time-varying nature of tension–resolution patterns opens new research opportunities for investigating the dynamic, temporal aspects of emotional experience, thus complementing extant emotion research which mostly focuses on static aspects of brief emotional episodes.

In this chapter, we explore tension–resolution patterns and their relation to aesthetic experience from a psychological and neuroscientific perspective. We begin the chapter with a discussion of general psychological principles underlying tension and resolution phenomena which are relevant to a wide variety of art forms. We will argue that experiences of tension are closely related to processes of expectancy, prediction, and anticipation, and reflect a general urge to resolve dissonances or uncertainty and reach a stable state of equilibrium. We will mainly focus on temporal art such as music, film, or literature, in which tension–resolution patterns are most apparent (interestingly, these art forms also happen to have the biggest "mass appeal," suggesting that they address very basic aspects of human emotion). However, many of the principles discussed also transfer to static visual art forms (e.g. painting, sculpture, or architecture). Following the general discussion of tension–resolution patterns, we consider the special case of music in more detail and discuss theoretical and empirical research on musical tension. We close the chapter with considerations on possible brain mechanisms mediating experiences of tension and resolution. Again, we mainly focus on evidence from music research.

14.2 Tension–resolution patterns: psychological principles

For a long time, writers, playwrights, choreographers, film directors, and composers have known the importance of creating tension and suspense[2] to capture the attention of their audiences and excite emotional responses. Western classical music, for example, can be understood as a continuous flow of consonance and dissonance, stability and instability—patterns of tension and resolution that orchestrate the emotional response of the listener. Likewise, tension and resolution phenomena play a key role in film and literature, and the appeal of many popular books and movies seems to directly derive from their power to evoke feelings of tension and suspense. Famous suspense movies like *Psycho*, *Jaws*, and *The Silence of the Lambs* have attracted millions of moviegoers, and suspenseful novels have hit bestseller lists around the world. Similarly, an abstract piece of visual art can create tension by putting the spectator in a state of incomprehension that is resolved when a meaningful interpretation has been found. Apart from works of art, various other instances, in which tension and resolution phenomena play a role and potentially evoke strong emotions can be identified. In many forms of play and games, experiences of tension and resolution are important (in sports events, for example, the collective emotional reactions of spectators are often framed by episodes of tension and resolution). Similarly, tension and resolution phenomena are omnipresent in everyday life: the fearful or hopeful expectation experienced before significant life events with uncertain outcomes, or the relaxation and contentment experienced after a desired goal has been reached can all be conceived of as feelings of tension and resolution. Although these examples at first glance appear to describe very different phenomena, the common occurrence of tension–resolution patterns—not only in art but in virtually all aspects of human life—suggests very basic underlying psychological principles, which we will explore in the following sections.

14.2.1 Uncertainty

Previously, we described tension as a state of dissonance, instability, or fearful/hopeful expectation that is associated with a yearning for resolution; that is, an urge for consonance, stability, harmony, or equilibrium. When dissecting the mechanisms by which feelings of tension or suspense are evoked, one major factor that can be identified as a mediator of tension is *uncertainty*, and many art forms systematically use uncertainty to evoke feelings of tension or suspense. Obvious examples are films or novels, in which uncertainty about events of the narrative plot creates suspense. In narratives, this uncertainty often takes the form of an explicit or implicit question in the mind of the reader or viewer ("Where was the purloined letter hidden?," "Who killed Laura Palmer?", etc.). Such questions arouse curiosity and create "the force that draws us through a narrative" (Rabkin 1973). The eventual resolution of uncertainty then results in a state of understanding and provides a sense of closure that is associated with a positive affective response. The effectiveness of so-called cliff-hangers in television series provides a vivid example of the potential power of using uncertainty as a plot device: ending an episode in a moment of high uncertainty—for example, by putting the protagonist in a life-threatening situation without providing resolution as to whether (s)he survives or not—can create a strong, almost pathological desire to watch the next episode, in which the uncertainty and the associated suspense is hopefully resolved.

Uncertainty in narrative plots comes in different forms: there can be uncertainty about *what* will happen, *when* it will happen, *how* it will happen, or *if* it will happen. When Gabriel García Márquez starts his novella *Chronicle of a Death Foretold* with the sentence, "On the day they were going to kill him, Santiago Nasar got up at five-thirty in the morning to wait for the boat the bishop was coming on," there is a high level of certainty about what will eventually happen (i.e. the murder of Santiago Nasar). However, suspense is created because the exact circumstances leading to the murder (i.e. how it will happen) remain unclear, thus capturing the interest and curiosity of the reader.

It may be argued that suspense in narrative plots is a very special case of tension that is only relevant to representational art forms such as films or novels, which usually refer to (hypothetical) events in the real world. In contrast, tension–resolution phenomena in more abstract non-representational art may be based on entirely different principles. However, investigating tension–resolution patterns in music (which is arguably less representational than a movie or a novel) reveals strikingly similar mechanisms. Referring to suspense in music, music philosopher Leonard Meyer states that "[s]uspense is essentially a product of ignorance as to the future course of events" (Meyer 1956, p. 27). Note that in music, uncertainty about *when* a certain event will happen is often more important than *what* will happen. For example, in most pieces it is very likely that harmonic tension is eventually resolved by returning to the tonic. However, it is not clear when this resolution occurs, and many musical pieces are sophisticated examples of skillfuly delaying the anticipated resolution. Likewise, abstract non-representational works of visual art can use uncertainty to evoke feelings of tension and resolution by putting the spectator in a psychological state of non-understanding that triggers a process of interpretation which may culminate in a rewarding state of resolution as soon as a sensible interpretation has been found. As noted by Fitch and colleagues (2009), some artworks leave the tension partly unresolved, stimulating repeated viewings. Uncertainty thus works as a mediator of tension, whereas the reduction of uncertainty resolves tension.

14.2.2 Expectancy, prediction, and anticipation

Despite its importance for mediating tension, uncertainty alone is not sufficient for evoking feelings of tension (otherwise, tension would be highest for random sequences of events that are completely unpredictable). Other key elements of tension are processes of expectancy, prediction, and anticipation.[3] As events unfold in time (e.g. in a piece of music, or in a narrative plot), they are constantly evaluated against a background of expectancies that are continuously updated during the temporal evolution of events. Events generating strong expectancies are then potential triggers of tension because they create an urge for resolution. To allow the build-up of tension, there is usually some temporal distance between the event generating the expectancy and the event fulfilling the expectancy, and the most effective tension–resolution patterns often span large structures (e.g. the plot of a novel, or the music of an entire symphony).

Importantly, in order to create tension, expected events must have some kind of significance. If the outcome of an expected event is irrelevant—that is, if it does not matter whether expectancies are fulfilled or violated—tension fails to build up. This means that expected events usually have either positive or negative valence. In narrative plots, impending danger conjures images of death or physical harm to the protagonist, which the reader hopes will be resolved by a seemingly unlikely, but highly desired happy ending. In music, tension can be driven by a yearning for consonant and stable sounds that are experienced as pleasurable.

In its most elementary form, tension can thus be conceptualized as a divergence between a desired goal state and the actual state of the world. This divergence between the present state and the desired goal state creates a psychological discomfort that is experienced as tension. The apparent paradox between the psychological discomfort and the feelings of pleasure derived from tension–resolution experiences will be addressed later in this chapter.

To make the formation of predictions possible, some kind of information or formal structure is necessary on which predictive inferences can be based. In music, the generation of predictions is afforded by the "rules" of a given musical system (e.g. Western tonal music). For example, certain chords bear strong implications (e.g. a dominant seventh chord tends to be followed by a tonic chord), thus creating expectancies against which subsequent events are evaluated (for an overview of predictive processing in music, see Rohrmeier and Koelsch 2012). Musical events with a strong implicative function generating clear expectancies about subsequent events then give rise to an experience of tension. In narrative plots, expectancies can be based on genre-specific conventions (the expected happy ending of a mainstream Hollywood movie) or rules of everyday life that activate specific cognitive schemata or scripts (Bartlett 1932; Rumelhart 1980; Schank and Abelson 1977). Of course, expectancies need not always be fulfilled but can be violated to generate surprise and make a piece of art more interesting. In fact, the right "mix" between fulfillment and violation of expectancies is essential: if an artwork is too predictable (i.e. if it always fulfills expectancies) it runs the risk of being boring, whereas a piece of art that is completely unpredictable can easily lead to frustration by overcharging the perceiver's cognitive capacities (of course, in some artworks, this may be exactly what the artist intended).

The predictive processes relevant for generating tension and resolution in works of art build on very basic aspects of human cognition. The philosopher Daniel Dennett stated: "[a] mind is fundamentally an anticipator, an expectation-generator" (Dennett 1996), and anticipation has been suggested as a basic emotion category (Plutchik 1980). Furthermore, the ability to form predictions has been proposed as a fundamental principle of human brain functioning (Bar 2007; Bubic et al. 2010; Friston 2010), which will be discussed in more detail in the last part of this chapter. The importance of anticipation processes as a general feature of human cognition is also plausible from an evolutionary perspective because the ability to make accurate predictions about the future promotes fast and adequate behavioral responses, and increases survival chances. Similarly, the urge to resolve uncertainties discussed previously has biological advantages because behavioral responses in a state of certainty and understanding usually lead more directly to a specific goal than in an uncertain environment. Thus, any art form that affords predictions and creates expectancies takes advantage of this general feature of human cognition. Likewise, the episodes of uncertainty and non-understanding that are deliberately employed in works of art to create tension address a basic psychological need for uncertainty reduction and understanding. Similar to the concept of homeostasis in biology (i.e. the tendency of biological organisms to maintain a stable state in a constantly changing environment), feelings of tension and resolution thus seem to reflect a kind of "psychological homeostasis"—a general tendency to resolve dissonances, conflicts, and uncertainty, and strive towards a stable state of equilibrium.[4]

Interestingly, the fact that art takes advantage of these basic features of human cognition to arouse emotions indicates that these "aesthetic" emotions are not qualitatively different from the emotional experiences of everyday life (for an opposing view distinguishing aesthetic emotions from utilitarian emotions see Scherer 2004). Instead, the omnipresence of tension–resolution phenomena suggests that the concomitant emotions build on general psychological principles that apply to a wide variety of seemingly different modalities.

14.2.3 The emotional valence of tension

If tension is associated with dissonance, instability, and uncertainty—that is emotional states that are usually associated with psychological discomfort, unease, and negative emotional valence—this raises an obvious question: why would anyone deliberately seek out experiences that are associated with tension? Works of art potentially evoke feelings of tension, yet their popularity apparently contradicts the hypothesis that tension causes psychological discomfort and should therefore be avoided. To solve this contradiction, it is helpful to distinguish between experiences of tension in art and tension in everyday life. Unlike tension experiences in real life, often considered as negative in Western societies because they are associated with physiological or psychological strain, art provides an environment in which tension can be experienced without compromising one's physical or psychological health. While tension in everyday life often induces negative emotions and is therefore generally avoided, looking for tension experiences in the arts thus provides a safeguard against potential physical or psychological harm (for a similar discussion focusing on the excitement aroused by sport events see Elias and Dunning 1986). However, even in art, tension is often only positively valued because it is tacitly assumed that it will eventually resolve. Not resolving tension (e.g. in a movie or a piece of music) is usually experienced as unsatisfying. The pleasure derived from tension–resolution experiences thus appears to be a function of the ratio between the discomfort associated with the tension and the comfort experienced during resolution: the higher the discomfort during the tension phase, the higher the pleasure in the moment of resolution. This view is also expressed in Berlyne's "arousal jag" theory (Berlyne 1960), which postulates that a drop from unusually high levels of arousal (that are experienced as unpleasant) to a low level of arousal is associated with feelings of pleasure (however, see Zillmann 1980 for an alternative account). Thus, the pleasure derived from this arousal jag is a motivator for engaging in activities that are associated with unusually high levels of arousal, such as fairground attractions or extreme sports. In a more moderate form, similar mechanisms appear to underlie the emotional reactions evoked by tension–resolution patterns in works of art.

14.2.4 The paradox of repeated exposure

Previously, we discussed uncertainty and processes of expectancy, prediction, and anticipation as key factors mediating experiences of tension and resolution. This raises the question of how it is possible that tension can be experienced over repeated exposures to a piece of art. Assuming that uncertainty is resolved after the first exposure, and expectations during subsequent exposures are largely accurate—thus rendering the expectancy violations or fulfillments experienced during the first exposure ineffective—experiencing a piece of art more than once should make it dull and uninteresting. However, this contradicts common experience. The film philosopher Noël Carroll states that after having seen *King Kong* at least 50 times, he still feels "the irresistible tug of suspense" (Carroll 1996, p. 71), and a piece of music can be listened to hundreds of times while still conveying feelings of tension and resolution.

Various theories accounting for this "paradox of suspense," "anomalous suspense," or "resilience of suspense" have been proposed (for the special case of narrative plots, see Brewer 1996, for an overview). These include the postulation of a willing suspension of memory, or a shift of motivation (repeated readings of a literary text, for example, are often not motivated by evoking a state of suspense but may have other reasons such as focusing on more subtle details of the story, the writing style of the author, etc.). In part, the paradox of repeated exposure may also be accounted for by imperfect memory because second exposures often occur after a period of time when considerable parts of the information of the first exposure may have been lost (and discovering what one still remembers might be a pleasurable experience). Furthermore, even if the expectancy violations and the uncertainty of the first exposure are less effective during subsequent exposures, some aspects of an artwork can create tension because they are intrinsically associated with psychological discomfort (e.g. due to sensory dissonance in a piece of music or an unpleasant image in a piece of visual art) that is resistant to repeated exposures.

In music, the paradox that unexpected musical events can maintain their quality of "unexpectedness" even after repeated exposures has been referred to as Wittgenstein's paradox (Dowling and Harwood 1986; Huron 2006). To solve this paradox, Bharucha (1994) distinguishes between *veridical* and *schematic* expectations. Whereas veridical expectations refer to the expectations concerning actual events in a familiar piece of music (e.g. a wrong note during the performance of a familiar piece violates a veridical expectation), schematic expectations are driven by the rules of a musical system (e.g. a dominant seventh chord followed by a chord that is not the tonic violates a schematic expectation). Although veridical and schematic expectations often coincide, they can diverge in familiar pieces, in which a musical event can be veridically expected but schematically unexpected, thus accounting for the remaining experience of unexpectedness over repeated exposures.

Finally, repeated exposure increases the predictive power of a listener, which might be experienced as pleasurable, and various studies have shown that repeated exposure to a musical piece increases liking (Lieberman and Walters 1968; Peretz et al. 1998; Verveer et al. 1933). This effect of "mere exposure" (Zajonc 1968) may to a certain degree compensate for the lower effectiveness of expectancy violations after repeated exposures.

14.2.5 Tension–resolution patterns and the aesthetic trajectory

The aesthetic experience associated with patterns of tension and resolution can be conceived of as a trajectory that unfolds in time and comprises different temporal stages. Fitch and colleagues (2009) proposed a minimum of three stages for this *aesthetic trajectory*: "an initial stage of familiarity, recognition, or stability; a second stage of surprise, ambiguity, or tension; and a third stage of integration, resolution, or synthesis" (p. 61). These three stages closely correspond to the mechanisms mediating experiences of tension and resolution described earlier. Previously we stated that the expectation and anticipation processes related to tension experiences require a formal structure that makes it possible to form predictions; that is, they presuppose some kind of familiarity and recognition reflecting the initial stage of Fitch's aesthetic trajectory. Second, the initial stage of familiarity is challenged by expectancy violations generating surprise, ambiguity, or tension. Finally, during the last stage, tension is resolved by integration and synthesis.

Similar to Fitch's aesthetic trajectory, the expectation processes generating tension can also be related to different stages of the *tension arc* proposed by Koelsch (2012): during the initial build-up of expectancy, tension is created; this tension may be modulated by the harmonic stability and the extent of the established structure, and further increased by breaches of expectancy generating an anticipation for resolution; when the resolution is finally reached, tension is released, giving rise to an experience of pleasure and reward. Usually, several of these tension arcs are superimposed and intertwined on different time scales (such as a phrase embedded in a part of a movement of a sonata). Koelsch's tension arc specifically refers to musical tension. However, considering the relevance of expectation processes as a general feature of human cognition, the different stages of the tension arc can also be generalized to other tension and resolution experiences.

14.3 Musical tension

The continuous ebb and flow of tension and resolution—a waxing and waning of episodes of dissonance and instability versus passages of consonance and stability—is one of the hallmarks of Western tonal music, and constitutes a major aspect of emotional experience during music listening. Tension and resolution are related to various features of the music. Different notes of a musical scale are associated with different degrees of stability and instability, different chords, or combinations of chords, convey different degrees of consonance or dissonance, metrical features like syncopation can delay or anticipate certain musical events, and expressive features such as dynamics and agogics create fluctuations in loudness or tempo, thus contributing to experiences of tension and resolution. Tension and resolution can be related to purely acoustic features (e.g. loudness or sensory dissonance being associated with an increase in experienced tension; Farbood 2012; Pressnitzer et al. 2000), or they can involve higher cognitive processes (e.g. expectation processes building on the rules of a musical system; Lerdahl and Krumhansl 2007). As already mentioned, tension–resolution patterns occur at different levels: they can be small-scale, changing from one musical event to the next as well as large-scale, encompassing, for example, the global structure of an entire symphony. These global and local tension–resolution patterns possibly interact with one another, thus maximizing the emotional effect of the aesthetic experience.

As discussed previously, expectancy and anticipation processes as well as a general striving for consonance, stability, and resolution play a key role for inducing experiences of tension. The important contribution of expectation processes as a mediator of music-evoked emotions has prominently been put forward in the seminal work *Emotion and meaning in music* by Leonard Meyer (1956), in which expectation and the associated experiences of tension and resolution were proposed as a key mechanism for creating musical meaning and emotion. According to Meyer, the continuous play with expectations in music (i.e. the fulfillment or violation of expectancies) is paramount to the associated affective and aesthetic response. Similarly, Huron (2006) proposes a general psychological theory of expectation illustrated for the special case of music. Like Meyer, he stresses the important role of expectation for generating tension, which he defines as a physiological reaction "that arises from changes in arousal and attention, in preparation for some expected event" (Huron 2006, p. 305).

Expectation processes in music build on the implicative structure of musical events (a dominant seventh chord, for example, is implicative of the subsequent tonic chord). For the special case of melodic expectations, Narmour (1990) devised a theory, identifying a set of general principles mediating expectation processes in melodies (summarized in Krumhansl 1995), which have been largely confirmed by experimental research (Cuddy and Lunney 1995). Subsequent research (Schellenberg 1997) has shown that the principles postulated by Narmour for melodic expectations can be reduced to two basic predictor variables, namely *pitch proximity* and *pitch reversal*. The pitch proximity principle states that listeners expect a melodic sequence to continue on a note that is proximate in pitch to the last note. Extending the pitch proximity principle, the pitch reversal principle claims that listeners expect the next note to be close in pitch to the note before the last note. For example, when a melody features a large leap (violating the pitch proximity principle), listeners expect the melody to return to a note close to the note before the leap.

According to Narmour (1990), melodic expectation is governed by innate bottom-up as well as learned top-down processes. This view has been challenged by computational modeling approaches showing that melodic expectation can be accurately modeled based on general learning mechanisms without assuming innate representational rules (Pearce and Wiggins 2012). The learning mechanisms built into the model distinguish between long-term and short-term learning processes. Whereas long-term learning reflects listeners' experiences acquired over previously encountered musical pieces, short-term learning reflects the learning that occurs during exposure of the current piece. Interestingly, the model has begun to be extended to harmony (Whorley et al. 2008) and even language (Wiggins 2011), indicating that it reflects general modality-independent cognitive processes rather than music-specific rules or principles.

Although musical expectancies have mostly been investigated in tonal contexts (i.e. in melodies or chord progressions), expectation processes are equally important in the processing of musical rhythm. Studying rhythms of the "Black Atlantic" diaspora (the rhythms underlying popular music genres such as jazz, blues, reggae, funk, rock, or hip-hop), Pressing (2002) argues that the emotional effect of these rhythms crucially relies on prediction and expectation processes.

Importantly, the expectation processes relevant to mediating musical tension can happen on different cognitive levels and involve different degrees of previously acquired knowledge. They can be knowledge-free (i.e. not relying on long-term knowledge) or they can be based on information stored in long-term memory. For example, expectation generation related to small-scale structures (e.g. the grouping of notes) may depend exclusively on auditory sensory memory and the processing according to Gestalt principles. On the other hand, the formation of expectations can be informed by (implicit) knowledge of music-syntactic structures stored in long-term memory (see Koelsch 2012, for a more detailed discussion of different levels of music processing). It remains to be investigated whether the different levels of cognitive processing involved in the generation of expectancies are associated with tension experiences that are qualitatively different from each other.

14.3.1 Empirical research

Various experimental studies have tapped into the subjective experience of tension. Most research investigating musical tension employs continuous self-report methods (Schubert 2010) to measure the subjective experience of tension. In studies using continuous self-report methods, the tension experience of participants is tracked online during the actual listening experience. In the first empirical study of musical tension, for example, Nielsen (1983, cited in Krumhansl 1996) used a set of spring-loaded tongs that participants pressed together according to their degree of experienced tension while listening to excerpts from Joseph Haydn's *Symphony No. 104* and Richard Strauss' *Also sprach Zarathustra*. The deflection of the tongs was recorded and put into relation to different structural features of the music (see below). The tension profiles obtained by Nielsen were replicated by Madsen and Fredrickson (1993) using the so-called continuous-response digital interface (CRDI), a potentiometer that records dial-movements of participants while they listen to a piece of music. An alternative method for acquiring tension ratings is to play music to participants until a certain event of interest is reached, then stop the music and ask participant to rate how much tension they experienced or how conclusive the last musical event felt to them (e.g. Bigand et al. 1996; Bigand and Parncutt 1999; Lerdahl and Krumhansl 2007). Other studies use sliders depicted on a screen that can be adjusted with a computer mouse depending on the degree of experienced tension (e.g. Farbood 2012; Krumhansl 1996; Lehne et al. 2013). In a web-based survey by Farbood (2012), tension of short musical excerpts was rated by selecting graphical representations that best corresponded to the excerpt (e.g. an upward-rising line symbolizing an increase in tension). Note that when acquiring subjective tension ratings (and emotion ratings in general) one has to distinguish between the tension that is subjectively experienced by participants (*felt* musical tension) and the tension that participants think that the music expresses (*perceived* musical tension). This distinction is important because felt and perceived emotion in music do not necessarily coincide (Gabrielsson 2002).

Musical expectancies have been studied using *priming paradigms* and the *probe tone technique*. In priming experiments, participants are presented with a musical event (the prime) that is followed by a target event. When the target is expected, it is processed faster than when it is unexpected. By having participants perform a simple decision task (e.g. deciding whether the target is in or out of tune), response times can be recorded that indicate how expected a specific musical event is in relation to a preceding prime (see, for example, Bharucha and Stoeckig 1986). In probe-tone paradigms, a musical context is played to participants that is followed by a tone, for which participants have to judge how well it fits the preceding musical context. Using this procedure, it is possible to construct a "tonal hierarchy" that ranks different pitches according to their perceived stability in a given context (see Krumhansl and Shepard 1979; Krumhansl and Kessler 1982). For example, notes of a musical scale are perceived as more or less stable, with the first note of the scale (the tonic) usually being experienced as most stable.

Experimental studies also investigated how different musical features influence the subjective experience of tension. The previously mentioned study by Nielsen (1983) found dynamics, tonal aspects (melodic contour and harmony), timbral features, and repetition to influence ratings of musical tension. Other features that were found to affect tension ratings are phrase structure, note density (Krumhansl 1996), pitch height, loudness, onset frequency, and tempo (Farbood 2012). Discarding expressive features of the music (i.e. dynamics and agogics) appears to preserve the overall tension–resolution patterns of a musical piece (Krumhansl 1996; Lehne et al. 2013), however, expressive features can enhance the experience of tension, and the contribution of different musical features depends on their relative importance in the specific musical piece (Farbood 2012; Lehne et al. 2013). Moreover, tension ratings correlate highly within individuals over different exposures (Krumhansl 1996) as well as between different groups of persons such as musicians and non-musicians or school children of different ages (Fredrickson 1997, 1999, 2000), suggesting relatively consistent and stable underlying cognitive and affective processes. Tension ratings also have been found to correlate with ratings of discrete emotions (happiness, sadness, and fear) and psychophysiological measures such as finger-pulse amplitude, respiration rate, blood pressure, skin conductance, and temperature (Krumhansl 1997), thus supporting its role in affective processing.

Musical tension has been modeled based on the tonal structure of a musical piece (Lerdahl 1996; Lerdahl and Krumhansl 2007). In these models, tension is estimated by calculating the psychological "distance" between chords of a musical piece (chords that are "further away" from the tonal center of a piece are associated with higher tension values) based on the theory of tonal pitch space (Lerdahl 2004) and the hierarchical syntactic structure as described by Lerdahl and Jackendoff (1983). A more recent model of musical tension (Farbood 2012) includes other musical parameters such as loudness, melodic expectation, onset frequency, and pitch height, and also accounts for dynamic cognitive processes (memory and attention) that affect the experience of tension.

Altogether, the empirical studies of musical tension indicate that a plethora of features affects the experience of tension, ranging from purely acoustic features such as sensory dissonance or loudness to aspects that require more complex cognitive processing (e.g. the processing of musical syntax). To gain a better understanding of musical tension, future research may profit from a closer investigation of the underlying mechanisms by which different musical features create tension. In particular, it may be useful to investigate to what extent the influence of different features on musical tension can be accounted for by the processes of expectation and prediction discussed above, and to what degree other mechanisms (e.g. sensory dissonance) play a role.

14.3.2 Musical tension and emotion

How do the tension–resolution patterns of a musical piece relate to the listener's subjective emotional experience? First, emotional qualities of musical events mediating tension and resolution experiences may partly depend on their specific psycho-acoustic properties. For example, musical chords conveying tension often are more dissonant than chords associated with resolution (regardless of the musical context). The experience of dissonance is, at least partly, determined by the sensory processing of sound in the ear. Chords perceived as dissonant often feature frequencies that are close together causing interferences between the auditory sensory receptors of the basilar membrane in the inner ear, which are perceptually experienced as roughness (Helmholtz 1913). Due to this roughness, chords associated with tension may thus intrinsically be experienced as more unpleasant than consonant chords.[5]

Apart from the intrinsic emotional quality of isolated musical events, expectation processes again play a key role in inducing emotional experiences related to tension and resolution during music listening. Previously, we referred to the works by Meyer (1956) and Huron (2006) who proposed expectation processes as the driving force behind music-evoked emotions. Likewise, Juslin and Västfjäll (2008) identify musical expectancy as one of six mechanisms by which music can induce emotions (the other five are brain stem reflexes, evaluative conditioning, emotional contagion, visual imagery, and episodic memory). The power of expectation processes as a mediator of music-evoked emotions is further corroborated by empirical research on so-called chill experiences during music listening. Chills (also referred to as thrills, shivers, or frisson) are intensely pleasurable responses to music that often are associated with specific physical sensations such as shivers down the spine or goosebumps (for an overview see Panksepp 1995). Sloboda (1991) found that these chill experiences often are evoked by new or unexpected harmonies, thus supporting the close connection between expectancy violations and emotional responses.

14.3.3 Underlying brain mechanisms

The advent of modern neuroimaging methods, in particular functional magnetic resonance imaging (fMRI), has opened new avenues for investigating brain processes related to aesthetic experience by providing a non-invasive method for studying cognitive and affective processes in the human brain. Although neuroscientific research that directly investigates the neural correlates of tension–resolution patterns is scarce, various studies investigating expectancy violations in music, as well as studies investigating chill experiences point to brain areas relevant for mediating feelings of tension. Furthermore, there is evidence that the pleasure derived from music listening is driven by neural mechanisms of prediction and expectation associated with dopaminergic neuronal activity in the midbrain.

A large body of research has investigated predictive processing in music (reviewed in Rohrmeier and Koelsch 2012), and neuroscientific studies have begun to identify underlying brain mechanisms. In particular, cognitive processing of expectancy violations in music appears to be associated with specific signatures in the electroencephalogram (EEG). These signatures differ with regard to the level of cognitive processing involved. At a relatively low level of cognitive processing, for example, occasional deviants in a sequence of tones give rise to the so-called mismatch negativity (MMN), a characteristic negativity of event-related scalp potentials (ERPs) peaking around 150–250 ms after the deviating event (for a review, see Näätänen et al. 2007). On a higher level of music-syntactic processing, expectancy violations in short chord sequences manifest in an early right anterior negativity (ERAN), which peaks around 150–200 ms after an unexpected chord function (Koelsch et al. 2000). In a study using magnetoencephalography (MEG), the neural sources of this brain potential were localized in the inferior frontal gyrus (IFG, Brodmann area 44; Maess et al. 2001). This finding has been corroborated by subsequent fMRI studies (e.g. Koelsch et al. 2005; Tillmann et al. 2006), indicating a functional role of the IFG in music-syntactic processing. Importantly, however, apart from activating cortical areas involved in syntactic processing, expectancy violations have also been found to activate areas related to emotive processing in the orbitofrontal cortex (Koelsch et al. 2005; Tillmann et al. 2006), as well as the amygdala (Koelsch et al. 2008a). The amygdala activations in response to irregular musical events are consistent with the observation that expectancy violations in music also evoke autonomic responses such as increases in electrodermal activity (Koelsch et al. 2008b), which are mediated by the central nucleus of the amygdala (LeDoux 2000).

A recent fMRI study of our own group (Lehne et al. 2013) investigated neural correlates of musical tension more directly by relating fMRI data to behavioral tension ratings of four piano pieces by Mendelssohn, Mozart, Schubert, and Tchaikovsky. It was found that tension increases (as rated by the participants) were related to signal increases in the orbitofrontal cortex (Brodmann area 47) and the amygdala (consistent with the fMRI studies mentioned; Koelsch et al. 2005; Tillmann et al. 2006; Koelsch et al. 2008a). This supports the involvement of orbitofrontal cortex and amygdala in musical tension.

There are also hints that certain brain areas are related to the different stages of the tension arc proposed by Koelsch (2012). Studying neural correlates of chill responses, Salimpoor and colleagues (2011) found that anticipation of a chill response and the actual chill response itself were related to different brain areas. Whereas actual chill responses were related to neuronal activity in the ventral striatum, presumably the nucleus accumbens (a structure involved in reward processing), anticipation of chill responses activated more dorsal parts of the striatum. Activations in the same compartment of the dorsal striatum were also evoked by unexpected chords in short chord sequences (Koelsch et al. 2008a). Assuming that irregular chords evoke an anticipation for resolution and that the resolution is experienced as pleasurable, it appears that two distinct stages of the aesthetic trajectory, *anticipation* and *resolution*, map to specific brain areas in the striatum. However, to disentangle the brain mechanisms underlying the different phases of the aesthetic trajectory, in particular the different stages of the tension arc proposed by Koelsch (2012), further research is required.

In previous sections we mentioned that the predictive processes related to experiences of tension and resolution constitute a fundamental aspect of human cognition and brain functioning. In this regard, the theory of predictive coding (Friston 2010) is especially relevant. This theory has been proposed as a unified brain theory, according to which action, perception, and learning can be accounted for by a basic principle: the minimization of prediction errors (i.e. the minimization of surprise). Considering the role of predictive processes in tension–resolution phenomena, it seems likely that their appeal and emotional power, at least partly, derive from the fact that they tap into this basic principle of brain functioning. For music, the close relation between the predictive coding theory and music-evoked emotions has been put forward in an article by Gebauer and colleagues (2012), suggesting that the fulfillment or violation of musical expectancies is associated with activity of the brain's dopaminergic reward system. According to this view, musical expectancies activate dopaminergic neurons in the midbrain, coding information about a listener's expectations concerning upcoming musical events. If these expectations are violated, a prediction error ensues that is associated with emotional experiences. Although the exact neuronal mechanisms by which prediction processes trigger affective responses during music processing remain largely unclear, the theory of predictive coding provides a promising framework from which new insights into the neural mechanisms underlying music-evoked emotions and affective experience in general can be expected in the near future.

14.4 Summary

Tension–resolution patterns are an important artistic device that drives emotional experience in a variety of different art forms (especially in temporal art forms such as music, film, or literature). In this chapter, we discussed general psychological mechanisms underlying tension and resolution phenomena. In particular, we argued that experiences of tension build on processes of expectation and anticipation, as well as a general urge to resolve uncertainty, dissonance and instability and strive towards a stable state of equilibrium. We discussed emotional aspects of tension and resolution phenomena and highlighted their important role in the aesthetic experience of art and their relation to different stages of an "aesthetic trajectory." The special case of musical tension was considered in more detail, giving an overview of empirical research that has identified various features mediating experiences of tension and resolution. Finally, we discussed neuroimaging research that has identified limbic and paralimbic structures of the brain related to expectancy violations and "chill" experiences (and their anticipation), thus pointing to possible neural correlates of tension and resolution experiences. The predictive coding theory was discussed as a promising theoretical framework that may help to integrate aesthetic emotions into a general neuro-cognitive model of human affect.

Acknowledgments

The preparation of this chapter was supported by the Cluster of Excellence "Languages of Emotion" of the Freie Universität Berlin. We would like to thank John DePriest and Philipp Engel for helpful comments on earlier versions of the manuscript.

Notes

1 Note that seemingly negative emotions such as fear or sadness are often experienced as positive when encountered in works of art.

2 The words *tension* and *suspense* will be largely used synonymously, and the following paragraphs will try to show that both concepts build on similar psychological mechanisms.

3 Note that there are subtle conceptual differences between the terms *expectancy*, *expectation*, *prediction*, and *anticipation* (for accounts on terminology, see Bubic et al. 2010; Rohrmeier and Koelsch 2012). However, there is no consensus regarding the definition of the terms, and they are often used interchangeably in the literature.

4 Note the parallels to Festinger's theory of cognitive dissonance (Festinger 1957), which states that humans have a strong urge to resolve psychological dissonances and strive towards cognitive states in which beliefs and behavior are consistent (even if this involves entertaining irrational beliefs).

5 In addition to sensory processes, the experience of dissonance is also influenced by listening habits and can change over time. For example, some chords that may have been regarded as relatively dissonant a century ago, are nowadays perceived as more consonant (e.g. major seventh chords).

References

Bar, M. (2007). The proactive brain: using analogies and associations to generate predictions. *Trends in Cognitive Sciences* **11**, 280–9.

Bartlett, F.C. (1932). *Remembering: A Study in Experimental and Social Psychology*. Cambridge: Cambridge University Press.

Berlyne, D.E. (1960). *Conflict, arousal, and curiosity*. New York, NY: McGraw-Hill.

Bharucha, J. (1994). Tonality and expectation. In R. Aiello (ed.), *Musical Perceptions*. Oxford: Oxford University Press, pp. 213–39.

Bharucha, J. and Stoeckig, K. (1986). Reaction time and musical expectancy: Priming of chords. *Journal of Experimental Psychology: Human Perception and Performance* **12**, 403–10.

Bigand, E. and Parncutt, R. (1999). Perceiving musical tension in long chord sequences. *Psychological Research* **62**, 237–54.

Bigand, E., Parncutt, R., and Lerdahl, F. (1996). Perception of musical tension in short chord sequences: the influence of harmonic function, sensory dissonance, horizontal motion, and musical training. *Perception and Psychophysics* **58**, 124–41.

Brewer, W.F. (1996). The nature of narrative suspense and the problem of rereading. In P. Vorderer, H.J. Wulff, and M. Friedrichsen (eds), *Suspense: Conceptualizations, Theoretical Analyses, and Empirical Explorations*. London: Routledge, pp. 107–27.

Bubic, A., Cramon, D.Y. von, and Schubotz, R.I. (2010). Prediction, cognition and the brain. *Frontiers in Human Neuroscience* **4**, 25.

Carroll, N. (1996). The paradox of suspense. In P. Vorderer, H.J. Wulff, and M. Friedrichsen (eds), *Suspense: Conceptualizations, Theoretical Analyses, and Empirical Explorations*. London: Routledge, pp. 71–91.

Cuddy, L.L. and Lunney, C.A. (1995). Expectancies generated by melodic intervals: perceptual judgments of melodic continuity. *Perception and Psychophysics* **57**, 451–62.

Dennett, D. (1996). *Kinds of Minds: Toward an Understanding of Consciousness*. New York, NY: Basic.

Dowling, W.J. and Harwood, D.L. (1986). *Music Cognition*. San Diego, CA: Academic Press.

Elias, N. and Dunning, E. (1986). *Quest for Excitement: Sport and Leisure in the Civilizing Process*. Oxford: Blackwell.

Farbood, M. (2012). A parametric, temporal model of musical tension. *Music Perception* **29**, 387–428.

Festinger, L. (1957). *A Theory of Cognitive Dissonance*. Stanford, CA: Stanford University Press.

Fitch, W.T., von Graevenitz, A., and Nicolas, E. (2009). Bio-aesthetics, dynamics and the aesthetic trajectory: a cognitive and cultural perspective. In M. Skov and O. Vartanian (eds), *Neuroaesthetics*. Amityville, NY: Baywood Publishing Company, Inc., pp. 59–102.

Fredrickson, W.E. (1997). Elementary, middle, and high school student perceptions of tension in music. *Journal of Research in Music Education* **45**, 626–35.

Fredrickson, W.E. (1999). Effect of musical performance on perception of tension in Gustav Holst's first suite in e-flat. *Journal of Research in Music Education* **47**, 44–52.

Fredrickson, W.E. (2000). Perception of tension in music musicians versus nonmusicians. *Journal of Music Therapy* **37**, 40–50.

Friston, K. (2010). The free-energy principle: a unified brain theory? *Nature Reviews Neuroscience* **11**, 127–38.

Gabrielsson, A. (2002). Emotion perceived and emotion felt: same or different? *Musicae Scientiae* Special issue 2001–2002, 123–47.

Gebauer, L., Kringelbach, M., and Vuust, P. (2012). Ever-changing cycles of musical pleasure: the role of dopamine and anticipation. *Psychomusicology: Music, Mind, and Brain* **22**, 152–67.

Helmholtz, H.L.F. von (1913). *Die Lehre von den Tonempfindungen als physiologische Grundlage für die Theorie der Musik*. Braunschweig: Friedrich Vieweg & Sohn.

Huron, D. (2006). *Sweet Anticipation: Music and the Psychology of Expectation*. Cambridge, MA: MIT Press.

Juslin, P.N. and Västfjäll, D. (2008). Emotional responses to music: the need to consider underlying mechanisms. *Behavioral and Brain Sciences* **31**, 559–621.

Koelsch, S. (2012). *Brain and Music*. Chichester: John Wiley & Sons, Ltd.

Koelsch, S., Fritz, T., and Schlaug, G. (2008a). Amygdala activity can be modulated by unexpected chord functions during music listening. *NeuroReport* **19**, 1815–9.

Koelsch, S., Fritz, T., Schulze, K., et al. (2005). Adults and children processing music: an fMRI study. *NeuroImage* **25**, 1068–76.

Koelsch, S., Gunter, T., Friederici, A., et al. (2000). Brain indices of music processing: "nonmusicians" are musical. *Journal of Cognitive Neuroscience* **12**, 520–41.

Koelsch, S., Kilches, S., Steinbeis, N., et al. (2008b). Effects of unexpected chords and of performer's expression on brain responses and electrodermal activity. *PLoS One* **3**, e2631.

Krumhansl, C.L. (1995). Music psychology and music theory: problems and prospects. *Music Theory Spectrum* **17**, 53–80.

Krumhansl, C.L. (1996). A perceptual analysis of Mozart's piano sonata K. 282: Segmentation, tension, and musical ideas. *Music Perception* **13**, 401–32.

Krumhansl, C.L. (1997). An exploratory study of musical emotions and psychophysiology. *Canadian Journal of Experimental Psychology* **51**, 336–53.

Krumhansl, C.L. and Kessler, E.J. (1982). Tracing the dynamic changes in perceived tonal organization in a spatial representation of musical keys. *Psychological Review* **89**, 334–68.

Krumhansl, C.L. and Shepard, R.N. (1979). Quantification of the hierarchy of tonal functions within the diatonic context. *Journal of Experimental Psychology: Human Perception and Performance* **5**, 579–94.

LeDoux, J.E. (2000). Emotion circuits in the brain. *Annual Review of Neuroscience* **23**, 155–84.

Lehne, M., Rohrmeier, M., Gollmann, D., et al. (2013). The influence of different structural features on felt musical tension in two piano pieces by Mozart and Mendelssohn. *Music Perception* **31**(2), 171–85.

Lehne, M., Rohrmeier, M., and Koelsch, S. (2013). Tension-related activity in the orbitofrontal cortex and amygdala: an fMRI study with music. *Social Cognitive and Affective Neuroscience* **9**(10), 1515–23.

Lerdahl, F. (1996). Calculating tonal tension. *Music Perception* **13**, 319–63.

Lerdahl, F. (2004). *Tonal pitch space*. New York, NY: Oxford University Press.

Lerdahl, F. and Jackendoff, R. (1983). *A Generative Theory of Tonal Music*. Cambridge, MA: MIT Press.

Lerdahl, F. and Krumhansl, C.L. (2007). Modeling tonal tension. *Music Perception* **24**, 329–66.

Lieberman, L.R. and Walters, W.M. (1968). Effects of repeated listening on connotative meaning of serious music. *Perceptual and Motor Skills* **26**, 891–5.

Madsen, C.K. and Fredrickson, W.E. (1993). The experience of musical tension: a replication of Nielsen's research using the continuous response digital interface. *Journal of Music Therapy* **30**, 46–63.

Maess, B., Koelsch, S., Gunter, T.C., et al. (2001). Musical syntax is processed in Broca's area: an MEG study. *Nature Neuroscience* 4(5), 540–45.

Meyer, L.B. (1956). *Emotion and meaning in music*. Chicago, IL: University of Chicago Press.

Näätänen, R., Paavilainen, P., Rinne, T., et al. (2007). The mismatch negativity (MMN) in basic research of central auditory: A review. *Clinical Neurophysiology* **118**, 2544–90.

Narmour, E. (1990). *The Analysis and Cognition of Basic Melodic Structures: The Implication-Realization Model*. Chicago, IL: Chicago University Press.

Nielsen, F.V. (1983). *Oplevelse of musikalsk spaending*. Copenhagen: Akademisk Forlag.

Panksepp, J. (1995). The emotional sources of "chills" induced by music. *Music Perception* **13**, 171–207.

Pearce, M.T. and Wiggins, G.A. (2012). Auditory expectation: the information dynamics of music perception and cognition. *Topics in Cognitive Science* **4**, 625–52.

Peretz, I., Gaudreau, D., and Bonnel, A.-M. (1998). Exposure effects on music preference and recognition. *Memory* **26**(5), 884–902.

Plutchik, R. (1980). *Emotion: Theory, research, and experience: Vol. 1. Theories of emotion*. New York, NY: Academic Association.

Pressing, J. (2002). Black Atlantic rhythm: its computational and transcultural foundations. *Music Perception* **19**, 285–310.

Pressnitzer, D., McAdams, S., Winsberg, S., et al. (2000). Perception of musical tension for nontonal orchestral timbres and its relation to psychoacoustic roughness. *Perception and Psychophysics* **62**, 66–80.

Rabkin, E.S. (1973). *Narrative suspense: "When Slim turned sideways . . ."* Ann Arbor, MI: University of Michigan Press.

Rohrmeier, M. and Koelsch, S. (2012). Predictive information processing in music cognition. A critical review. *International Journal of Psychophysiology* **83**, 164–75.

Rumelhart, D.E. (1980). *Schemata: the building blocks of cognition.* In R.J. Spiro, B.C. Bruce, and W.F. Brewer (eds), *Theoretical Issues in Reading Comprehension: Perspectives From Cognitive Psychology, Linguistics, Artificial Intelligence, and Education.* Hillsdale, NJ: Lawrence Erlbaum Associates, pp. 33–58.

Salimpoor, V., Benovoy, M., Larcher, K., et al. (2011). Anatomically distinct dopamine release during anticipation and experience of peak emotion to music. *Nature Neuroscience* **14**, 257–62.

Schank, R.C. and Abelson, R. (1977). *Scripts, Plans, Goals, and Understanding.* Hillsdale, NJ: Lawrence Erlbaum Associates.

Schellenberg, E.G. (1997). Simplifying the implication-realization model of melodic expectancy. *Music Perception* **14**, 295–318.

Scherer, K.R. (2004). Which emotions can be induced by music? What are the underlying mechanisms? And how can we measure them? *Journal of New Music Research* **33**, 239–51.

Schimmack, U. and Reisenzein, R. (2002). Experiencing activation: energetic arousal and tense arousal are not mixtures of valence and activation. *Emotion* **2**, 412–7.

Schubert, E. (2010). Continuous self-report methods. In P. Juslin and J.A. Sloboda (eds), *Handbook of Music and Emotion—Theory, Research, Applications.* Oxford: Oxford University Press, pp. 223–53.

Sloboda, J.A. (1991). Music structure and emotional response: some empirical findings. *Psychology of Music* **19**, 110–20.

Thayer, R.E. (1989). *The biopsychology of mood and arousal.* New York, NY: Oxford University Press.

Tillmann, B., Koelsch, S., Escoffier, N., et al. (2006). Cognitive priming in sung and instrumental music: activation of inferior frontal cortex. *NeuroImage* **31**, 1771–82.

Verveer, E.M., Barry, H., and Bousfield, W.A. (1933). Change in affectivity with repetition. *The American Journal of Psychology* **45**, 130–4.

Watson, D., Wiese, D., Vaidya, J., et al. (1999). The two general activation systems of affect: structural findings, evolutionary considerations, and psychobiological evidence. *Journal of Personality and Social Psychology* **76**, 820–38.

Whorley, R.P., Pearce, M.T., and Wiggins, G.A. (2008). Computational modelling of the cognition of harmonic movement. In K. Miyazaki, M. Adachi, Y. Hiraga, et al. (eds), *Abstracts of the 10th international conference on music perception and cognition.* Tokyo: Sapporo, p. 212.

Wiggins, G.A. (2011). "I let the music speak": cross-domain application of a cognitive model of musical learning. In P. Rebuschat and J. Williams (eds), *Statistical learning and language acquisition.* Amsterdam, The Netherlands: Mouton De Gruyter, pp. 463–94.

Wundt, W. (1896). *Grundriss der Psychologie.* Leipzig: Engelmann.

Wundt, W. (1911). *Grundzüge der physiologischen Psychologie.* Leipzig: Engelmann.

Zajonc, R.B. (1968). Attitudinal effects of mere exposure. *Journal of Personality and Social Psychology* **9**, 1–27.

Zeki, S. (1999). Art and the brain. *Dædalus* **127**, 71–103.

Zillmann, D. (1980). Anatomy of suspense. In P.H. Tannenbaum (ed.), *The Entertainment Functions of Television.* Hillsdale, NJ: Lawrence Erlbaum Associates, pp. 133–63.

Chapter 15

From pleasure to liking and back: Bottom-up and top-down neural routes to the aesthetic enjoyment of music

Elvira Brattico

15.1 Music and aesthetic emotions

In daily life we are surrounded by music; in the shopping mall, on TV, or on the radio. Nowadays, even while surfing the Internet pop-up commercials might start playing a tune. For the adult layperson, situations where music is listened to intentionally (such as when playing favorite songs on a mobile device or from the hi-fi equipment at home or when attending a concert) happen less frequently compared to instances of casual listening as just described. According to a review by Sloboda (2010), intentionally listening to music (as the main activity) amounts to only 2–12 per cent of the total time spent listening to music. Nevertheless, intentional musical activities can be considered as fairly common in daily life (compared to physical activities, for instance), even when a person does not have, or plan to have, a career in music. This is even more common in adolescence, when music assumes a special role in daily life (Saarikallio and Erkkila 2007). Intentional listening and active musical activities represent typical aesthetic experiences of music, generating affective and evaluative outcomes. Recently, the aesthetic experience of music has been defined as a situation in which an individual approaches a musical event or involves oneself in a musical activity—possibly with an aesthetic attitude, intentionality, affective expectations, and dedicated attention (Brattico et al. 2013; Brattico and Pearce 2013). According to the framework developed by Brattico and colleagues (2013), an aesthetic musical experience in an optimal situation (such as listening to music in a concert hall) would engender three different kinds of aesthetic responses: (1) aesthetic judgments ("this music is so beautiful"), (2) aesthetic emotions (such as when goosebumps appear on the skin or when one bursts into tears while listening to a song), and (3) preference ("I love that song!"). Although the chronological succession of these responses is yet to be determined, Brattico and her team (2013) proposed that aesthetic emotions and preference follow aesthetic judgments of beauty or of other formal aspects of the music heard.

But can an objective property in music (or even in other art forms) directly engender aesthetic pleasure? How do external stimuli, affective state, and prior listening experience coalesce to create idiosyncratic pleasurable responses to music? The paradox of a stimulus which begets a positive aesthetic response despite its intrinsic negative properties exists both in music and in visual arts. Conway and Rehding (2013, p. 1) remind us that "throughout history, artists have created deeply moving artwork that is emphatically not beautiful; Goya's *Saturn Devouring One of His Sons* provides a famous historical example." For instance, a stimulus might originate unpleasant sensations at peripheral organs and hence be associated with withdrawal reactions, and yet simultaneously produce strong feelings of pleasure mediated neurally by higher-level mechanisms and psychologically by associations with positive past events or by drives for compliance with peer sets of values. Here, we explore this paradox between sensory pleasure and conscious enjoyment. In some cases the two experiences coincide; in others they are drastically opposite.

Several factors modulate the affective and evaluative responses to music, adding to the complexity of the relationship between sensory pleasure and conscious enjoyment. Juslin and Västfjäll (2008) identified six psychological mechanisms by which music induces emotions identified by brainstem reflexes, evaluative conditioning, emotional contagion, visual imagery, episodic memory, and expectancy, and according to Konecni (2008) these mechanisms can all be termed extra-musical associations. Vuust and Kringelbach (2010) listed two mechanisms in addition to extra-musical associations: hardwired responses and anticipation (which respectively are reminiscent of the brainstem reflexes and expectancy mechanisms of Juslin and Vastfjäll's proposal). The latter mechanism can be considered the only music-specific one, since it predicts that emotions derive from former knowledge of the temporal course of the music heard, acquired from previous listening. This chapter will analyse the neural mechanisms generating the aesthetic emotion of enjoyment, rather than the psychological ones. First, it will focus on the sensory mechanisms determining the succession of neural events from the periphery to the central nervous system and leading to musical enjoyment. This can be called the sensory hypothesis of musical pleasure. Next, the chapter will introduce the cognitive mechanisms determining musical pleasure that originate from prefrontal and associative cortices and modulate peripheral reactions. This can be considered the conceptual hypothesis of musical pleasure. While reviewing these mechanisms, the paradoxical case of cognitive pleasure or enjoyment that does not match sensory pleasure will be discussed. Finally, the chapter will offer the first formal description of how these two types of neural mechanisms are influenced by internal contextual factors, for example by expertise, mood, personality, and genes, and external contextual factors like the presence of peers and the environment.

15.2 The bottom-up neural route to sensory pleasure

A musical aesthetic experience might lead to conscious judgments of beauty concerning a musical piece or a particular performance. These judgments are often accompanied by hedonic feelings, although not always. A number of psycho-acoustic and behavioral studies demonstrated that some perceptual features, like sensory dissonance or consonance, innately trigger pleasant or unpleasant sensations to sounds across cultures and ages, with empirical evidence obtained even from infants as young as a few months of age (Trainor et al. 2002). Functional magnetic resonance imaging (fMRI) data with infants only a few days old also showed asymmetric activity over the supratemporal plane in the right hemisphere, where the primary auditory cortex is located, while they listened to consonant music as compared with altered dissonant music; furthermore, altered music (containing out-of-key chords) activated the ventral striatum (which consists of the nucleus accumbens and the ventromedial parts of the caudate and putamen, and in which the neural signal is transmitted mainly by using the neurotransmitter dopamine) and the left amygdala (Perani et al. 2010). These latter findings diverge from what was observed in adults in a pioneering positron emission tomography (PET) study (Blood et al. 1999): ratings of increasing dissonance were proportionally correlated with neural activity in the right parahippocampal gyrus (not the amygdala) and precuneus, whereas increasing ratings of consonance correlated with activity in the orbitofrontal cortex and subcallosal cingulate. In subsequent neuropsychological studies, patients with lesions in the parahippocampal gyrus (and not a patient with a selective amygdala lesion) exhibited reduced sensitivity to musical unpleasantness with intact emotional recognition of happiness and sadness in (consonant) music (Gosselin et al. 2006). Notably, all of these studies reviewed defined dissonance as the alteration of the chord accompaniment of a melody.

Although these previous studies seem to rule out a contribution of the amygdala in the perception of musical dissonance, others show that the amygdala does play a role in the perception of negative auditory sensations. For instance, Gosselin and colleagues (2005) provided evidence that 16 patients with unilateral (left or right) temporal lobe excisions encompassing the amygdala could not perceive fear in music as distinctly as controls could; this impairment in emotional recognition was specific to fear, since the patients' recognition of happy and sad music was normal (for a replication of these findings, see Gosselin et al. 2011). Furthermore, Kumar and colleagues (2012) measured the fMRI responses of 16 subjects listening to 74 sounds (including not just musical but also animal and environmental sounds) and rating them according to their pleasantness. Sounds varied greatly in perceived unpleasantness, from very high (e.g. chalk scratching on a blackboard, a knife scraped on a bottle) to very low (e.g. bubbling water). Results revealed the importance of the functional coupling (based on anatomical connections; LeDoux 2012) between the right basolateral nucleus of the amygdala and the left primary auditory cortex for experiencing aversive, unpleasant sensations from sounds. This finding indicates that the amygdala provides the auditory cortex with information about the pleasant or aversive quality of sounds. A recent study by Koelsch and colleagues (2013) also revealed stronger functional connectivity between the left superficial amygdala and the auditory cortex resulting from higher auditory cortex activity during positive emotions from music (joy compared to fear). Similarly, with intracranial electroencephalography (EEG) recordings on a male patient, Liégeois-Chauvel and colleagues (2014) observed activations of the right amygdala to sad and peaceful emotions induced by 20-second piano pieces and of the left amygdala to only happy pieces. Based on these findings we can venture to propose that the coupling between the auditory cortex and the amygdala represents a central neural mechanism for evoking musical emotions. In contrast, a decoupling between those areas, as observed by Liégeois-Chauvel and co-workers (2014) for angry-sounding music that was strongly disliked by the patient measured might underlie negative aesthetic experiences of music.

The lateralized activity of the amygdala according to emotional valence has previously been noticed in a meta-analysis conducted mainly on neuroimaging studies using visual stimuli (Fusar-Poli et al. 2009). However, contrarily to the data obtained with music as stimulus, the left amygdala was activated by negative emotions whereas the right amygdala was activated by positive emotions. Discrepant findings are also present within the literature and in one meta-analysis focused on the effects of gender: the left amygdala in females was recruited more for negative emotions and in males more for positive emotions (Stevens and Hamann 2012). Based on the few studies with music reviewed here, it seems that the picture for music is more straightforward than for visual stimuli. Nevertheless, the neuroimaging studies on music described above demonstrate the importance of coupling between the amygdala and auditory cortices for the elicitation of an emotional experience from music.

While the amygdala, especially in the right hemisphere, and the nearby parahippocampal gyrus might be the necessary relay stations for an aversive, unpleasant sensory experience directly deriving from the acoustic features of the music heard, the striatum, particularly the right nucleus accumbens (see Koelsch 2014) and the caudate nucleus, are the subcortical brain structures important for experiencing very pleasurable sensations from music. To study these subcortical structures, researchers have isolated chills or frissons, which are replicable and measurable bodily responses, associated with shivers down the spine or goosebumps in the skin, triggered by music or other artistic stimuli (also climax passages in a speech might trigger chills). Several studies have demonstrated that chills are associated with a strong pleasurable sensation and hence index highly positive emotions induced, for example, by music. Studying the neural correlates of the chill response to familiar, highly pleasurable music, there was evidence that these are associated with activity in reward areas of the brain, such as the ventral striatum (including the nucleus accumbens) and ventral tegmental area, which are regulated by endogenous dopamine transmission (Blood and Zatorre 2001; Pereira et al. 2011; Menon and Levitin 2005; Salimpoor et al. 2011). Furthermore, in a recent paper it was found that the bodily reactions of chills are anticipated by neural activity in the dorsal striatum (particularly, in the right caudate) followed by activation of the ventral striatum, and particularly of the nucleus accumbens, a subcortical nucleus of the mesodopaminergic pathway using mainly dopamine as neurotransmitter (Salimpoor et al. 2011). Additionally, Salimpoor and colleagues (2013) noticed that activity in the dorsal and ventral striatum was concurrent during listening to unfamiliar music, indicating temporal overlap between expectations of the music heard and its positive evaluation (as indexed by the subjects' bids mimicking realistic purchasing of a tune on the Internet). The apparent discrepancy in the chronological succession of the responses might be explained by the nature of the expectations to familiar and unfamiliar music, and by the differences in the paradigms (the second study did not measure chill responses). Indeed the basal ganglia, of which the dorsal striatum is a part, seem to have a role in the online prediction of beat and temporal regularities in music (Grahn and Rowe 2013). These studies, though, left open the question of how this pleasurable response originates, be it from bottom-up auditory processing from the periphery to the auditory cortex and then the subcortical nuclei (perhaps even bypassing the auditory cortex), or top-down from higher-order cognitive structures like the prefrontal cortex, posing attitudes and expectations that modulate peripheral responses.

15.3 The top-down neural route to conscious liking

The neural mechanisms described so far rely mainly on subcortical structures and on the auditory cortex. They might operate in a very fast manner and without conscious awareness by the subject. For instance, a very loud and rough sound triggers the acoustic startle reflex via subcortical nuclei in the brainstem (particularly the caudal pontine reticular nucleus) without fine-feature processing by the auditory cortex (Koch 1999). The passage from sensory pleasure to conscious hedonic feeling leading to the liking of a musical piece (or even a sound) involves the mediation of higher-order structures. By using apposite experimental designs it is possible in the laboratory to isolate the implicit, subconscious neural processes related to an emotional stimulus from those related to its conscious processing. Here we focus on implicit versus explicit tasks that have been used first in visual research. In a recent study, we contrasted individuals who preferred to listen to (and are fans of) the musical genre heavy metal with those individuals who enjoyed listening mainly to the musical genre Latin American music. These preferences might be considered opposites, since fans of one are often critical of the other. In a recent EEG study we took advantage of this feature of musical preferences with a between-subjects design in which individuals with clear preferences for heavy metal or Latin American music listened to excerpts of musical tunes belonging to one of those genres (Istók et al. 2013). The findings evidenced that the late positive event-related brain responses to the music stimuli from the favorite genre were enhanced compared to those to the stimuli from the disliked genre, but only when subjects were performing a descriptive task of classifying the timbre and not when they were consciously deciding whether or not they liked the music. This counterintuitive effect was explained by the authors of the study as due to the phase-locking of evoked neural responses, which highlights the preference modulation of auditory processing (locked to the onset of the stimuli) but not the more time-varying evaluative processing associated with conscious liking judgments. In other words, the hedonic responses to favorite music influence perceptual and cognitive processing of music in an implicit way, even when attention is focused on performing a non-affective task.

According to some authors (for a review, see Hargreaves and North 2010), arousal and valence (or sensory consonance/dissonance) are central components of an aesthetic experience. Here, however, we consider these—particularly valence, conceived as sensory pleasure—as only subcomponents of the musical aesthetic experience. This is similar to how the psychological constructionist account of the brain basis of emotion considers valence and arousal, the two dimensions of "core affect," realized by the visceral and peripheral sensory systems and defined as the "mental representation of the bodily changes that are sometimes (but not always) experienced as feelings of hedonic pleasure and displeasure with some degree of arousal" (Lindquist et al. 2012, p. 125). "Core affect" is thus viewed as an initial, mainly unconscious stage of affective processing, giving emotional nuance to stimuli and events and largely utilizing subcortical brain structures related to the limbic system. That nuance becomes a conscious emotion, such as sadness, happiness, and conscious enjoyment, only after a conceptual act requiring language, executive attention, episodic memory, and categorization processes, and is controlled mainly by frontal and parietal lobe structures. For instance, the dorsomedial and ventromedial prefrontal cortex, the retrosplenial cortex/posterior cingulate cortex, and the medial temporal lobe are all active when categorizing an emotion or focusing on the emotional content of a stimulus. In the framework proposed by Lindquist and colleagues, these brain structures are hypothesized to store representations of prior experiences to attribute a meaning to a core affective state coming from the self or from others. In turn, the anterior temporal lobe and the ventrolateral prefrontal cortex are involved in conceptualization since they control language functions, which are necessary for categorizing an emotion. The dorsolateral prefrontal cortex, associated with goal-directed control of attention and working memory, is instead recruited for explicitly holding affective information about a stimulus in mind in order to categorize it. In the current proposal, sensory pleasure is viewed as an early affective reaction to music similar to "core affect." To become a conscious emotion of enjoyment and engender a liking judgment for a musical piece, sensory pleasure must be followed by value attribution mediated by personal associations, knowledge, social constructs, and other top-down processes, and all controlled by the prefrontal cortex and associative temporoparietal structures of the brain. Salimpoor and colleagues (2013) provided the first direct evidence of the interplay between frontal and subcortical structures originating musical pleasure. They revealed that the connectivity between the nucleus accumbens and the auditory cortex represents the core neural mechanism predicting a rewarding response to music, indexed by the highest bid given during an auction task. The authors also observed increased connectivity between the nucleus accumbens and the inferior prefrontal cortex, serving to explain the role played by matched or mismatched temporal expectations for harmonic structures in the generation of musical pleasure.

An explanatory construct, not much explored within the domain of music emotions, is represented by cognitive dissonance (not to be confused with musical dissonance), which is a very influential theory in social psychology. According to this theory, cognitive dissonance takes place when our actions are in conflict with our prior attitudes and the latter hence change to match the former, avoiding the negative affect resulting from the cognitive conflict. In a study by van Veen and colleagues (2009), recruitment of the dorsal anterior cingulate cortex and the anterior insula were found to predict the subsequent changes in attitude by the subjects (in this case, a positive change towards the experimental setting of the fMRI scanner). Cognitive dissonance might also be a psychological mechanism for pleasure derived from the resolution of mismatching expectations while listening to music. A rare example of a study focusing on the effects of attitudes on auditory processing of the same musical stimuli was conducted by Steinbeis and Koelsch (2009). In this block design study, participants were convinced that 60 excerpts of musical pieces lasting from 8 to 13 seconds (originally composed by Schönberg and Webern) were written either by a composer or by a computer while their fMRI signal was measured. When the piece was thought to be the product of a composer, the anterior medial frontal cortex, the superior temporal sulcus, and the temporal poles—namely those brain structures that are typically used when subjects attribute a mental state to another human—were active. Notably, the strength of the activity in the anterior medial frontal cortex even correlated positively with the extent to which the participants thought an intention was being expressed in the musical pieces (as reported in a post-experimental questionnaire). Hence, when an intentional stance is attributed to a musical piece, anterior frontal structures are recruited and assist the listener in interpretation. One could assume that such a search for an intention in the composition is an important stage towards a complete aesthetic experience of music, including a conscious decision of liking.

A peculiar case of discrepancy between "sensory" pleasure and "conscious" enjoyment occurs when the hedonic response derives from understanding the formal structure of a piece, for example, of contemporary classical music (think of *Sonata No. 1* by Pierre Boulez composed according to the principles of serial music). The affective sensations deriving from the acoustic processing of such a piece might be negative, with strong activity in the parahippocampal gyrus and the (right) amygdala, and yet a listener who is able to identify the repetitive notes and to identify the neatness of the construction might still give a positive evaluative judgment of liking and even have hedonic responses with neural activity in the reward system or at least the drive to listen to the same piece again. In the domain of abstract visual art, Leder and colleagues (2004) named this psychological process "cognitive mastering," a crucial stage of the information processing sequence leading to the aesthetic outcomes of judgments and emotions. Perlovsky (2010a, 2010b) also talks of "knowledge instinct" as a mechanism generating the aesthetic pleasure which stems from knowledge and understanding. According to this theory, aesthetic emotions are viewed as needs of the mind to understand the ever-changing world by thinking conceptually whereas basic emotions are considered as bodily instincts. The cognitive consonance or dissonance between the knowledge and the world generates aesthetic emotions of satisfaction or dissatisfaction.

Finally, as discussed earlier, cognitive mastering might change the affective state of an individual when facing a musical (or other artistic) activity. Indeed, familiar music has been shown to activate more limbic subcortical areas than unfamiliar music and even liked music. The positive affect derived from understanding might be related to such a brain response. Overall, in the study by Pereira and colleagues (2011), subjects liked familiar music more than unfamiliar music. In another fMRI study by Chapin and colleagues (2010) investigating expressivity in music with a real performance of a piano piece and a digital one (produced with Musical Instrument Digital Interface or MIDI), musicians showed higher activity in the ventral striatum than non-musicians while listening to an expressive musical performance. In other words, the study indicated that musicians are better able to master expectations of expressivity and consequently identify the agogic and dynamic variations in a musical performance, which are typically aimed at conveying emotions. This is still an isolated finding, since most previous studies on music expertise have concentrated on cognitive hypotheses rather than on affective ones, but the finding of modulation of limbic and reward activity is expected to be replicated if the hypothesis that cognitive mastering produces positive affect is valid.

15.4 The role of context: social and person-related factors

The aesthetic emotion of musical enjoyment is, however, much more variable across individuals than the mechanistic view of how sensory pleasure arises would predict—that neural mechanisms causally link a stimulus with certain acoustic features (e.g. a dissonant sound) to a physiological response, namely the recruitment of parahippocampal gyrus, amygdala, and ventral striatum, and their connectivity with auditory cortex. For instance, in the previous section, it was demonstrated that "sensory" pleasure might clash with "cognitive" pleasure, such as when one receives very positive feelings from listening to the very dissonant and loud music of a death metal band. The origins of this paradox might be related to the person or to the social and external context. The physical environment during a musical activity might by itself determine the emotional responses, including enjoyment. For instance, the proximity of a friend while attending a concert, and noticing her facial expressions and bodily reactions while listening to music might trigger mimicry, thanks to the properties of the mirror neuron system (Gallese and Freedberg 2007), and generate similar emotions via the psychological mechanism of emotional contagion. An alternative neural mechanism that has been proposed to underlie emotional contagion is the temporal synchronization of neural activity in the same brain structures between individuals experiencing the same stimulus, such as a movie or a musical piece (Nummenmaa et al. 2012; see also Alluri et al. 2012). The most recent demonstration of the effects of the presence of peers on an aesthetic response to music has been thus far produced by Egermann and colleagues (2011) using 14 members of an amateur orchestra. By using skin conductance responses, questionnaires, and ten one-minute classical music excerpts, they reported that both the intensity of emotions prompted by the music and the frequency of chills experienced (accompanied by a change in skin conductance) were higher when participants listened to the music alone than when in a group. This study did not support the hypothesis of social facilitation by presence of peers, which instead found validation in Lamont's study (2011), which was not restricted to the classical music genre; in this case, strong experiences of music occurred most typically during a live concert with other listeners present.

An important social psychological mechanism affecting musical enjoyment is that of compliance, the desire to comply with what peers do. In a web-based study with more than

5000 participants, Egermann and colleagues (2013) noticed that online music listeners changed their ratings of arousal more than their ratings of valence for five music excerpts of various styles (lasting 30 seconds each) based on social feedback in the form of information from previous music listeners (represented in this experiment by the position of the rating sliders during listening and/or rating). Additionally, the authors included a control condition with a second group of participants where the feedback came directly from the computer predicting the rating responses of the participants. The findings showed that social feedback was again more effective than informational feedback on changing arousal ratings, indicating that the emotional effects of music might be modulated by normative and social drives to conform to the majority. According to the seminal research by Asch (1955), this type of conformity based on normative influence does not necessarily change one's own attitudes or perception but only the public compliance. According to Egermann and colleagues (2013), their results diverge from this prediction since they showed changes of felt excitement deriving particularly from social feedback. However, it is not known how long-lasting this effect is nor whether it involves physiological emotional responses as well as changes in the activity of the limbic system.

In adolescence, the effects of these social factors (and possibly even others, such as the construction of self-identity through specific behavior) on various kinds of behavior seem to be boosted. In a behavioral study it was found that adolescents took three times more risks in a driving game when in the presence of their friends than when performing the game alone; in contrast, peer presence had no influence on risk-taking in the same driving game in adults (Gardner and Steinberg 2005). A follow-up study utilizing fMRI further revealed that adults exhibited higher activity in the lateral prefrontal cortex during the driving game than adolescents whereas adolescents showed higher activity in the reward structures of the ventral striatum and the orbitofrontal cortex during driving decisions in the presence of their friends than when alone (Chein et al. 2011). Again, interconnectivity between striato-frontal structures might be at the core of the neural mechanism responsible for sensory and conscious musical pleasure and preference.

A pioneering study by Berns and colleagues (2010) asked subjects to rate the liking of songs twice, once just after the first listening and then a second time, with or without popularity ratings shown for each song. Subjects changed their liking ratings by 12 per cent in their second listening and by 22 per cent when popularity ratings were shown. FMRI data showed that activity in the bilateral anterior insula and frontal poles was positively correlated with popularity scores, whereas negative correlations were found in the posterior cingulate cortex. Hence, popularity ratings did not affect conscious enjoyment (called "intrinsic preferences" by the authors of the study), since no modulation of reward circuits, such as ventral striatum and orbitofrontal cortex, was found. Instead, the drive to match social consensus was associated with a change in internal arousal states, suggesting neural costs in changing prior judgments to conform to those of peers. In other words, the psychological mechanism of cognitive dissonance might have been at play here.

In psychology, there is accumulating evidence for a relationship between the aesthetic response of musical preference and person-related factors. For instance, by using a questionnaire survey of about 3500 individuals, Rentfrow and Gosling (2003) noticed that the four main types of preferred music were associated with the personality traits of the Big Five questionnaire (Extroversion, Agreeableness, Conscientiousness, Emotional Stability, and Openness), self-views, and cognitive ability. However, more recently these associations of musical preferences with personality traits were proven to be not so robust whereas a larger proportion of observed variance was explained by gender and age; for instance, older individuals preferred increasingly complex music, females showed higher liking for classical, mainstream, folk, alternative rock, Latino, and music of black origin (MOBO), and men for jazz, dance, rock, and functional music (North 2010).

Even the induction of strong musical pleasure can be modulated by internal context. McCrae (2007) found that Item 188 of the NEO Personality Inventory, which reads: "Sometimes when I am reading poetry or looking at a work of art, I feel a chill or wave of excitement," was found to be the best indicator of final "Openness to Experience" scores in American individuals. Openness is conceived as "the breadth, depth, and permeability of consciousness, and in the recurrent need to enlarge and examine experience" (McCrae and Costa 1997, p. 826); individuals who score high in Openness are curious, innovative, willing to create unusual associations, and also sensitive to art and beauty. Additionally, Grewe and colleagues (2007) found an association between chill experience and Reward Dependence (Cloninger 1994), a scale related to Agreeableness in the NEO Personality Inventory. According to Panksepp (1995, 2009–2010), the reason why individuals who experience more chills are also those with high sensitivity and a tendency towards nurturing and agreeable behavior is to be found in the evolutionary origin of the chill response, which in his opinion must be related to the sensation (and its concurrent strong emotion) of physical separation of the offspring from the mother.

In the field of neuroimaging, the relations between personality and brain responses to music (and even more so to music emotions) have been little explored thus far. Montag and colleagues (2011) were among the first to explore a link between personality traits and brain responses to musical pleasure with fMRI. They asked 33 subjects to bring their favorite and most disliked musical pieces to the lab and then listen to 3 minutes of each song during an fMRI session. As expected from previous studies, favorite music activated the ventral striatum (caudate and nucleus accumbens) as well as the insula and the cuneus. Remarkably, the activity in the nucleus accumbens negatively correlated with self-transcendence, a subscale of the Temperament and Character Inventory (TCI), which explained 25 per cent of its activity. Tellegen absorption scores were also negatively correlated with striatal activity, but did not substantially improve the model when included in the same model as self-transcendence. The growing behavioral evidence supporting the relevance of person-related and social factors in affecting emotional responses to music calls for further neuroimaging investigations to identify the neural underpinnings of these effects.

15.5 The role of genes in idiosyncratic aesthetic emotions

In the broader field of affective neuroscience, variability in emotional responses has often been regarded as random noise or else has substantiated relativistic views of emotion processing (like the conceptual act model), reducing any emotional experience to an emergent state associated with dynamically varying (and hard to measure) neural networks and to a variety of cognitive functions (Lindquist et al. 2012). In the music domain, Konecni (2008, p. 118) pointed out the following methodological problem: "the confusion of the music's expressiveness and induced emotion goes further and enters the instructions to the research participants." According to this author, such confusion not only weakens the efficacy of data collection but also lead to the formation of unsubstantiated theories of musical emotions. The picture becomes even more complex when considering the neuroimaging literature, exhibiting discrepancies in the brain structures associated with certain musical emotions, and particularly with sadness and happiness responses to music (Mitterschiffthaler et al. 2007; Brattico et al. 2011; Khalfa et al. 2005). For musical pleasure, the picture seems to be more homogenous with the ventral striatum, anterior cingulate cortex, and orbitofrontal cortex often recruited. However, the approaches for studying the neural correlates of musical pleasure are varied. Instructions can range from ratings of pleasantness (Koelsch et al. 2006) to ratings of likeability (Montag et al. 2011), or brain responses could be selected based on coinciding bodily changes (Salimpoor et al. 2011), leading to some inconsistency in the conclusions that can be drawn from these studies. For instance, chills can be used as markers for high musical pleasure (Grewe et al. 2009), but they occur in a minority of the population and involve substantial bodily reactions whose neural generators have not yet been directly identified.

However, taking into consideration that music-related emotions, like enjoyment, are highly variable and subjective, this idiosyncrasy cannot be ascribed only to the extra-musical associations and the mutable (hard to predict) life circumstances determining the internal and external contexts of a musical experience, nor to the discrepancies between stimulation paradigms and experimental procedures. Low-level sensory processes already vary between individuals and within the same individual in quite a stable way due to inborn, genetic predispositions for micro- and macro-structural physiological properties of critical brain regions such as the auditory cortex and amygdala, which are both crucial for musical emotions. The novel avenue of imaging genetics permits the tracing of variability in emotions and their neural correlates to an individual's biology, to gender, personality traits, and genetic variants (Munoz et al. 2009). For instance, the functionality of the amygdala has been linked to genetic polymorphisms—for example, in serotonin-related (Hariri et al. 2002) or in dopamine-related genes (Blasi et al. 2009)—as well as correlated to personality traits such as Emotional Stability or Openness to Experience (Pecina et al. 2013). In a study by Blasi and colleagues (2009), the functional DRD2 rs1076560 polymorphism was associated with greater amygdala activity during the implicit processing of facial expressions and greater dorsolateral prefrontal cortex activity during explicit emotional processing. In addition, Hariri and colleagues (2002) discovered that individual variability in the processing of fearful stimuli was associated with a functional polymorphism of the serotonin transporter (5-HTT) gene and with amygdala reactivity, a finding at least partially replicated in recent meta-analyses (Murphy et al. 2013). Consequently, the individual variations in mesolimbic and cortico-striato-cortical circuits during emotion processing might not merely be the result of measurement noise, experimental protocols, or contextual factors, but might systematically result from dopaminergic or serotonergic genes, for example. Those predispositions might in turn explain part of the behavior involving experiencing strong pleasurable emotions derived from music.

Furthermore, the human population contains a small percentage of people with a peculiar inborn and hereditable disorder specific to music, named "tone deafness" (implying, for instance, the inability to recognize even famous tunes) or congenital amusia (Peretz et al. 2002, 2005, 2009), accompanied by normal cognitive functioning and other auditory and speech abilities. This disorder, originating from dysfunctional connectivity between the auditory cortex and the inferior prefrontal cortex, hints at the genetic foundations of brain-based music functions. Recent genetic studies involving twins and families indeed revealed that there exist predispositions for musical abilities enabling some individuals to be more or less hardwired for musical activities than others (Ukkola et al. 2009; Ukkola-Vuoti et al. 2011, 2013; for a review, see Levitin 2012). Such candidate genes with variations associated with superior musical skills were the arginine vasopressin receptor gene AVPR1A (associated with scores in an audio-structuring test and listening behavior), as well as its epistatis (interaction) with the human serotonin transporter gene SLC6A4, UNC5C (adenylate cyclase 8) connected to synaptic plasticity and highly expressed in the central nervous system gene UGT8 (Ukkola-Vuoti et al. 2013; Park et al. 2012). With a complex behavioral trait like musicality, more than one gene is possibly in play ("music is likely to be the result of thousands of small-effect loci, which together can produce significant heritability quotients," p. 636), as Levitin recently proposed (Levitin 2012). However, even if musicality skills have been correlated with the morphology and functionality of Heschl's gyrus (Schneider et al. 2002), thus far no study has directly tested any links between the auditory system or limbic system functions and genotypes previously associated with musicality skills. Such links are of particular interest here because they would establish whether genes predispose an individual for musicality merely through boosting perceptual skills via the enhancement of neural discrimination abilities in the primary auditory cortex and/or via increasing the functional connectivity between the auditory cortex and the subcortical emotion-related structures (such as the amygdala, nucleus accumbens and the basal ganglia), as studies on musical emotions seem to indicate.

15.6 **Conclusion**

This chapter reviewed brain-based evidence supporting two hypotheses previously tested only within the domain of visual arts (e.g. Cupchik et al. 2009; Thakral et al. 2012): first, the sensory hypothesis founding a pleasurable experience of music in bottom-up mechanisms, and second, the conceptual hypothesis centering the aesthetic enjoyment of music on prefrontal structures and the internal and social factors modulating them. To borrow Thakral and colleagues' words (2012):

> this aesthetic experience [of beauty] has been hypothesized to intimately depend on visual sensory details in the art, such as the wavy lines and colors that constitute the modern painting Stairway at

Auvers by van Gogh. It has also been hypothesized that aesthetic experience depends on concepts or thoughts associated with viewing art to a greater degree than visual sensory details, such as a contemporary sculpture Brillo Soap Pads Box by Andy Warhol that is nearly identical—perceptually—to those that were once found on supermarket shelves.

In the current chapter, the cognitive routes in the brain leading to the conscious enjoyment of music have been illustrated from the perspective of both internal and social factors. These factors can sometimes be so powerful as to detach the sensory response coming from the inner ear and reaching the brainstem and the amygdala/parahippocampal gyrus (e.g. loudness or sensory dissonance) from the conscious enjoyment mediated by internal and social determinants, and accompanied by the tendency to an approach action exemplified by the drive to listen repeatedly to the same musical piece in spite of the aversive, almost painful, sensations it induces. Rock music concerts, often loud enough to cause tinnitus in the attendees (e.g. Bogoch et al. 2005), provide an example of stimulating this behavior. Future research should pay more attention to the underlying social and internal mechanisms generating emotional reactions to music and to how they change our brain processing of it.

References

Alluri, V., Toiviainen, P., Jaaskelainen, I.P., et al. (2012). Large-scale brain networks emerge from dynamic processing of musical timbre, key and rhythm. *NeuroImage* **59**, 3677–789.

Asch, S.E. (1955). Opinions and social pressure. *Scientific American* **193**, 1–7.

Berns, G.S., Capra, C.M., Moore, S., et al. (2010). Neural mechanisms of the influence of popularity on adolescent ratings of music. *NeuroImage* **49**, 2687–96.

Blasi, G., Lo Bianco, L., Taurisano, P., et al. (2009). Functional variation of the dopamine D2 receptor gene is associated with emotional control as well as brain activity and connectivity during emotion processing in humans. *Journal of Neuroscience* **29**, 14812–9.

Blood, A.J. and Zatorre, R.J. (2001). Intensely pleasurable responses to music correlate with activity in brain regions implicated in reward and emotion. *Proceedings of the National Academy of Sciences* **98**, 11818–23.

Blood, A.J., Zatorre, R.J., Bermudez, P., et al. (1999). Emotional responses to pleasant and unpleasant music correlate with activity in paralimbic brain regions. *Nature Neuroscience* **2**, 382–7.

Bogoch, I.I., House, R.A., and Kudla, I. (2005). Perceptions about hearing protection and noise-induced hearing loss of attendees of rock concerts. *Canadian Journal of Public Health* **96**, 69–72.

Brattico, E., Alluri, V., Bogert, B., et al. (2011). A functional MRI study of happy and sad emotions in music with and without lyrics. *Frontiers in Psychology* **2**, 308. doi: 10.3389/fpsyg.2011.00308.

Brattico, E., Bogert, B., and Jacobsen, T. (2013). Toward a neural chronometry for the aesthetic experience of music. *Frontiers in Psychology* **4**, 206.

Brattico, E. and Pearce, M.T. (2013). The neuroaesthetics of music. *Psychology of Aesthetics, Creativity, and the Arts* **7**, 48–61.

Chapin, H., Jantzen, K., Kelso, J.A., et al. (2010). Dynamic emotional and neural responses to music depend on performance expression and listener experience. *PLoS One* **5**, e13812.

Chein, J., Albert, D., O'Brien, L., et al. (2011). Peers increase adolescent risk taking by enhancing activity in the brain's reward circuitry. *Developmental Science* **14**, F1–10.

Cloninger, C.R. (1994). Temperament and personality. *Current Opinion in Neurobiology* **4**, 266–73.

Conway, B.R. and Rehding, A. (2013). Neuroaesthetics and the trouble with beauty. *PLoS Biology* **11**, e1001504.

Cupchik, G.C., Vartanian, O., Crawley, A., et al. (2009). Viewing artworks: contributions of cognitive control and perceptual facilitation to aesthetic experience. *Brain and Cognition* **70**, 84–91.

Egermann, H., Kopiez, R., and Altenmüller, E. (2013). The influence of social normative and informational feedback on musically induced emotions in an online music listening setting. *Psychomusicology. Music, Mind, and Brain* **23**, 21–32.

Egermann, H., Sutherland, M.E., Grewe, O., et al. (2011). The influences of a group setting on the experience of music: A physiological and psychological perspective on emotion. *Musicae Scientiae* **15**, 307–23.

Fusar-Poli, P., Placentino, A., Carletti, F., et al. (2009). Laterality efect on emotional faces processing: ALE meta-analysis of evidence. *Neuroscience Letters* **452**, 262–67.

Gallese, V. and Freedberg, D. (2007). Mirror and canonical neurons are crucial elements in esthetic response. *Trends in Cognitive Science* **11**, 411.

Gardner, M. and Steinberg, L. (2005). Peer influence on risk taking, risk preference, and risky decision making in adolescence and adulthood: an experimental study. *Developmental Psychology* **41**, 625–35.

Gosselin, N., Peretz, I., Hasboun, D., et al. (2011). Impaired recognition of musical emotions and facial expressions following anteromedial temporal lobe excision. *Cortex* **47**(9), 1116–25.

Gosselin, N., Peretz, I., Noulhiane, M., et al. (2005). Impaired recognition of scary music following unilateral temporal lobe excision. *Brain* **128**, 628–40.

Gosselin, N., Samson, S., Adolphs, R., et al. (2006). Emotional responses to unpleasant music correlates with damage to the parahippocampal cortex. *Brain* **129**, 2585–92.

Grahn, J.A. and Rowe, J.B. (2013). Finding and feeling the musical beat: striatal dissociations between detection and prediction of regularity. *Cerebral Cortex* **23**, 913–21.

Grewe, O., Kopiez, R., and Altenmueller, E. (2009). The chill parameter: Goose bumps and shivers as promising measures in emotion research. *Music Perception* **27**, 61–74.

Grewe, O., Nagel, F., Kopiez, R., et al. (2007). Listening to music as a re-creative process: Physiological, psychological, and psychoacoustical correlates of chills and strong emotions. *Music Perception* **24**, 297–314.

Hargreaves, D.J. and North, A.C. (2010). Experimental aesthetics and liking for music. In P.N. Juslin and J.A. Sloboda (eds), *Handbook of Music and Emotion.* Oxford: Oxford University Press, pp. 513–46.

Hariri, A.R., Mattay, V.S., Tessitore, A., et al. (2002). Serotonin transporter genetic variation and the response of the human amygdala. *Science* **297**, 400–3.

Istok, E., Brattico, E., Jacobsen, T., et al. (2013). "I love Rock 'n' Roll"—music genre preference modulates brain responses to music. *Biological Psychology* **92**, 142–51.

Juslin, P.N. and Västfjäll, D. (2008). Emotional responses to music: The need to consider underlying mechanisms. *Behavioral and Brain Sciences* **31**, 559–621.

Khalfa, S., Schon, D., Anton, J.L., et al. (2005). Brain regions involved in the recognition of happiness and sadness in music. *NeuroReport* **16**, 1981–4.

Koch, M. (1999). The neurobiology of startle. *Progress in Neurobiology* **59**, 107–28.

Koelsch, S. (2014). Brain correlates of music-evoked emotions. *Nature Reviews Neuroscience* **15**, 170–80.

Koelsch, S., Fritz, T., von Cramon, D.Y., et al. (2006). Investigating emotion with music: an fMRI study. *Human Brain Mapping* **27**, 239–50.

Koelsch, S., Skouras, S., Fritz, T., et al. (2013). The roles of superficial amygdala and auditory cortex in music-evoked fear and joy. *Neuroimage* **81C**, 49–60.

Konecni, V.J. (2008). Does music induce emotion? A theoretical and methodological analysis. *Psychology of Aesthetics, Creativity, and the Arts* **2**, 115–29.

Kumar, S., von Kriegstein, K., Friston, K., et al. (2012). Features versus feelings: dissociable representations of the acoustic features and valence of aversive sounds. *Journal of Neuroscience* **32**, 14184–92.

Lamont, A. (2011). University students' strong experiences of music: Pleasure, engagement, and meaning. *Musicae Scientiae* **15**, 229–49.

Leder, H., Belke, B., Oeberst, A., et al. (2004). A model of aesthetic appreciation and aesthetic judgments. *British Journal of Psychology* **95**, 489–508.

LeDoux, J. (2012). Rethinking the emotional brain. *Neuron* **73**, 653–76.

Levitin, D.J. (2012). What does it mean to be musical? *Neuron* **73**, 633–7.

Liégeois-Chauvel, C., Bénar, C., Krieg, J., et al. (2014). How functional coupling between the auditory cortex and the amygdala induces musical emotion: A single case study. *Cortex* **60**, 82–93.

Lindquist, K.A., Wager, T.D., Kober, H., et al. (2012). The brain basis of emotion: a meta-analytic review. *Behavioral and Brain Sciences* **35**, 121–43.

McCrae, R.R. (2007). Aesthetic chills as a universal marker of openness to experience. *Motivation and Emotion* **31**, 5–11.

McCrae, R.R. and Costa, P.T. (1997). Conceptions and correlates of openness to experience. In R. Hogan and J.A. Johnson (eds), *Handbook of Personality Psychology*. San Diego, CA: Academic Press, pp. 825–47.

Menon, V. and Levitin, D.J. (2005). The rewards of music listening: response and physiological connectivity of the mesolimbic system. *NeuroImage* **28**, 175–84.

Mitterschiffthaler, M.T., Fu, C.H., Dalton, J.A., et al. (2007). A functional MRI study of happy and sad affective states induced by classical music. *Human Brain Mapping* **28**, 1150–62.

Montag, C., Reuter, M., and Axmacher, N. (2011). How one's favorite song activates the reward circuitry of the brain: personality matters! *Behavioral Brain Research* **225**, 511–4.

Munoz, K.E., Hyde, L.W., and Hariri, A.R. (2009). Imaging genetics. *Journal of the American Academy of Child and Adolescent Psychiatry* **48**, 356–61.

Murphy, S.E., Norbury, R., Godlewska, B.R., et al. (2013). The effect of the serotonin transporter polymorphism (5-HTTLPR) on amygdala function: a meta-analysis. *Molecular Psychiatry* **18**, 512–20.

North, A.C. (2010). Individual differences in musical taste. *American Journal of Psychology* **123**, 199–208.

Nummenmaa, L., Glerean, E., Viinikainen, M., et al. (2012). Emotions promote social interaction by synchronizing brain activity across individuals. *Proceedings of the National Academy of Sciences* **109**, 9599–604.

Panksepp, J. (1995). The emotional sources of "chills" induced by music. *Music Perception* **13**, 171–207.

Panksepp, J. (2009–10). The emotional antecedents to the evolution of music and language. *Musicae Scientiae* Special issue 2009–2010, 229–59.

Park, H., Lee, S., Kim, H.J., et al. (2012). Comprehensive genomic analyses associate UGT8 variants with musical ability in a Mongolian population. *Journal of Medical Genetics* **49**, 747–52.

Pecina, M., Mickey, B.J., Love, T., et al. (2013). DRD2 polymorphisms modulate reward and emotion processing, dopamine neurotransmission and openness to experience. *Cortex* **49**, 877–90.

Perani, D., Saccuman, M.C., Scifo, P., et al. (2010). Functional specializations for music processing in the human newborn brain. *Proceedings of the National Academy of Sciences* **107**, 4758–63.

Pereira, C.S., Teixeira, J., Figueiredo, P., et al. (2011). Music and emotions in the brain: familiarity matters. *PLoS One* **6**, e27241.

Peretz, I., Ayotte, J., Zatorre, R.J., et al. (2002). Congenital amusia: A disorder of fine-grained pitch discrimination. *Neuron* **33**, 185–91.

Peretz, I., Brattico, E., Jarvenpaa, M., et al. (2009). The amusic brain: in tune, out of key, and unaware. *Brain* **132**, 1277–86.

Peretz, I., Brattico, E., and Tervaniemi, M. (2005). Abnormal electrical brain responses to pitch in congenital amusia. *Annals of Neurology* **58**, 478–82.

Perlovsky, L. (2010). Musical emotions: functions, origins, evolution. *Physics of Life Reviews* **7**(1), 2–27.

Perlovsky, L.I. (2010). Neural mechanisms of the mind, Aristotle, Zadeh, and fMRI. *IEEE Transactions on Neural Network* **21**, 718–33.

Rentfrow, P.J. and Gosling, S.D. (2003). The do re mi's of everyday life: The structure and personality correlates of music preferences. *Journal of Personality and Social Psychology* **84**, 1236–56.

Saarikallio, S. and Erkkila, J. (2007). The role of music in adolescents' mood regulation. *Psychology of Music* **35**, 88–109.

Salimpoor, V.N., Benovoy, M., Larcher, K., et al. (2011). Anatomically distinct dopamine release during anticipation and experience of peak emotion to music. *Nature Neuroscience* **14**, 257–62.

Salimpoor, V.N., van den Bosch, I., Kovacevic, N., et al. (2013). Interactions between the nucleus accumbens and auditory cortices predict music reward value. *Science* **340**, 216–9.

Schneider, P., Scherg, M., Dosch, H.G., et al. (2002). Morphology of Heschl's gyrus reflects enhanced activation in the auditory cortex of musicians. *Nature Neuroscience* **5**, 688–94.

Sloboda, J.A. (2010). Music in everyday life: The role of emotions. In P.N. Juslin and J.A. Sloboda (eds), *Handbook of Music and Emotion: Theory, Research, Applications*. New York, NY: Oxford University Press, pp. 493–514.

Steinbeis, N. and Koelsch, S. (2009). Understanding the intentions behind man-made products elicits neural activity in areas dedicated to mental state attribution. *Cerebral Cortex* **19**, 619–23.

Stevens, J.S. and Hamann, S. (2012). Sex differences in brain activation to emotional stimuli: a meta-analysis of neuroimaging studies. *Neuropsychologia* **50**, 1578–93.

Thakral, P.P., Moo, L.R., and Slotnick, S.D. (2012). A neural mechanism for aesthetic experience. *NeuroReport* **23**, 310–3.

Trainor, L.J., Tsang, C.D., and Cheung, V.H.W. (2002). Preference for sensory consonance in 2- and 4-month-old infants. *Music Perception* **20**, 187–94.

Ukkola, L.T., Onkamo, P., Raijas, P., et al. (2009). Musical aptitude is associated with AVPR1A-haplotypes. *PLoS One* **4**, e5534.

Ukkola-Vuoti, L., Kanduri, C., Oikkonen, J., et al. (2013). Genome-wide copy number variation analysis in extended families and unrelated individuals characterized for musical aptitude and creativity in music. *PLoS One* **8**, e56356.

Ukkola-Vuoti, L., Oikkonen, J., Onkamo, P., et al. (2011). Association of the arginine vasopressin receptor 1A (AVPR1A) haplotypes with listening to music. *Journal of Human Genetics* **56**, 324–9.

van Veen, V., Krug, M.K., Schooler, J.W., et al. (2009). Neural activity predicts attitude change in cognitive dissonance. *Nature Neuroscience* **12**, 1469–74.

Vuust, P. and Kringelbach, M.L. (2010). The pleasure of making sense of music. *Interdisciplinary Science Reviews* **35**, 166–82.

Chapter 16

Effects of expertise on the cognitive and neural processes involved in musical appreciation

Marcus T. Pearce

16.1 Introduction

There is relatively little empirical research that specifically addresses the question of expertise effects on musical appreciation, but an anecdote will serve to make the importance of the issue clear. Recently, I attended a concert with a friend and, afterwards, discussing the performance, mentioned that I had found the timing a little erratic. My companion replied, "Well, I'm no expert but I enjoyed the concert!" Is musical training or expertise important for the appreciation of music? Or can anyone appreciate music? Does it depend on the type of music? If training is important, which aspects are crucial? Which components of aesthetic experience do those aspects of training affect? These questions have deep implications for the importance of musical training and experience in education and development of artistic appreciation. The goal of this chapter is to establish a framework for generating and testing hypotheses about the effects of musical expertise on the appreciation of music.

Before we embark, it's worth pausing to consider a few important methodological points (see Schellenberg 2006). The vast majority of psychological and neuroscientific research on the effects of musical training has taken a cross-sectional approach by comparing musicians and non-musicians on some particular measure. However, this does not demonstrate a *causal* effect of musical training, since it cannot be known whether the measure in question differed between the groups before commencement of musical training (it may even have influenced the initiation and maintenance of musical training leading to expertise). Second, it is difficult to know how specific the effect is to musical training per se unless a control group of experts in another domain are also studied (e.g. Dick et al. 2011; Schellenberg 2004), which is rare in the literature. Third, there are many demographic factors (e.g. socio-economic status) that correlate with musical education, so it is difficult to isolate any effects of musical training unless control groups matched for these factors are included (usually they are not). With those caveats in mind, we begin our enquiry by locating music and musical expertise in more precise terms.

16.2 What is music?

Music is a complex phenomenon that belies definition, leading Edgard Varèse to describe his own compositions very generally as *organized sound* (Varèse and Wen-chung 1966). Virtually omnipresent across cultures, music consists of patterns of human-initiated sound[1] extended in time, sometimes associated with visual displays, and used in a bewildering variety of contexts for an enormous range of purposes.

We might have more success if we follow a cognitive-scientific approach and attempt to pin down the ways in which a musical object may be instantiated in different kinds of representation. Music can be represented in at least three ways (see Babbitt 1965):

1 physically, as an acoustic phenomenon (traditionally resulting from human movement);
2 in symbolic encodings usually designed for recording compositions and giving directions for their performance (e.g. scores, vocal representations, piano-roll representations in sequencers and other software);
3 in various kinds of cognitive and neural representation (e.g. procedural memory, perceptual, and cognitive representations).

Representations of music may vary across multiple dimensions: pitch, timbre, timing, duration, loudness, spatial location, and other more complex characteristics relating to harmonic movement, metrical structure, and overall form. In trying to identify "objective" properties of musical sounds, one quickly finds oneself moving from the acoustic domain, represented by the first category above, into the psychological domain, represented by the third category.[2] Meter, for example, has no well-defined interpretation as an objective property of sounds, without a (cognitive) model to interpret metrical structure from those sounds.

Nonetheless, this characterization of music is useful for our present purposes since it allows us to identify different kinds of expertise afforded by the phenomenon of music. A given individual might not have expertise in all three domains: a piano tuner, for example, is an expert primarily in the acoustic domain while a medieval scribe would have been an expert in the symbolic domain of music notation. However, it is difficult to distinguish music from non-music unless one considers the psychological domain, which is the proper place to start investigating musical appreciation. In fact, both the piano tuner and the scribe, as listeners, also have a psychological experience of sound and surely draw on that experience in plying their different trades. Later, we will be interested in the relationship between sound as an acoustic phenomenon and music as a psychological phenomenon experienced by a listener. As a psychological phenomenon, music involves a communicative process in which an audience receives, cognitively/neurally represents, and processes sensory input (whether from acoustic or visual input or both) and appreciates it *as music*. We might argue that organized sound only becomes music when it is psychologically represented as such by a listener.

Musical communication typically involves a composer, a performer, and a listener (Kendall and Carterette 1990), who may be different individuals or groups depending on the situation (e.g. in improvisation the distinction between the composer and performer is blurred; a performer can be his or her own audience). Individuals may develop expertise in any or all of these areas: one may be an expert in composition, improvisation, conducting, performance, listening (e.g. as a critic, analyst, theorist, producer, DJ), among other things. Within each of these areas expertise may relate to different compositional tools, instruments, musical genres, symbolic encodings, and so on.

16.3 Frameworks for understanding expertise

If music itself is a slippery concept to define, then *musical expertise* is even more so.

A long tradition of research in cognitive science and artificial intelligence focuses on the study of *experts* in various domains (e.g. board games, science, medicine, the arts, sport) (Ericsson 2006). This research has tended to characterize expertise as unusually high performance in specialized skills, acquired over many years of focused training and deliberate practice. Feltovich, Prietula, and Ericsson (2006) identify the following generalizable characteristics of expertise:

(1) limited in scope and non-transferable;

(2) comprises a large amount of knowledge;

(3) features larger and more integrated cognitive units;

(4) exhibits functional, abstracted representations of presented information;

(5) automation of basic strokes or movements;

(6) shows selective access to relevant information;

(7) includes an ability to reflect consciously on the expert domain;

(8) comprises efficient and effective adaptation to the task;

(9) requires more than simple experience for its development.

Many of these can be found in musical experts (Lehmann and Gruber 2006). One prediction that can be made is that the effects of musical expertise should be limited in scope and non-transferable to other domains and items 5–9 are likely to be characteristic of expertise in music performance. However, items 3 and 4 are especially relevant to perception and appreciation, since they concern cognitive representations rather than the generation of expert performance. On this basis, we might hypothesize that musical experts re-present music to themselves:

(1) in larger chunks or longer-lasting excerpts, both in short-term memory, but also to facilitate efficient retrieval from long-term memory;

(2) as deep, abstract, generalized structures, suited to functional transfer across musical situations, rather than surface characteristics of the music.

These hypotheses are partially congruent with research on *perceptual expertise* (either pre-existing or trained), primarily using visual discrimination paradigms (Bukach et al. 2006). First, experts show *holistic processing* of visual images; for example, *the inversion effect* describes the observation that experts show impaired recognition of vertical inversions of objects taken from the expert domain (Rossion and Curran 2010). Second, experts have stronger *relational processing* in which spatial relations between the component parts are encoded to a greater extent than non-experts (Gauthier and Tarr 2002; Gauthier et al. 1998). These imply a global kind of *configural* processing, which is compatible with larger chunk sizes of information and abstract representations. However, this research also finds evidence for more detailed representation of the stimulus. For example, *the entry-level shift*, where expertise leads to faster discrimination responses to differences at specific levels of detail than more general levels of detail. According to the expertise hypothesis, general-purpose brain regions become specialized for objects of expertise indexed by neural responses such as the N170 (Hole 1994) and blood oxygen-level-dependent (BOLD) responses in the inferior temporal cortex (Gauthier et al. 2000). It is worth noting that there is considerable controversy about this hypothesis (Robbins and McKone 2007).

Can these approaches be generalized to musical expertise? If so, we might look for the following attributes in musicians:

(1) musical representations present in larger chunks than in non-experts;

(2) abstract, generalized representations (holistic processing);

(3) greater sensitivity to relations between musical structures;

(4) greater sensitivity to more specific levels of detail in musical structure.

The last item might seem paradoxical in combination with the first three items. However, if we allow for multiple levels of cognitive representation (Marr 1982; McClamrock 1991) there is no reason why expertise should not lead to both more abstract and more detailed representations. We return to these hypothesized effects below.

16.4 **Musical expertise**

It is important to distinguish musical *training* from musical *expertise*, since one can exist without the other. Musical training, like any form of training, does not always lead to expertise. Less obviously, expertise can sometimes be attained without explicit, formal training (Sloboda 1991). Musical training might refer to a number of different things depending on the kind of expertise targeted and reflecting the fact that, as we have seen, musical practice involves a diverse range of skills, each of which might be emphasized to different degrees in any one programme of training. However, in contemporary Western cultures, the primary goal of musical training is usually to attain a high level of performance in playing a musical instrument, which may be the voice itself. In some cases, this is combined with training in other art forms such as dance, theater, and so on. Because training is easier to quantify than expertise, in much of the research performed to date, questions about the effects of expertise in musicianship are often reduced to questions about the effects of musical training. Furthermore, training is often quantified in terms of number of years of formal lessons but this may not be a good proxy for the extent of actual practice. Even when hours of practice are assessed, the effects of practice on expertise may vary depending on the individual concerned and the nature of the practice.

It is remarkable how much implicit understanding of music can be acquired merely through exposure without the need for explicit musical training. Indeed, research has shown non-musicians to possess sophistication in such a wide range of abilities related to musical perception that they have been characterized as "musically experienced listeners" (Bigand and Poulin-Charronnat 2006). Sloboda (1991) goes further, characterizing musical expertise itself as an emotional sensitivity to the music of a given culture possessed by many (or perhaps, most) individuals within a culture. According to Sloboda, this sensitivity depends on:

(1) the existence in a culture of forms that have perceptible structures of certain kinds;

(2) frequent informal exposure to examples of these forms over a lifetime;

(3) the existence of a normal range of human emotional responses;

(4) the opportunity to experience those emotions mediated through perceived musical structures, which itself requires

(5) the opportunity to experience music in contexts free of externally imposed constraints or negative reinforcements.

We might start our enquiry, therefore, by investigating whether musical training has any effect at all on the different components of this conception of musical expertise as emotional sensitivity. Taking a slightly broader perspective than Sloboda (1991) (and combining the related points 4 and 5 above), this entails asking the following four questions:

(1) Are there effects of musical training on the perception and cognition of musical structure, separable from mere exposure?

(2) Are there effects of musical training on the frequency of informal exposure to music?

(3) Are there effects of musical training on the experience of human emotion?

(4) Are there effects of musical training on the emotional and aesthetic experience of music?

These questions will be addressed in the following sections.

16.5 **Expertise effects on perception and cognition of musical structure**

The most striking features of musical expertise are the perceptual motor skills necessary to physically perform highly complex sequences of actions on a musical instrument. For example, in highly trained musicians, we observe changes in neural organization in the anatomical regions of motor cortex responsible for controlling the relevant effectors (e.g. the left hand for violinists, Amunts et al. 1997; Bangert and Schlaug 2006). Furthermore, there is some evidence that a sensitive period exists, prior to the age of seven, during which musical training has particularly marked effects on behavioral and neural markers of auditory and motor development (Penhune 2011).

What about perceptual skills? Based on the literature on expertise, we might predict that musical training would lead to more abstract representation of music, with larger chunks of musical material held in memory and greater integration between the component parts. To address this question, researchers have studied the aesthetic judgment of musicians and non-musicians for pieces of music in which the large-scale tonal form has been disrupted by rearranging or rewriting certain parts. The results consistently suggest that musicians are no more affected than non-musicians by these disruptions (Cook 1987; Karno and Konecni 1992; Marvin and Brinkman 1999), suggesting that both rely more on local representation than on high-level, integrated representations of large-scale tonal closure.

At the other end of the scale, it seems likely that extensive periods of time spent in focused listening to music would sharpen auditory perception. There is evidence suggesting that this is the case. Professional musicians show smaller pitch discrimination thresholds for pure and complex tones than non-musicians (Micheyl et al. 2006). These effects may increase with number of years of musical training (Kishon-Rabin et al. 2001) and there is some evidence that pitch discrimination is better in musicians who habitually tune their instruments (Micheyl et al. 2006). Musicians are also better than non-musicians at discriminating temporal intervals (Banai et al. 2012) as well as a range of other auditory temporal tasks (Rammsayer and Altenmüller 2006).

These effects suggest that musical training has a positive influence on the efficiency with which sounds are encoded: musicians show optimized perception and discrimination of pitch and timbre, in some cases related specifically to their particular instrument. How do these effects of musical training on low-level auditory processing influence the higher-level cognitive processing of music? In processing rhythm and meter, for example, there is evidence that musicians represent deeper metrical hierarchies than non-musicians (Palmer and Krumhansl 1990). Musicians are sensitive to a greater range of features (e.g. slurs, rests, articulation, and timbre) in identifying grouping boundaries in music (Delige 1987). This presumably allows musicians to make finer-grained distinctions in their interpretations of metrical and grouping structure, facilitating temporal prediction of events. More predictable music in turn is better remembered (Agres et al. 2013; Schmuckler 1997), and evidence suggests that, musicians show better recognition memory for melodies than non-musicians (Dowling and Bartlett 1981; Dowling 1978; Orsmond and Miller 1999), perhaps because they have more accurate predictive models.

Research directly examining predictive processing of music is relevant here. General theories of sensory processing in cognitive science and cognitive neuroscience include top-down prediction as a central component in constructing coherent representations of incoming sensory input (Barlow 1959; Dayan et al. 1995; Friston 2005; Gregory 1980; Helmholtz 1866). Recent incarnations of this approach take a probabilistic approach in a hierarchical framework where information is processed over successively larger time windows at higher scales in the hierarchy (Friston 2005; Friston and Kiebel 2009). According to this theory of perception, learning is driven by *prediction errors*, occurring when bottom-up sensory input conflicts with top-down prediction. Therefore, violations of expectation are of central importance in understanding how the brain constructs optimized models of sensory input through experience. If musical training optimizes those top-down predictive models, then we should observe differences between experts and non-experts in terms of prediction error.

Recent research suggests that expectations in music are acquired through a process of statistical learning in which listeners construct implicit probabilistic models of the next element in a musical sequence given the preceding context both at the psychological (Huron 2006; Meyer 1957; Oram and Cuddy 1995; Pearce and Wiggins 2006; Tillmann et al. 2000) and neural levels (Kim et al. 2011; Loui et al. 2009; Pearce et al. 2010). The majority of behavioral research on expectation in music perception has found that the expectations of non-musicians look very similar to those of musicians (Bigand et al. 2003; Schellenberg 1996), although some research has identified differences (Pearce et al. 2010). However, neuroscientific studies on musical expectations, using electroencephalography (EEG) and magnetoencephalography (MEG), have revealed differences in neural responses of experts and non-experts to violations of musical expectation (i.e. prediction errors).

Research introducing low-probability auditory events in simple melodic tone sequences has found larger event-related potentials (ERPs) in auditory cortical regions for musicians compared to non-musicians. In an MEG study, for example, Fujioka, Trainor, Ross, Kakigi, and Pantev (2004) examined neural responses to notes forming deviant contours or intervals in short five-note melodies, finding that the mismatch negativity (MMN) response was significantly larger in musicians than non-musicians for both interval and contour deviants, while no difference was observed for a pitch-deviant control condition. There is also evidence that instrument-specific musical training affects sensitivity to timbre. Pantev, Roberts, Schulz, Engelien, and Ross (2001) presented trumpeters and violinists with oddball sequences reproduced on the trumpet or the violin during EEG recording. The results indicated a larger N1 response to the oddball presented in the timbre of their own instrument. In addition to instrument-specific differences in neural processing of prediction errors, additional research has shown style-specific differences between jazz, classical, or pop musicians processing deviants in terms of pitch mistuning, intensity, timbre, sound-source location, rhythm, and pitch slide (Vuust et al. 2012).

Effects of musical training are also observed for violations of temporal expectations. In an MEG study, jazz musicians showed a left-lateralized MMN to violations of metrical expectation, while in non-musicians the response was right-lateralized (Vuust et al. 2005). Using elegantly manipulated drum patterns, Geiser, Sandmann, Jäncke, and Meyer (2010) examined the MMN to meter-congruent and meter-incongruent deviants. Trained percussionists showed a greater difference in MMN amplitude between these conditions than non-musicians, suggesting a greater sensitivity to violations of metrical structure.

Related effects of musical training have been found in neural processing of harmonic movement. Stylistically unexpected chords generate characteristic neural responses, including an early right anterior negativity (ERAN) peaking at around 180 ms post-stimulus (Koelsch et al. 2000). Furthermore, the ERAN has been observed both in artificial grammars (Loui et al. 2009) and in real musical examples (Koelsch et al. 2008; Steinbeis et al.

2006). Although the ERAN is present in non-musicians, it has a greater amplitude in trained musicians (Koelsch et al. 2002).

It seems, then, that musicians show larger neural prediction errors to violations of musical syntax. One plausible explanation is that musical training produces more specific, and more accurate, expectations for forthcoming musical structures, resulting from an optimized predictive model. Recent research has examined this question using a probabilistic cognitive model of auditory expectation which acquires pitch expectations through statistical learning of sequential dependencies in music (Pearce 2005). The trained model generates a probability distribution, whose *entropy* reflects how uncertain the model is about the continuation of the current context, and whose components reflect the likelihood of each possible continuation, given the preceding context, in terms of *information content* (MacKay 2003). Using this model to select melodic contexts that differ systematically in entropy, Hansen and Pearce (2014) found that musicians showed lower levels of uncertainty overall but also were better able to distinguish high- and low-entropy contexts than non-musicians. In particular, the musicians were able to make better use of the predictive cues in low-entropy contexts to generate more accurate expectations and showed greater prediction errors to low-probability continuations in those contexts. Furthermore, the degree to which listeners' expectations matched those of the probabilistic model increased linearly with extent of musical training.

These results suggest that musical training can lead to optimized predictive models of musical structure. However, the question arises of whether the crucial factor is musical training per se, or style-specific training and experience. In a follow-up study, Hansen et al. (2013) examined predictive uncertainty for contexts taken from solos by Charlie Parker in professional jazz and classical musicians as well as non-musicians. The results suggest that the predictive advantage in low-entropy contexts, shown by greater prediction errors to low-probability continuations, is greater for jazz musicians than classical musicians, although both groups outperform non-musicians. While style-specific training and experience appears to be important, the question of whether these effects are driven by experience or training remains unanswered.

16.6 Expertise effects on frequency of exposure to music

As in other domains, expertise in music requires enormous amounts of time spent in deliberate practice. In fact, hours of deliberate practice is a significant predictor of level of expertise attained (Sloboda et al. 1996). However, it is highly likely that musicians listen to more music throughout their lives than non-musicians, through individual practice, group practice, critical listening to music, concert attendance, and other activities. Perhaps more importantly, they experience music in more focused and, crucially, more emotionally charged situations than non-musicians. Might this more focused and emotionally engaged exposure, rather than explicit training per se, have any effect on musical appreciation?

We know that exposure to music (as with other sensory stimuli) has a strong influence on its perception and appreciation. For example, passive exposure to music appears to be fundamental in the development of metrical perception (Hannon and Trehub 2005; Hannon et al. 2012) and increasing exposure to and familiarity with a stimulus (e.g. a musical style) increases subjective preference for it (Bornstein and D'Agostino 1992). The effect of mere exposure on aesthetic experience may, according to one theory (Huron 2006; Reber et al. 1998), be directly related to the predictive processing discussed in the previous section if pleasure results from the perceptual fluency (facilitated processing and reduced cognitive load) associated with more accurate predictive models (acquired through exposure and familiarity). As noted earlier, there is a thorny methodological problem in distinguishing the effects of explicit musical training from the effects of increased exposure, since musicians are likely to spend more time listening to music than non-musicians. Furthermore, musicians are more likely to engage in focused listening to music and we know that the efficiency of statistical learning is affected by attentional mechanisms (Toro et al. 2005).

At this stage, therefore, it is at least plausible that the greater amount of time musicians spend in focused listening to music, in combination with the effects of musical training per se, leads to more accurate predictive models which, in turn, may impact on aesthetic experience through increased perceptual fluency (we return to this point later).

16.7 Expertise effects on emotional experience

Our third question is whether musical training affects the general experience of emotion. In a study of schoolchildren, Schellenberg (2004) found that 36 weeks of weekly music lessons increased ability in full-scale IQ and its index scores (Verbal comprehension, Perceptual organization, Freedom from distractability, Processing speed) more than did taking equivalent lessons in drama. However, subsequent studies have shown that musical training does not affect emotional intelligence in adults (Schellenberg 2011) or emotion comprehension in children (Schellenberg and Mankarious 2012).

There is evidence that musicians and 7-year-old children randomly assigned to one year of musical training show better emotion recognition for speech prosody (Thompson et al. 2004) and that this persists even when groups are matched for cognitive ability (Lima and Castro 2011). However, it is important to distinguish between the experience of emotion and the perception or recognition of emotion in a stimulus such as language or music (Juslin and Västfjäll 2008). These findings do not necessarily indicate a difference in the experience of emotion itself, only a better ability to recognize emotion in speech prosody. Therefore, there is no evidence to suggest that musicians differ from non-musicians in terms of their experience of emotion as a result of their musical training (although this does have other effects on general cognitive abilities).

16.8 Expertise effects on emotional and aesthetic experience of music

If emotional experiences of music depend on understanding of high-level musical structure, then we would expect any effects of musical training on musical perception and cognition to have concomitant effects on the emotional experience of music. Again, it is important here to distinguish the recognition of emotion *in* music from the induction of emotion *by* music (Juslin and Västfjäll 2008). Regarding the effects of musical training, the majority of research has addressed emotion recognition in Western instrumental music. In general, the evidence suggests that musical training has very little effect on the perception of emotion in music in adults (Bigand et al. 2005; Edmonston 1966; Hevner 1935; Robazza et al. 1994; Waterman 1996) or children (Robazza et al. 1994; Terwogt and Van Grinsven 1991). Even the training-related differences in the perception of emotion in speech prosody discussed earlier are weaker or absent when tone sequences mimicking speech prosody are used (Thompson et al. 2004; Trimmer and Cuddy 2008).

There is, however, evidence to suggest that musicians may make less use of emotion in making aesthetic judgments of music. In an ERP study of responses to chord sequences in which the final chord varied in terms of its harmonic congruency with the context, Müller, Höfel, Brattico, and Jacobsen (2010) found enhanced emotion-related neural processing (the late positive potential (LPP); Hajcak et al. 2006) for beauty judgments compared with judgments about how correct each chord sequence sounded in non-musicians but not in musicians. It is possible that musicians make less use of emotion and rely on other factors in making aesthetic judgments of music.

What might these factors be? Scientific theories of aesthetic experience emphasize a number of different cognitive components involved (Brattico and Pearce 2013; Leder et al. 2004; Leder 2013) in addition to emotional experience. These range from basic perceptual processing, through effects of attention, memory, knowledge, and judgment. Perhaps the most studied of these are aesthetic judgments for liking, pleasantness, or preference. Here we examine two potential factors affecting aesthetic judgment: musical complexity and dissonance.

16.8.1 Complexity

Wilhelm Wundt, a founder of experimental psychology and one-time assistant of Hermann Helmholtz in Heidelberg, demonstrated that physiological arousal is related to stimulus complexity and argued that aesthetic pleasure is maximal at intermediate degrees of complexity (Wundt 1874). In his new experimental aesthetics, Berlyne (1974) developed this idea further, positing an inverted U-shaped function (sometimes called the Wundt curve) linking the "arousal potential" of a stimulus with its "hedonic value." Berlyne's research attempted to identify how stimulus properties (such as complexity, familiarity, novelty, uncertainty) influence aspects of the aesthetic experience such as arousal, pleasure, and level of interest.

Support for a U-shaped relationship between complexity and liking has been found in artificial auditory stimuli (Crozier 1974; Heyduk 1975; Vitz 1966) and in musical styles ranging from baroque, romantic, and serial (Burke and Gridley 1990), jazz (Gordon and Gridley 2013), bluegrass (Orr and Ohlsson 2001), and pop (North and Hargreaves 1995). If musical expertise facilitates more sophisticated cognitive representations of music, then we would predict that musical training would shift the U-shaped function towards optimal preference for greater degrees of objective stimulus complexity. Importantly, this prediction hinges on an objective definition of complexity. With changes of expertise or familiarity, we would expect the relationship between objective and subjective complexity to change, while the relationship between subjective complexity and liking remains constant (with moderate degrees of subjective complexity producing optimal liking; see Orr and Ohlsson 2005).

There is some evidence that this effect of expertise on the relationship between objective complexity and aesthetic experience does, in fact, hold. Rubin-Rabson (1940) reported a positive correlation between degree of musical training and liking for "modern" music. In an investigation of preferences for world music, Fung (1996) found that musicians preferred excerpts with very complex texture, whereas non-musicians preferred moderately complex textures. Burke and Gridley (1990) showed that musically trained individuals exhibit greater relative preference for complex musical works (by Debussy and Boulez) over less complex compositions (by Bach, Grieg) than non-musicians. In a verbal association task, Istok and colleagues (2009) found that musicians used more adjectives related to originality and variety, and fewer terms related to mood or mood induction, than non-musicians, to describe the aesthetic value of music. These results are somewhat difficult to interpret since the verbal responses may reflect cultural or professional norms related to expertise rather than psychological experience per se.

Other research has empirically tested the effects of expertise on the inverted-U shaped relationship in music more directly. North and Hargreaves (1995), for example, examined excerpts of popular music, finding that the inverted-U relationship between liking and subjective complexity peaked at a slightly lower complexity level for non-musicians than for individuals with intermediate or high levels of musical training. The relationship between complexity, preference, and expertise may even reflect more specific effects of stylistic musical training. Using a paradigm developed in research on music education, Coggiola (2004) examined preferences for four jazz tunes varying in complexity (as defined by the experimenter) by jazz musicians studying at university. The ratings of the two groups differed only for the most complex music, for which the jazz musicians reported stronger aesthetic experience than non-jazz musicians.

Other results apparently contradict this overall picture. Using specially composed jazz and bluegrass improvisations, Orr and Ohlsson (2001) found an inverted-U shaped relationship between subjective complexity and liking in non-musicians for bluegrass but weaker support for the jazz excerpts. In a subsequent study, Orr and Ohlsson (2005) asked professional jazz and bluegrass musicians to rate their liking for short jazz and bluegrass compositions. Comparing the responses of the musicians with non-musicians from an earlier study (Orr and Ohlsson 2001), the results suggested that musical training dissolves the U-shaped relationship between complexity and liking. This may be because experts focus on different aspects of musical structure, such as harmony (Conley 1981), when making complexity judgments. It is also likely that other factors besides complexity, such as expressive performance (Clarke 1993; Repp 1992) play a greater role in experts' aesthetic experience of music.

Running through this work is the problem of obtaining a broad enough range of objective complexity to observe the complete inverted-U shaped function. In most of these studies, complexity is defined subjectively, either by the experimenter or by the participants in the studies. Therefore, we lack insight into which aspects of musical structure contribute to the perception of complexity and, therefore, liking. One way of avoiding this problem is to bypass the complexities of real music and isolate the structures of interest with objective measures of complexity. A successful approach has been to create stimuli that systematically vary in information content (an objective information-theoretic measure of complexity). Research on the pleasantness of artificial tone sequences (Crozier 1974; Vitz 1966) and piano themes (Simon and Wohlwill 1968) suggests that individuals with musical training show greater preferences for more complex stimuli than non-musicians. Smith and Melara (1990) assessed aesthetic responses to harmonic progressions (sequences of chords) varying in degree of syntactic prototypicality to test the hypothesis that unpredictable musical events are experienced as pleasant. The results indicated that all subjects (even novices) found atypical progressions more interesting/complex but non-musicians and undergraduate musicians preferred harmonic prototypes, while only music graduate students preferred atypical progressions.

16.8.2 Dissonance

Another aspect of the acoustic structure of music that contributes to aesthetic experience (Brattico and Pearce 2013) is dissonance. Here, however, the evidence suggests that expertise sharpens the perception of dissonance and accentuates its negative effect on liking. For example, Bigand, Parncutt, and Lerdahl (1996) found the greater tension-inducing effect of dissonant chords over consonant chords was more pronounced in musicians than in non-musicians. Other research has examined the pleasantness of melodic and harmonic intervals varying in dissonance, finding that musicians distinguish perfect consonance, imperfect consonance, and dissonance to a greater extent than non-musicians who tend to classify all intervals as moderately pleasant (Schön et al. 2005). Similarly, dissonant chords are experienced as more unpleasant and minor chords as more sad by musicians than by non-musicians (Pallesen et al. 2005), although fMRI showed no group differences in neural processing.

Effects of musical expertise on responses to dissonance have also been found in physiological data. Dellacherie, Roy, Hugueville, Peretz, and Samson (2011) found that dissonant piano music induces more unpleasant feelings (valence) and stronger physiological responses (skin conductance responses (SCR) and zygomatic EMG) in musicians than in non-musicians, while musical training was a good predictor of valence. The conclusion that musical training reinforces the aversion to dissonance is supported by EEG and MEG research. Brattico and colleagues (2009) report that the MMN response to dissonant and mistuned chords presented in the context of major chords is stronger for musicians than non-musicians (and this varies as a function of degree of musical training), while no difference is observed in the response to minor chords in the same context. Schön and colleagues (2005) also report qualitative differences between musicians and non-musicians in terms of neural processing of harmonic intervals varying in dissonance.

These studies have focused on musical dissonances rather than sensory dissonance per se. Because musically dissonant intervals are used (less frequently than consonant intervals) in the music that people listen to, these studies cannot discount the possibility that the effects are related to top-down cognitively-driven expectations rather than bottom-up sensory dissonance. However, this is unlikely to be the case for the mistuned chords used by Brattico and her research team (2009). Furthermore, at least one study has found that the effects of chord dissonance (an augmented triad) on ERPs is clearly distinguished from the top-down cognitive effects of harmonic function discussed above (Regnault et al. 2001).

16.9 **Conclusion**

Drawing inspiration from Sloboda's (1991) definition of musical expertise as an emotional sensitivity to the music of a given culture, we have examined whether musical training affects aesthetic sensitivity to music. Contrary to predictions from the general literature on expertise, there is little evidence that musicians have more abstract representations of music, storing larger, more integrated chunks of music in memory, than non-musicians. There is also little evidence for any effects of musical training on emotional experience in general or on emotion recognition in music per se. However, there is evidence that musical training leads to sharpened perceptual representation and processing and, in particular, the optimization of predictive models of musical structure. Importantly, much of the evidence for more accurate predictive models of musical structure in musicians comes from EEG and MEG studies, whose temporal precision allows detailed analysis of responses to individual musical events. Given that musicians are likely to spend significantly more time in focused listening to music than non-musicians, it is unclear whether this is an effect of music training per se or simply increased musical exposure.

What appears to be clear, however, is that musicians do tend to find pleasure in more complex music than non-musicians (at least in some circumstances), consistent with a Wundt curve shifted to greater levels of objective complexity. It is noteworthy that this effect of expertise on aesthetic experience occurs without any changes to emotional experience of music. It is tempting to associate this preference for increased complexity with the findings of more accurate predictive models, although robust, empirical evidence has yet to be developed. Somewhat paradoxically, it seems that musicians show greater sensitivity to dissonance in their aesthetic judgments than non-musicians. This enhanced sensitivity to perceptual qualities runs counter to a suggestion that experts might make greater use of knowledge-based criteria and less use of perceptual criteria than non-experts in their aesthetic judgments (Juslin 2013). It is, however, somewhat reminiscent of research in which visual experts show both greater configural processing and sensitivity to finer levels of detail than non-experts.

One potential solution to this apparent paradox might be to suggest that dissonance introduces uncertainty that cannot be resolved through greater experience and training while other aspects of objective complexity (e.g. harmonic syntax, metrical sophistication) can be rendered more certain with more accurate predictive models derived through experience and training. However, this would not explain why non-musicians appear to be less sensitive to dissonance than musicians. Another possibility, therefore, is that musicians have more accurate predictive models of musical consonance and dissonance and, as a result, generate greater prediction errors to dissonant tones and intervals. This would be consistent with their preference for greater complexity resulting from more accurate predictive models.

There is clearly much work to be done. It is fundamental to link the development of accurate predictive models during musical training specifically to changes in aesthetic experience of music. Do differences in predictive uncertainty underlie the differential effects of training on different aspects of complexity (e.g. dissonance versus melodic complexity)? How can we develop more general predictive computational models of musical structure that integrate melodic, harmonic, rhythmic, and timbral structure? It is puzzling that musicians fail to show greater sensitivity to high-level tonal form, but perhaps we are asking the question in the wrong way; it is possible that neurophysiological recording might prove more suitable to the task. Finally, the framework predicts that effects of training on predictive coding and aesthetic experience should be experience-specific, therefore we predict limited transfer between areas of musical expertise and to other domains. As already noted, this is consistent with general findings in expertise research (Feltovich et al. 2006).

We have developed a framework for understanding the effects of expertise on aesthetic appreciation of music, facilitating hypothesis generation. Testing these hypotheses is likely to require a sophisticated combination of psychological study, computational modeling, and neuroimaging in which great care is taken to distinguish the causes and effects of musical training and distinguish these from general training effects and other factors that covary with musical training.

Acknowledgments

I am extremely grateful to Dan Carey, Niels Chr. Hansen, and John Sloboda for their careful reading and insightful comments during the development of this chapter.

Notes

1 Or, *pace* John Cage, in at least one case, it is absence, although notice here that the absence is still human-produced and extended for a prescribed period of time.

2 Note that Babbit (1965) uses different terms to describe the three kinds of representation.

References

Agres, K., Abdallah, S., and Pearce, M.T. (2013). An information-theoretic account of musical expectation and memory. In M. Knauff, M. Pauen, N. Sebanz, and I. Wachsmuth (eds), *Proceedings* *of the 35th Annual Conference of the Cognitive Science Society*. Austin, TX: Cognitive Science Society, pp. 127–32.

Amunts, K., Schlaug, G., Jäncke, L., et al. (1997). Motor cortex and hand motor skills: structural compliance in the human brain. *Human Brain Mapping* **5**, 206–15.

Babbitt, M. (1965). The use of computers in musicological research. *Perspectives of New Music* **3**(2), 74–83.

Banai, K., Fisher, S., and Ganot, R. (2012). The effects of context and musical training on auditory temporal-interval discrimination. *Hearing Research* **284**(1–2), 59–66. doi:10.1016/j.heares.2011.12.002.

Bangert, M. and Schlaug, G. (2006). Specialization of the specialized in features of external human brain morphology. *European Journal of Neuroscience* **24**, 1832–4.

Barlow, H.B. (1959). Sensory mechanisms, the reduction of redundancy, and intelligence. In *Proceedings of a Symposium on the Mechanisation of Thought Processes*. National Physical Laboratory, Teddington: Her Majesty's Stationery Office, **Vol. 2**, pp. 537–59.

Berlyne, D.E. (1974). The new experimental aesthetics. In D.E. Berlyne (ed.), *Studies in the New Experimental Aesthetics: Steps Towards an Objective Psychology of Aesthetic Appreciation*. Washington, DC: Hemisphere Publishing Co., pp. 1–25.

Bigand, E., Parncutt, R., and Lerdahl, F. (1996). Perception of musical tension in short chord sequences: the influence of harmonic function, sensory dissonance, horizontal motion, and musical training. *Perception Psychophysics* **58**, 124–41.

Bigand, E., Poulin, B., Tillmann, B., et al. (2003). Sensory versus cognitive components in harmonic priming. *Journal of Experimental Psychology: Human Perception and Performance* **29**, 159–71.

Bigand, E. and Poulin-Charronnat, B. (2006). Are we "experienced listeners"? A review of the musical capacities that do not depend on formal musical training. *Cognition* **100**, 100–30.

Bigand, E., Vieillard, S., Madurell, F., et al. (2005). Multidimensional scaling of emotional responses to music: The effect of musical expertise and of the duration of the excerpts. *Cognition and Emotion* **19**(8), 1113–39. doi:10.1080/02699930500204250.

Bornstein, R.F. and D'Agostino, P.R. (1992). Stimulus recognition and the mere exposure effect. *Journal of Personality and Social Psychology* **63**(4), 545–52.

Brattico, E., Pallesen, K.J., Varyagina, O., et al. (2009). Neural discrimination of nonprototypical chords in music experts and laymen: an MEG study. *Journal of Cognitive Neuroscience* **21**(11), 2230–44. doi:10.1162/jocn.2008.21144.

Brattico, E. and Pearce, M. (2013). The neuroaesthetics of music. *Psychology of Aesthetics, Creativity, and the Arts* **7**(1), 48–61. doi:10.1037/a0031624.

Bukach, C.M., Gauthier, I., and Tarr, M.J. (2006). Beyond faces and modularity: the power of an expertise framework. *Trends in Cognitive Sciences* **10**(4), 159–66. doi:10.1016/j.tics.2006.02.004.

Burke, M.J. and Gridley, M.C. (1990). Musical preferences as a function of stimulus complexity and listeners' sophistication. *Perceptual and Motor Skills* **7**, 687–90.

Clarke, E.F. (1993). Imitating and evaluating real and transformed musical performances. *Music Perception* **10**, 317–41.

Coggiola, J.C. (2004). The effect of conceptual advancement in jazz music selections and jazz experience on musicians' aesthetic response. *Journal of Research in Music Education* **52**(1), 29. doi:10.2307/3345523.

Conley, J.K. (1981). Physical correlates of the judged complexity of music by subjects differing in musical background. *British Journal of Psychology* **72**(4), 451–64.

Cook, N. (1987). The perception of large-scale tonal closure. *Music Perception* **5**(2), 197–206.

Crozier, J.B. (1974). Verbal and exploratory responses to sound sequences varying in uncertainty level. In D.E. Berlyne (ed.), *Studies in the New Experimental Aesthetics: Steps Towards an Objective Psychology of Aesthetic Appreciation*. Washington, DC: Hemisphere Publishing Co., pp. 27–90.

Dayan, P., Hinton, G.E., Neal, R.M., et al. (1995). The Helmholtz Machine. *Neural Computation* **7**, 889–904.

Deliège, I. (1987). Grouping conditions in listening to music: An approach to Lerdahl and Jackendoff's grouping preference rules. *Music Perception* **4**(4), 325–60.

Dellacherie, D., Roy, M., Hugueville, L., et al. (2011). The effect of musical experience on emotional self-reports and psychophysiological responses to dissonance. *Psychophysiology* **48**(3), 337–49. doi:10.1111/j.1469-8986.2010.01075.x.

Dick, F., Lee, H.L., Nusbaum, H., et al. (2011). Auditory-motor expertise alters "speech selectivity" in professional musicians and actors. *Cerebral Cortex* **21**(4), 938–48.

Dowling, W.J. (1978). Scale and contour: Two components in a theory of memory for melodies. *Psychological Review* **85**, 341–54.

Dowling, W.J. and Bartlett, J.C. (1981). The importance of interval information in long-term memory for melodies. *Psychomusicology* **1**(1), 30–49.

Edmonston, W.E. (1966). The use of the semantic differential technique in the esthetic evaluation of musical excerpts. *American Journal of Psychology* **79**, 650–2.

Ericsson, K.A. (2006). The influence of experience and deliberate practice on the development of superior expert performance. In K.A. Ericsson, N. Charness, P.J. Feltovich, and R.R. Hoffman (eds), *The Cambridge Handbook of Expertise and Expert Performance*. Cambridge: Cambridge University Press, pp. 683–703.

Feltovich, P.J., Prietula, M.J., and Ericsson, K.A. (2006). Studies of expertise from psychological perspectives. In K.A. Ericsson, N. Charness, P.J. Feltovich, and R.R. Hoffman (eds), *The Cambridge Handbook of Expertise and Expert Performance*. Cambridge: Cambridge University Press, pp. 41–67.

Friston, K. (2005). A theory of cortical responses. *Philosophical Transactions of the Royal Society Series B* **360**, 815–36.

Friston, K. and Kiebel, S. (2009). Predictive coding under the free-energy principle. *Philosophical Transactions of the Royal Society of London, Series B* **364**, 1211–21. doi:10.1098/rstb.2008.0300.

Fujioka, T., Trainor, L.J., Ross, B., et al. (2004). Musical training enhances automatic encoding of melodic contour and interval structure. *Journal of Cognitive Neuroscience* **16**(6), 1010–21. doi:10.1162/0898929041502706.

Fung, C.V. (1996). Musicians' and nonmusicians' preferences for world musics: relation to musical characteristics and familiarity. *Journal of Research in Music Education* **44**(1), 60. doi:10.2307/3345414.

Gauthier, I., Skudlarski, P., Gore, J.C., et al. (2000). Expertise for cars and birds recruits brain areas involved in face recognition. *Nature Neuroscience* **3**, 191–7.

Gauthier, I. and Tarr, M.J. (2002). Unraveling mechanisms for expert object recognition: bridging brain activity and behavior. *Journal of Experimental Psychology Human Perception and Performance* **28**, 431–46.

Gauthier, I., Williams, P., Tarr, M.J., et al. (1998). Training "greeble" experts: a framework for studying expert object recognition processes. *Vision Research* **38**, 2401–28.

Geiser, E., Sandmann, P., Jäncke, L., et al. (2010). Refinement of metre perception—training increases hierarchical metre processing. *European Journal of Neuroscience* **32**(11), 1979–85. doi:10.1111/j.1460-9568.2010.07462.x.

Gordon, J. and Gridley, M.C. (2013). Musical preferences as a function of stimulus complexity of piano jazz. *Creativity Research Journal* **25**(1), 143–6. doi:10.1080/10400419.2013.752303.

Gregory, R.L. (1980). Perceptions as hypotheses. *Philosophical Transactions of the Royal Society B* **290**(1038), 181–97.

Hajcak, G., Moser, J.S., and Simons, R.F. (2006). Attending to affect appraisal strategies modulate the electrocortical response to arousing pictures. *Emotion* **6**, 517–22.

Hannon, E.E. and Trehub, S.E. (2005). Metrical categories in infancy and adulthood. *Psychological Science* **16**, 48–55.

Hannon, E.E., Vanden Bosch Der Nederlanden, C.M., and Tichko, P. (2012). Effects of perceptual experience on children's and adults' perception of unfamiliar rhythms. *Annals of the New York Academy of Sciences* **1252**, 92–9. doi:10.1111/j.1749-6632.2012.06466.x.

Hansen, N C. and Pearce, M.T. (2014). Predictive uncertainty reflects statistical learning of sensory input. *Frontiers in Psychology* **5**, 1052.

Hansen, N.C., Vuust, P., and Pearce, M.T. (2013). Predictive processing of musical structure: Effects of genre-specific expertise. In *Poster presented at the 35th Annual Conference of the Cognitive Science Society*. Berlin, Germany.

Hevner, K. (1935). The affective value of the major and minor modes in music. *American Journal of Psychology* **47**, 103–18.

Heyduk, R.G. (1975). Rated preference for musical compositions as it relates to complexity and exposure frequency. *Perception Psychophysics* **17**, 84–90.

Hole, G.J. (1994). Configurational factors in the perception of unfamiliar faces. *Perception* **23**, 65–74. doi:10.1068/p230065.

Huron, D. (2006). *Sweet Anticipation: Music and the Psychology of Expectation*. Cambridge, MA: MIT Press.

Istok, E., Brattico, E., Jacobsen, T., et al. (2009). Aesthetic responses to music: A questionnaire study. *Musicae Scientiae* **13**, 183–206. doi:10.1177/102986490901300201.

Juslin, P.N. (2013). From everyday emotions to aesthetic emotions: Toward a unified theory of musical emotions. *Physics of Life Reviews* **10**, 235–66.

Juslin, P.N. and Västfjäll, D. (2008). Emotional responses to music: The need to consider underlying mechanisms. *Behavioral and Brain Sciences* **31**, 559–75.

Karno, M. and Konecni, V.J. (1992). The effects of structural interventions in the First Movement of Mozart's Symphony in G-Minor K. 550 on aesthetic preference. *Music Perception* **10**, 63–72.

Kendall, R.A. and Carterette, E.C. (1990). The communication of musical expression. *Music Perception* **8**, 129–63.

Kim, S.-G., Kim, J.S., and Chung, C.K. (2011). The effect of conditional probability of chord progression on brain response: an MEG study. *PLoS One* **6**, e17337.

Kishon-Rabin, L., Amir, O., Vexler, Y., et al. (2001). Pitch discrimination: are professional musicians better than non-musicians? *Journal of Basic and Clinical Physiology and Pharmacology* **12**, 125–43.

Koelsch, S., Gunter, T., and Friederici, A.D. (2000). Brain indices of music processing: "Nonmusicians" are musical. *Journal of Cognitive Neuroscience* **12**(3), 520–41.

Koelsch, S., Kilches, S., Steinbeis, N., et al. (2008). Effects of unexpected chords and of performer's expression on brain responses and electrodermal activity. *PLoS One* **3**(7), e2631.

Koelsch, S., Schmidt, B.H., and Kansok, J. (2002). Effects of musical expertise on the early right anterior negativity: an event-related brain potential study. *Psychophysiology* **39**, 657–63.

Leder, H. (2013). Next steps in neuroaesthetics: Which processes and processing stages to study? *Psychology of Aesthetics, Creativity, and the Arts* 7(1), 27–37. doi:10.1037/a0031585.

Leder, H., Belke, B., Oeberst, A., and Augustin, D. (2004). A model of aesthetic appreciation and aesthetic judgments. *British Journal of Psychology* **95**(Pt 4), 489–508. doi:10.1348/0007126042369811.

Lehmann, A.C. and Gruber, H. (2006). Music. In K.A. Ericsson, N. Charness, P.J. Feltovich, and R.R. Hoffman (eds), *The Cambridge Handbook of Expertise and Expert Performance*. Cambridge: Cambridge University Press, pp. 457–70.

Lima, C.F. and Castro, S.L. (2011). Speaking to the trained ear: Musical expertise enhances the recognition of emotions in speech prosody. *Emotion* **11**, 1021–31. doi:10.1037/a0024521.

Loui, P., Wu, E.H., Wessel, D.L., et al. (2009). A generalized mechanism for perception of pitch patterns. *Journal of Neuroscience* **29**, 454–9.

MacKay, D.J.C. (2003). *Information Theory, Inference, and Learning Algorithms*. Cambridge: Cambridge University Press.

Marr, D. (1982). *Vision*. San Francisco, CA: W.H. Freeman.

Marvin, E.W. and Brinkman, A. (1999). The effect of modulation and formal manipulation on perception of tonic closure by expert listeners. *Music Perception* **16**, 389–408.

McClamrock, R. (1991). Marr's three levels: A re-evaluation. *Minds and Machines* **1**, 185–96.

Meyer, L.B. (1957). Meaning in music and information theory. *Journal of Aesthetics and Art Criticism* **15**(4), 412–24.

Micheyl, C., Delhommeau, K., Perrot, X., et al. (2006). Influence of musical and psychoacoustical training on pitch discrimination. *Hearing Research* **219**(1–2), 36–47. doi:10.1016/j.heares.2006.05.004.

Müller, M., Höfel, L., Brattico, E., et al. (2010). Aesthetic judgments of music in experts and laypersons—an ERP study. *International Journal of Psychophysiology* **76**(1), 40–51. doi:10.1016/j.ijpsycho.2010.02.002.

North, A.C. and Hargreaves, D.J. (1995). Subjective complexity, familiarity and liking for popular music. *Psychomusicology* **14**, 77–93.

Oram, N. and Cuddy, L.L. (1995). Responsiveness of Western adults to pitch-distributional information in melodic sequences. *Psychological Research* **57**(2), 103–18.

Orr, M.G. and Ohlsson, S. (2001). The relationship between musical complexity and liking in jazz and bluegrass. *Psychology of Music* **29**, 108–27. doi:10.1177/0305735601292002.

Orr, M.G. and Ohlsson, S. (2005). Relationship between complexity and liking as a function of expertise. *Music Perception* **22**, 583–611. doi:10.1525/mp.2005.22.4.583.

Orsmond, G.I. and Miller, L.K. (1999). Cognitive, musical and environmental correlates of early music instruction. *Psychology of Music* **27**, 18–37.

Pallesen, K.J., Brattico, E., Bailey, C., et al. (2005). Emotion processing of major, minor, and dissonant chords: a functional magnetic resonance imaging study. *Annals of the New York Academy of Sciences* **1060**, 450–3. doi:10.1196/annals.1360.047.

Palmer, C. and Krumhansl, C.L. (1990). Mental representations for musical metre. *Journal of Experimental Psychology: Human Perception and Performance* **16**(4), 728–41.

Pantev, C., Roberts, L.E., Schulz, M., et al. (2001). Timbre-specific enhancement of auditory cortical representations in musicians. *Neuroreport* **12**(1), 169–74.

Pearce, M.T. (2005). *The Construction and Evaluation of Statistical Models of Melodic Structure in Music Perception and Composition*. London: Department of Computing, City University.

Pearce, M.T., Ruiz, M.H., Kapasi, S., et al. (2010). Unsupervised statistical learning underpins computational, behavioural and neural manifestations of musical expectation. *NeuroImage* **50**, 302–13.

Pearce, M.T. and Wiggins, G.A. (2006). Expectation in melody: The influence of context and learning. *Music Perception* **23**(5), 377–405.

Penhune, V.B. (2011). Sensitive periods in human development: Evidence from musical training. *Cortex* **47**, 1126–37.

Rammsayer, T. and Altenmüller, E. (2006). Temporal information processing in musicians and nonmusicians. *Music Perception* **24**, 37–48. doi:10.1525/mp.2006.24.1.37.

Reber, R., Winkielman, P., and Schwarz, N. (1998). Effects of perceptual fluency on affective judgments. *Psychological Science* **9**(1), 45–8. doi:10.1111/1467-9280.00008.

Regnault, P., Bigand, E., and Besson, M. (2001). Different brain mechanisms mediate sensitivity to sensory consonance and harmonic context: evidence from auditory event-related brain potentials. *Journal of Cognitive Neuroscience* **13**(2), 241–55.

Repp, B.H. (1992). A constraint on the expressive timing of a melodic gesture: Evidence from performance and aesthetic judgment. *Music Perception* **10**, 221–41.

Robazza, C., Macaluso, C., and D'Urso, V. (1994). Emotional reactions to music by gender, age, and expertise. *Perceptual and Motor Skills* **79**, 939–44.

Robbins, R. and McKone, E. (2007). No face-like processing for objects-of-expertise in three behavioural tasks. *Cognition* **103**, 34–79.

Rossion, B. and Curran, T. (2010). Visual expertise with pictures of cars correlates with RT magnitude of the car inversion effect. *Perception* **39**, 173–83.

Rubin-Rabson, G. (1940). The influence of age, intelligence, and training on reactions to classic and modern music. *Journal of General Psychology* **22**(2), 413–29. doi:10.1080/00221309.1940.9710043.

Schellenberg, E.G. (1996). Expectancy in melody: Tests of the implication-realisation model. *Cognition* **58**(1), 75–125.

Schellenberg, E.G. (2004). Music lessons enhance IQ. *Psychological Science* **15**, 511–4.

Schellenberg, E.G. (2006). Long-term positive associations between music lessons and IQ. *Journal of Educational Psychology* **98**(2), 457–68. doi:10.1037/0022-0663.98.2.457.

Schellenberg, E.G. (2011). Music lessons, emotional intelligence, and IQ. *Music Perception* **29**, 185–94.

Schellenberg, E.G. and Mankarious, M. (2012). Music training and emotion comprehension in childhood. *Emotion* **12**, 887–91.

Schmuckler, M.A. (1997). Expectancy effects in memory for melodies. *Canadian Journal of Experimental Psychology* **51**(4), 292–305.

Schön, D., Ystad, S., Regnault, P., et al. (2005). Sensory Consonance: An ERP Study. *Music Perception* **23**(2), 105–17.

Simon, C.R. and Wohlwill, J.F. (1968). An Experimental Study of the Role of Expectation and Variation in Music. *Journal of Research in Music Education* **16**(3), 227. doi:10.2307/3344079.

Sloboda, J.A. (1991). Musical Expertise. In K.A. Ericsson and J. Smith (eds), *Toward a General Theory of Expertise: Prospects and Limits*. Cambridge: Cambridge University Press, pp. 153–71.

Sloboda, J.A., Davidson, J.W., Howe, M.J.A., et al. (1996). The role of practice in the development of performing musicians. *British Journal of Psychology* **87**, 287–309. doi:10.1348/135910799168452.

Smith, J.D. and Melara, R.J. (1990). Aesthetic preference and syntactic prototypicality in music: tis the gift to be simple. *Cognition* **34**, 279–98.

Steinbeis, N., Koelsch, S., and Sloboda, J.A. (2006). The role of harmonic expectancy violations in musical emotions: Evidence from subjective, physiological and neural responses. *Journal of Cognitive Neuroscience* **18**(8), 1380–93.

Terwogt, M.M. and Van Grinsven, F. (1991). Musical expression of moodstates. *Psychology of Music* **19**, 99–109. doi:10.1177/0305735691192001.

Thompson, W.F., Schellenberg, E.G., and Husain, G. (2004). Decoding speech prosody: do music lessons help? *Emotion* **4**, 46–64. doi:10.1037/1528-3542.4.1.46.

Tillmann, B., Bharucha, J.J., and Bigand, E. (2000). Implicit learning of music: A self-organizing approach. *Psychological Review* **107**, 885–913.

Toro, J.M., Sinnett, S., and Soto-Faraco, S. (2005). Speech segmentation by statistical learning depends on attention. *Cognition* **97**, B25–B34.

Trimmer, C.G. and Cuddy, L.L. (2008). Emotional intelligence, not music training, predicts recognition of emotional speech prosody. *Emotion* **8**, 838–49. doi:10.1037/a0014080.

Varèse, E. and Wen-chung, C. (1966). The liberation of sound. *Perspectives of New Music* **5**, 11–9. doi:10.2307/832385.

Vitz, P.C. (1966). Affect as a function of stimulus variation. *Journal of Experimental Psychology* **71**, 74–9.

von Helmholtz, H. (1866). Concerning the perceptions in general. Treatise on Physiological Optics (3rd edn.), **Vol. III** (1866) (translated by J.P.C. Southall 1925 Opt. Soc. Am. Section 26, reprinted New York: Dover, 1962).

Vuust, P., Brattico, E., Seppänen, M., et al. (2012). The sound of music: Differentiating musicians using a fast, musical multi-feature mismatch negativity paradigm. *Neuropsychologia* **50**, 1432–43.

Vuust, P., Pallesen, K.J., Bailey, C., et al. (2005). To musicians, the message is in the meter pre-attentive neuronal responses to incongruent rhythm are left-lateralized in musicians. *NeuroImage* **24**(2), 560–4. doi:10.1016/j.neuroimage.2004.08.039.

Waterman, M.W. (1996). Emotional responses to music: implicit and explicit effects in listeners and performers. *Psychology of Music* **24**, 53–67.

Wundt, W. (1874). *Grundzüge der physiologischen Psychologie*. Leipzig: Wilhelm Engelmann.

Section 5

Neuropsychology of art and aesthetics

Chapter 17

The neuropsychology of visual art

Anjan Chatterjee

17.1 Introduction

Katherine Sherwood was 44 years old when she had a massive left hemisphere stroke. This catastrophic event left her with an aphasia and right-sided weakness. She was (and is, as at the time of writing this chapter) an artist and a professor. After recovering from the psychological impact and some of the neurological deficits from her stroke, she began to paint again using her left hand. She laid canvasses horizontally, and moved around them in a mobile chair. She also noticed a curious thing. The style and content of her painting changed (Waldman 2000). She previously regarded herself as a particularly cerebral painter, approaching her work in an over-intellectual manner. After her stroke, she felt that images flowed from her more easily and expressively (Sherwood 2012) (Figure 17.1). Something changed about her process of making art (Chatterjee 2008). What had changed?

More generally, cases like Professor Sherwood raise questions about what we can learn from the neuropsychology of art. Neuropsychology has been critical to advancing our knowledge of various cognitive systems, including perception, memory, and language, as shown by other chapters in this book. However, its impact on the study of aesthetics has been limited. Why? Certainly, an important factor is that as a domain of inquiry neuroaesthetics is relatively immature (Chatterjee 2004b, 2011). The data from which one might draw inferences are sparse. Beyond the constraints of limited data, fundamental questions about proper methods remain. It is easy to be seduced by the striking phenomenology and to construct interesting "just so" stories to account for these observations. Cognitive neurology and neuropsychology has had a long tradition of starting with careful clinical observations before applying rigorous quantitative methods to the phenomena under consideration. The neuropsychology of art is at this threshold and seems poised to make the quantitative shift (Chatterjee 2009).

One could speculate that artists, by virtue of their developed visuo-motor skills, might be relatively protected from the visuo-motor deficits associated commonly with brain damage. Such speculation appears not to be true when we consider the effects of brain damage on artistic production (Bogousslavsky and Boller 2005; Chatterjee 2004a, 2004b; Rose 2006; Zaidel 2005). Unilateral spatial neglect is a disorder in which patients are unaware of one side of space (Chatterjee 2003; Heilman et al. 1993). Artists with unilateral spatial neglect demonstrate neglect in their artwork (Jung 1974). Visual agnosia is the inability to perceive or recognize objects visually. Artists with visual agnosias have difficulty depicting objects in their art (Wapner et al. 1978). Achromatopsia is the inability to apprehend color.

(a)
(b)

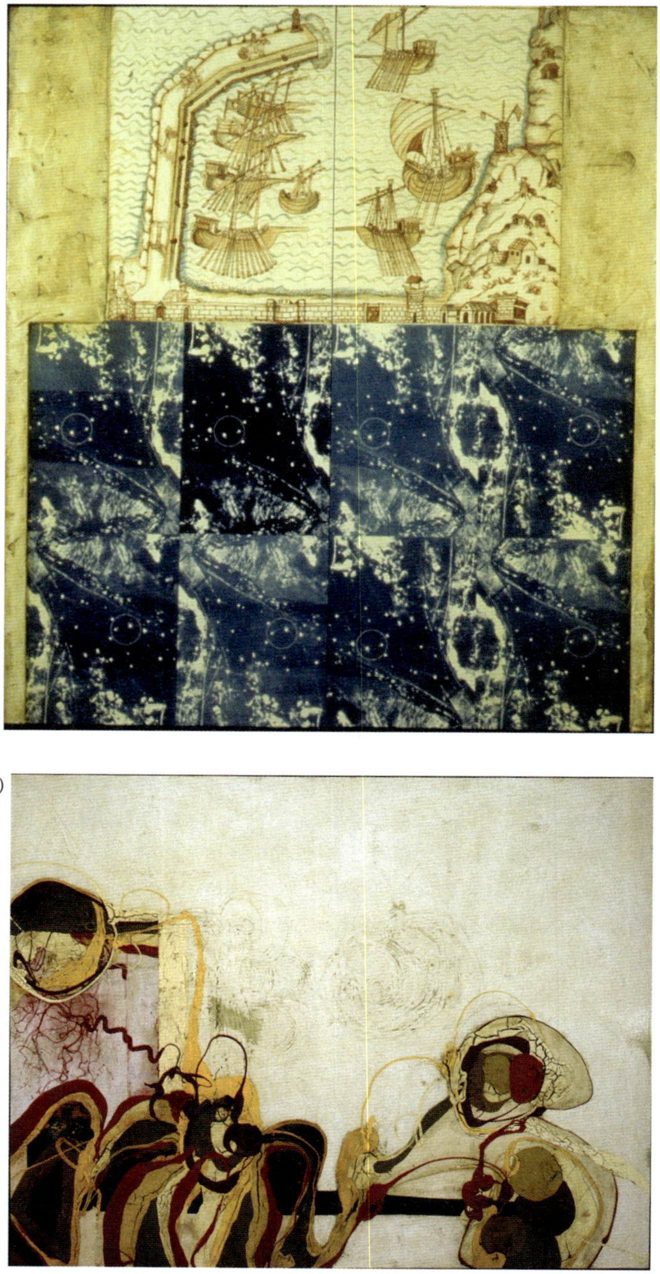

Figure 17.1 Examples of Katherine Sherwood's paintings. Permission obtained from Katherine Sherwood. (a) *Test Sites*, painted before her stroke. (b) *The Cart Before the Horse*, painted after her stroke. This material was originally published as Visual Art by Anjan Chatterjee in *The Roots of Cognitive Neuroscience: Behavioural Neurology and Neuropsychology*, edited by Anjan Chatterjee and H. Branch Coslett and has been reproduced by permission of Oxford University Press <http://ukcatalogue.oup.com/product/9780195395549.do>. For permission to reuse this material, please visit <http://www.oup.co.uk/academic/rights/permissions>.

Artists with achromatopsia may be unable to use color properly in their images (Sacks 1995a). In these cases, the artwork illustrates the nature of neuropsychological deficits. While the depictions may be beautiful and captivating, they do not offer insight into the nature of artistic production itself.

In this chapter, I focus on two aspects of the neuropsychology of art. First, I will examine situations in which brain damage produces a paradoxical facilitation (Kapur 1996) of artistic output. The effects of brain damage on the ability to produce visual art stand in sharp contrast to virtually all other complex human abilities. Diseases of the brain can impair our ability to communicate, coordinate movements, recognize objects, apprehend emotions, and make rational decisions. By contrast, while diseases of the brain often alter the ability to produce art, the alterations are sometimes interesting and occasionally even regarded as improvements (Chatterjee 2006), but they can still be regarded as not impeding an individual's ability to create artworks. Second, I will discuss the issue of measurement in the experience of viewing art. One might approach the idea of measuring aesthetic experiences with some trepidation. The fear is that quantifying these experiences robs them of that which is fundamental to them. However, if neuropsychology of art is to advance as a science, it must incorporate some form of quantification in its methods (Chatterjee et al. 2009). I will describe our attempts to do so.

17.2 The paradoxical facilitation of visual art

Implicit in a discussion of this paradoxical facilitation of art is the view that there is no single "art center" in the brain. Nor does one hemisphere play a privileged role in art production per se. Rather, the production of art is highly complex with different components mediated by different parts of the brain. The final artistic output emerges from a coordination of these different components. Brain damage alters the available parts of the brain dedicated to the overall artistic output that becomes the product of a different coordination of components. By analogy, we might think of these neural systems like a suspended mobile. The mobile rests still in the equilibrium of its weights. If a particular weight is removed, the entire configuration might collapse. However, the configuration might also find itself in a new resting state that is different from the original but which is also appealing. Similarly, brain damage may render an artist incapable of continuing to work, or it may create a new equilibrium in which the individual's artistic work proceeds in new and interesting configurations.

There are four ways in which neurological disorders might change and sometimes improve art production (Chatterjee 2006). These are the (1) disposition to produce visual art, (2) provision of a unique visual vocabulary, (3) aids to descriptive accuracy, and (4) changes in expressive powers.

17.2.1 Disposition to produce art

Fronto-temporal dementias (FTD) are a group of degenerative neurological diseases which cause people to undergo profound changes in their social interactions. They can become disinhibited and disorganized. They can have problems with language, attention span, and the ability to make decisions. In addition to these alterations in personality and cognition, a few people with FTD develop a propensity to produce art for the first time. Miller and colleagues (Miller et al. 1998) note that these people's art is realistic, not abstract or symbolic. It is highly detailed and has an obsessive quality. The patients themselves seem intensely preoccupied with their work. The artistic output of people with FTD appears to be a consequence of changes in their personalities. The acquired obsessive–compulsive traits find graphic expression and they produce striking visual images as a part of their repetition and attention to detail.

Several other examples show that obsessive–compulsive traits rendered by neurological diseases can predispose people to produce art. Sacks (1995b) described the Italian painter, Franco Magnani, living in San Francisco. Magnani painted hundreds of realistic scenes of an Italian town, Pontito, where he grew up. At the age of 31, Magnani probably had an encephalitic illness. Following that illness he painted compulsively. Pontito preoccupied his thoughts and conversations. Sacks speculated that he had partial complex seizures and was in part demonstrating an obsessive "sticky" personality sometimes associated with temporal lobe epilepsy (Waxman and Geschwind 1975). Such patients are also sometimes "hypergraphic," meaning they write incessantly. Magnani's hypergraphia was being expressed visually rather than verbally.

Lythgoe and colleagues (Lythgoe et al. 2005) reported the case of a builder with a subarachnoid hemorrhage. He also had no interest in art premorbidly but became an obsessive artist after recovery from the initial injury. After the hemorrhage, he had a normal verbal and performance IQ and exhibited behavioral patterns except for some degree of verbal disinhibition. He did well on most neuropsychological tasks except for those that involved mental flexibility. He also began to draw hundreds of sketches, mostly of faces. He then moved to large-scale drawings sometimes covering entire rooms, while confining his art to a few themes.

Finally, about 10 per cent of children with autism develop savant-like abilities (Rimland and Fein 1988). A sub-set of these children produce striking visual images (Sacks 1995c). The most detailed description of such a case was Nadia, reported by Selfe (1977). As a baby, Nadia did not respond to her mother and as she got older she lacked social empathy. As a child, she was obsessively concerned with the presence of other children without forming any substantial bonds with them. Her acquisition of language was delayed. Despite these developmental abnormalities, she was remarkable skilled at drawing. By the age of three years she was drawing life-like horses. She drew intensively for a few moments at a time, always copying images. She also focused on specific images, like horses, of which she drew hundreds of examples. While Nadia's abilities were striking, she was not unique. Autistic children with these striking drawing skills seem to focus on specific subjects and draw them repeatedly.

Thus, neurological disorders that produce obsessive–compulsive traits can also dispose people to produce art. These artists produce realistic images and focus on a narrow range of themes. While the neural basis for obsessive–compulsive disorders is not completely understood, dopamine and alterations of reward circuits are probably involved. We reported a patient with Parkinson's disease who produced art obsessively after being placed on dopamine agonists (Chatterjee et al. 2006). Interestingly, despite profound impairments in his motor control, which were evident when he wrote, he could make graceful sinuous movements when drawing. Obsessive–compulsive traits are also associated with dysfunction of the orbito-frontal and medial temporal cortices and fronto-striatal circuits (Kwon et al. 2003; Saxena et al. 1999; Ursu et al. 2003). Notably, in cases such as those described in this section, these regions may have been damaged, leaving posterior occipito-temporal cortices intact. The preservation of posterior cortices ensures that the neural substrate that represents faces, places, and objects are preserved and are available as the focus of these patients' obsessions.

17.2.2 Visual vocabulary

Neurological disorders such as migraine and epilepsy are associated with productive visual phenomena. Of over 200 entries submitted to the first National Art Competition sponsored by the British Migraine Association and WB Pharmaceuticals, 70 per cent showed spectral appearances, 48 per cent showed fortifications, 16 per cent showed areas of visual loss, and 2.5 per cent showed mosaic visions (Wilkinson and Robinson 1985).

The artist, Ignatius Brennan, (Podoll and Robinson 2000) eloquently expressed how migrainous auras inspired his art. His migraines, beginning at the age of 11 years, were experienced as frightening episodes of visual loss, often with a zigzag cloud obscuring much of his visual field. As he got older, he saw triangles and rounded forms as well as mosaics. He also experienced visual distortions of things getting larger or smaller. He recounts the effects of migraines on his art as follows:

> I started with pictures of my migraine experiences unconsciously rather than deliberately, when I was in art school. I used to do a lot of drawings of landscapes at that time and often found that I would be drawing clouds not just in the sky, but everywhere, which I think was a reference to the visual voids experienced during visual loss. I also used serrated zigzag shapes in my drawings, symbolizing the experience of a whole being broken up. . . . Clouds, zigzags and other imagery are part of my own personal visual vocabulary, but which certainly has come out of migraine experiences. I'm absolutely sure. I don't tend to do that deliberately, but when it suits a particular subject, e.g. to represent a feeling or an emotion, I make use of these images in different ways. . . .

Such patients offer insights into how artists generally have a visual vocabulary at their disposal. Greater analyses of these cases may inform us about how artists develop this visual vocabulary and how they develop a visual grammar in concatenating this visual vocabulary.

17.2.3 Descriptive accuracy

For centuries visual artists have been preoccupied with rendering objects and their surroundings accurately. Underlying the problem of descriptive accuracy in drawing and painting is the role of knowledge. Visual agnosias are a class of disorders in which patients have difficulty recognizing objects (Farah 1990). Since Lissauer's classic descriptions of visual agnosias, object recognition deficits are known to lie on a continuum between perceptual and conceptual deficits (Lissauer 1890). Perceptually based agnosias, called apperceptive agnosias, impair the ability to process the visual information into a coherent object. Conceptually based agnosias, called associative agnosias, involve impairments of semantic knowledge of the object (Farah 1990).

Wapner, Judd, and Gardner (1978) described an artist with an apperceptive agnosia who had difficulty copying images despite being able to convey depth and shading in drawings that were otherwise fragmented. His preserved semantic system did not help guide his artistic production. Thus, when asked to draw a telephone, he constructed images by reasoning, "It needs a base for it to stand on, a place to speak into, something to hear with a wire to plug in for communication and a place to dial." This verbal strategy was not effective in rendering accurate images. Semantic knowledge by itself does not help render objects accurately.

This patient contrasts with two people with associative agnosias (Franklin et al. 1992; Schwartz and Chawluck 1990). In both cases, when asked to draw objects from verbal labels these people drew crude, simplified images similar to those drawn by a young child. However, when drawing from complex visual images, the results were strikingly different. For example, one of these people could copy a portrait originally painted by Botticelli or draw a beautiful portrait of a staff worker (Franklin et al. 1992). People are thus able to render objects accurately and beautifully without semantic knowledge.

The case of a Polish aphasic artist also informs us about the relationships between semantics and art (Kaczmarek 1991). Profoundly influenced by the events of the Second World War, his premorbid paintings were anti-war statements. They often included numbers, letters, and ideograms. Following his stroke, he could only produce a few words. At its core, our semantic system functions to abstract and generalize. The use of abstracted symbols could be considered a marker for a preserved semantic system. Although testing of his semantic system was not reported, from the descriptions of his aphasia one might infer that it was impoverished. His inability to make use of verbal symbols extended to his artwork. He was no longer able to produce paintings in his previous style of using symbols to communicate his anti-war sentiments. However, he was still able to draw realistic landscapes and portraits well, despite having lost his ability to manipulate symbols!

Does semantic knowledge of an object hinder artistic production? The art historian Ernst Gombrich (1960) observed that even trained artists impose knowledge of what they are looking at in their depictions in a way that can compromise their accuracy. Thus, impaired knowledge, provided visual-motor systems are intact, might aid in the ability to depict objects and scenes accurately. Perhaps such impaired knowledge accounts for cases of autistic children with savant-like artistic abilities.

Autistic artists need only to look at an object for a few minutes before being able to draw them rapidly and accurately (Sacks 1995c; Selfe 1977). Nadia's abilities were not an accelerated version of other children's drawing development. She did not first pass through a phase of drawing simple schematic images before learning to draw realistically. Rather her skills developed very early and did not change much over time. Nadia initially drew horses deftly and without hesitation. Two observations suggest that she treated horses as visual patterns rather than as objects. First, she would start drawing anywhere on the page. Rather than trying to squeeze the whole horse into a page, she would stop drawing when she came to the paper's edge even if it meant only drawing part of the horse. Secondly, most people draw horses by starting at the head. Nadia could start her drawing at the neck and seemed unaffected by critical features associated with a specific object, such as the head of a horse. Her remarkable skill at drawing horses, and later pelicans, was not hindered by semantic associations that interfere with the ability to "see" the visual object. As Nadia eventually acquired language her drawings became more prosaic. Presumably, the acquisition of language indicated development of a richer semantic system and detracted from her artistic skills.

17.2.4 Changed expressivity

Visual art is of course not restricted to rendering objects and scenes accurately. Perhaps driven by the advent of photography, visual art diverged into many forms of expression. Among the most intriguing effects of brain damage on artists are a class of phenomena in which the inability to make accurate depictions results in surprisingly appealing stylistic changes. These changes can occur in the use of color and form and in the content of images.

Sacks (1995a) described an artist with an acquired achromatopsia following a traumatic brain injury. His earlier paintings were colorful and abstract. After the accident everything appeared "dirt gray" to him. His initial attempts to use color did not work and he resigned himself to painting in black and white. Eventually he incorporated a limited set of colors in his paintings. After an initial sense of helplessness, he considered his new way of seeing as a strange gift. He thought he saw the world as pure form, uncluttered by color, and endowed with a new range of expressions.

Right hemisphere damage can produce left spatial neglect and artists with neglect omit the left side of images that they draw or paint (Blanke et al. 2003; Cantagallo and Sala 1998; Halligan and Marshall 1997; Jung 1974; Marsh and Philwin 1987; Schnider et al. 1993). As they recover from their neglect, their use of line may still be altered. Two examples show how this change in the use of line can produce art that is highly regarded. Lovis Corinth, an important German artist, had a right hemisphere stroke in 1911. As he recovered, he resumed painting. His self-portraits and those of his wife showed clear changes in style, with details on the left sometimes omitted and textures on the left blending with the background. Alfred Kuhn characterized this work as follows (quoted in Gardner 1975):

> He [Corinth] had become prescient for the hidden facets of appearance. . . . The contours disappear, the bodies are often as ript asunder, deformed, disappeared into textures . . . also the faithfulness of portraits had ceased almost entirely. . . . With wide stripes the person is captured in essence. Characterization is now exaggerated, indeed, often to caricature . . . Corinth always seems to be painting a picture behind the picture, one which he alone sees . . . at this point Corinth shifted from the ranks of the great painters into the circle of the great artists.

Heller (1994) reported the experience of the artist Loring Hughes, who after a right hemisphere stroke had difficulty coordinating the spatial relationship between lines. This forced her to abandon her premorbid style of accuracy in realistic depiction. Instead she turned to her own imagination and emotions. Initially, she was too ashamed to display her paintings, but she began to show publicly once she became comfortable with her new style. The artistic community responded well to these distorted images. The critic Eileen Watkins described her work as now delivering "an emotional wallop" that was not present previously.

When they occur as a result of left brain damage, stylistic changes might be different from those observed stemming from right brain damage. The specific changes reported are the introduction of more vivid colors and a change in content. These changes are exemplified in the Bulgarian painter, Zlatio Boiyadjiev, the Californian artist Katherine Sherwood who was mentioned in the introduction, and a Swiss painter reported by Annoni and colleagues.

Boiyadjiev's premorbid artistic style was natural and pictorial and he tended to use earth tones in his paintings. Following the onset of his aphasia, Boiyadjiev's paintings have been described as richer, more colorful, and containing more fluid and energetic lines (Brown 1977; Zaimov et al. 1969). The imagery in his work became more inventive and at times even bizarre and fantastical. Similarly, Katherine Sherwood suffered a left hemisphere hemorrhagic stroke that left her with an aphasia and right-sided weakness (Waldman 2000). She trained herself to paint with her left hand, and since then her career has flourished. Premorbidly, her images were described as "highly cerebral" incorporating a range of esoteric images such as cross-dressers, medieval seals, and spy photos. After her stroke she felt that she could not produce these images even if she tried. Her new style is described as "raw" and "intuitive," with large irregular circular movements. She says her left hand enjoys an ease and a grace with the brush that her right hand never had, and describes it as "unburdened" (Sherwood 2012). Finally, Annoni and colleagues (2004) recently described a Swiss landscape painter whose art was described as being "figurative–impressionist." He had a small stroke in the left thalamus. His wife thought he had mild emotional dysfunction after his stroke. He felt that he was more sensitive to the hidden beauty of images and used bolder colors. He switched from realistic to more impressionistic images. He thought that he was less likely to use lines, contours, and perspective clearly and felt that he was more creative.

A final stylistic change rendered by brain damage is a move towards simplicity. Annoni and colleagues (2004) described a person with a left occipital lesion. A month after his stroke he resumed drawing and painting. His new artwork was simplified, stylized, increasingly abstract, and confined to a limited use of colors. It is not known if damage to visual association cortices would consistently result in simplification and abstraction. A few artists with Alzheimer's disease have continued to paint (Crutch et al. 2001; Maurer and Prvulovic 2004; Miller and Hou 2004) and demonstrate a similar pattern. William Utermohlen painted a series of self-portraits several years into the course of his illness. As these portraits became increasingly simplified and distorted, they appear as haunting psychological self-expressions.

Willem de Kooning is probably the best-known artist who continued to paint after the onset of Alzheimer's (Storr 1995). De Kooning's ex-wife and students provided the structure for him to continue to work. They stretched his canvases and mixed his colors. Experts think this late period represents a new and coherent style. His paintings were simpler and he confined his palette to primary colors. Traces of shapes from earlier works are evident, but are pared down. Gary Garrels (1995), the senior curator at the San Francisco Museum of Modern Art, thought "the vocabulary of forms was retained, but clarified. . . . the results are paintings of an openness and freedom not seen before, paintings that are extraordinarily lyrical, immediately sensual, and exhilarating."

17.3 Quantification in the neuropsychology of art

Despite the fascinating observations recounted here, much needs to be done if neuropsychology is to make substantive contributions to empirical aesthetics. For the field to mature, it must move beyond recounting interesting anecdotes. Methodologically it is critical to quantify features of artwork in order to assess change. Most of the cases reported are devoid of measurement. Claims about changes in art after brain damage are usually post-hoc comments supported by a few illustrative examples of artwork.

We need an instrument to assess art quantitatively. Such an instrument should have at least two characteristics. First, it should assess artwork via a comprehensive set of attributes which would include formal properties (use of line, color, composition, etc.) and content properties (abstract, realism, etc.). Second, it should distinguish between descriptive and evaluative judgments. With such an instrument one could begin to impose quantitative structure on otherwise casual observations about works of art.

With these considerations in mind, we developed the Assessment of Art Attributes (AAA) (Chatterjee et al. 2010). The AAA assesses 12 descriptive attributes that would apply to any piece of visual art. Six attributes refer to formal/perceptual properties and six to content/conceptual properties. The six formal/perceptual attributes are: *Balance*, *Color Saturation*, *Color Temperature*, *Depth*, *Complexity*, and *Stroke Style*. The six content/conceptual attributes are: *Abstractness*, *Animacy*, *Emotionality*, *Realism*, *Representational Accuracy*, and *Symbolism*. The stimuli in the AAA consist of 24 paintings from the Western canon. These are paintings by well-known artists, but not their most famous works and encompass a range in the six formal/perceptual and six content/conceptual attributes of interest. The paintings are also evaluated for preference or liking and their level of interest for the viewer.

17.4 Art production

We have used the AAA to examine changes in art production as well as art perception following brain damage. For art production, we examined the work of three artists that have already been mentioned (Chatterjee et al. 2011). These are Sherwood and Boiyadjiev, both with left brain damage, and Corinth with right brain damage. In each case we sampled several paintings completed before and after the stroke occurred. Average ratings on each attribute were obtained for each painting. Then the average ratings for all the pre- and post-morbid paintings were obtained for each participant for each artist.

We examined Sherwood and Boiyadjiev's work to determine if left brain damage produces consistent changes in art. Sherwood describes her approach to her work premorbidly as over-intellectualized (Chatterjee 2008). Post-morbidly, her work has been described as raw and more expressive, and less forced (Sherwood 2012). In concordance with these descriptions, raters found her work following brain injury to be more abstract, more symbolic, more distorted, more vibrant, less realistic, and depicted with looser strokes. Raters also found her work to be flatter and to contain warmer colors, changes not mentioned previously by critics describing her work. Boiyadjiev's work has been described as becoming fantastic and sometimes bizarre with richer, more colorful forms, and having fluid, energetic lines (Brown 1977). Consistent with these descriptions, raters found his work to be more abstract, more symbolic, less realistic, more distorted as well as having a looser stroke style. They also found his work to be flatter and less animated, changes also not mentioned by critics before.

Could motor deficits for Sherwood and Boiyadjiev have caused these changes? Both were right-handed artists who began to paint with their left hands after their strokes. From these data, it is not clear if some pictorial aspects of the artwork, such as the coarser brushstrokes, changed because of hemispheric brain damage or because they began to use their left hand. However, a shift to the left hand is unlikely to explain the use of increasingly vibrant colors or changes in conceptual attributes of their artwork, such as greater symbolism or abstraction.

A priori, one could be pessimistic about the prospects of identifying systematic effects of brain damage on art production. After all, artistic styles and content vary so much across different artists that one might be comparing changes in qualitatively different kinds of objects. Our observations of Sherwood and Boiyadjiev's paintings suggest that this pessimism might not be warranted. Sherwood and Boiyadjiev's artistic styles are quite different from each other. For example, Sherwood's paintings started out being substantially flatter than Boiyadjiev's. Critically, both artists' paintings were judged as becoming flatter following their strokes, despite the fact that Sherwood's paintings executed *before* her stroke were more similar to Boiyadjiev's paintings made *after* his stroke in depicting depth. Thus, it would be a mistake to think that people with left brain damage produce a *prototypic style* of painting. Rather, it appears that left brain damage produces a *prototypic shift* in style of painting.

Sherwood and Boiyadjiev's paintings became more abstract, symbolic, and distorted as well as less realistic, and they were painted with looser strokes, more vibrant colors, and less depth (Chatterjee et al. 2011). Examining Corinth's work would test the specificity of these changes. Corinth's paintings after his stroke were described as "deformed," and attempts at producing faithful portraits "ceased almost entirely". Blanke (2006) describes a broadening of brushstrokes, a lack of depth, less spatial detail, and several deformities depicted on the left side of his self-portrait paintings. Researchers found Corinth's paintings to be more abstract, more distorted, less realistic, and also to exhibit looser strokes and a flatter perspective.

When Sherwood, Boiyadjiev, and Corinth's paintings are considered together, the following changes were found in all three artists' works. Their paintings became more abstract and distorted and less realistic and accurate. They were also rendered with looser strokes, less depth, and with more vibrant colors. Thus, none of these changes can be ascribed to laterality of brain functions. Also, their paintings did not change in complexity or the level of emotion being conveyed. It remains to be seen whether these attributes would be susceptible to change with other kinds of neurological illness. For example, we might expect the artworks of people with Alzheimer's disease to become less complex over time.

All the changes observed in Corinth's paintings were also observed in works by both Sherwood and Boiyadjiev (Chatterjee et al. 2011). Thus, chronic right hemisphere damage does not appear to produce specific patterns of change in artistic production. By contrast, both Sherwood and Boiyadjiev's subjects became more symbolic. This could mean that artists with left brain damage, because of an unfettered right hemisphere, engage meaning more loosely, specifically in the use of symbolism. These data show that the popular idea that right hemisphere is the artistic hemisphere is wrong. Clearly, both hemispheres participate in artistic production as shown by the fact that these artists' paintings changed regardless of which hemisphere was damaged. If anything, damage to the left hemisphere produced more significant alterations in artistic production than did damage to the right hemisphere.

We have also used the AAA to assess gradual changes in artistic styles in two patients with Alzheimer's disease (AD). Investigating artists with AD allows us to look for gradual changes in art production rather than sudden changes brought about by a stroke.

William Utermohlen was diagnosed with AD after a long, successful painting career (Crutch et al. 2001). He was born in Philadelphia and studied at the Pennsylvania Academy of Fine Arts from 1951 to 1957. He moved to England in 1957 where he enrolled at the Ruskin School of Art in Oxford before settling in London. His early work was characterized by linear expressionism, with frequent inclusion of Pop imagery and styles, including a use of strong colors. His earliest symptoms began four years before diagnosis, and involved difficulties with tying a necktie, calculating household finances, and memory of day-to-day events. Formal neuropsychological examination revealed a moderate degree of global cognitive deterioration, while MRI indicated generalized cerebral atrophy. Following the initial diagnosis at the age of 61, regular clinical assessments documented the expected gradual decline in cognitive function (Crutch and Rossor 2006). We looked at five self-portraits executed by William Utermohlen in the four-year period following his diagnosis.

The second artist, Lester Potts, was diagnosed with Alzheimer's disease around age 70. He had worked in a rural Alabama sawmill during the Great Depression. He served in the Korean War and was regarded as a dependable civic leader. He had not painted before his enrolment at an adult day-care center. At the center he was taught to use watercolors by a retired artist as part of a community outreach program. By the time he died at age 78 Lester Potts had painted over 100 original watercolors.

Using the AAA, we found that both artists' paintings became more abstract and more symbolic, less realistic and less accurate in depicting figurative detail. Potts' paintings also showed more color saturation, shifts in hues, less complexity, and less emotion as his disease progressed. Notably, changes were not seen with balance, depth, or stroke quality in either person's paintings.

Our observations were in general accord with other observations of the relationship between AD and artistic production and offered greater detail in the nature of change observed. For example, Fornazzari (2005) describe changes in the portraiture of a woman who suffered from AD. These changes include trends toward "unusual figure fond, loss of proportion in the facial features, and loss of proportionality. . . ." The changes mentioned might parallel those observed in Utermohlen and Potts. "Unusual figure fond" and "loss of proportionality" might correspond to a decline in depictive accuracy. However, it is not clear if loss of proportionality could refer to loss of balance, which we did not see in either person. This uncertainty highlights one of our points, which is that clear operational definitions of art attributes are needed if we are to compare results across studies.

Our findings also validate anecdotal observations of changes in abstraction and increased symbolism among artists with AD (Crutch et al. 2001; Cummings and Zarit 1987; Rankin et al. 2007; Chatterjee 2004a). For example, art historians describe a heightened trend towards abstraction in paintings executed by Willem de Kooning during the progression of his AD (Garrels 1995). Maurer and Prvulovic (2004) thought that in the later works of the painter Carolus Horn, "ornamental symbols and mythical creatures appeared, which were derived from a conjunction of different species." Cummings and Zarit (1987) reported that an artist with AD over a period of two-and-a-half years moved towards simplicity and distortion. Miller and Hou (2004) observed AD artists to produce works with less precision and attention to spatial relationships.

17.5 **Art perception**

We know virtually nothing about the neuropsychology of the perception of art. One might expect that sensitivity to different attributes within a work of art would be affected by different kinds of brain damage. Recently, we used the AAA to examine the effects of brain damage on art perception (Bromberger et al. 2011). Our investigation focused on the role of the right hemisphere. The right hemisphere participates prominently in visuospatial attention and representation (Chatterjee 2003; Heilman et al. 1993). As a first step, we wished to avoid confounding language comprehension with perceptual judgments in our study. For example, if a participant does not understand what the word "symbolic" means, it would not be difficult to assess their perception of symbolism in any painting. We also use contemporary lesion analysis methods in our study. Voxel-based lesion-symptom mapping (VLSM) techniques allow us formally to assess the way in which damage to a brain area correlates with behavioral scores (Bates et al. 2003; Kimberg et al. 2007; Wu et al. 2007) with the advantage that one does not have to establish a deficit cut-off. Rather, behavior in VLSM considered a continuous variable.

In this study, the judgments of 20 people with right brain damage were compared to 30 healthy age-matched control participants. Of all the content/conceptual qualities, damage to the right frontal lobe was associated with differences in judgments of abstractness, realism, animacy, and symbolism in visual art. In addition, damage to the right parietal lobe was related to deviations in judgments of animacy and symbolism. There were no brain–behavior relationships for differences in evaluative judgments, those of preference or level of interest for the viewer.

Any aesthetic experience is built upon at least three components (Chatterjee 2004b; Chatterjee 2011). These components are the experience of sensory qualities, the associated sets of meanings, and the emotional responses evoked by the aesthetic object. Broadly, one might regard the formal-perceptual attributes of the AAA as probing the sensory experience, the content-conceptual attributes as probing the meaning, and the evaluative questions as probing the emotional response to these paintings. The data from this study suggests that these three components of visual aesthetic experiences might segregate broadly in the organization of the brain. Most of our participants had damage in the distribution of the right middle cerebral artery. This distribution of brain damage involving lateral frontal-parietal and temporal cortices was more likely to affect judgments of conceptual attributes. We would predict that damage in the posterior cerebral artery distribution affecting ventral occipital and temporal cortices might be more likely to affect the perceptual attributes. Furthermore, given the extensive data implicating the ventral striatum and orbitofrontal cortex in subjective rewards (Kable and Glimcher 2009), damage to ventro-medial prefrontal cortices is plausibly more likely to affect people's evaluation of paintings.

These recent results demonstrate the feasibility of conducting systematic quantitative studies in the neuropsychology of art. This point by no means implies that theoretical analyses or qualitative approaches to art are not useful. It does mean that if the neuropsychology of art is to mature *as a science*, it needs to incorporate quantitative approaches, and formal tests of hypotheses as part of its research program.

17.6 Conclusion

In reviewing the literature of the neuropsychology of art production, I emphasized the possible enhancing effects of brain damage on artistic production. Of course, these observations are based on selected artists. It is likely that many more artists are devastated by their brain injury and these cases are not reported. We simply do not know the base rate of facilitation effects. Yet, the examples of changed and possibly improved art following brain damage point to the multifaceted nature of art. In my view, exploring the mechanisms underlying the alterations and gaining insights into these facets is one way that neuropsychology can advance neuroaesthetics. Looking forward, unless the neuropsychology of art can incorporate the basic rigors of any experimental science, we will not get past the description of seductive artistic phenomenology under consideration. We are in a position to augment qualitative insights with quantitative approaches. The observations from patients can also serve to generate hypotheses to be tested using functional imaging techniques. For example, the hypothesis that the perception of art dissociates from the evaluation of art is amenable to functional imaging methods. We can be guardedly optimistic that the neuropsychology of art may mature and serve as a fundamental anchor in the new field of neuroaesthetics.

Acknowledgment

This material was originally published as the chapter "Visual Art" by Anjan Chatterjee, pp. 349–66 in *The Roots of Cognitive Neuroscience: Behavioural Neurology and Neuropsychology* edited by Anjan Chatterjee and H. Branch Coslett and has been reproduced by permission of Oxford University Press, <http://ukcatalogue.oup.com/product/9780195395549.do>. For permission to reuse this material, please visit <http://www.oup.co.uk/academic/rights/permissions>.

References

Annoni, J., Devuyst, G., Carota, A., et al. (2004). Changes in artistic style after minor posterior stroke. *Journal of Neurology, Neurosurgery and Psychiatry* **76**, 797–803.

Bates, E., Wilson, S., Saygin, A., et al. (2003). Voxel-based lesion symptom mapping. *Nature Neuroscience* **6**, 448–50.

Blanke, O. (2006). Visuo spatial neglect in Lovis Corinth's self portraits. *International Review of Neurobiology* **74**, 193–214.

Blanke, O., Ortigue, S., and Landis, T. (2003). Color neglect in an artist. *Lancet* **361**, 264.

Bogousslavsky, J. and Boller, F. (2005). *Neurological Disorders in Famous Artists.* Basel: Karger.

Bromberger, B., Sternschein, R., Widick, P., et al. (2011). The right hemisphere in aesthetic perception. *Frontiers in Human Neuroscience* **5**. doi: 10.3389/fnhum.2011.00109.

Brown, J. (1977). *Mind, Brain, and Consciousness. The Neuropsychology of Cognition.* New York, NY: Academic Press.

Cantagallo, A. and Sala, S.D. (1998). Preserved insight in an artist with extrapersonal spatial neglect. *Cortex* **34**, 163–89.

Chatterjee, A. (2003). Neglect. A disorder of spatial attention. In M. D'Esposito (ed.), *Neurological Foundations of Cognitive Neuroscience.* Cambridge, MA: The MIT Press, pp. 1–26.

Chatterjee, A. (2004a). The neuropsychology of visual artists. *Neuropsychologia* **42**, 1568–83. doi: http://dx.doi.org/10.1016/j.neuropsychologia.2004.03.011.

Chatterjee, A. (2004b). Prospects for a cognitive neuroscience of visual aesthetics. *Bulletin of Psychology and the Arts* **4**, 55–9.

Chatterjee, A. (2006). The neuropsychology of visual art: conferring capacity. *International Review of Neurobiology* **74**, 39–49. doi: http://dx.doi.org/10.1016/S0074-7742(06)74003-X.

Chatterjee, A. (2008). Apoplexy and personhood in Katherine Sherwood's paintings. In K. Sherwood (ed.), *Golgi's Door.* Washington, DC: National Academy of Sciences, pp. 44–52.

Chatterjee, A. (2009). Prospects for a neuropsychology of art. In M. Skov and O. Vartanian (eds), *Neuroaesthetics.* Amityville, New York: Baywood Publishing Company, pp. 131–43.

Chatterjee, A. (2011). Neuroaesthetics: a coming of age story. *The Journal of Cognitive Neuroscience* **23**, 53–62. doi: doi:10.1162/jocn.2010.21457.

Chatterjee, A., Bromberger, B., Smith, W.B., et al. (2011). Artistic production following brain damage: a study of three artists. *Leonardo* **44**, 405–10. doi: http://dx.doi.org/10.1162/LEON_a_00240.

Chatterjee, A., Hamilton, R.H., and Amorapanth, P.X. (2006). Art produced by a patient with Parkinson's disease. *Behavioural Neurology* **17**, 105–8.

Chatterjee, A., Thomas, A., Smith, S.E., et al. (2009). The neural response to facial attractiveness. *Neuropsychology* **23**, 135–43.

Chatterjee, A., Widick, P., Sternschein, R., et al. (2010). The assessment of art attributes. *Empirical Studies of the Arts* **28**, 207–22. doi: 10.2190/EM.28.2.f.

Crutch, S., Isaacs, R., and Rossor, M. (2001). Some workmen can blame their tools: artistic change in an individual with Alzheimer's disease. *Lancet* **347**, 1096–8. doi: 10.1016/S0140–6736(00)05187–4.

Crutch, S.J. and Rossor, M.N. (2006). Artistic changes in Alzheimer's disease. *International Review of Neurobiology* **74**, 147–61. doi: 10.1016/S0074–7742(06)74012–0.

Cummings, J.L. and Zarit, J.M. (1987). Probable Alzheimer's disease in an artist. *Journal of the American Medical Association* **258**, 2731–34. doi: 10.1001/jama.1987.03400190113039.

Farah, M.J. (1990). *Visual Agnosia*. Cambridge, MA: The MIT Press.

Fornazzari, L.R. (2005). Preserved painting creativity in an artist with Alzheimer's disease. *European Journal of Neurology* **12**, 419–24. doi: 10.1111/j.1468–1331.2005.01128.x.

Franklin, S., van Sommers, P., and Howard, D. (1992). Drawing without meaning?: dissociations in graphic performance of an agnosic artist. In R. Campbell (ed.), *Mental Lives. Case Studies in Cognition*. Cambridge, MA: Blackwell, pp. 9–37.

Gardner, H. (1975). *The Shattered Mind. The Person After Brain Damage*. New York, NY: Alfred A. Knopf.

Garrels, G. (1995). *Three toads in the garden: line, color, and form* In J. Jenkins and S. Engberg (eds), *Willem de Kooning. The Late Paintings, The 1980s*. Minneapolis: San Francisco Museum of Modern Art and Walker Arts Center, pp. 9–37.

Gombrich, E. (1960). *Art and Illusion*. Princeton, NJ: Princeton University Press.

Halligan, P.W. and Marshall, J.C. (1997). The art of visual neglect. *Lancet* **350**, 139–40.

Heilman, K.M., Watson, R.T., and Valenstein, E. (1993). Neglect and related disorders. In K.M. Heilman and E. Valenstein (eds), *Clinical Neuropsychology* (3rd edn). New York, NY: Oxford University Press, pp. 279–336.

Heller, W. (1994). Cognitive and emotional organization of the brain: Influences on the creation and perception of art. In D. Zaidel (ed.), *Neuropsychology*. New York, NY: Academic Press, pp. 271–92.

Jung, R. (1974). Neuropsychologie und neurophysiologie des kontur-und formsehens in zeichnerei und malerei. In H. Weick (ed.), *Pyschopathologie Mususcher Gestaltungen*. Stuttgart: F.K. Shattauer, pp. 27–88.

Kable, J.W. and Glimcher, P.W. (2009). The neurobiology of decision: consensus and controversy. *Neuron* **63**, 733–45.

Kaczmarek, B. (1991). Aphasia in an artist: a disorder of symbolic processing. *Aphasiology* **5**, 361–71.

Kapur, N. (1996). Paradoxical functional facilitation in brain-behavior research. *Brain* **119**, 1775–90.

Kimberg, D.Y., Coslett, H.B., and Schwartz, M.F. (2007). Power in Voxel-based Lesion-Symptom Mapping. *Journal of Cognitive Neuroscience* **19**, 1067–80.

Kwon, J., Kim, J., Lee, D., et al. (2003). Neural correlates of clinical symptoms and cognitive dysfunctions in obsessive-compulsive disorder. *Psychiatry Research* **122**, 37–47.

Lissauer, H. (1890). Ein Fall von Seelenblindheit Nebst Einem Beitrage zur Theori derselben. *Archiv fur Psychiatrie und Nervenkrankheiten* **21**, 222–70.

Lythgoe, M., Polak, T., Kalmus, M., et al. (2005). Obsessive, prolific artistic output following subarachnoid hemorrhage. *Neurology* **64**, 397–98.

Marsh, G.G. and Philwin, B. (1987). Unilateral neglect and constructional apraxia in a right-handed artist with a left posterior lesion. *Cortex* **23**, 149–55.

Maurer, K. and Prvulovic, D. (2004). Paintings of an artist with Alzheimer's disease: visuoconstructive deficits during dementia. *Journal of Neural Transmission* **111**, 235–45. doi: 10.1007/s00702-003-0046-2.

Miller, B.L., Cummings, J., Mishkin, F., et al. (1998). Emergence of artistic talent in frontotemporal dementia. *Neurology* **51**, 978–82.

Miller, B. and Hou, C. (2004). Portraits of artists: Emergence of visual creativity in dementia. *Archives of Neurology* **61**, 842–4.

Podoll, K. and Robinson, D. (2000). Migraine experiences as artistic inspiration in a contemporary artist. *Journal of the Royal Society of Medicine* **93**, 263–5.

Rimland, B. and Fein, D. (1988). Special talents of autistic savants. In L. Obler and D. Fein (eds), *The Exceptional Brain*. New York, NY: Guilford, pp. 474–92.

Rose, F.E. (2006). *The Neurobiology of Painting*. London: Academic Press.

Sacks, O. (1995a). *The case of the colorblind painter: An Anthropologist on Mars*. New York, NY: Alfred A. Knopf, pp. 3–41.

Sacks, O. (1995b). *The landscape of his dreams: An Anthropologist on Mars*. New York, NY: Alfred A. Knopf, pp. 153–87.

Sacks, O. (1995c). *Prodigies: An Anthropologist on Mars*. New York, NY: Alfred A. Knopf, pp. 188–243.

Saxena, S., Brody, A., Maidment, K., et al. (1999). Localized orbitofrontal and subcortical metabolic changes and predictors of response to paroxetine treatment in obsessive–compulsive disorder. *Neuropsychopharmacology* **21**, 683–93.

Schnider, A., Regard, M., Benson, D.F., et al. (1993). Effect of a right-hemisphere stroke on an artist's performance. *Neuropsychiatry, Neuropsychology and Behavioral Neurology* **6**, 249–55.

Schwartz, M. and Chawluck, J. (1990). Deterioration of language in progressive aphasia: a case study. In M. Schwartz (ed.), *Modular Deficits in Alzheimer-Type Dementia*. Cambridge, MA: The MIT Press, pp. 245–96.

Selfe, L. (1977). *Nadia. A case of extraordinary drawing ability in an autistic child*. New York, NY: Academic Press.

Sherwood, K. (2012). How a cerebral hemorrhage altered my art. *Frontiers in Human Neuroscience* 6. doi: 10.3389/fnhum.2012.00055/abstract.

Storr, R. (1995). At last light. In J. Jenkins and S. Engberg (eds), *Willem de Kooning. The Late Paintings, The 1980s*. Minneapolis, MN: San Francisco Museum of Modern Art and Walker Arts Center, pp. 37–79.

Ursu, S., Stenger, V., Shear, M., et al. (2003). Overactive action monitoring in obsessive-compulsive disorder. *Psychological Science* **14**, 347–53.

Waldman, P. (2000). Master stroke: A tragedy transforms a right-handed artist into a lefty—and a star. *Wall Street Journal*, Friday 12 May.

Wapner, W., Judd, T., and Gardner, H. (1978). Visual agnosia in an artist. *Cortex* **14**, 343–64.

Waxman, S. and Geschwind, N. (1975). The interictal behavior syndrome associated with temporal lobe epilepsy. *Archives of General Psychiatry* **32**, 1580–6.

Wilkinson, M. and Robinson, D. (1985). Migraine art. *Cephalalgia* **5**, 151–7.

Wu, D.H., Waller, S., and Chatterjee, A. (2007). The functional neuroanatomy of thematic role and locative relational knowledge. *Journal of Cognitive Neuroscience* **19**, 1542–55.

Zaidel, D. (2005). *Neuropsychology of Art*. New York, NY: Psychology Press.

Zaimov, K., Kitov, D., and Kolev, N. (1969). Aphasie chez un peintre. *Encephale* **58**, 377–417.

Chapter 18

The creation of art in the setting of dementia

Indre V. Viskontas and Suzee E. Lee

18.1 Introduction

While many researchers studying the neuroscience of aesthetics are primarily interested in describing how healthy brains create, produce, and appreciate art, there is much to learn from individuals whose neurological illnesses alter their artistic output. In particular, there are rare cases of patients whose art gains greater critical acclaim as their neurodegenerative disease progresses (Miller et al. 1998; Chakravarty 2011). These unique patients provide illuminating glimpses into the brain regions that underlie creativity, as well as insights into what aspects of art are appreciated by specialists and the general public.

Increasingly, neuroscientists are conceptualizing mind–brain relationships in terms of circuits of brain regions that work together to create thoughts, feelings, experiences, and actions, rather than focusing on a single area of the brain. In line with this trend, clinician-scientists characterize neurodegenerative disorders with respect to which circuits are affected by disease. This characterization is also useful when it comes to a discussion of the neural underpinnings of artistic production because producing art is a multifaceted task, demanding the participation of many distinct brain regions.

A comprehensive model of the neural basis of creativity remains elusive in part because defining and measuring the value of art is subjective. While most researchers agree that creativity involves producing ideas or products that are both novel and useful, this definition is more difficult to apply to the production of visual art than to creative cognition or other manifestations of creativity. Here, the goals of science and art clash: while science is used to extract general rules or principles about the world, art is by its very nature subjective—designed to use the individual experience to demonstrate what is universal. Therefore, this chapter will focus both on studies of groups of patients and healthy individuals and exemplary case histories that provide additional insights into our capacity to produce art.

18.2 The path of degeneration in dementia

Neurodegenerative diseases that cause dementia begin with the aggregation of misfolded proteins in specific parts of the brain (Pick 1892; Alzheimer 1911). These misfolded proteins cause neurons to malfunction and die, and there is evidence to suggest that degeneration spreads via functionally connected brain areas (Zhou et al. 2012) such that each neurodegenerative disease targets a different network of these regions (Seeley et al. 2009). Thus, each disease recapitulates unique constellations of symptoms that reflect the functions of the affected neural circuits (Seeley et al. 2009; Zhou et al. 2010).

Clinical examination results in classification of neurodegenerative diseases based on symptoms and physical examination findings. Neuropathological examination enables neurodegenerative diseases to be classified based on characteristics of disease proteins seen under microscopy. Diagnoses of clinical and neuropathological examinations, however, do not always concur in individual patients. Since the majority of the scientific literature on neurodegenerative disease and creativity is based on clinical classifications, this chapter will primarily use clinical characterizations.

We will focus on Alzheimer's disease (AD), the most common neurodegenerative disease and frontotemporal dementia (FTD), a common dementia in those under 65 years of age. Common diseases with parkinsonism include Lewy Body disease and Parkinson's disease; progressive supranuclear palsy and corticobasal degeneration are rare. FTD is further classified into behavioral variant FTD (bvFTD) and two progressive disorders of language known as primary progressive aphasias (PPAs), including semantic (svPPA) and nonfluent (nfvPPA) variants (Gorno-Tempini et al. 2011). Before inferences can be drawn regarding the general effect of neurodegeneration on artistic proclivities, a characterization of each disease is useful.

Alzheimer's disease most commonly starts with short-term memory loss. As the disease progresses, other cognitive functions decline, including planning and organizational abilities (executive functions), language, and visual perception. People with early-onset AD may initially have relative sparing of short-term memory, with language or visual skills affected first. AD affects the medial temporal, parietal, and occipital lobes. In the earlier stages of AD, cognition for social norms and emotional function remains intact.

By contrast, bvFTD is characterized by prominent, early behavioral, or personality changes. Apathy, loss of empathy, rigid personality, obsessions, and compulsions occur. Patients with bvFTD harbor right frontoinsula, anterior cingulate, and orbitofrontal damage. The PPAs both manifest in progressive loss of language function. On the one hand, in svPPA, speech is fluent and grammatical, however loss of word and object meaning occurs. Patients with svPPA initially have early degeneration in left anterior temporal lobes; however, disease eventually spreads to the right anterior temporal lobe, resulting in personality changes and trouble recognizing faces. On the other hand, nfvPPA shows agrammatical, effortful speech with vowel distortions; thus disruptions in language and speech production occur. Most affected regions in nfvPPA include left inferior frontal cortex (see Table 18.1).

Alzheimer's disease is sometimes referred to by neurologists as targeting the posterior brain, since early degeneration is often seen in parietal and occipital lobes, particularly in people with early-onset disease, and one of the early signs of the disease is an inability to copy complex figures. Thus, visual art, as produced by patients who suffer from AD, is often degraded, in terms of visual complexity and technique (Rankin et al. 2007). In contrast, patients with neurodegenerative disease whose artistic production increases or seems to improve with

Table 18.1 Anatomical systems involved in earlier stage AD and FTD

Neurodegenerative disease	Anatomical regions affected
Alzheimer's disease	Entorhinal cortex, hippocampus, parietal and occipital lobes
Behavioral variant FTD	right frontoinsula, anterior cingulate cortex, orbitofrontal cortex
Semantic variant PPA	left > right anterior temporal poles
Nonfluent variant PPA	left inferior frontal gyrus (operculum)

disease progress, tend to have posterior regions intact, and damage, at least initially, localized in the anterior portion of the brain (Miller et al. 1998).

18.3 The emergence of art in the setting of dementia

The study by Miller and colleagues (1998) was among the first observations that certain neurodegenerative diseases might lead patients to engage in creative endeavors. The study describes five patients suffering from FTD who developed artistic skills during the early stages of the disease. The authors found that while AD is a far more common dementia, most, if not all the patients they observed who developed artistic prowess after the onset of dementia, were diagnosed with FTD. In addition, patient artistry was manifested in the visual, and never the verbal domain. These patients developed skills in painting, photography, or sculpture and most of the work they produced was realistic, rather than abstract or symbolic (see Figure 18.1). Miller and colleagues (1998) observed that the art seemed to result from previously seen images that were recalled even when patients had lost the ability to describe the images using language. The artists, while losing most other cognitive and social functioning, showed increased attention and fascination with details of faces and objects, as well as sounds. This study sparked an interest in exploring how degeneration of particular brain regions might release or enhance latent artistic abilities.

Spurred on by the discovery of the five patients mentioned above, Miller and colleagues (2000) conducted a larger study on the emergence of artistic talent in patients with FTD. They included 69 patients diagnosed with FTD and found 12 who acquired or maintained artistic skills in both musical and visual domains. Like their original five patients, these patients showed skills that were not reliant on language, and their work "lacked a symbolic or abstract component." Eight of the twelve patients had svPPA comprising almost 50 per cent (8/17) of all patients studied with this variant. Three more had predominantly temporal degeneration as evidenced by imaging or perfusion, although they had frontal degeneration later on in the disease. The authors note that degeneration of temporal cortex coupled with sparing of frontal functioning was a characteristic feature of all patients who showed artistic ability. Nine of the twelve patients showed predominantly left hemisphere pathology, while only one showed bilateral degeneration.

Indeed, patients with asymmetrical degeneration resulting from frontotemporal lobar degeneration (FTD) seem to be more likely to show changes in creative behavior with

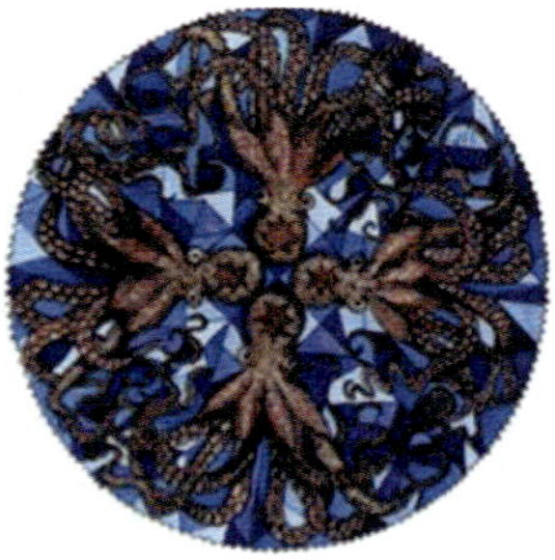

O is for Octopus
(from *An ABC Book of Invertebrates*)
-Anne Adams 2000

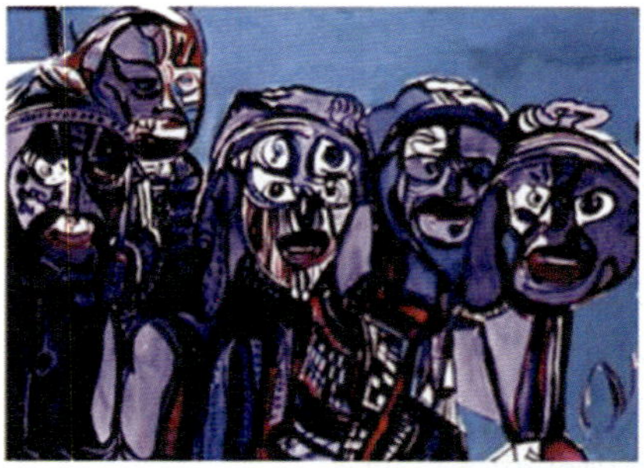

Four Masks
-Jancy Chang 2000

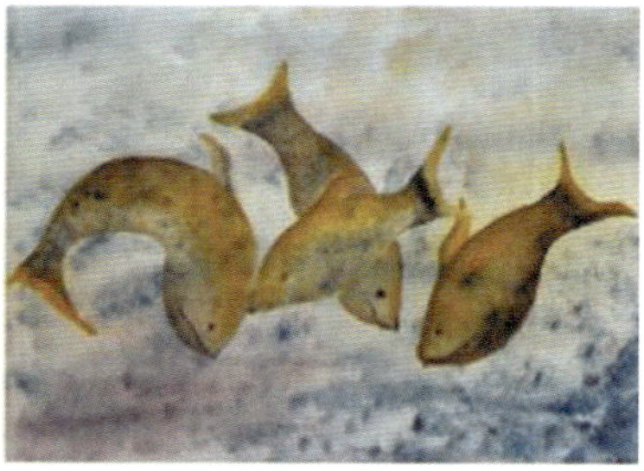

Fishes
-Carol Franz 1999

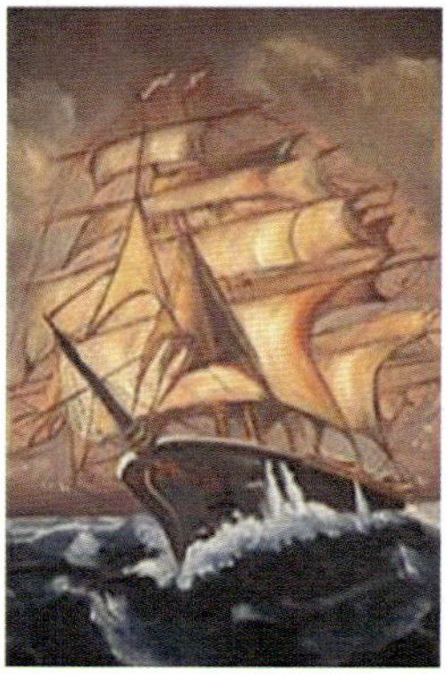

Untitled
-Fountiene Lee Duda 1970

Figure 18.1 Samples of artwork by patients towards the end stages of dementia. Reproduced from <http://memory.ucsf.edu>.

disease than those who have symmetrical damage. That is, bvFTD patients tend to show somewhat more symmetrical involvement of disease processes across regions of the frontal lobes, whereas patients with svPPA tend to show atrophy that begins unilaterally, only spreading to the contralateral hemisphere late in the disease (Johnson et al. 2005). Interestingly, four out of the five artistic patients that Miller and his team (1998) studied had svPPA, not bvFTD. Of the four patients diagnosed with svPPA, three showed more degeneration of the left than the right temporal cortex either in SPECT or post-mortem pathology.

18.4 Left versus right brain thinking

Deciphering the neural basis of creativity by studying neurological or psychiatric patients has a long history, although the focus has been on loss rather than on gain or release of latent talent. This same history has led to the proliferation of several oversimplifications and myths. Arguably, the most popular myth to date is the hemispheric asymmetry hypothesis. Simply put, proponents of the hemispheric asymmetry hypothesis suggest that creativity is localized primarily in the right hemisphere, while the left hemisphere impedes or blocks creative thought. This hypothesis relies on the idea that the two cerebral hemispheres are functionally dissociable, with different cognitive processes either wholly or largely localized to one or the other side.

The hemispheric asymmetry hypothesis with respect to the brain basis of creativity also has a long history, one that stretches back to the turn of the twentieth century. At this time, modern psychology was in its infancy: psychiatrists were investigating the possibility of multiple personalities residing in different brain regions and neurologists were finding that certain functions were localized in one hemisphere or the other, as patients with unilateral neurological insults were demonstrating. The dual nature of the brain inspired fiction, such as the famous tale of *Dr Jekyll and Mr Hyde* by Robert Louis Stevenson.

It has long been known that aspects of visuospatial cognition are localized in the right hemisphere, including the perception and manipulation of forms and their spatial relationships (Hecaen et al. 1962). Deficits in visuospatial cognition are typically demonstrated by examples of visual agnosia, the inability to recognize or identify visual stimuli. A particularly striking example of visual agnosia is prosopagnosia, or the inability to recognize faces. In short, visual agnosias are not thought to be due to deficits in primary sensory capabilities, but rather in secondary and associative processing of visual information. That is, agnosias reflect deficits in visuospatial processing: the affected individual is aware of the visual stimulus but cannot process the relationships between its features, thereby causing an inability to recognize the object.

Hécaen and colleagues (1962) looked at over 400 patients with lesions of the posterior parts of the brain and found that in most cases, disorders of visuospatial abilities were associated with right hemisphere insults: 51 of 59 cases of visual agnosia were right hemisphere lesions, while 16 of 22 cases of prosopagnosia were in patients with right hemisphere damage. Similar conclusions can be drawn from several studies conducted by Brenda Milner (1978) on temporal lobectomy patients. She found that patients with left temporal lobectomies suffered impairment in certain verbal abilities, whereas patients with right temporal lobectomies suffered from disruptions in the ability to comprehend pictures. Similarly, the right hemisphere has become strongly associated with various aspects of music perception and music production (Liegeois-Chauvel et al. 1998). Studies by Wilder Penfield, who stimulated parts of the neocortex during epilepsy surgery, suggested that activity in the right temporal lobe could evoke the auditory hallucinations of musical melodies (Penfield and Perot 1963).

When Roger Sperry and Michael Gazzaniga began publishing their findings from studies of epilepsy patients in the 1960s, the idea that the two hemispheres might largely operate independently of one another gained momentum once again (Gazzaniga and Sperry 1966). Their patients had undergone a radical surgical procedure in which the corpus callosum, the fiber tract that joins the two hemispheres, was severed to prevent the spread of seizures from one side of the brain to the other. Because certain aspects of sensory processing are localized to a single hemisphere, the investigators were able to query one side of the brain independently of the other. For example, information from the left side of the visual field is represented in the right hemisphere and vice-versa. Therefore, by presenting a stimulus only in the left or right visual field, Sperry and colleagues were able to observe how each of the hemispheres responded to it (Gazzaniga and Sperry 1966). These studies confirmed previous suspicions that the left hemisphere is much better at communicating verbally than the right hemisphere, and that the right side knows how to use and manipulate objects presented to it, even though it cannot name them.

These studies led to a re-evaluation of previous observations in patients with unilateral neurological insults as well as spurring new studies of such patients. In the following decades, extensive research has indicated that the right hemisphere excels in global, holistic and intuitive processing, whereas the left hemisphere specializes in sequential and analytical tasks. In line with this view, Kaplan and Gardner (1989; Gardner 1982) noted that when asked to draw a still life, patients with right-sided brain insults tended to produce drawings that sacrificed the gestalt for details, whereas patients with left-sided injury concentrated efforts on form and contour, rather than details.

18.5 A study of drawing in patients with dementia

As a follow-up to the larger study of art production in the setting of dementia, Rankin and colleagues (2007) conducted a study of drawing comparing 18 patients with FTD, 18 patients with AD and 15 age-matched healthy individuals. Strikingly, patients with svPPA faired poorly on the Picture Completion sub-test, part of the standardized Torrance Test of Creative Thinking (TTCT) of non-verbal artistic creativity. In the Picture Completion sub-test, participants are given a page with ten incomplete abstract figures and asked to use the figures to draw interesting objects or pictures. They are told to make the pictures or objects as unique and unusual as possible. SvPPA patients produced drawings that were rated as more conventional and less original in content, containing fewer details and some evidence of perseverative behavior (see Figure 18.2). They also produced fewer drawings overall compared with AD patients, bvFTD patients, and healthy controls. BvFTD patients showed a greater perseverative tendency to complete the incomplete figures and patients with AD showed performance that was no different from healthy controls. Replicating these findings, another recent study also showed that patients with bvFTD are impaired on a creative thinking test compared with both healthy controls and patients with Parkinson's disease (de Souza et al. 2010).

In the second part of the study, the participants were asked to draw three representations of common artistic subjects: a still-life, which consisted of a vase holding multicolored flowers, an imagined room, and a self-portrait. When still-life drawings were compared across patient groups and controls, those produced by bvFTD patients tended to show fewer details, no reference to the background, less variety, and less color. Drawings by svPPA patients, in contrast, were richly detailed, multicolored, and often focused on a small segment of the image, such as one sprig of blue flowers, rather than on the whole. While drawings of the imagined room did not differ in number of details between groups, svPPA patients were the least likely to include personal decorations (38 per cent), although bvFTD and AD patients also drew rooms with less personal detail

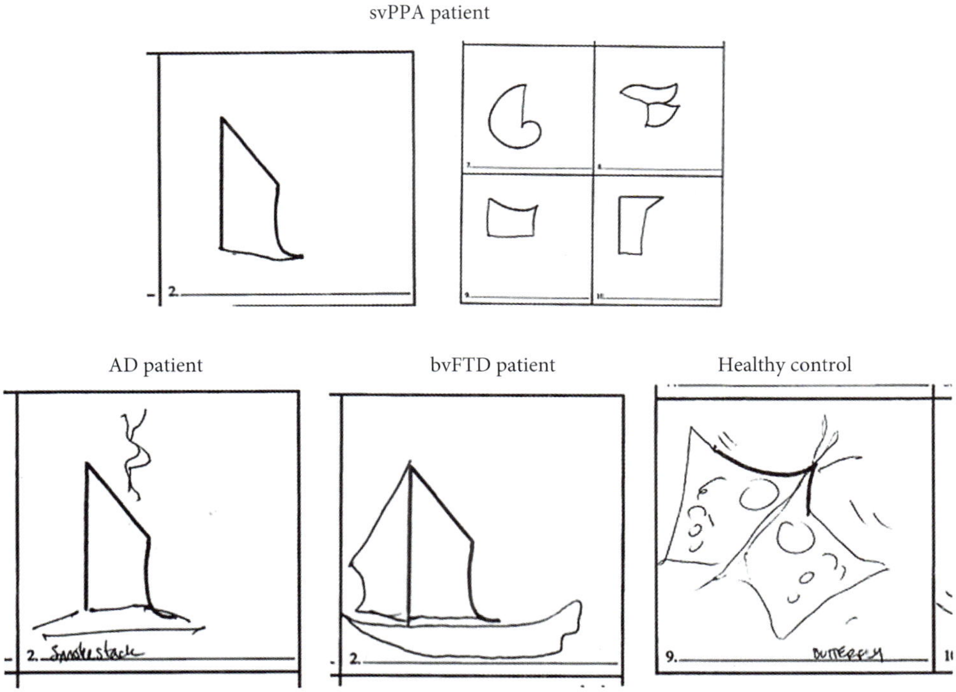

Figure 18.2 Representative samples of patient responses on the Picture Completion Subtest of the Torrance Test of Creative Thinking (TTCT). The top panel shows five trials completed by a patient with svPPA, while trials completed by AD, bvFTD patients, and a healthy control are shown below. Note that the svPPA images are devoid of recognizable or nameable objects while the other images are clearly specific objects. Data from Rankin, Katherine P., Liu, Anli A., Howard, Sara, Slama, Hilary, Hou, Craig E., Shuster, Karen, and Miller, Bruce L., A case-controlled study of altered visual art production in Alzheimer's and FTLD, *Cognitive and Behavioral Neurology*, 20 (1), pp. 48–61, doi: 10.1097/WNN.0b013e31803141dd Wolters Kluwer Health.

included than healthy controls (50 per cent and 53 per cent, respectively). Given the greater demands of this task, including internal visualization, composition, and cognitive creativity, it is not surprising that all dementia groups drew less rich, more homogeneous rooms.

Perhaps the most interesting results, however, came from the self-portraits. Whereas all patient groups tended to draw more distorted and disordered images than healthy controls, bvFTD and svPPA images were strikingly bizarre, with geometric facial features, unnatural color palettes, and disembodied heads or other body parts, seemingly floating in space (see Figure 18.3). These drawings were not recognizable as portraits of the individuals who drew them, compared with those of healthy controls. Given the fact that much of contemporary art is abstract, disordered, and often bizarre, it is easy to see how the art produced by these patients might be considered more creative and critically acclaimed than the drawings produced by their healthy but non-artistic counterparts. Of course,

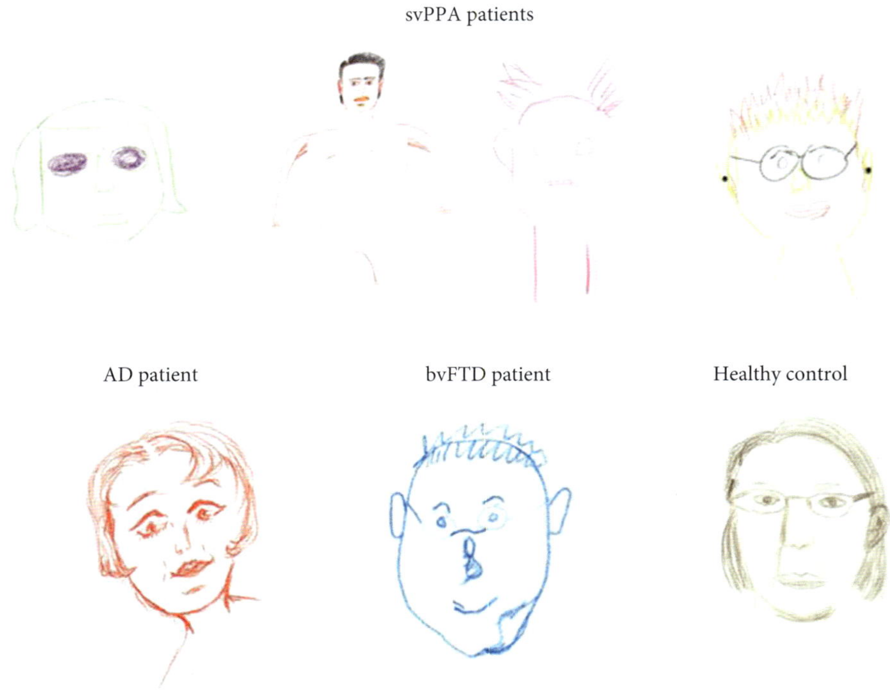

Figure 18.3 Self-portraits by patients and a healthy control drawn during the study published by Rankin and colleagues (2007). Note that the drawings by patients with svPPA show fewer personal details or characteristics, use bizarre colors and strange expressions, compared with the drawings by AD and bvFTD patients, as well as the healthy control participant. Reproduced from Rankin, Katherine P., Liu, Anli A., Howard, Sara, Slama, Hilary, Hou, Craig E., Shuster, Karen, and Miller, Bruce L., A Case-controlled Study of Altered Visual Art Production in Alzheimer's and FTLD, *Cognitive and Behavioral Neurology*, 20 (1), pp. 48–61, doi: 10.1097/WNN.0b013e31803141dd (c) 2007, Wolters Kluwer Health.

although the patients' drawings might be hailed as more creative by the artistic community, the brain networks underlying art production in patients are unlikely to be the same as those used by healthy artists. Whereas artists might choose to steer away from traditional representations, patients might not be able to generate realistic images and might be less able to bring an artistic vision to life.

The enhancements in creativity with a particular degenerative pattern are also often domain-specific. Skills in one domain (e.g. music) may improve creative output in that domain but may not necessarily transfer to other artistic realms. Similarly, the paradoxical facilitation of creativity with neurological disease is generally limited to a particular domain, be it gardening, painting, sculpture, music, or something else. But what is likely to be common among acclaimed healthy artists and svPPA patients who are also artists, is perseveration or an obsession to create art.

18.6 **Perseverative, obsessive, and visually dominated behaviors in svPPA patients**

Perhaps as a result of the decline in the ability to communicate and perceive the world through language, svPPA patients tend to prefer activities that are primarily visual. Perseverative or obsessive behaviors are commonplace in patients with svPPA. Together, both the dominance of the visual and an obsessive nature are manifest in the large proportion (~25 per cent) of patients for whom jigsaw puzzles become the most frequent, or even the primary activity of daily living (Green and Patterson 2009). Green and Patterson (2009) found that svPPA patients have above-average jigsaw completion skills, especially in "reality-disrupted" puzzles, where expectations of the real world can interfere with puzzle completion, and in abstract jigsaw puzzles, where conceptual knowledge does not benefit performance. These patients often exhibit flat emotional affect and demeanor, but were seen to visibly enjoy the completion of jigsaw puzzles, and take pride in their skills, suggesting that an obsession can bring pleasure. This observation is encouraging both from a clinical standpoint and as a putative reason underlying the proliferation of art in this patient group: producing art requires effort, time, and commitment, which is more easily given when the activity is experienced as fun.

Inspired by this work, Viskontas and colleagues (2011) conducted a study of dementia patients, whose purpose was to investigate whether svPPA patients might show enhanced abilities in a complex visual search task, given their tendency towards pursuing visual activities. Because the number of objects that humans are able to perceive in detail at any given time is limited, where and how attention is paid can greatly influence our experience of the world. As William James has noted, "each of us literally chooses, by his way of attending to things, what sort of universe he shall appear himself to inhabit" (James 1890). Therefore, by observing how and where svPPA patients direct their visual attention, we might glimpse the processes underlying their art production.

To investigate where and how visual attention is paid in our patient groups, we adapted a task pioneered by Anne Treisman and colleagues (1980) in which participants are asked to search an array of letters for a specific target letter (Viskontas et al. 2011). When the target and distracters have at least one prominent distinguishing feature, the search is conducted in parallel; that is, increasing the number of distracters has little or no effect on response time or accuracy, and the target letter seems to "pop-out" of the display. When the features of the target and distracters overlap, however, the search is conducted by investigating each item one at a time, causing the response time to increase with the number of items in the array. During this serial search, participants are more likely to make errors as the number of distracters increases and/or the time to complete each trial is limited (Treisman and Gelade 1980). Because we were primarily interested in where and how visual attention is paid in patient groups, we monitored participants' eye movements, in addition to recording accuracy and response times. Finally, to understand the brain basis of this behavior, we used a technique called voxel-based morphometry

(VBM) that quantified differences in brain volume that correlate with performance on the task.

We found that as their performance on the jigsaw puzzle tasks suggested, svPPA patients did, in fact, show enhanced performance on the most difficult visual search task: they were notably faster than healthy controls during the serial search trials with the largest number of distracters. These patients were just as accurate as healthy controls during both pop-out and serial search. AD patients, in contrast, showed significant impairments even under the easiest conditions and their performance was no different than chance during the most difficult serial search trials. BvFTD patients were significantly worse than controls only under the most difficult serial search conditions, and, like controls, their response times increased with the number of distracters in serial search. Strikingly, svPPA patients showed no increase in response time when the display size was doubled from 15 to 30 items in the serial search condition, and found the target more quickly than all other groups even while looking at a smaller proportion of the image (see Figure 18.4). We interpreted these findings as suggestive of the fact that svPPA patients may have a more efficient visual search strategy that allows them to ignore distracters and find the target more quickly than healthy people.

Next, we were interested in determining which brain regions were associated with better performance on this task. To that end, we conducted a VBM analysis which demonstrated that better serial search performance correlated with gray matter volume in a bilateral dorsal frontoparietal network across all participants (including a region in the superior parietal lobe, the precuneus, the middle frontal gyrus, and higher visual regions in the occipital lobe). This network is known to support visual attention and is spared in patients with svPPA.

Because the serial search task relies heavily on the process of integrating separate visual features into one coherent image, it is not surprising that the region of the parietal lobe that correlated with better performance in this task has also been shown to result in visual binding problems when it is damaged, as is the case in patients with Balint's syndrome or hemispatial neglect (Damasio 1985). On the other end of the spectrum, hyperactive binding, or the melding of information from different sensory association areas as is seen in individuals with synaesthesia, has been correlated with neural activity and more gray matter in the parietal lobe (Weiss and Fink 2009), and greater connectivity in a region of the parietal lobe that nearly overlaps with the part of the superior parietal lobe shown to correlate with better performance on our task (Rouw and Scholte 2007). In our study, we found no significant differences in gray matter volume in this superior parietal region between patients with svPPA and their healthy counterparts. This observation implies that the enhanced visual search function in our svPPA patients might reflect a change in functional rather than structural brain differences. In summary, the results of this comprehensive study of visual search in people with svPPA suggest that the paradoxical enhancement of visual artistry seen in certain patients might reflect a shift in attentional space from language to vision, which may reflect changes in connectivity in the dorsal parietal network.

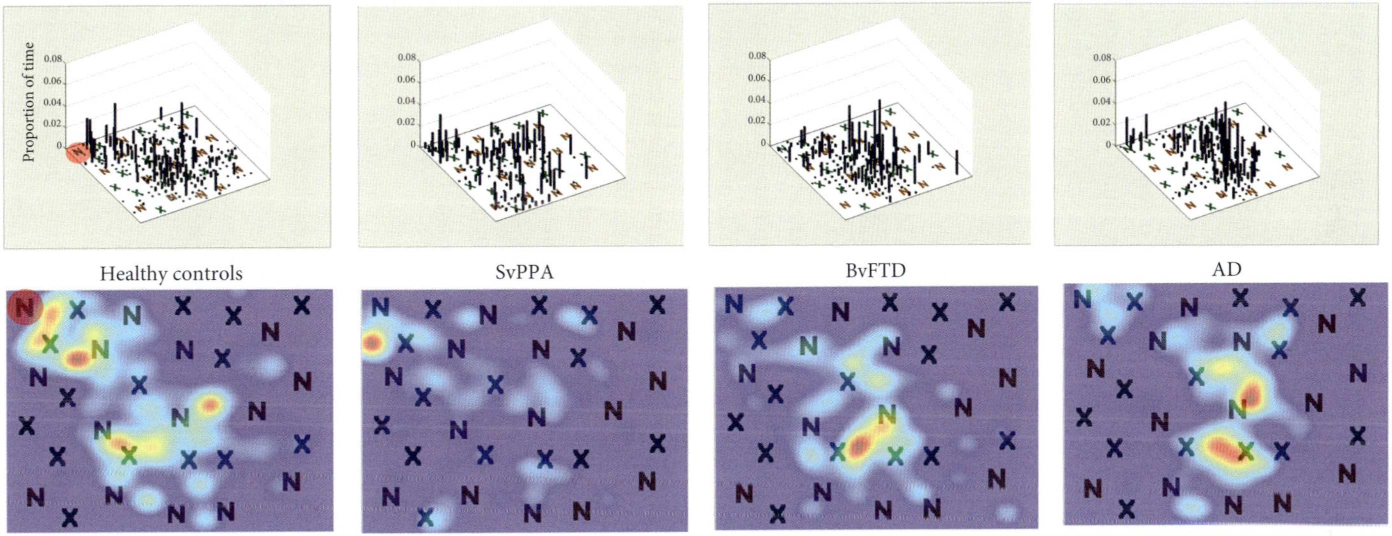

Figure 18.4 Examples of visual search patterns during conjunction search. Representative plots of eye movements during a conjunction search trial with a display size of 30 items. Target is indicated by the arrows in the NC figures. The top panel shows a histogram of the proportion of time spent in each part of the figure, with a density map (smoothed Gaussian distribution) of the same data shown below. The data shown here are representative of 10–14 trials in each patient group. Reprinted from *Neuropsychologia*, 49 (3), Indre V. Viskontas, Adam L. Boxer, John Fesenko, Alisa Matlin, Hilary W. Heuer, Jacob Mirsky, and Bruce L. Miller, Visual search patterns in semantic dementia show paradoxical facilitation of binding processes, pp. 468–78, Copyright (2011), with permission from Elsevier.

18.7 A comprehensive case study

After the initial discovery by Miller and colleagues (1998) of patients whose artistic talents emerge or proliferate after disease onset, one case in particular stood out as exceptional (Seeley et al. 2008). Anne Adams was a cell biologist until, at age 46, her son was in a serious motor-vehicle accident. She took a leave of absence from her academic career to care for him and, around the same time, began to paint. Once her son recovered, she decided not to return to academic life and spent more and more of her time in her studio. Although she had a lifelong interest in art and music, her earlier paintings were primarily simple drawings and architectural watercolors, characterized by muted color schemes and classical perspective. Between ages 46 and 52, however, as she was spending more and more time painting, her work became much more vibrant, colorful, and multifaceted. She synthesized her love of music with her artistic output by painting representations of favorite musical pieces.

At the same time, she began to show an increased tendency to pun, which was also reflected in her artistic choices. When she was 53, she began to paint what would become one of her signature masterpieces: "Unraveling Bolero," an homage to Maurice Ravel's most famous piece. She laid out the music, measure by measure, into an intricate, exacting, and compulsively detailed diptych, which took her four years to complete. Oddly, Maurice Ravel (1875–1937) himself suffered from a progressive aphasia, and composed his bolero in the early stages of his disease, at age 53. The piece is a study of perseverations (Amaducci et al. 2002), with the reiteration of a simple melodic theme, accompanied by a simple, incessantly repetitive bass line. Despite the fact that it became his best-known work, Ravel himself considered the "Bolero" a rather trivial opus, describing it simply as "a piece for orchestra without music". He explains, "I'm going to try and repeat [the theme] a number of times without any development, gradually increasing the orchestra as best I can" (Orenstein 1991). After writing "Bolero," Ravel's illness gradually progressed until he died eight years later. Unaware of Ravel's illness, Anne Adams painted her "Bolero," in a precise and compulsive fashion, with the height of each of her rows signifying the increasing texture and volume of the orchestra.

After "Bolero," the subject of her paintings shifted from music to even more abstract concepts such as *pi*, and a series of paintings of invertebrates arranged as letters that together created a complete alphabet. Shortly after her completed "Bolero," her language symptoms, harbingers of nfvPPA, were first noticed. Initially, as her language declined and she showed no other cognitive impairments, she turned towards photographic realism and her paintings became vividly realistic and intricately detailed. By age 64, she was nearly mute.

In nvfPPA, atrophy begins in the left fronto-opercular cortex, leaving patients with effortful, non-fluent speech (Gorno-Tempini et al. 2004), and difficulties with grammar and articulation. In Anne Adams' case, structural and functional imaging demonstrated enhanced gray matter and activation in heteromodal associative (IPS/SPL) and polymodal (STS) neocortex in the right hemisphere (Seeley et al. 2008). These areas are involved in visuomotor search and attentional control (Corbetta and Shulman 2002; Seeley et al. 2007) and sensory transcoding, necessary for sight-reading music (Sergent 1992; Schon et al. 2002). These regions are also part of the same network that seems to underlie enhanced performance on the serial visual search task described above. VBM has also demonstrated enhancement in these same regions in professional musicians (Gaser and Schlaug 2003). Taken together, these studies point to the possibility that as one brain region degenerates and loses functionality, another region might be gaining functionality and even gray matter volume as a result of changes in behavior and/or compensatory activities. Then the question becomes one of the chicken or the egg: does the creative impulse lead to the behavior, which causes rewiring over time? or, does rewiring occur as a result of degeneration elsewhere, leading to changes in behavior? Assuredly, the answer is likely both.

18.8 **Theory of creativity in patients with dementia: release of inhibition**

Despite its popularity, the idea that creative thought is restricted to the right hemisphere while scientific or logical thought is restricted to the left hemisphere is patently false. If this supposition was true, then one would find many examples of people with neurological damage to the right hemisphere with no loss of rational or logical thought, and people with left hemisphere damage with no loss of creativity. No scientist would ever argue that science does not involve creativity and no artist would contend that art does not demand rigorous analysis or meticulous attention to detail. Furthermore, even the neurologist Joseph Bogen, who participated in early studies of split-brain patients and was an active promoter of the hemispheric asymmetry perspective, conceded that the neural basis of creativity resulted from the functional specialization of the two hemispheres *and* the combinatorial interaction of the hemispheres which requires the corpus callosum (Bogen and Bogen 1988). Both hemispheres are involved in different aspects of creative cognition. However, an insult or degeneration localized to or disproportionately affecting one hemisphere or the other can have profound effects on a person's perceptions and focus, altering creative thinking and execution.

There is some evidence that in the context of focal cerebral infarctions, regional blood flow to unaffected areas can be greater than in healthy control participants (Weiller et al. 1992). Not surprisingly, the pattern of functional reorganization as the patient recovers does not seem to follow a universal map, but differs between individuals (Weiller et al. 1993). Miller and colleagues (2000) suggested that the paradoxical emergence of visual artistic talent in patients with predominantly left-sided anterior temporal lobe degeneration, as is the case in patients with svPPA, might reflect a release from inhibition of the more right-sided posterior and dorsolateral frontal regions involved in visual art production and creativity.

Miller and colleagues (1998) also proposed that the accompanying deficits in social and (other) cognitive functioning that the patients experienced may have led to the facilitation of artistic skills through the release from inhibition of these skills. While their planning, organization, and motivation (processes relying on intact frontal regions) remained intact, and they were able to access fairly remote memories from which they produced realistic reproductions of visual images, they showed marked deficits in semantic memory and verbal prowess. As a case in point, Mell, Howard, and Miller (2003) described an accomplished visual artist who developed FTD. At the age of 49, this woman began to develop her own unique painting technique, producing complicated patterns with Chinese horoscope icons. This development coincided with the onset of illness and a progressive decline in her ability to perform her duties as a high-school teacher. Decline in language function coincided with enhanced creativity in her paintings: they grew more emotionally intense and impressionistic, and less realistic and detailed. Similarly, Kaczmarek (1991) described a painter who suffered a left hemisphere stroke that resulted in aphasia. Unlike the woman described above, this painter retained the ability to draw realistic pictures. In contrast, the painter lost the ability to "create [the] highly symbolic pictures" that characterized the paintings that he had completed before his stroke.

The idea that creativity stems from the release of baseline inhibitory activity is at the heart of another prominent theory of creative cognition. This theory has been championed and studied most thoroughly by Colin Martindale, but also has a long history as reflected in the nineteenth-century notions regarding the relationship between degeneration and psychoticism advocated by Morel as far back as 1857. Supporting this notion is the observation that atrophy of higher inhibitory centers of the brain resulted in a constellation of symptoms such as criminality and over-emotionality, traits that overlap considerably with those used to describe eccentric, highly creative individuals. Martindale's theory emphasizes cognitive disinhibition, or the ability to shed schematic constraints and biases that impede creative thoughts and/or actions (Martindale 1999; Martindale 1971). In the context of this theory, degeneration in the dominant hemisphere, as seen in svPPA patients, might result in greater functional activity and rewiring between posterior visual areas and prefrontal cortex over time, such that previously-untapped creative potential emerges in the setting of disease.

18.9 Conclusion and future directions

Although neurodegenerative diseases cause profound loss of memory, personality, and other aspects of thinking, the emergence of visual artistic abilities provides a window into understanding brain networks that serve visual creativity. A long history of studies and theories regarding left versus right brain functional specialization frames an understanding of how patients with neurodegenerative disease asymmetrically affecting one brain hemisphere may develop prolific visual artistic ability. With the advent of technologies such as functional MRI, scientists have been equipped to parse visual perception into tractable components, revealing possible mechanisms underlying the neural basis of visual perception and creativity. Neuroscientists will probe further into how brain networks for various functions work in concert with each other. It may be that specific brain networks take "center stage" as the functional demands of the human change from moment to moment. Understanding these relationships will shed light on whether the release of inhibition as a theory for the emergence of visual creativity in patients with neurodegenerative disease holds true.

References

Alzheimer, A. (1911). Über eigenartige Krankheitsfälle des späteren Alters. *Zeitschrift fur die Gesamte Neurologie und Psychiatrie* **4**, 356–85.

Amaducci, L., Grassi, E., and Boller, F. (2002). Maurice Ravel and right-hemisphere musical creativity: influence of disease on his last musical works? *European Journal of Neurology* **9**, 75–82.

Bogen, J.E. and Bogen, G.M. (1988). Creativity and the corpus callosum. *Psychiatric Clinics of North America* **11**, 293–301.

Chakravarty, A. (2011). De novo development of artistic creativity in Alzheimer's disease. *Annals of Indian Academy of Neurology* **14**, 291–4.

Corbetta, M. and Shulman, G.L. (2002). Control of goal-directed and stimulus-driven attention in the brain. *Nature Reviews Neuroscience* **3**(3), 201–15.

Damasio, A.R. (1985). Disorders of complex visual processing: agnosias, achromatopsia, Balint's syndrome, and related difficulties of orientation and construction. In M-M. Mesulam (ed.), *Principles of behavioral neurology*. Philadelphia, PA: F.A. Davis, pp. 259–88.

De Souza, L., Volle, E., Bertoux, M., et al. (2010). Poor creativity in frontotemporal dementia: a window into the neural basis of the creative mind. *Neuropsychologia* **48**, 3733–42.

Gardner, H. (1982). *Art, Mind, and Brain: A Cognitive Approach to Creativity*. New York, NY: Basic Books, pp. 318–35.

Gaser, C. and Schlaug, G. (2003). Brain structures differ between musicians and non-musicians. *Journal of Neuroscience* **23**(27), 9240–5.

Gazzaniga, M.S. and Sperry, R.W. (1966). Simultaneous double discrimination response following brain bisection. *Psychonomic Science* **4**(7), 262–3.

Gorno-Tempini, M.L., Dronkers, N.F., Rankin, K.P., et al. (2004). Cognition and anatomy in three variants of primary progressive aphasia. *Annals of Neurology* **55**(3), 335–46.

Gorno-Tempini, M.L., Hillis, A. E., Weintraub, S., et al. (2011). Classification of primary progressive aphasia and its variants. *Neurology* **76**(11), 1006–14.

Green, H.A. and Patterson, A.E. (2009). Jigsaws-A preserved ability in semantic dementia. *Neuropsychologia* **47**(2), 569–76.

Hécaen, H., Angelerges, T., and Houllier, S. (1962). Les varieties cliniques des acalculies au cours des lesions retrorolandiques. *Revue Neurologique* **105**, 85–103.

James, W. (1890). *Principles of psychology*. Boston, MA: Harvard University Press.

Johnson, J.K., Diehl, J., Mendez, M.F., et al. (2005). Frontotemporal lobar degeneration: demographic characteristics of 353 patients. *Archives of Neurology* **62**(6), 925–30.

Kaczmarek, B. (1991). Aphasia in an artist: A disorder of symbolic processing. *Aphasiology* **5**, 361–71.

Kaplan, J.A. and Gardner, H. (1989). Artistry after unilateral brain disease. In F. Boller and J. Grafman (eds), *Handbook of Neuropsychology*. Amsterdam: Elsevier, pp. 141–55.

Liegeois-Chauvel, C., Peretz, I., Babai, M., et al. (1998). Contribution of different cortical areas in the temporal lobes to music processing. *Brain* **121**(10), 1853–67.

Martindale, C. (1971). Degeneration, disinhibition, and genius *Journal of the History of the Behavioral Sciences* **7**, 177–82.

Martindale, C. (1999). Biological bases of creativity. In R.J. Sternberg (ed.), *Handbook of Creativity*. Cambridge, UK: Cambridge University Press, pp. 137–52.

Mell, J.C., Howard, S.M., and Miller, B.L. (2003). Art and the brain: the influence of frontotemporal dementia on an accomplished artist. *Neurology* **60**(10), 1707–10.

Miller, B.L., Boone, K., Cummings, J.L., et al. (2000). Functional correlates of musical and visual ability in frontotemporal dementia. *British Journal of Psychiatry* **176**, 458–63.

Miller, B.L., Cummings, J., Mishkin, F., et al. (1998). Emergence of artistic talent in frontotemporal dementia. *Neurology* **51**, 978–82.

Milner, B. (1978). Clues to the cerebral organization of memory. In P.A. Bluser and A. Rougeul-Buser (eds), *Cerebral Correlates of Conscious Experience*. Holland: Elsevier, pp. 139–53.

Morel, B.A. (1857). *Traité des dégénérescence physiques, intellectuelles, et morales de l'espèce humaine*. Paris: J.B. Balliere.

Orenstein, A. (1991). *The ballets of Maurice Ravel: creation and interpretation*. Burlington, VT: Ashgate.

Penfield, W. and Perot, P. (1963). The brain's record of visual and auditory experience: A final summary and discussion. *Brain* **86**, 595–696.

Pick, A. (1892). Ueber die Beiehungen der senilen Hirnatrophie zur Aphasie. *Prager Medicinische Wochenschrift* **17**, 165.

Rankin, K.P., Liu, A.A., Howard, S., et al. (2007). A case-controlled study of altered visual art production in Alzheimer's and FTLD. *Cognitive Behavioral Neurology* **20**, 48–61.

Rouw, R. and Scholte, H.S. (2007). Increased structural connectivity in grapheme-color synesthesia. *Nature Neuroscience* **10**, 792–7.

Schon, D., Anton, J.L., Roth, M., et al. (2002). An fMRI study of music sight-reading. *Neuroreport* **13**(17), 2285–9.

Seeley, W.W., Allman, J.M., Carlin, D.A., et al. (2007). Divergent social functioning in behavioral variant frontotemporal dementia and Alzheimer disease: reciprocal networks and neuronal evolution. *Alzheimer Disease and Associated Disorders* **21**(4), S50–7.

Seeley, W.W., Crawford, R.K., Zhou, J., et al. (2009). Neurodegenerative diseases target large-scale human brain networks. *Neuron* **62**(1), 42–52.

Seeley, W.W., Matthews, B.R., Crawford, R.K., et al. (2008). Unravelling Bolero: progressive aphasia, transmodal creativity and the right posterior neocortex. *Brain* **131**(Pt 1), 39–49.

Sergent, J.Z., Terriah, S., and MacDonald, B. (1992). Distributed neural network underlying musical sight-reading and keyboard performance. *Science* **257**, 106–9.

Treisman, A.M. and Gelade, G. (1980). A feature integration theory of attention. *Cognitive Psychology* **12**, 97–136.

Viskontas, I.V., Boxer, A.L., Fesenko, J., et al. (2011). Visual search patterns in semantic dementia show paradoxical facilitation of binding processes. *Neuropsychologia* **49**(3), 468–78.

Weiller, C., Chollet, F., Friston, K.J., et al. (1992). Functional reorganization of the brain in recovery from striatocapsular infarction in man. *Annals of Neurology* **31**(5), 463–72.

Weiller, C., Ramsay, S.C., Wise, R.J., et al. (1993). Individual patterns of functional reorganization in the human cerebral cortex after capsular infarction. *Annals of Neurology* **33**(2), 181–9.

Weiss, P.H. and Fink, G.R. (2009). Grapheme-colour synaesthetes show increased grey matter volumes of parietal and fusiform cortex. *Brain* **132**(Pt 1), 65–70.

Zhou, J., Gennatas, E.D., Kramer, J.H., et al. (2012). Predicting regional neurodegeneration from the healthy brain functional connectome. *Neuron* **73**(6), 1216–27.

Zhou, J., Greicius, M.D., Gennatas, E.D., et al. (2010). Divergent network connectivity changes in behavioural variant frontotemporal dementia and Alzheimer's disease. *Brain* **133**(Pt 5), 1352–67.

Chapter 19

Hemispheric specialization, art, and aesthetics

Dahlia W. Zaidel

19.1 Introduction

Hemispheric specialization is a hallmark of human cognition when compared to all other animals. It describes the fact that the left and right cerebral hemispheres separately control different functions of the mind (Corballis 2009). For this reason, functional brain lateralization is a source of much scientific curiosity and research. The notion of functional lateralization for art, in particular, gained a solid foothold in neurological and neuroscientific thinking in the 1960s and 1970s following the dramatic findings of the commissurotomy cases in Roger Sperry's lab at Caltech (Sperry 1974; Trevarthen 1990; Zaidel 1993). These studies were particularly suited to the discovery of which cognitive functions are specialized in one or the other hemisphere, and this explains the excitement surrounding the findings. Before that time, the nature of hemispheric specialization was not adequately characterized, even though functional cognitive asymmetry was long understood to be a unique feature of the human brain.

By now, countless empirical findings, conducted with both normal subjects and brain-damaged patients, have supported the notion that the left and right hemispheres have separate cognitive and thinking styles (Corballis 2012; Springer and Deutsch 2001). The left hemisphere specializes in the main language functions, detail, and focused cognition, and applies logical, deductive thinking styles. The right specializes in visuospatial functions, including topographical knowledge, face perception, and memory, and holistic, global, and gestalt thinking styles. The asymmetrical hemispheric specializations are now viewed as complementary to each other and controlled by interhemispheric dynamics shaped by adaptive evolution (Sperry 1974, 1981). Despite the specialization, in the normal brain, the two cerebral hemispheres work in tandem; this is what enables humans, more than any other biological organism, to communicate with others through a sophisticated combinatorial language as well as through art, to navigate expertly in their environment, to build lasting constructions, to create, and to innovate.

At the time when the increased interest in art and its functional brain lateralization was sparked, art was "art"; it was lumped into a single unitary entity. The myriad of art types were not elaborated upon, paintings and drawings received the greatest attention as instances of art, there was no meaningful discussion of the brain of the artist as distinguished from that of the viewer of the art, the components of art were not fractionated with respect to their possible neurological components, and, importantly, the lateralization notions for art were applied exclusively to right hemisphere specialization. What was emphasized outside Sperry's laboratory, by scholars and non-scholars alike, was that art by its very nature is antagonistic to language and "must" be supported by a non-language system in the brain, namely the system specialized in the right hemisphere (Zaidel 2013). However, this was an inference, not a scientific fact. No empirical studies directly measuring aspects of art were published by the Sperry laboratory, nor were any of the commissurotomy patients professional artists. But, sadly, this association between art and hemispheric specialization has since dominated interpretations of scientific data dealing with art and brain, by both scientists and laypersons alike. This chapter reviews the background as well as discusses findings on brain laterality in art.

19.2 Historical background for hemispheric lateralization in art

As noted above, the association between hemispheric specialization and art was jump-started by the work on the complete commissurotomy patients of neurosurgeons Bogen and Vogel who were studied in Roger Sperry's psychobiology laboratory at Caltech in Pasadena, California (Bogen and Vogel 1962; Sperry 1968, 1981). These patients underwent surgical midline disconnection of fibers in the forebrain commissures—namely, in the corpus callosum, anterior, and hippocampal commissures—in order to relieve severe, intractable epilepsy that no longer responded to medication. Following surgery, they were tested and studied in Sperry's laboratory because the surgical disconnection allowed, for the first time, direct assessment of hemispheric competence. This was a greatly improved and reliable method for determining hemispheric specialization. Moreover, previously held certainties about the right hemisphere, mostly derived from the study of patients with unilateral focal damage, were challenged through empirical tests and shown to be wrong in that more abilities were controlled by the disconnected right hemisphere than was thought to be the case (Sperry 1981).

Prior to the galvanizing effects of the studies with the complete commissurotomy patients, the scientific findings on groups of neurological brain-damaged patients, where the damage was unilateral and focal, provided initial evidence for functional brain asymmetries. Those results were known and discussed in the confines of neurology and neurosurgery. The lateralization of the language brain areas to the left hemisphere after the mid 1800s, notably by Paul Broca and Carl Wernicke, inspired countless of functional localization observations in the brain. The preferred strategy was, as noted earlier, the study of neurological patients with unilateral damage confined to a single hemisphere. If patients were unable to carry out a particular task after the damage, which they were able to do beforehand, the damaged area was inferred to control (was responsible for) that function in the intact brain. However, with the complete commissurotomy patients in Sperry's laboratory, direct testing of hemispheric competence as opposed to negatively inferring competence from brain damage, as is done in the case of focal, unilateral damage, became possible. Consequently, there was much enthusiasm in scientific circles for results and ideas emerging from Sperry's laboratory, so much so that assumptions and inferences were made about hemispheric specialization, particularly that of the right, without empirical evidence upon which to base them.

In general, one major reason art and brain had remained an enigma for so long, despite the classic 1948 publication by Alajouanine on a few artists with brain damage (Alajouanine 1948; Gardner 1975, 1982), is that precious few talented individuals in society are capable of producing art, and of those, very few come to the attention of neurologists. From a neurological perspective, pairing art with its neuroanatomical underpinnings would not have universal application whereas agnosia (loss of previous knowledge about specific things), apraxia (loss of voluntary purposeful movements), or aphasia (acquired language impairments) would apply universally. Consequently most neurological, neuropsychological, and neuroscience laboratories devoted efforts into elucidating those conditions, leaving brain control of art as esoteric and mysterious issues. Only in the past 15 to 20 years or so, with the advent of neuroimaging techniques, have we seen a growth in the study of the neural underpinning of art, albeit mostly in visual and the musical arts (Zaidel 2005; Nadal 2013).

19.3 The right hemisphere and art?

Several factors in scientific and intellectual thinking contributed to the notion that art is an expression that can be lateralized in the brain. The lateralization applied specifically to the right hemisphere. The most significant of the factors concerned the nonverbal nature of right hemisphere cognition and its specialization in visuospatial cognition.

In neuropsychology and neurology investigating left hemisphere cognition has been a straightforward endeavor, because the main language centers are represented in the left, and failures in language communication are relatively easy to detect and measure. The right hemisphere, by contrast, had been an enigma and a challenge to researchers because it lacks language's communication channels. The publications from the Sperry laboratory pursued the mystery of the right hemisphere by attempting to understand its functions and cognitive style. Consequently several empirical facts and reasonable conclusions emerged: right hemisphere cognition (a) is indeed, for the most part, nonverbal in nature, (b) specializes in visuospatial tasks, and (c) in holistic, gestalt thinking styles.

Against this background, conceptual characterization of what other functions might be specialized in the right hemisphere has led to inferences noted earlier that ended up guiding data interpretation, namely that the right hemisphere is, (a) artistic, because visuospatial abilities were assumed to be the very essence of art, (b) emotional, because emotions are non-verbal, and (c) creative, because, again, creativity is nonverbal. But they were mere conjectures, born out of speculation, and they lacked empirical support. At the heart of the logic was the nonverbal aspect of right hemisphere cognition together with the nonverbal nature of visual art (as well as of emotions and creativity). It was assumed then, and now, based on personal subjective observations, that thoughts, ideas, emotions, and creativity reflect a neural source lacking language, and thus must be formed in the right hemisphere. Subjectively, they appear language-less. In fact, we do not know how all of these are formed and cannot be certain that they do not arise in the left hemisphere.

The ensuing decades did not support these notions about the lateralization of art to the right hemisphere; indeed, they may be viewed as derailing research into brain, art, and creativity. Researchers were swayed by fashionable notions and felt they had to interpret their empirical findings in light of those unsubstantiated conjectures, while being largely unaware of what gave rise to them in the first place.

In reality, there was no evidence that Sperry's artistically unskilled complete commissurotomy patients sketched or made drawings of superior quality when using the left hand (controlled by the right hemisphere). They did, however, copy two-dimensional representations of three-dimensional figures, such as a cube, better with the left than with the right hand. This was in agreement with neurological patients with damage in the right hemisphere, particularly in the parietal lobe, who could not, on the whole, accurately copy a cube or any other pictorial representations (Bogen 1992). This type of performance contributed to the conclusion that the normal right hemisphere specializes in visuospatial perception. However, because art was viewed as the conglomerate "art," irrespective of whether or not it was pictorial, musical, literary, and so on, the realization that spatial ability is not key to artistic expression was not reached. Even one of the important researchers into brain and art, Howard Gardner, was influenced by the general findings about the commissurotomy patients in Sperry's laboratory (Gardner 1975, 1982). Fashion in science exerts powerful influence on scientific opinions and data interpretation.

19.4 Perceptual effects of right hemisphere damage in artists

Damage in the right hemisphere, in artists and non-artists alike, results in commonly observable deficits in perception, and this leads to alterations in the way they depict their art. Most of what we know is based on visual art, most of it pictorial. It is not as if the damage has led to changes in artistic expression per se; rather, the alterations reflect the inability of the brain to process visual information in a complete way. In other words, observable alterations reflect broken perception, and the perceptual processes themselves do not define artistic talent, creativity, and expression.

One example of a perceptual deficit following brain damage is hemi-neglect of the left half of space, whether the space is personal (on and around the person's body) or extrapersonal (away from the person's body). With artists, the neglect manifests in leaving out details in the left half of the canvas, for example. Some of the well-known artists who displayed such effects in their works following right hemisphere damage are Anton Raederscheidt, Otto Dix, and Lovis Corinth (see Bogousslavsky and Boller 2005; Zaidel 2005). It would seem that a central invisible midline axis divides space into left or right, and the half that is ignored is typically left of the midline. The symptoms consist of failing to attend, orient, draw, paint, eat, comb, and so on in the patient's left space. The damage is typically in the right parietal lobe, although damage in the thalamus or even subcortical areas in the right hemisphere can give rise to hemi-neglect.

The prevalent interpretation for the hemi-neglect is an imbalance in hemispheric control of space such that the left hemisphere becomes abnormally dominant in "grabbing" attentional resources to the right half of space (Heilman et al. 2000). In most patients the neglect symptoms dissipate in six to eight weeks and this contributes to the proposal of an imbalance, albeit a temporary one. It is important to emphasize that the effects are observed in most of the cases only in the left half of space, that they have right hemisphere damage, and that the effects apply to both artists and non-artists alike. Thus, the perceptual alterations observed in the artists' works are not restricted to how the brain controls the production of pictorial art.

Another outcome of right hemisphere damage, particularly when it is localized in the parietal lobe, is spatial perception disturbances such as spatial agnosia and topographical agnosia. In the former, the understanding of relationships among objects or among abstract geometrical lines is lost, and the latter, knowledge of previously known routes, maps, layouts, and so on is no longer accessible. The whole notion that the right hemisphere specializes in visuospatial perception and knowledge originated with observations of patients suffering from such damage. In addition, when the damage is localized in the fusiform region (roughly the inferior junction of the occipital and temporal lobes) it leads to severe disturbance in recognizing previously known faces, a condition known as prosopagnosia. The condition also impairs recognition of newly encountered faces.

When considering the effects of left or right hemisphere damage in more than 50 brain-damaged artists, several conclusions emerged (Zaidel 2005). The first is that neither unilateral hemispheric damage leads to specific, characteristic alterations in art. The second is that any alterations that do emerge post-damage (alterations do not always emerge) are not unique to art; they merely reflect perceptual deficits of the kind observed in non-artists as well. The third is that the only aspect of an artist's work that seems to change following the damage is the technique, not the skill or the talent, due to acquired motoric problems; artists adjust their artistic expression with different techniques.

19.5 Perception of paintings in the left and right hemispheres

Most studies on art and brain assume that the artist's production and the viewer's observation are one and the same. They are not. To learn the underlying neural mechanisms of the production we need to look to the artist, while to learn the neural underpinning in the observer we need to study the observer. The former aspect can be determined through the study of professional artists with unilateral focal brain damage (discussed in subsequent sections of this chapter). The latter can be studied through the use of the hemi-field technique in normal subjects. The first study to put to a direct test the question of right hemisphere superiority in art perception was a hemi-field study in normal subjects (Zaidel and Kasher 1989). It investigated hemispheric differences in the perception of surrealistic versus realistic paintings. Subjects viewed the pictures of a series of 24 paintings

(12 surrealistic and 12 realistic) in the center of vision and then had their recognition memory of the paintings tested in a hemi-field paradigm. In this recognition phase a total of 48 pictures, of which 24 were the original to-be-remembered-pictures and 24 were decoys (half were surrealistic and half realistic) were projected briefly (100 milliseconds) in the left or right visual half fields. They pressed a key to indicate "yes," they remembered seeing the painting in the to-be-remembered series, and a "no" key to indicate they had not seen the painting in the memory series. The results showed superior memory for the surrealistic paintings in the right visual half-field (left hemisphere) and no difference between the left and right visual half-fields for the realistic paintings.

The above hemi-field study indicated not only absence of right hemisphere superiority in remembering paintings, but also that not all paintings are the same. The surrealistic paintings depicted the world through violations of reality rules, some representing logical violations, while some representing physical violations, and this particular type of reality distortions favors left hemisphere cognitive processing. The content itself influences the hemispheric role in processing the painting. When it comes to realistic depictions of the known physical world, it seems both hemispheres have equal cognitive approaches. Other painting genres were not tested, unfortunately, for that may have provided a continuum on which to determine hemispheric asymmetry.

Similarly, in a different study, pictures generated by a computer to depict "abstract" style paintings and "realistic" style paintings were applied in a hemi-field paradigm to normal subjects (Vogt and Magnussen 2005). Some of the subjects were professional painters and some were laypersons. All subjects were asked to remember the pictures. The results revealed that the abstract pictures were better remembered in the right hemisphere. The pictures depicting the realistic style were remembered better by the artists in the left hemisphere, and in the right hemisphere by the non-artists. These results support the notion that art is not a single entity. Rather, that art can be fractionated into separate different *genres*, and how viewers process the art depends on their expertise. Thus, again, the idea that a single hemisphere, the right, can specialize in art is untenable.

19.6 **Drawings and the right hemisphere**

What is known about the right hemisphere and drawings? Neuropsychologists in Britain asked their brain-damaged patients to draw a bicycle from memory. The damage was in the right or left parietal and occipital lobes. The drawings had characteristic errors. The right-brain-damaged group omitted the contour frame but left in the details of the bicycle parts; the left-brain-damaged group drew the contour frame and omitted the bicycle parts details (McFie and Zangwill 1960; Paterson and Zangwill 1944; Warrington and James 1966). The damage in the right hemisphere abolished the ability to depict the gestalt, whole, global bicycle configuration and the damage in the left hemisphere abolished the ability to depict the details and pieces of the bicycle. These findings prompted Paterson and Zangwill (1944) to propose that the right hemisphere uses a cognitive strategy heavily dependent on global and gestalt configurations while the left applies a strategy that emphasizes local details. However, with regards to drawing and the brain (by non-professionals) these findings show that neither hemisphere alone controls drawing abilities.

Subsequently, Kirk and Kertesz (1993) asked 125 patients with unilateral, focal brain-damaged to draw seven figures from memory (cube, circle, square, clock, tree, house, a person). The patients with right hemisphere damage displayed spatial-relations errors, and those with the left damage had numerous errors of various kinds. The key finding is that the ability to draw was not abolished in either group. Thus, we can assume that in the normal brain, the control of drawing is present in each hemisphere.

19.7 Lateralized and bilateral brain activation with EEG and fMRI

Laterality effects have been tested mainly in non-artists. They have been reported in both electroencephalography (EEG) and functional magnetic resonance imaging (fMRI) studies, but those effects were not limited exclusively to a single hemisphere. Neurophysiological responses such as EEG uncovered some lateralized effects in viewing paintings. Hofel and Jacobsen (2007) recorded evoked potentials from surface skull electrodes in normal subjects while they were making aesthetic judgments of graphic patterns. They found pronounced right hemisphere lateralization in late positive potentials, particularly 600 milliseconds after the initial exposure of the graphic pattern.

On the other hand, fMRI studies in which subjects made aesthetic judgments of paintings revealed a predominant left hemisphere activation as follows: (1) Jacobsen and colleagues (Jacobsen et al. 2006) found maximal bilateral activation of the inferior frontal lobe gyrus (Brodmann areas 45 and 47), lateralized increased activation of the left lateral orbitofrontal region, in the left temporal pole region, as well as in the temporoparietal junction region. They also reported activation of the right Brodmann area 10 (frontal lobe) as well as the right posterior cingulate gyrus. (2) The fMRI work of Cupchik and colleagues (Cupchik et al. 2009) revealed bilateral maximal activation of the insula (Brodmann area 13) during aesthetic judgments and the left lateral prefrontal cortex (Brodmann area 10), the left superior parietal lobe, left cingulate (Brodmann areas 24 and 32), and left lingual gyrus region (Brodmann area 19).

Similarly, Ishizu and Zeki (2011) explored explicit aesthetic reactions to pictures of paintings and measured responses to how beautiful they were on a nine-point scale. They found numerous hemispheric regions that were maximally activated in judgments of "beautiful"; they were in the left medial orbitofrontal cortex, left caudate nucleus, and left inferior occipital gyrus. For judgments of "ugly" they found maximal bilateral activation of the amygdala as well as the left postcentral gyrus (somatosensory cortex). Vessel and colleagues (Vessel et al. 2012) found maximal activation in the left hemisphere, specifically in the anterior medial prefrontal cortex, hippocampus, substantia nigra, and posterior cingulate gyrus; they found a large area of activation in the medial area of the left hemisphere, including the thalamus. Barely any activation in the right hemisphere was noted.

Reviews of fMRI results in studies measuring subjects' responses to pictorial material have been reported (Nadal 2013; Nadal et al. 2008). Together, these results support the notion that art evaluation lacks a consistent lateralized localization in the brain.

19.8 Painted portraits: lateralization of the face in art and viewers' aesthetic reactions

For hundreds of years, particularly since the Renaissance, Western artists have painted single sitters with a lateralized pose. Previously, profile views of faces were the norm. From the Renaissance onward, they showed more of the left side of the face than of the right, and this was especially true for women's faces, more so than for men's (McManus and Humphrey 1973). McManus and Humphrey found that more women than men sitters were painted so that the left side of the face was more exposed (68 per cent versus 56 per cent, respectively). On the surface, there was no reason why artists would show such bias in their works.

The background for this outcome is described by Zaidel (2005) thus:

> An empirical attempt to understand the basis for the side bias in Western portraiture was launched many years after those results were first published (Zaidel and FitzGerald 1994). The first key conjecture was that the principal determinant in the bias was a preference by the observer, particularly the paying observer (for hundreds of years, artists made a living through commissioned portraits). The working hypothesis was that observers liked to view women's left side of the face more than the right, and that artists wanted to accommodate them. They wanted their works to be liked. They spent days, weeks, months, and even years on a given portrait (e.g. Leonardo's "Mona Lisa"). The bias could not have been some random and haphazard event. Thus, it was important to find out what subjects think by way of liking a portrait.
>
> Subjects in a laboratory were shown painted portraits of single sitters and asked to judge on a five-point scale the degree to which they liked the painting as a whole. A separate group of subjects were shown these portraits and asked to judge them on a five-point scale according to how attractive they considered the sitter. In each type of judgment requirement, one group of subjects saw the original orientation of the paintings and another group saw a mirror reversal version of the paintings. The unexpected findings were that in each type of judgment, regardless of whether the original orientation or its lateral reversal were shown, women sitters whom the artist originally painted with a right-side emphasis were preferred: The painting was preferred as a whole and the sitter was considered much more attractive than paintings where the left side was emphasized. Moreover, no significant left emphasis versus right emphasis was found for the men sitters, in each type of judgment. One would have expected at the very least that portraits of women with left side emphasis to be the most preferred, since this is the Western trend in painted portraits. However, the fact that gender difference emerged in the portrayed face indicated that the subjects picked up (unconsciously) an artistic bias after all. (There was no statistically significant sex difference in the judging subjects.)
> Zaidel 2005, pp. 166–7

These results indicate that artists' bias in depicting faces is more a reflection of their own aesthetic, perceptual, and cognitive preference than that of naive observers. To test whether or not facial attractiveness is organized asymmetrically on human faces, another experiment in which head-on photographs of people's faces were used to determine attractiveness preference was carried out. After digitizing the photographs, which were taken with straight-on views with neutral expressions, each photograph was made into a left–left face and a right–right face with the aid of Adobe Photoshop (Chen et al. 1997; Zaidel et al. 1995). When subjects were shown these perfectly symmetrical left–left and right–right faces and asked to decide which one is the most attractive, they chose the right–right for women's faces and revealed no significant difference between the two for men's faces. These results illustrate the difference between the artist's artistic decision and the viewer's artistic preference. Thus, when it comes to aesthetic judgment, we need to distinguish between the brain of the artist, who is the producer of the art, and the viewer's brain.

19.9 Conclusion

The emphasis in brain and art research thus far has been on pictorial art. With this in mind, the behavioral data, physiological brain recordings, functional imaging techniques, measured in the observers of the art, as well as effects of unilateral brain damage in professional artists, together point to an absence of hemispheric specialization for art. Instead, the collective data indicate that art is not a unitary entity, that multiple art-related components are involved in producing it; some are features of the producing artist's brain, some are features of the expressive medium of art (e.g. pictorial, musical, literary, dance), and some are features of the observer's brain. From a purely logical point of view, it is not reasonable to assume that there is a hemispheric specialization for producing and observing art. It is more parsimonious to assume that sub-components of art are specialized in different hemispheres and that both hemispheres contribute to the final product.

References

Alajouanine, T. (1948). Aphasia and artistic realization. *Brain* **71**, 229–41.

Bogen, J.E. (1992). The callosal syndromes. In K.M. Heilman and E. Valenstein (eds), *Clinical Neuropsychology* (4th edn). New York, NY: Oxford University Press, pp. 337–407.

Bogen, J.E. and Vogel, P.J. (1962). Cerebral commissurotomy in man. Preliminary case report. *Bulletin of the Los Angeles Neurological Society* **27**, 169–72.

Bogousslavsky, J. and Boller, F. (eds) (2005). *Neurological disorders in famous artists. Frontiers in Neurological Neuroscience*. Basel: Karger.

Chen, A.C., German, C., and Zaidel, D.W. (1997). Brain asymmetry and facial attractiveness: beauty is not simply in the eye of the beholder. *Neuropsychologia* **35**, 471–6.

Corballis, M.C. (2009). The evolution and genetics of cerebral asymmetry. *Philosophical Transactions of the Royal Society Biological Sciences* **364**, 867–79.

Corballis, M.C. (2012). Lateralization of the human brain. *Progress in Brain Research* **195**, 103–21.

Cupchik, G.C., Vartanian, O., Crawley, A., et al. (2009). Viewing artworks: Contributions of cognitive control and perceptual facilitation to aesthetic experience. *Brain and Cognition* **70**, 84–91.

Gardner, H. (1975). *The shattered mind*. New York, NY: Vintage Books.

Gardner, H. (1982). *Art, mind, and brain*. New York, NY: Basic Books.

Heilman, K.M., Valenstein, E., and Watson, R.T. (2000). Neglect and related disorders. *Seminars in Neurology* **20**, 463–70.

Hofel, L. and Jacobsen, T. (2007). Electrophysiological indices of processing aesthetics: Spontaneous or intentional processes? *International Journal of Psychophysiology* **65**, 20–31.

Ishizu, T. and Zeki, S. (2011). Toward a brain-based theory of beauty. *PLoS One* **6**, e21852.

Jacobsen, T., Schubotz, R.I., Hofel, L., et al. (2006). Brain correlates of aesthetic judgment of beauty. *Neuroimage* **29**, 276–85.

Kirk, A. and Kertesz, A. (1993). Subcortical contributions to drawing. *Brain and Cognition* **21**, 57–70.

McFie, J. and Zangwill, O.L. (1960). Visual-constructive disabilities associated with lesions of the left cerebral hemisphere. *Brain* **83**, 243–60.

McManus, I.C. and Humphrey, N.K. (1973). Turning the left cheek. *Nature* **243**, 271–2.

Nadal, M. (2013). The experience of art: Insights from neuroimaging. In S. Finger, D.W. Zaidel, F. Boller, et al. (eds), *The fine arts, neurology, and neuroscience: History and modern perspectives*. Oxford: Elsevier, pp. 135–58.

Nadal, M., Munar, E., Capó, M.A., et al. (2008). Towards a framework for the study of the neural correlates of aesthetic preference. *Spatial Vision* **21**, 379–96.

Paterson, A. and Zangwill, O.L. (1944). Disorders of visual space perception associated with lesions of the right cerebral hemisphere. *Brain* **67**, 331–58.

Sperry, R.W. (1968). Hemisphere deconnection and unity in conscious awareness. *American Psychologist* **23**, 723–33.

Sperry, R.W. (1974). Lateral specialization in the surgically separated hemispheres. In F.O. Schmitt and F.G. Worden (eds), *Neuroscience third study program*. Boston, MA: MIT Press, pp. 5–19.

Sperry, R.W. (1981). Some effects of disconnecting the cerebral hemispheres. *Science* **217**, 1223–6.

Springer, S.P. and Deutsch, G. (2001). *Left brain, right brain: Perspectives from cognitive neuroscience* (5th edn). Chicago, IL: W. H. Freeman and Company.

Trevarthen, C.B. (ed.) (1990). *Brain circuits and functions of the mind: Essays in honor of Roger W. Sperry*. New York, NY: Cambridge University Press.

Vessel, E.A., Starr, G.G., and Rubin, N. (2012). The brain on art: Intense aesthetic experience activates the default mode network. *Frontiers in Human Neuroscience* **6**, 1–17.

Vogt, S. and Magnussen, S. (2005). Hemispheric specialization and recognition memory for abstract and realistic pictures: A comparison of painters and laymen. *Brain and Cognition* **58**, 324–33.

Warrington, E.K. and James, M. (1966). Drawing disability in relation to laterality of cerebral lesion. *Brain* **89**, 53–82.

Zaidel, D.W. (1993). View of the world from a split-brain perspective. In E.M.R. Critchley (ed.), *Neurological boundaries of reality*. London: Farrand Press, pp. 161–74.

Zaidel, D.W. (2005). *Neuropsychology of art: Neurological, cognitive, and evolutionary perspectives*. London: Psychology Press.

Zaidel, D.W. (2013). Split-brain, the right hemisphere, and art: Fact and fiction. In S. Finger, D.W. Zaidel, F. Boller, et al. (eds), *The fine arts, neurology, and neuroscience: History and modern perspectives*. New York, NY: Elsevier, pp. 3–17.

Zaidel, D.W., Chen, A.C., and German, C. (1995). She is not a beauty even when she smiles: Possible evolutionary basis for a relationship between facial attractiveness and hemispheric specialization. *Neuropsychologia* **33**, 649–55.

Zaidel, D.W. and FitzGerald, P. (1994). Sex of the face in western art: Left and right in portraits. *Empirical Studies of the Arts* **12**, 9–18.

Zaidel, D.W. and Kasher, A. (1989). Hemispheric memory for surrealistic versus realistic paintings. *Cortex* **25**, 617–41.

Section 6

The evolution of art, aesthetics, and the brain

Chapter 20

Towards a comparative approach to empirical aesthetics

Gesche Westphal-Fitch and W. Tecumseh Fitch

20.1 Introduction

Aesthetic traditions are ubiquitous in human societies and seem to be a human cultural universal, akin to language and music. Empirical aesthetics is emerging as an area of cognitive research on a par with musicology and linguistics in its potential to reveal fundamental properties of the human mind. Empirical aesthetics as we understand it focuses on the processes underlying the interaction of one or more individuals with an aesthetic stimulus (often visual), during both perception and production of the object. An important aim of this research should be to describe and understand the worldwide human drive to create and surround ourselves with objects that are pleasing to one or more of the senses, in a way that goes beyond their immediate utilitarian function.

There is a rich potential for comparative work in aesthetics. First, differentiating cultural-specific features from universally shared elements necessitates studying aesthetic traditions across a wide range of human cultures. This is analogous to searching for and defining universal features present in all languages (e.g. different word classes such as verbs and nouns), versus language-specific features, such as the Chinese tonal system. In addition to cross-cultural comparisons, comparisons across species are required to distinguish those components of the aesthetic process that are uniquely human from those shared with other species and to discover possible examples of convergent evolution.

In this chapter, we consider some common features of human aesthetic traditions with a particular emphasis on early artifacts incorporating non-representational patterns, and we discuss their possible evolutionary origins. Non-representational patterns and ornaments are present in most, if not all, human cultures. However, they are not a typical research focus in empirical aesthetics, where the main focus has been on Western "high art." From a cross-cultural view, however, it makes sense to study abstract ornamental patterns due to their ubiquity and relative simplicity (cf. Westphal-Fitch et al. 2012; Westphal-Fitch et al. 2013). We therefore emphasize the importance of geometrical and non-representational patterns here.

Biologically, it seems likely that many low-level perceptual components of the aesthetic process are shared with other species, while certain aspects (e.g. production of repeated hierarchically organized patterns) may well be unique to our own. Such features may also be present in other human domains such as music or language, with important implications for our understanding of the nature of the human mind. In this chapter, we review literature on symmetry perception and on nest-building in birds, bees, and primates, two areas where we think known similarities and differences between humans and animals concerning aesthetics can be usefully pursued. It is vital to study not only chimpanzees and other primates (Zaidel et al. 2013), but also species that may have evolved traits convergently, for example birds. In particular, we highlight the need to take the ecology and sociobiology of a species into account in order to understand the mechanisms by which pre-existing traits can become exaggerated and gain social importance beyond their immediate function.

20.2 Pervasive issues in aesthetic inquiry

We begin with a short discussion of the historical development of aesthetics, observing that the sharp distinction between ornament and "high art" is a relatively new phenomenon idiosyncratic to Western European cultures. We also consider a tension at the heart of all aesthetic inquiry, namely that between the broad pan-human universality of aesthetic activity and the wide variety of superficial fashions and traditions that may change rapidly in individual cultures and time periods.

20.2.1 Natural and acquired aesthetics: beyond Kant

Although aesthetics began as a purely philosophical discipline, with the advent of Gustav Fechner's writings introducing an empirical approach to aesthetics (Fechner 1871, 1876) scientists began to gather data on aesthetic behaviors. This makes empirical aesthetics one of the oldest branches of psychology, dating back to the 1870s.

Our modern Western conception of the fine arts as "high art" phenomena, distinct from ornament and craft, is a relatively new one which emerged in eighteenth-century Europe (Kristeller 1990; Shiner 2001). Prior to that, and hence for the vast majority of human history, no strict distinction was made between art and craft, and in many cultures this distinction does not exist (Anderson 1979). The concept of "aesthetics" as a distinctive kind of appreciation specific to art (as opposed to "good taste" which might apply to any number of things) arose at roughly the same time in Europe.

The word "aesthetics" derives from the Greek "aisthetiko," meaning sentient or sensory (cf. the medical term "anaesthetic"). Aesthetics was first used as a term to describe appreciation of beauty in 1750 by Alexander Baumgarten (Gregor 1983; Shiner 2001). Baumgarten proposed aesthetics as a second cognitive mode that relies on the senses (*cognitio sensitiva*) in addition to rational thought. In his model, rational thought is independent of sensory input, but *cognitio sensitiva* relies mainly on input through the senses. Additionally, Baumgartner differentiates between the natural and artificial (acquired) aesthetic sense. While humans are born with a natural aesthetic sense—Baumgarten calls it a natural disposition of the soul for beautiful thought: *dispositio naturalis animi ad pulcre cogitandum* (Groß 2001, p. 97)—enculturation leads to a refinement of this sense.

We interpret this natural aesthetic sense that Baumgarten describes as the aesthetic sense that is shared between all humans, regardless of cultural surroundings, while traits specific to a culture are acquired in a later enculturation process (or refinement, in Baumgarten's terms) during ontogeny.

Beauty for Baumgarten is a cognitive phenomenon in an aesthetically minded perceiver—beauty is perfect perception, and not a property of the object being viewed (Groß 2001). This is reminiscent of David Hume's famous statement (Hume 1757): "Beauty is no quality in things themselves. It exists merely in the mind that contemplates them; and each mind perceives a different beauty."

Baumgarten's original concept of aesthetics crucially attributes the capacity for aesthetic perception to all humans, with the proviso that enculturation can refine this ubiquitous natural aesthetic sense. This contrasts sharply with the very influential concept of aesthetics proposed by Immanuel Kant, presented in *Kritik der Urtheilskraft* (*Critique of Judgement*; Kant 1872).

The key concept for Kant is "disinterested contemplation." Kant posits that a perceiver needs to lack all interest in the object (e.g. the desire to possess it) for a pure aesthetic judgment of the object to be possible (Kant 1872, p. 50):

> *. . . so sagt Jedermann: Hunger ist der beste Koch, und Leuten*
> *von gesundem Appetit schmeckt Alles, was nur essbar ist: mithin*
> *beweist ein solches Wohlgefallen keine Wahl nach Geschmack.*
> *Nur wenn das Bedürfnis befriedigt ist, kann man unterscheiden,*
> *wer unter Vielen Geschmack habe oder nicht . . . Geschmack ist*
> *das Beurtheilungsvermögen eines Gegenstandes oder einer Vorstellungsart*
> *durch ein Wohlgefallen oder Missfallen ohne alles Interesse.*
> *Der Gegenstand eines solchen Wohlgefallens heisst schön.*
> Our translation: . . . people say that hunger is the best cook, and
> people with a healthy appetite enjoy everything that is edible. Such
> pleasure does not evidence a choice based on taste. Only when
> the needs are satisfied can one discern who amongst the many
> has taste or not . . . Taste is the ability to judge an object or
> a performance with pleasure or displeasure without any interest.
> The object of such a pleasure is called beautiful.

Kant's view takes a highly restrictive position on aesthetic judgment, differentiating between "pure" (what we today might call "objective") and "impure" judgments, effectively excluding most perceivers. It is also elitist: if such a pure and disinterested judgment is only available to the few people on this planet whose needs are met entirely, then aesthetic judgment is not available to the majority of humans. For an empirical approach to aesthetics seeking broad commonalities between humans, Kant's view implies an overly restricted working model, and Baumgarten's previous, more inclusive approach is preferable.

A later movement, influenced by the reports of explorers and adventurers describing the art practices of other cultures from around the world, explicitly argued that the aesthetic capacity is fully shared by all humans, as stated by Franz Boas:

> In one way or another esthetic pleasure is felt by all members of mankind. No matter how diverse the ideals of beauty may be, the general character of the enjoyment of beauty is of the same order everywhere. The very existence of song, dance, painting and sculpture among all the tribes known to us is proof of the craving to produce things that are felt as satisfying through their form, and of the capability of man to enjoy them. (Boas 1955, p. 9)

Indeed, the pervasiveness of ornamental objects and patterns and their similarities in symmetry, repetition, and structure gave rise to the notion that common, describable elements might underlie ornament and that ornament could usefully be described in "grammars" (see e.g. Jones 1856). This approach has been greatly expanded in more recent work by Washburn and Crowe (1988).

20.2.2 The multiple purposes of art

Alois Riegl, in his posthumously published *Historische Grammatik der bildenden Künste* (*Historical grammar of the visual arts*) (Riegl 1966) emphasized that humans naturally possess a *Kunstschaffenstrieb* (a drive to create art) (Riegl 1966, p. 217). Riegl differentiates between three purposes ("*Zweck*") characterizing all human artifacts: 1. *Schmückungszweck* (decoration), 2. *Gebrauchszweck* (use), 3. *Vorstellungszweck* (imaginative purpose) (Riegl 1966, p. 217). There may be mainly functional creations with little artistic content (e.g. an arrowhead), and almost purely artistic creations with no actual functional purpose other than decoration and/or representation. The intertwining of function and aesthetics in tools is particularly interesting. Some quite early hominid tools, so-called bifaces (or handaxes) first produced about 1.4 million years ago (Mithen 1996), have a high degree of bilateral symmetry which would have required considerable planning and careful execution. The largest specimens weigh well over a kilogram and are 30 centimeters long (Wenban-Smith 2004), and it is not clear that they could have functioned usefully as tools. Both the size and symmetry of these Achulean handaxes have stimulated discussion about aesthetic or social functions (Kohn and Mithen 1999; Mithen 2003; Nowell and Chang 2009) and the concomitant cognitive abilities required to create them (Mithen 1996) early in human cultural development.

20.2.3 Fechner and non-representational art

Fechner (1876) distinguished between direct and associative factors in aesthetics. Associative factors correspond roughly to Riegl's *Vorstellungszweck*, the ideas and associations that an object triggers in the perceiver's mind. Direct factors are more closely tied to properties inherent in the object, such as symmetry, color, and other formal or material properties. Associative factors can potentially intensify or counteract the effect of the direct factors. Therefore, Fechner suggested that in order to identify aesthetic principles, non-representational ornament may be more useful than representational art, because the associative factors in the objects portrayed in representational art can occlude or override the more subtle organizing principles. This is not the case in ornament, where formal organizing principles take center stage. The associations that a portrayed object triggers are very specific to a certain place and time, and thus hinder a comprehensive historical and cross-cultural study of aesthetics. Furthermore, direct factors are likely to be the only ones that can be meaningfully studied across species.

Concerning data collection, Fechner proposed three different methods for empirical inquiry in aesthetics: *Wahl* (choice), *Herstellung* (production), and *Verwendung* (usage in the real world). The underlying assumption is again that aesthetic proclivities can be active in multiple modalities, and thus can potentially be broken down into component parts, some of which are shared across all cultures and potentially across species (although to our knowledge, Fechner does not discuss the latter possibility). Fechner's three-way approach takes into account that aesthetics is not a simple stimulus/response activity captured by a sender/receiver model. Inevitably, the producer(s) of an artwork/aesthetic object also engage their perceptual proclivities during production. Very often, the finished product differs from what was originally envisioned, presumably due to the effects of feedback from perception influencing production, leading to unexpected results and "happy accidents."

Biological approaches to aesthetics have only recently been proposed and empirical data from animals is rarely included. Eibl-Eibesfeldt (1988), writing about the biological underpinnings of aesthetics, distinguishes between three layers of perceptual biases that we might find: those that are shared with other species, specifically other vertebrates (level 1), those that are unique to the human species but shared across humankind (level 2), and those that are unique to a specific human culture (level 3). These are important distinctions worth keeping in mind when developing a comparative approach in aesthetics and can, in our opinion, be usefully extended to production biases as well. We hypothesize that Fechner's associative factors are located at level 3, while direct factors are situated at level 2. If true, this alignment again underscores the relevance of non-representational abstract art to cross-cultural and cross-species inquiry.

The contents of level 1 (aesthetic proclivities shared with other species), remain little studied and poorly understood. However, an increasing body of research on birds, insects, and primates (reviewed below), allows us to form tentative hypotheses about what may be shared among animals.

20.3 Some implications of non-representational human artifacts

Some of the oldest human aesthetic artifacts known are patterned markings on ochre dating back about 70 000 years (Henshilwood et al. 2002; Kuhn and Stiner 2007). Contemporaneous perforated marine shells seem to have been collected and worn specifically for their visual appeal (Henshilwood et al. 2004; Vanhaeren et al. 2006; Bouzouggar et al.

2007). Both pre-date paleolithic representations of humans or animals by roughly 30 000 years (Hodgson 2006; Verpooten and Nelissen 2010). This is surprising, because geometric patterns are far rarer than animate beings in the natural world, and animals and fellow humans would have had a high relevance for early human artists. The fact that abstract patterns and ornaments can be found across cultures, including those that have not developed representational art, further supports the idea that decorative abstraction is a basic and direct outlet for the human aesthetic drive, with no general need for representation.

Ochre has been used by humans for at least 100 000 years: quite sophisticated ochre-processing tools from that time have been found at Blombos cave, South Africa, including an abalone shell containing a well-fitting grindstone (Henshilwood et al. 2011). This "toolkit" contained a residue of the mixture used, consisting of ochre, bone and marrow, charcoal and other minerals, indicating that these humans were experienced in producing pigments. Ochre can be used for body painting (and still is in many cultures), but also for tanning, hafting, and other things, so its presence alone does not demonstrate aesthetic activities.

Gastropod shells were used very early on as body decorations; some of the oldest examples known date back 70 000–82 000 years (Henshilwood et al. 2004; Vanhaeren et al. 2006; Bouzouggar et al. 2007). Some of these oldest examples show evidence of having been covered with or come into contact with ochre (Henshilwood et al. 2004; Bouzouggar et al. 2007). Wear marks around perforations in the shell suggest that the shells were suspended from cords, like beads (Bouzouggar et al. 2007). Ornamental shells have been found in locations several kilometres from the nearest beach, and due to their small size they were unlikely to represent a food source. Unlike the shells of molluscs that were used for food, shells used as beads show abrasion marks suggesting that they were collected after they washed up on the beach, rather than caught fresh (Kuhn et al. 2001).

Around 30 000–19 500 years ago, beads began to be fashioned from teeth, bone, stones, and other materials, with further embellishments and patterns engraved into the surface (Dubin 1997). Beads from 38 000–10 000 years ago have been found in Australia, Africa, Russia, India, China, and Europe. Even today, beads are popular as jewellery, continuing one of the oldest aesthetic traditions of humankind.

Darwin (1874) discussed body ornament as the most basic form of art in *The Descent of Man*, and interprets it mainly as a means to enhance physical features and to increase physical beauty. He (Darwin 1874, p. 577) noted the universality of ornament, suggesting that it is due to a shared cognitive architecture:

> Lastly, it is a remarkable fact . . . that the same fashions . . .
> now prevail, and have long prevailed, in the most distant quarters
> of the world. It is extremely improbable that these practices,
> followed by so many nations, should be due to tradition from any
> common source. They indicate the close similarity of the mind of
> man, to whatever race he may belong.

Ornaments are produced and appreciated by men and women alike. In contrast, most famous painters, sculptors, architects, musicians, and poets in Western culture were men, which has led some to hypothesize that art production is mainly a male trait (Voland 2003). However, a male preponderance in the arts is most likely due to socio-historical factors specific to Western cultures, where the aesthetic objects produced by women occupied the less-valued domestic and craft domains. Everyday objects that may be subject to aesthetic modification such as weaving, carving, or sewing are produced by both sexes (Shiner 2001), although the allocation to the sexes may vary from culture to culture.

Decorating the body with beads, tattoos, scars, or paint is extremely common (Gröning 1997; Dubin 1997) and is found in men and women alike, as well as in children. Interestingly, symmetrical paintings on the face increase the perceived attractiveness of faces (Cárdenas and Harris 2006). Swaddle and Cuthill (1995) have shown that artificial manipulations of photographs of individual faces to make them symmetrical actually decreases their attractiveness. However, symmetry is preferred in images of faces created from many averaged faces (Perrett et al. 1999). It may be the case that symmetrical face markings are a means to make a perceiver focus on a general global symmetry (akin to averaging facial properties as in Perrett et al. 1999), while drawing attention away from characteristics that define an individual face, and individual facial asymmetries. Cárdenas and Harris (2007) found that women do not rate symmetry of the face or face decorations more highly during the fertile phase of their cycle, suggesting that at least symmetrical ornamentation of the face does not act as a straightforward cue allowing females to assess mate quality or good genes. In addition to looking attractive, body decoration probably always has been a social activity (it is difficult to tattoo or scar oneself, particularly when symmetry and regularity is the goal; see e.g. Figure 20.1), thus weakening the obvious parallel between body ornament and animal courtship displays. Ornaments may thus serve a group cohesion function (see Figure 20.2), or as a means for rapid broadcast of diverse social information to a large number of people beyond immediate associates (Kuhn and Stiner 2007).

Verpooten and Nelissen (2010) have proposed sensory exploitation in combination with social learning as a mechanism by which representational art evolved in humans. Sensory exploitation is a mechanism on the part of the sender of a signal, manipulating the signal to heighten its salience to the receiver due to pre-existing perceptual biases (Ryan 1998). Verpooten and Nelissen argue that the early rise of non-representational art and geometric patterns can be explained by stimulation of lower (e.g. primary visual cortex) visual areas. It required social learning and rituals, in addition to an emerging mental bias for iconic images (e.g. readiness to detect faces or figures in inanimate matter), for iconic art to emerge in human society. They thus attempt to map the cultural evolution of artistic tradition onto neural processing in the visual system, with abstract patterns being both earlier in history and lower in the hierarchy.

Although intriguing, we find this argument unconvincing. Not only has it been shown that symmetry perception does not primarily activate early visual cortex (Sasaki et al. 2005), but as already observed, abstract art is often a social art form. The postulated "basic"

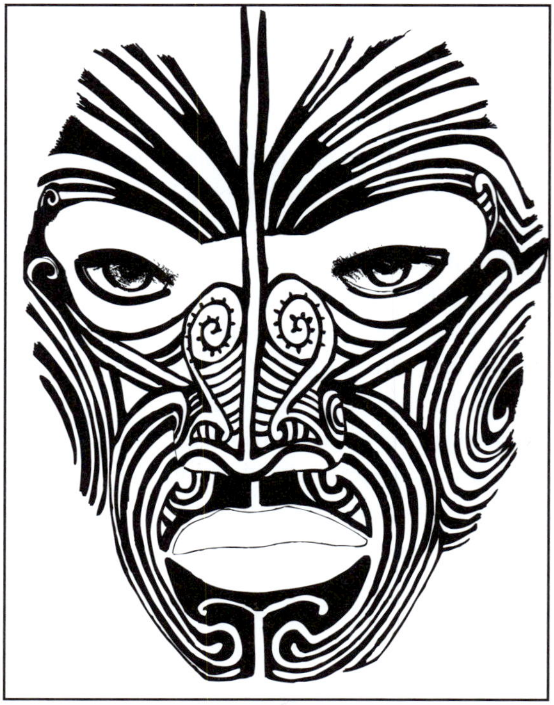

Figure 20.1 An example of the intricate and highly symmetrical facial tattoos typical of Maori tradition in New Zealand.

manifestations of aesthetic behaviors may be more sophisticated than previously thought, and already involve higher level visual processing and social interactions. Research with animals, whose basic visual system closely resembles that of humans, provides one way to evaluate this hypothesis.

In another attempt to understand the origins of art, entoptic phenomena have been proposed as the trigger for early non-representational art (Lewis-Williams and Dowson

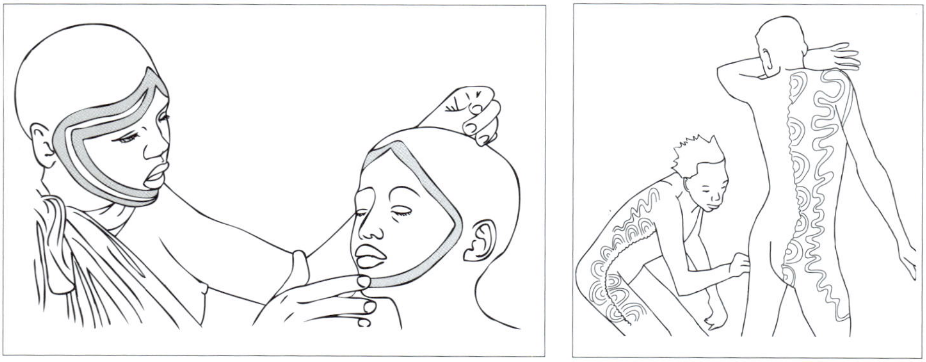

Figure 20.2 Body ornament is often by necessity applied in a social setting, for example between mothers and children (left) or same-sex dyads, suggesting that aesthetics has a strong social core.

1988; Dronfield 1996). These are visual phenomena that arise not from stimulation of the retina by light but from activity within the visual system itself—for example, the flashes and geometrical patterns perceived when pressure is applied to the eyeballs, during exposure to stroboscopic flicker, or the visual aura perceived by many migraine sufferers prior to an attack. Given the similarities in the visual systems across species, this would predict that entoptic phenomena are not unique to humans (although testing this prediction is non-trivial). Thus the question of why only humans have developed the urge to produce abstract geometrical patterns remains open.

Taking a broader view on the motivation to produce aesthetic objects, Deacon (2006) suggests that aesthetics may be part of a general novel cognitive style that evolved in humans, involving a modified motivational system. Such an internal reward system would make it inherently enjoyable to perceive and manipulate certain visual stimuli in a manner that may also apply to music and language. Additionally, Deacon (2006, p. 30) posits that humans have a unique "representational stance" which biases perception towards detecting a symbolic relation between an object and another referent (e.g. interpreting cloud shapes as objects). This would explain the origin of representational art.

Certainly, aesthetic experiences are rewarding, and it seems likely that the brain's reward system has a reinforcing effect on the production and perception systems underlying aesthetics. However, we suspect that the representational stance did not play an important role initially, given that geometrical patterns seem to be the oldest art forms and lack any obvious iconic representational function (although symbolic reference may arise later—e.g. meander patterns representing Greek culture).

While it may be impossible to reconstruct the pathways by which geometrical patterns arose, the fact that they are expressed so ubiquitously across virtually all cultures as well as in the oldest artifacts suggests that they are part of our core aesthetic sense, and thus make them a good departure point for considering research in other species.

20.4 **Animal aesthetics?**

At least since the 1950s, claims have been made that chimpanzees and other primates produce "artistic" paintings when provided with paper, paint, and brushes. Other animals which have been claimed to have artistic abilities include elephants, horses, dolphins, and rhinoceros. We see these performances as having limited relevance. Animal paintings are only produced in a captive or domesticated setting, with human encouragement and provisioning of materials, suggesting that these paintings are more likely due to encouragement and rewards by humans than a natural artistic or aesthetic inclination of the animals.

Unfortunately, little solid empirical work in the lab has been done specifically on the types of aesthetic perception that animals might have. However, using training and food rewards, Watanabe, Sakamoto, and Wakita (1995) have shown that pigeons can discriminate between Monet and Picasso paintings and generalize the distinction to new paintings by other painters in a similar style. Pigeons can also be trained to distinguish between children's drawings that had been classified as "good" or "bad" by humans prior to the experiment, based on their own subjective judgment, and generalize to further, novel examples of these categories (Watanabe 2010). Even so, there is no indication that the birds are enjoying these experiences beyond the immediate food reward they receive for successful answers or that they preferred "good" over "bad" paintings. Indicators of enjoyment might be that the bird paused for a longer period of time near a given artwork without receiving a reward, viewed it for a longer time, or might be physiological measures such as lower cortisol levels or lower heart rates when exposed to art. Nonetheless, because pigeons have excellent vision this research is valuable in showing that visual tasks relying on food rewards, measuring actual behavior and using artworks as stimuli can be successfully applied to animals without involving aesthetic appreciation as motivation, as would inevitably be the case with human participants.

Perhaps more convincingly, Watanabe and Nemoto (1998) provided Java sparrows (*Padda oryzivora*) a choice between perches which either elicited playbacks of Bach and Schönberg or silence. Two of the four birds spontaneously spent more time on the perch that triggered Bach's music. Accepting that this may reflect some preference on the birds' part, the question of what aspect of the stimulus is preferred (loudness, tempo, pitch, range, etc.) remains open (cf. Fitch 2006).

Early evidence suggested that non-human primates prefer regular to irregular visual shapes. Rensch (1957) studied visual preferences in a capuchin (*Cebus apella*) and a vervet monkey (*Cercopithecus aethiops*, now *Chlorocebus aethiops*) with stimuli that either contained symmetry on one or two axes ("regular") or were asymmetrical ("irregular"), and reported a bias in both species towards the symmetrical shape. Anderson, Kuwahata, Kuroshima, Leighty, and Fujita (2005) later conducted a similar study with four capuchin and four squirrel monkeys (*Saimiri sciureus*) with additional stimuli including (a) images of bilaterally symmetrical versus scrambled faces, and (b) geometrical shapes that were arranged regularly but not in a bilaterally symmetrical fashion versus scrambled shape arrangements. Regular or irregular images were pasted on cards and several of them presented simultaneously to the animal. The first card to be picked up was interpreted as the preferred image. The animals did not receive food rewards. There was a slight preference for regular over irregular patterns, but this was not consistent for all individuals of a species and not always statistically significant. However, even when not significant, the trend tended to be against irregular and for regular patterns. Again, it is not clear that the animals gained any aesthetic pleasure from this task, and there is no evidence that these species produce such regular patterns. Nonetheless these results may indicate a subtle perceptual bias for regularity in other primates that might be utilized and reinforced in human aesthetic artifacts.

Walker (1970) exposed rats to backgrounds ranging from monochromatic gray to black-and-white geometric patterns of varying complexity and found "a general tendency for the animals to move from less complex to more complex stimuli during the day" (Walker 1970, p. 643). As the order of the backgrounds was not counterbalanced, this effect may have merely been due to familiarity or novelty-seeking behavior in the animals.

Artificial body ornaments in the form of leg bands change preferences for conspecifics in zebra finches (*Taeniopygia guttata*) and can have a positive effect on both males and females. Zebra finches have orange-red beaks, and males additionally have orange-red patches in the face. Females prefer males with red bands over unbanded individuals or those with blue and green bands, while males prefer females with black or pink bands (Burley et al. 1982). When males are kept in dyads, those equipped with red bands monopolized food sources and displaced individuals with green bands (Cuthill et al. 1997). Recently, Pariser, Mariette, and Griffith (2010) reported that when wild-caught zebra finches were kept in male-only aviaries, those individuals with red bands gained more weight and sang more than those individuals with green or neutral bands. Seguin and Forstmeier (2012), however, failed to replicate this effect with zebra finches bred in captivity. We note that ultraviolet light seems to play a crucial role in the preference for red, and that artificial lighting may distort the UV information available to the birds (Hunt et al. 1997). This may explain the difference in these two studies.

Studying the aesthetic experience as a monolithic whole in animals and comparing it to that of humans can only provide the coarsest insights into the biology and evolution of the aesthetic sense. Particular species may have similar abilities and proclivities in some parts of the aesthetic process, but not others. We suggest a modular approach: in the visual domain, perception can be broken down into perception of color, symmetry (reflectional, rotational, translational), repetition, Gestalt principles (for an early study see Hertz 1928), and so on. Production could be broken down into production contexts, production self-reward systems (i.e. the effect that producing something has on the producer itself), demonstrating the ability to manipulate objects and use tools, self-adornment versus object adornment, etc. We now explore some commonalities and differences between bees, birds, and primates concerning production (nest-building) and symmetry perception.

20.4.1 Nest-building

Nest-building is an intriguing behavior to examine in the context of aesthetics because it characterizes all great apes, most birds, and many bees and wasps. Not only does nest construction entail a sophisticated use of external material, but in many species cooperation and quality assessment by perceivers play a role as well

Bees produce a wide range of nests, using sand, leaves, clay, feathers, or wax (von Frisch 1974). The honeycomb produced by honeybees (*Apis mellifera*) is particularly pleasing to our eye because of its astonishingly regular hexagonal pattern. Many bees work together to produce the comb; a single cell may be constructed by multiple bees by placing wax chips at 120 degrees to each other to form the walls of the cell (von Frisch 1974). The entire comb structure is independently aligned by many bees along a line determined by the earth's magnetic field. The production of the comb cells is dependent on motor-sensory feedback from bristles on the side of the bee's neck, and ceases when these are immobilized (von Frisch 1974). The hexagonal shape of the cells can also be found in other species of bees and wasps, which use non-wax building materials, so that the hexagons are unlikely to be due to properties of the wax, contra Pirk et al. (2004). This fact taken together with the experimental results reported by von Frisch refute the oft-repeated myth, originating with D'Arcy Thompson (Thompson 1948), that regular hexagonal cells emerge automatically due to physical principles. Instead, these structures seem to require active, precise control of the insect builders themselves.

The majority of bird species build nests and use them for parental care, although the complexity and materials of the structure as well as the social dynamics involved in nest-building and courtship vary considerably (Hansell 2000). While females are typically involved in nest-building, either alone or together with the male, there are some cases where females assess and choose between nests built by males; for example, weaver birds (Ploceidae). Male Village weaver birds (*Ploceus cucullatus*), native to Africa, construct nests later inspected by females (see Figure 20.3). The nests are attached to branches and consist of interwoven blades of grass. If the female approves, she will line it with grass and mate with the male (Hansell 2000). Thus, the female chooses her mate partially on the basis of the quality of the nest he can construct. Walsh, Hansell, Borello, and Healy (2011) have shown that male Southern Masked weaver birds (*Ploceus velatus*) become more efficient builders over time. Nests of both species built late in the season are shorter and lighter than early ones (Walsh et al. 2010). This suggests that experience interacts with the birds' inherent drive to build nests, demonstrating that males are not merely performing an invariant fixed-action pattern. Nest quality may thus reflect not only dexterity and physical prowess, but also the learning ability of the builder.

While nest structures are typically used to incubate eggs, in some cases, the male construction is not related at all to incubation. Bower birds are a small family of birds (Ptilonorhynchidae) native to New Guinea and Australia (Borgia 1986; Diamond 1982), consisting of roughly 20 species. In fifteen of these species (Hansell 2000), males construct bowers in specially cleared courts in the forest (see Figure 20.3). The bower, together with physical displays of the males, serves to attract females (their use or *Gebrauchszweck*, in Riegl's terms). A bower is a large structure made from grass and twigs, somewhat reminiscent of a nest, that is built on the forest floor and decorated with various, typically brightly colored, objects. The preferred colors of the objects and construction style of the bower varies widely between species. The bowers are not used as nests; these are built by the female after mating. Rather, bowers are a component of the courtship display, serving to entice and impress the female (Riegl's *Schmückungszweck*). There is also competition between males: neighboring males often try to steal objects from rivals' bowers and/or destroy the bower structure. The dimensions and construction principles of the bowers vary between species, as do the types and colors of the objects used to decorate the bowers.

While bower birds are famous for a behavior that appears purely ornamental, it is not unique to them alone. Other bird species are known to clear courts for their displays (Hansell 2000), and some species such as the superb lyrebird (*Menura novaehollandiae*) additionally modify the center of the clearing with a mound of earth (superb lyrebird) or a grass tuft (Jackson's widowbird, *Euplectes jacksonii*). In the bird-of-paradise species Lawe's

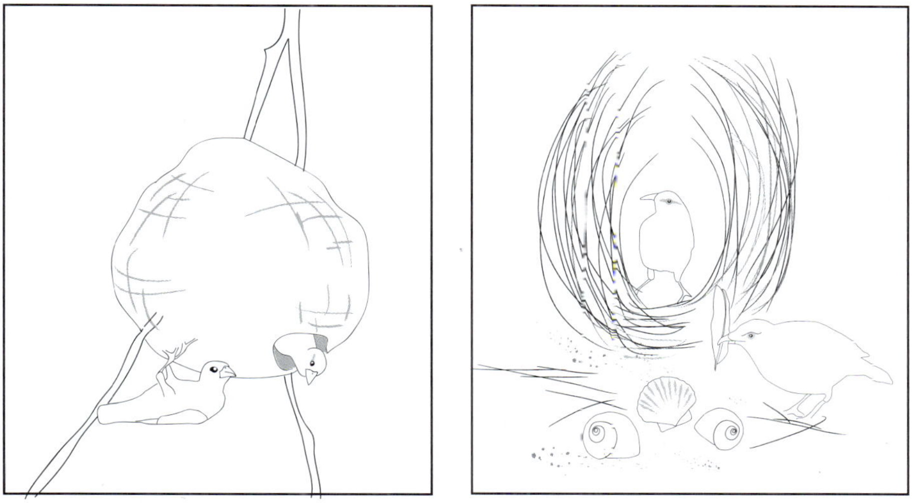

Figure 20.3 Male weaver birds (left) construct elaborate nests that are inspected by the female prior to mating. Male bower birds (right) build bowers that do not function as nests, but serve to attract females. The construction is enhanced with attractive, colorful objects, and the male performs a display for the female, holding objects in his beak.

parotia (*Parotia lawesii*), males place selected objects in the clearings, reminiscent of bowerbirds, with the difference that females subsequently remove the objects from the clearings (Pruett-Jones and Pruett-Jones 1988).

All adult great apes (gorillas, chimpanzees, bonobos, and orang-utans) build nests every night, and use them for sleeping and sometimes for social interactions such as grooming (Sabater Pi et al. 1997). The nests also offer protection against predation and wet ground. Individuals build nests that they rarely share with others, with the exception of mothers sharing nests with their infants. Nests are rarely reused and are built anew each evening. As these nests are mostly arboreal—with the exception of gorillas who usually build terrestrial nests (Tutin et al. 1995)—the nests need to be sturdily constructed. An analysis of orang-utan nests showed that the structures are complex, with sturdy branches preferentially used for support, and thinner branches used for interweaving (van Casteren et al. 2012). Nest construction in chimpanzees and bonobos is rapid, taking about a minute for day nests (used during midday rest) and no more than five minutes for night nests. Juveniles begin constructing nests in a playful manner, in preparation for their own "solo" nests that they must construct after weaning (Fruth and Hohmann 1996).

To summarize, nest-building is a behavior that in birds has the potential to be a crystallization point for the emergence of aesthetic production and perception due to its dual productive/perceptive nature, with competition in males driven by female perceptual judgment. In great apes, although nest-building is a universal trait, there is little social pressure for this behavior to become more elaborate or fulfil a courtship function because each individual builds their own nest. In bees, we see that the collaborative construction by many individuals biases the group towards a unified production process, leaving no room for individual variation or innovation if the combined effort is to converge on a well-formed structure. Each of these examples shows certain points of contact with human artistic practices, but none possesses all of the relevant features.

20.4.2 Symmetry perception

Symmetry—in particular, reflectional symmetry—is an important element in many visual designs and is a highly salient cue for humans (for an extensive review on human symmetry perception, see Treder 2010).

It has been hypothesized (Møller 1992; Penton-Voak et al. 1999) that symmetrical features and markings can act as indicators of high genetic fitness in humans and other animals (usually in males), with few deviations from a symmetrical ideal ("fluctuating asymmetry") being indicative of "good genes," meaning a high ability to deal with developmental stress and hence most desirable to females. However, fluctuating asymmetry has been criticized (e.g. by Palmer and Strobeck 2003), since the measurement of symmetry may be very sensitive to small measurement errors (especially in small traits), and asymmetries may arise for reasons other than developmental anomalies. Furthermore, Johnstone (1994) argues that a preference for symmetry in females can arise in the absence of a link between symmetry in males and genetic fitness.

Neural networks have been implemented to test whether a visual signaling process might inherently be biased towards symmetrical signals, similar to the symmetrical markings often found in animals (Enquist and Arak 1994; Enquist and Johnstone 1997). While bilateral reflectional symmetry does tend to emerge in these models with repeated iterations of initially random visual signals, critics have noted that this effect may be a byproduct of the simplified perceptual models implemented (Dawkins and Guilford 1995; Bullock and Cliff 1997; Kamo et al. 1998). Both biological approaches make interesting and valid points but fall short of capturing (human) symmetry processing in its full generality.

In animals, symmetry preferences have been studied extensively in the context of mate choice and sexual selection (Møller and Thornhill 1998). Far fewer studies have been published that examine symmetry preferences outside of a mate selection context. Since symmetry preferences in humans can be found in many contexts and feature prominently in many, if not most, human artifacts, we review research on symmetry perception in animals beyond the context of mate selection to see whether clues can be found in other species concerning the socio-ecological pressures that might have led to such an expansion of symmetry use in humans.

Honeybees can be trained to discriminate bilaterally symmetrical visual stimuli from asymmetrical stimuli and can generalize to new exemplars (Giurfa et al. 1996; Giurfa and Menzel 1997). Although there was no initial preference for symmetrical images, after exposure those trained to approach symmetrical stimuli performed consistently better and hovered longer and nearer to the stimuli than those trained to approach asymmetrical stimuli. Bees can also distinguish a vertical axis of symmetry from other orientations in bilateral symmetry (Horridge 1996). Bumblebees raised with no exposure to flowers or other symmetrical visual stimuli nonetheless show a marked preference (measured by approaches and staying times on the stimulus) for bilaterally symmetrical over asymmetrical images (Rodríguez et al. 2004). Given that these species obtain food from flowers, which are typically highly symmetrical, with more symmetrical flowers containing more nectar than less symmetrical flowers (Møller 1995), it is not surprising that symmetry is a salient cue to these insect species in a foraging context.

Turning to birds, extensive studies have been carried out on symmetry perception in both pigeons and starlings, with contradictory results. Pigeons can distinguish bilaterally symmetrical (with a vertical symmetry axis) from asymmetrical stimuli and generalize to new exemplars (Delius and Habers 1978; Delius and Nowak 1982). However, Huber and colleagues (1999) call into question whether this discrimination was based on an abstract concept of symmetry. Using three types of stimuli classes, they show that pigeons can learn to discriminate between symmetrical and asymmetrical items in two out of three stimuli classes. However, in the successful cases, the birds do not seem to acquire a general concept of symmetry to differentiate the two types of images. Instead, the authors suggest that pigeons use alternative discrimination strategies, the main one being rote learning. Results indicate that pigeons store each training exemplar together with the reward contingency. When viewing novel stimuli, the birds seem to compare the stored exemplars with the novel ones and respond according to a threshold based on stimulus similarity.

A mixed picture is also emerging for European starlings (*Sturnus vulgaris*). Starlings have a speckled breast, and the difference between the number of dots on each side averages about 9 per cent (Swaddle and Witter 1995). The birds can differentiate between artificial symmetrical and asymmetrical dot patterns where the asymmetry is achieved by moving the dots of one side of a bilaterally symmetrical pattern around randomly without removing them; that is, the number of spots on either side remains the same (Swaddle and Pruett-Jones 2001). However, if the number of dots is increased and the asymmetry is brought about by randomly removing dots from one side of a bilaterally symmetrical image and placing them randomly on the other side (the number of spots on either side differs) birds fail to discriminate between symmetrical and asymmetrical (Swaddle and Ruff 2004), corroborating the findings of Huber et al. (1999) with pigeons.

Similarly, young chicks (*Gallus gallus*) can be trained to discriminate bilaterally symmetrical stimuli from asymmetrical stimuli (Mascalzoni et al. 2012) but they have a preference for asymmetrical stimuli right after hatching. Furthermore, they preferentially peck on irregular dot arrays deviating from a straight line rather than arrays that are spaced along a straight line (Elliott et al. 2012), suggesting that a capacity to perceive mirror symmetry is not necessarily an indicator of a general preference for regularity and order. Taken together, the studies for pigeons, starlings, and chickens tell a cautionary tale: symmetry detection can be achieved with training and positive feedback under some circumstances, but it is not a robust capacity as it is in humans. The detection of symmetry, if present in a species, may be confined to narrowly delineated circumstances and stimuli types (e.g. mate choice or foraging), rather than indicating a generalized perceptual principle, as it seems to be in humans.

The literature on non-human primate symmetry perception is surprisingly sparse. A brain imaging study conducted by Sasaki and colleagues (2005) while participants were viewing radially symmetrical and asymmetrical dot patterns showed that the same areas of extrastriate visual cortex are activated in both humans and macaques (regions V3a, V4d), but that the activation was stronger in humans. There was little to no symmetry-specific activation in primary visual cortex (V1 and V2). These findings suggest that the visual processing of symmetrical stimuli in humans uses similar visual pathways as in other primates.

Waitt and Little (2006) showed macaques symmetrical and asymmetrical versions of conspecific faces, and the animals looked longer at symmetrical faces. Longer viewing time, at least in humans, is usually interpreted as a sign of preference (Langlois et al. 1987; Quinsey et al. 1996). In an eye-tracking study, Kano and Tomonaga (2009) showed chimpanzees and humans images of humans, chimpanzees, and other mammals. The basic scan patterns of images were very similar, focusing mainly on the head and face regions of the images and spending little time on the background. This again suggests that perceptual similarities exist between humans and other primates. To our knowledge no face preference studies have yet been conducted in chimpanzees. However, there is evidence that chimpanzees have no preference for symmetrical versus asymmetrical images of perineal swellings in female conspecifics (Breaux et al. 2012).

Based on the scant data available, one possibility is that the basic perceptual processes involved in symmetry perception may be quite similar in humans and other primates. So far, in all animal species studied symmetry recognition is restricted to particular symmetry types (typically reflectional symmetry) and contexts (mate selection in birds or foraging in bees), which may reflect a shared "canonical neural substrate" (Cohen and Zaidi 2013, p. 2) that relies on relatively simple perceptual mechanisms (Osorio 1996; Cohen and Zaidi 2013) that possibly evolved due to the biological relevance of symmetrical input in certain circumstances. The very broad appreciation of multiple types of symmetry that humans show, particularly involving both the perception and production of geometrical patterns, is to our knowledge unparalleled in the animal kingdom.

Despite the rarity of flexibly generated, creative geometrical patterns in the animal kingdom and their ubiquity in human society, surprisingly little empirical research has focused on human pattern production (although cf. Westphal-Fitch et al. 2012, 2013), and we think that the ability and proclivity of ordinary humans to generate structured and attractive patterns should be an important focus of future work in empirical aesthetics.

20.5 **Discussion**

Over its long history, aesthetics has grappled with the distinction between a "natural" aesthetic appreciation and further development (or refinement) of aesthetic sensitivities through cultural input.

The aesthetic traditions of a culture may vary over historical time, but while fashions come and go, we suggest that there are some core fundamental biases and proclivities of natural aesthetics that drive humans to produce aesthetic objects and remain unchanged.

In particular, a very general ability to both perceive and produce symmetrical stimuli appears to be a stable, and biologically unusual, feature of our species. Production of highly ordered but also extremely varied abstract geometrical patterns is, according to current knowledge, unique to our species, as is the production of representational art.

Although there are phenomena reminiscent of aesthetic behavior (e.g. bower building) in the animal kingdom, these nonetheless differ in important ways from those of humans (restriction to certain contexts, males only).The development of symmetrical signals in the animal kingdom is frequent but again typically restricted to specific contexts and forms. Little is known to what extent abstract patterns are perceived in animals, but it seems clear that patterns are not generated in a creative, flexible, and open-ended fashion in non-human species.

20.5.1 Music, language, and art as a cognitive triad

Art, music, and language are cultural phenomena that have a strong social component: language facilitates communication between humans and is typically used in a social setting (there are exceptions such as babbling and self-directed speech). Music is often played or sung in a group, and an audience may be present. Art is less obviously social: although artifacts may be produced and perceived in a solitary setting, often they are produced in a group, particularly in traditional societies. In particular, body art is highly social both in its production and in its later intended perception.

Besides their social functions, we suggest that a further common feature at the heart of the music/language/art triad is the ability to apply generative rules iteratively during their production (i.e. a generative syntax). In the case of visual art, we refer here in particular to abstract geometric patterns. Syntactic rules that govern the combinatorics of words are studied extensively in language, and it is a promising research field in musicology (Temperley 2010).

We argue that some sizeable subset of aesthetic behaviors involving abstract patterns in the visual domain also involves repeated application of generative rules with parallels to musical and linguistic structure (for examples see Figure 20.4).

Art, music, and language are all produced over time and the generative rules are applied serially during production. The ordering in the visual domain is less strict than in the auditory domain. Production may be restricted to a serial "one element at a time" production style, for example during sewing or beading, when only one joining operation can be executed at a time, but the order in which elements are joined is typically flexible.

A more striking difference between visual art on the one hand and music and language on the other becomes obvious during perception. While listening to speech or music, the stream must be parsed serially over time. Thus, the perceptual process must closely track the production process of the speaker or performer temporally. However, eye-tracking studies show that this is clearly not the case in visual art: perceivers can scan the two-dimensional array (or three dimensions in the case of sculptures) any way and in any order they like, without temporal restrictions. Humans (disregarding certain clinical groups) have a strong tendency to take in the entire artwork or stimulus first, focusing on details

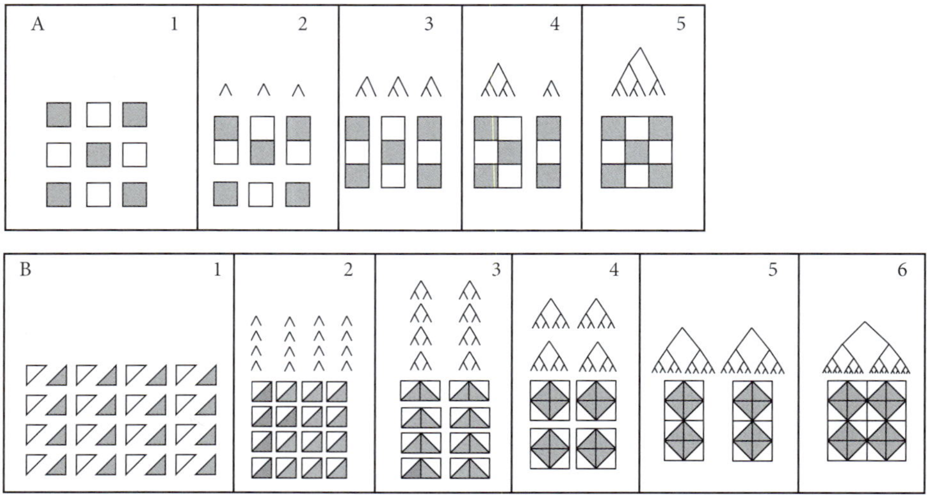

Figure 20.4 The production process underlying two visual patterns produced from basic shapes (squares, triangles) using a rule that only two edges may be joined at each step, as when sewing patchwork. The repeated application of the generative rule results in binary branching structures that show multiple levels of hierarchical embedding. A: The number of nodes is odd, leading to an asymmetrical binary tree, while the nodes in B are even, resulting in a symmetrical, recursive structure. The branching direction in A is arbitrary, but is kept consistent in the diagram for clarity.

at a later stage. Also, the perceiver can return their attention to regions that have already been looked at as often as they like. This relatively unconstrained aspect of visual perception versus the temporally bound perception of music and language offers a noteworthy freedom to the perceiver. Of course, this applies to static artwork but not to film or dance, which share the temporal flow of music or speech. It may be that such oppositions between visual and auditory, or static and temporally dynamic input have a profound effect on our aesthetic experience. However, it is also possible that all aesthetic experience shares certain common organizing principles, for example an "aesthetic trajectory" of recognition/surprise and resolution regardless of the medium or temporal dynamics (Fitch et al. 2009). Both possibilities provide issues for exploration in future empirical work.

Language is limited in its generativity by pragmatics and semantics. While it is possible to utter sentences such as "colorless green ideas sleep furiously," meaningless sentences are not conducive to successful communication, which is usually the goal of speaking. Music and abstract art are not similarly limited by semantics. Music is the most abstract case, and art can be representational (i.e. convey representational meanings) or abstract. Abstract patterns are particularly visually striking, have the longer developmental history (as we have argued above), and are much more widespread across cultures than representational art. Both music and visual art have an additional dimension of flexibility due to the freedom from semantic content (optional in the case of visual art) that language does not typically enjoy outside the narrow domains of linguistic examples or Dadaist poetry.

20.5.2 Outlook

Precisely because of the vast variance in the types of artwork that humans produce, empirical aesthetics needs to cast a wide net to achieve a comprehensive picture of the human aesthetic drive, and to discover features that all cultures have in common. Therefore, we maintain that in the future, empirical aesthetics should in principle incorporate the aesthetic judgments and activities of humans from a wide variety of cultures, and include manifestations of the aesthetic drive that can feasibly be studied in the lab. In particular, we strongly question the notion that it takes special knowledge or formal education to have a fully-fledged aesthetic experience (Fitch and Westphal-Fitch 2013).

In the same way that linguistics, and more recently, musicology have both rejected traditional prescriptive approaches to what is good or bad or right or wrong, empirical aesthetics needs to embrace a non-elitist, cross-cultural approach which recognizes and explores the aesthetic capacity of ordinary humans.

A broad comparative approach to aesthetics both across different human cultures and cross-species comparisons will allow the field to benefit from methodologies and research questions of linguistics and musicology, which have already mostly transitioned from a prescriptive (i.e. normative) to a descriptive approach (e g. Honing 2011). We suggest that a wholehearted embrace of such a broadly comparative perspective has rich insights to offer into aesthetics, and the intriguing hidden structural similarities of the human cognitive triad of music, language, and art.

Acknowledgments

An earlier version of this chapter was published as part of Gesche Westphal-Fitch's PhD thesis "Comparative Studies in Visual Pattern Processing," University of Vienna. We thank Ludwig Huber for helpful comments. This research was funded by ERC Advanced Grant SOMACCA (230604) and FWF grant W1234-G17 (to W. Tecumseh Fitch).

References

Anderson, R.L. (1979). *Art in Small-Scale Societies*. Englewood Cliffs, NJ: Prentice-Hall.

Anderson, J.R., Kuwahata, H., Kuroshima, H., et al. (2005). Are monkeys aesthetists? Rensch (1957) revisited. *Journal of Experimental Psychology: Animal Behavior Processes* **31**(1), 71–8.

Boas, F. (1955). *Primitive Art*. New York, NY: Dover.

Borgia, G. (1986). Sexual selection in bowerbirds. *Scientific American* **254**(6), 92–100.

Bouzouggar, A., Barton, N., Vanhaeren, M., et al. (2007). 82 000-year-old shell beads from North Africa and implications for the origins of modern human behavior. *Proceedings of the National Academy of Sciences of the USA* **104**(24), 9964–9.

Breaux, S.D., Watson, S.L., and Fontenot, M.B. (2012). A free choice task evaluating chimpanzees' preference for photographic images of sex swellings: effects of color, size and symmetry. *International Journal of Comparative Psychology* **25**, 118–36.

Bullock, S. and Cliff, D. (1997). The role of "hidden preferences" in the artificial co-evolution of symmetrical signals. *Proceedings of the Royal Society B* **264**, 505–11.

Burley, N., Krantzberg, G., and Radman, P. (1982). Influence of colour-banding on the conspecific preferences of zebra finches. *Animal Behaviour* **30**(2), 444–55.

Cárdenas, R.A. and Harris, L.J. (2006). Symmetrical decorations enhance the attractiveness of faces and abstract designs. *Evolution and Human Behavior* **27**, 1–8.

Cárdenas, R.A. and Harris, L.J. (2007). Do women's preferences for symmetry change across the menstrual cycle? *Evolution and Human Behavior* **28**, 96–105.

Cohen, E.H. and Zaidi, Q. (2013). Symmetry in context: Salience of mirror symmetry in natural patterns. *Journal of Vision* **13**(6), 1–9.

Cuthill, I.C., Hunt, S., Cleary, C., et al. (1997). Colour bands, dominance, and body mass regulation in male zebra finches (*Taeniopygia guttata*). *Proceedings of the Royal Society B* **264**, 1093–9.

Darwin, C. (1874). *The Descent of Man and Selection in Relation to Sex*. London: William Clowes and Sons.

Dawkins, M.S. and Guilford, T. (1995). An exaggerated preference for simple neural network models of signal evolution? *Proceedings of the Royal Society B* **261**, 357–60.

Deacon, T. (2006). The aesthetic faculty. In M. Turner (ed.), *The Artful Mind. Cognitive Science and the Riddle of Human Creativity*. Oxford: Oxford University Press, pp. 21–53.

Delius, J.D. and Habers, G. (1978). Symmetry: can pigeons conceptualize it? *Behavioral Biology* **22**(3), 336–42.

Delius, J.D. and Nowak, B. (1982). Visual symmetry recognition in pigeons. *Psychological Research* **44**(3), 199–212.

Diamond, J.M. (1982). Evolution of bowerbirds' bowers: animal origins of the aesthetic sense. *Nature* **297**, 99–102.

Dronfield, J. (1996). The vision thing: diagnosis of endogenous derivation in abstract arts. *Current Anthropology* **37**(2), 373–91.

Dubin, L.S. (1997). *Alle Perlen dieser Welt: eine Kulturgeschichte des Perlenschmucks*. Köln: Dumont Buchverlag.

Eibl-Eibesfeldt, I. (1988). The biological foundation of aesthetics. In I. Rentschler, B. Herzberger, and D. Epstein (eds), *Beauty and the Brain. Biological Aspects of Aesthetics*. Basel: Birkhäuser, pp. 29–68.

Elliott, M.A., Salva, O.R., Mulcahy, P., et al. (2012). Structural imbalance promotes behavior analogous to aesthetic preference in domestic chicks. *PLoS One* **7**(8), e43029.

Enquist, M. and Arak, A. (1994). Symmetry, beauty and evolution. *Nature* **372**, 169–72.

Enquist, M. and Johnstone, R.A. (1997). Generalization and the evolution of symmetry preferences. *Proceedings of the Royal Society B* **264**, 1345–8.

Fechner, G.T. (1871). *Zur Experimentalen Aesthetik*. Leipzig: Breitkopf & Härtel.

Fechner, G.T. (1876). *Vorschule der Ästhetik*, **Vol. 1**. Leipzig: Breitkopf & Härtel.

Fitch, W.T. (2006). The biology and evolution of music: a comparative perspective. *Cognition* **100**, 173–215.

Fitch, W.T., von Graevenitz, A., and Nicolas, E. (2009). Bio-aesthetics and the aesthetic trajectory: A dynamic cognitive and cultural perspective. In M. Skov and O. Vartanian (eds), *Neuroaesthetics*. Amityville, NY: Baywood, pp. 59–102.

Fitch, W.T. and Westphal-Fitch, G. (2013). Fechner revisited: towards an inclusive approach to aesthetics. *Behavioral and Brain Sciences* **36**(2), 140–1.

Fruth, B. and Hohmann, G. (1996). Comparative analyses of nest building behavior in bonobos and chimpanzees. In R.W. Wrangham, W.C. McGrew, F.B.M. deWaal, et al. (eds), *Chimpanzee Cultures*. Cambridge, MA: Harvard University Press, pp. 109–28.

Giurfa, M., Eichmann, B., and Menzel, R. (1996). Symmetry perception in an insect. *Nature* **382**, 458–61.

Giurfa, M. and Menzel, R. (1997). Insect visual perception: complex abilities of simple nervous systems. *Current Opinion in Neurobiology* **7**(4), 505–13.

Gregor, M.J. (1983). Baumgarten's "Aesthetica." *The Review of Metaphysics* **37**(2), 357–85.

Groß, S.W. (2001). *Felix aestheticus. Die Ästhetik als Lehre vom Menschen*. Würzburg: Königshausen & Neumann.

Gröning, K. (1997). *Decorated Skin. A World Survey of Body Art*. London: Thames & Hudson.

Hansell, M. (2000). *Bird Nests and Construction Behaviour*. Cambridge: Cambridge University Press.

Henshilwood, C., d'Errico, F., Vanhaeren, M., et al. (2004). Middle Stone Age shell beads from South Africa. *Science* **304**, 404.

Henshilwood, C.S., d'Errico, F., van Niekerk, K.L., et al. (2011). A 100 000-year-old ochre-processing workshop at Blombos Cave, South Africa. *Science* **334**, 219–22.

Henshilwood, C.S., d'Errico, F., Yates, R., et al. (2002). Emergence of modern human behavior: Middle Stone Age engravings from South Africa. *Science* **295**, 1278–80.

Hertz, M. (1928). Wahrnehmungspsychologische Untersuchungen am Eichelhäher I. *Zeitschrift für Vergleichende Physiologie* **7**(1), 144–94.

Hodgson, D. (2006). Understanding the origins of paleoart: the neurovisual resonance theory and brain functioning. *PaleoAnthroplogy* 54–67.

Honing, H. (2011). *Musical Cognition. A Science of Listening*. New Brunswick, NJ: Transaction Publishers.

Horridge, G.A. (1996). The honeybee (*Apis mellifera*) detects bilateral symmetry and discriminates its axis. *Journal of Insect Physiology* **42**(8), 755–64.

Huber, L., Aust, U., Michelbach, G., et al. (1999). Limits of symmetry conceptualization in pigeons. *Quarterly Journal of Experimental Psychology* **52**(4), 351–79.

Hume, D. (1757). *Four Dissertations*. London: A. Millar.

Hunt, S., Cuthill, I.C., Swaddle, J.P., et al. (1997). Ultraviolet vision and band-colour preference in female zebra finches, *Taeniopygia guttata*. *Animal Behaviour* **54**, 1383–92.

Johnstone, R.A. (1994). Female preference for symmetrical males as a byproduct of selection for mate recognition. *Nature* **372**, 172–5.

Jones, O. (1856). *The Grammar of Ornament*. London: Day & Son.

Kamo, M., Kubo, T., and Yoh, I. (1998). Neural network for female mate preference, trained by a genetic algorithm. *Philosophical Transactions of the Royal Society B* **353**, 399–406.

Kano, F. and Tomonaga, M. (2009). How chipanzees look at pictures: a comparative eye-tracking study. *Proceedings of the Royal Society B* **276**, 1949–55.

Kant, I. (1872). *Kritik der Urtheilskraft*. Berlin: L. Heimann's Verlag.

Kohn, M. and Mithen, S. (1999). Handaxes: products of sexual selection? *Antiquity* **73**, 518–26.

Kristeller, P.O. (1990). *Renaissance Thought and the Arts: Collected Essays*. Princeton, NJ: Princeton University Press.

Kuhn, S.L. and Stiner, M.C. (2007). Paleolithic ornaments: implications for cognition, demography and identity. *Diogenes* **214**, 40–8.

Kuhn, S.L., Stiner, M.C., Reese, D.S., et al. (2001). Ornaments of the earliest Upper Paleolithic: new insights from the Levant. *Proceedings of the National Academy of Sciences of the USA* **98**(13), 7641–6.

Langlois, J.H., Roggman, L.A., Casey, R., et al. (1987). Infant preferences for attractive faces: rudiments of a stereotype? *Developmental Psychology* **23**, 363–9.

Lewis-Williams, J.D. and Dowson, T.A. (1988). The signs of all times: Entoptic phenomena in Upper Paleolithic art. *Current Anthropology* **29**(2), 201–45.

Mascalzoni, E., Osorio, D., Regolin, L., et al. (2012). Symmetry perception by poultry chicks and its implications for three-dimensional object recognition. *Proceedings of the Royal Society B* **279**, 841–6.

Mithen, S. (1996). *The Prehistory of the Mind*. London: Thames and Hudson.

Mithen, S. (2003). Handaxes: The first aesthetic artefacts. In E. Voland and K. Grammer (eds), *Evolutionary Aesthetics*. Berlin: Springer, pp. 261–275.

Møller, A.P. (1992). Female swallow preference for symmetrical male sexual ornaments. *Nature* **357**, 238–40.

Møller, A.P. (1995). Bumblebee preference for symmetrical flowers. *Proceedings of the National Academy of Sciences of the USA* **92**, 2288–92.

Møller, A.P. and Thornhill, R. (1998). Bilateral symmetry and sexual selection: A meta-analysis. *The American Naturalist* **151**(2), 174–92.

Nowell, A. and Chang, M.L. (2009). The case against sexual selection as an explanation of handaxe morphology. *PaleoAnthroplogy* 77–88.

Osorio, D. (1996). Symmetry detection by categorization of spatial phase, a model. *Proceedings of the Royal Society B* **263**, 105–10.

Palmer, R.A. and Strobeck, C. (2003). Fluctuating asymmetry analyses revisited. In M. Polak (ed.), *Developmental Instability: Cause and Consequences*. Oxford: Oxford University Press, pp. 279–319.

Pariser, E.C., Mariette, M.M., and Griffith, S.C. (2010). Artificial ornaments manipulate intrinsic male quality in wild-caught zebra finches (*Taeniopygia guttata*). *Behavioral Ecology* **21**(2), 264–9.

Penton-Voak, I.S., Perrett, D.I., Castles, D.L., et al. (1999). Menstrual cycle alters face preference. *Nature* **399**, 741–2.

Perrett, D.I., Burt, D.M., Penton-Voak, I.S., et al. (1999). Symmetry and human facial attractiveness. *Evolution and Human Behavior* **20**, 295–307.

Pirk, C., Hepburn, H., Radloff, S., et al. (2004). Honeybee combs: construction through a liquid equilibrium process? *Naturwissenschaften* **91**, 350–3.

Pruett-Jones, S.-G., and Pruett-Jones, M. (1988). The use of court objects by Lawes' parotia. *The Condor* **90**, 538–45.

Quinsey, V.L., Kesetzis, M., Earls, C., et al. (1996). Viewing time as a measure of sexual interest. *Ethology and Sociobiology* **17**, 341–54.

Rensch, B. (1957). Ästhetische Faktoren bei Farb- und Formbevorzugungen von Affen. *Zeitschrift für Tierpsychologie* **14**(1), 71–99.

Riegl, A. (1966). *Historische Grammatik der bildenden Künste*. Graz: Hermann Böhlaus Nachf.

Rodríguez, I., Gumbert, A., Hempel de Ibarra, N., et al. (2004). Symmetry is in the eye of the "beeholder": innate preference for bilateral symmetry in flower-naïve bumblebees. *Naturwissenschaften* **91**, 374–7.

Ryan, M.J. (1998). Sexual selection, receiver biases, and the evolution of sex differences. *Science* **281**, 1999–2003.

Sabater Pi, J., Vea, J.J., and Serrallonga, J. (1997). Did the first hominids build nests? *Current Anthropology* **38**, 914–6.

Sasaki, Y., Vanduffel, W., Knutsen, T., et al. (2005). Symmetry activates extrastriate visual cortex in human and nonhuman primates. *Proceedings of the National Academy of Sciences of the USA* **102**(8), 3159–63.

Seguin, A. and Forstmeier, W. (2012). No band color effects on male courtship rate or body mass in the zebra finch: four experiments and a metaanalysis. *PLoS One* **7**(6), e37785.

Shiner, L. (2001). *The Invention of Art: A Cultural History*. Chicago, IL: University of Chicago Press.

Swaddle, J.P. and Cuthill, I.C. (1995). Asymmetry and human facial attractiveness: symmetry may not always be beautiful. *Proceedings of the Royal Society B* **261**, 111–6.

Swaddle, J.P. and Pruett-Jones, S. (2001). Starlings can categorize symmetry differences in dot displays. *American Naturalist* **158**(3), 300–7.

Swaddle, J.P. and Ruff, D.A. (2004). Starlings have difficulty in detecting dot symmetry: implications for studying fluctuating asymmetry. *Behaviour* **141**, 29–40.

Swaddle, J.P. and Witter, M.S. (1995). Chest plumage, dominance and fluctuating asymmetry in female starlings. *Proceedings of the Royal Society B* **260**, 219–23.

Temperley, D. (2010). *Music and Probability.* Cambridge, MA: MIT Press.

Thompson, D.W. (1948). *On Growth and Form.* New York, NY: Cambridge University Press.

Treder, M.S. (2010). Behind the looking-glass: a review of human symmetry perception. *Symmetry* **2**, 1510–43.

Tutin, C.E.G., Parnell, R.J., White, L.J.T., et al. (1995). Nestbuilding by lowland gorillas in the Lopé Reserve, Gabon: environmental influences and implications for censusing. *International Journal of Primatology* **16**(2), 53–76.

van Casteren, A., Sellers, W.I., Thorpe, S.K.S., et al. (2012). Nestbuilding orangutans demonstrate engineering know-how to produce safe, comfortable beds. *Proceedings of the National Academy of Sciences of the USA* **109**(18), 6873–7.

Vanhaeren, M., d'Errico, F., Stringer, C., et al. (2006). Middle Paleolithic shell beads in Israel and Algeria. *Science* **312**, 1785–8.

Verpooten, J. and Nelissen, M. (2010). Sensory exploitation and cultural transmission: the late emergence of iconic representations in human evolution. *Theory in Biosciences* **129**(2–3), 211–21.

Voland, E. (2003). Aesthetic preferences in the world of artifacts—adaptations for the evaluation of "honest signals?" In E. Voland and K. Grammer (eds), *Evolutionary Aesthetics.* Berlin: Springer, pp. 239–60.

von Frisch, K. (1974). *Animal Architecture.* London: Hutchinson.

Waitt, C. and Little, A.C. (2006). Preferences for symmetry in conspecific facial shape among *Macaca mulatta. International Journal of Primatology* **27**(1), 133–45.

Walker, E.L. (1970). Complexity and preference in animals and men. *Annals of the New York Academy of Sciences* **169**, 619–53.

Walsh, P.T., Hansell, M., Borello, W.D., et al. (2010). Repeatability of nest morphology in African weaver birds. *Biology Letters* **6**, 149–51.

Walsh, P.T., Hansell, M., Borello, W.D., et al. (2011). Individuality in nest building: do Southern Masked weaver (*Ploceus velatus*) males vary in their nest-building behaviour? *Behavioural Processes* **88**(1), 1–6.

Washburn, D.K. and Crowe, D.W. (1988). *Symmetries of Culture. Theory and Practice of Plane Pattern Analysis.* Seattle, WA: University of Washington Press.

Watanabe, S. (2010). Pigeons can discriminate "good" and "bad" paintings by children. *Animal Cognition* **13**(1), 75–85.

Watanabe, S. and Nemoto, M. (1998). Reinforcing property of music in Java sparrows (*Padda oryzivora*). *Behavioural Processes* **43**(2), 211–8.

Watanabe, S., Sakamoto, J., and Wakita, M. (1995). Pigeons' discrimination of paintings by Monet and Picasso. *Journal of the Experimental Analysis of Behavior* **63**, 165–74.

Wenban-Smith, F. (2004). Handaxe typology and Lower Paleolithic cultural development: ficrons, cleavers and two giant handaxes from Cuxton. *Lithics* **25**, 11–21.

Westphal-Fitch, G., Huber, L., Gómez, J.C., et al. (2012). Production and perception rules underlying visual patterns: Effects of symmetry and hierarchy. *Philosophical Transactions of the Royal Society B* **367**(1598), 2007–22.

Westphal-Fitch, G., Oh, J., and Fitch, W.T. (2013). Studying aesthetics with the method of production: effects of context and local symmetry. *Psychology of Aesthetics, Creativity, and the Arts* **7**(1), 13–26.

Zaidel, D.W., Nadal, M., Flexas, A., et al. (2013). An evolutionary approach to art and aesthetic experience. *Psychology of Aesthetics, Creativity and the Arts* **7**(1), 100–9.

Chapter 21

Art and brain co-evolution

Camilo J. Cela-Conde and Francisco J. Ayala

The capacity for appreciating aesthetic qualities in objects, movements, and sounds seems to be a human universal; all human groups possess this competence.

The capacity for *producing* aesthetic items is also universal: painters, dancers, and musicians are not restricted to any culture or historical epoch. However, *appreciating* aesthetic attributes—what we may call "beauty"—goes beyond producing them in at least two aspects. First, "artists" (producers) make up a small fraction of human groups; on the contrary, "spectators" are numerous. Second, it is possible to appreciate aesthetic qualities in natural objects and events, such as sunsets on a beach, whales' songs, or bird flight. These natural aesthetic items have no author.

We cannot establish the phylogenetic appearance of the human competence for appreciating beauty. Neither fossil nor archaeological records contain sufficient evidence to ascertain the appearance of this capacity. It is not possible to ascertain whether spectators with the ability for appreciating landscapes, dances, or songs existed in previous human species.

Producing beauty seems less elusive, and beauty's origin is also difficult to establish. Regarding artworks, Paleolithic polychromies, for instance, are an example of the early presence of artisans. Some traces of early artists' work should exist. How can we detect them?

We have extensively examined early evidences of decorative, artistic, or symbolic objects in our earlier work (Cela-Conde and Ayala 2007). We will not repeat here the arguments in favor of the symbolic function of burials, for instance. As we are here interested in the co-evolution of art and the brain, we will now focus on the search for items of proof of mental correlates that might speak in favor of a capacity for appreciating beauty.

21.1 Neuroaesthetics

Beyond some valuable precedents, such as Ramachandran and Zeki's ideas on art and the brain, the empirical field of neuroaesthetics started in 2004 when three different studies offered the first accounts of the activation of brain areas during aesthetic appreciation. Vartanian and Goel found brain activity related to preference for artworks in the right caudate nucleus, the left cingulate sulcus, and the bilateral fusiform gyri (Vartanian and Goel 2004). Kawabata and Zeki identified activity in the medial orbitofrontal cortex for beautiful, compared with ugly, stimuli, and also in the anterior cingulate gyrus in beautiful versus neutral stimuli (Kawabata and Zeki 2004). In turn, Cela-Conde and collaborators found increased activity for beautiful stimuli, compared with not-beautiful, in the left prefrontal dorsolateral cortex (Cela-Conde et al. 2004).

Since 2004 many related investigations have been published. Due to the different cognitive tasks asked from the participants, several brain regions have been identified as activated when aesthetic appreciation occurs (see Table 21.1).

Some early models of the cognitive activity of appreciation of beauty exist. Anjan Chatterjee (Chatterjee 2003) indicates the existence of a temporal sequence of processing within the nervous system. Chatterjee's (2003) model distinguishes between early, universal, and late, personal, stages of aesthetic perception. The point of departure is the visual attributes of the object whose perception is an act on the part of the viewer which is performed in the same way as perceptions otherwise unrelated to aesthetic appreciation. By means of a binding process, forms, colors, and spatial arrangements of movements—if there are any—are combined, engaging frontal-parietal circuits. These neural networks

Table 21.1 Brain areas activated in 20 neuroaesthetics experiments. The column "N" expresses the number of experiments mentioning each brain area.

Area	N	Cognitive processes
vMPFC: ventromedial prefrontal cortex	1	Resting state
aMPC: anterior medial prefrontal corte	5	
pCC: posterior cingulate cortex (L left, R right)	1	
Precuneus	2	
SN: substantia nigra	1	Reward and
Hippocampus	5	emotional
DS: dorsal striatum (caudate)	3	processing
VS (Nacc): ventral striatum (nucleus accumbens)	4	
Amygdala	2	
Insula	4	
ACC: anterior cingulate cortex	6	
OFC: orbitofrontal cortex	5	
Temp P: temporal pole	3	
DLPFC: dorsolateral prefrontal cortex	3	Judgment and
VLPFC: ventrolateral prefrontal cortex	4	decision making
Motor C: motor cortex	4	Perceptual
Occip C: occipital cortex	8	processing
P.hippo C: parahippocampal cortex	1	
TPJ: temporoparietal junction	1	
SPC: superior parietal cortex	2	
IPC: inferior parietal cortex	2	

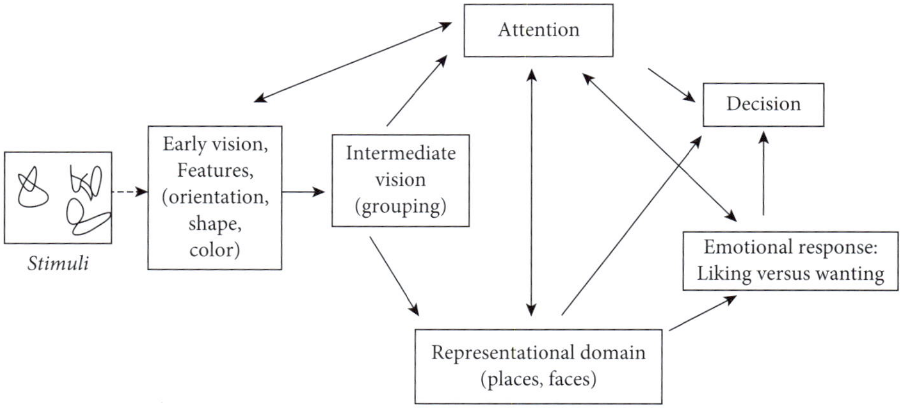

Figure 21.1 Chatterjee's model of the neural underpinnings of visual aesthetics. Reproduced from A. Chatterjee, Prospects for a cognitive neuroscience of visual aesthetics, *Bulletin of Psychology and the Arts*, 4, pp. 55–60 © 2003, American Psychological Association, with permission.

inherent to the attention lead, through a feed-forward system, to processing different attributes including mnemonic domains related to previous personal history (places, faces). An emotional component adds hedonic sensations by extending the feed-forward system. The complete phenomenon integrates what might be called the universal, innate components of visual perception (including the limits of the psychophysical as perceivable wavelengths) and the subjective aspects that relate to past experience of the subject (including its historical and cultural context).

Leder and colleagues (2004) have proposed a comprehensive model of visual aesthetic preference which considers the same elements as Chatterjee's (2003) scheme, but with some important additions. Aesthetic judgment is now included within the aesthetic experience per se. Crucial features of the context of perception are taken into account (to see an object in a museum is not the same as seeing it in the street or in a private home). Also, the training experience of the subject (expert/non-expert) forms part of the model.

21.2 Neural networks

The models by Chatterjee (2003) and Leder and colleagues (2004) theoretically imply the existence of connectivity among the different brain areas related to the perception of beauty, that is, the activity of neural networks.

Regarding aesthetics networks, Brown and collaborators proposed the existence of a "core circuit for aesthetic processing" by means of a model in which exteroceptive information passing through the orbitofrontal cortex (OFC), and interoceptive information passing through the anterior insula, are integrated to achieve aesthetic appraisal (Brown et al. 2011). However, this model is not based on any empirical studies of aesthetic appreciation. Rather, it is grounded in analyses of the reward system and valence processing.

Empirical verification of neural connectivity has also been addressed. The most interesting contribution with respect to putative aesthetic networks may have been the suggestion

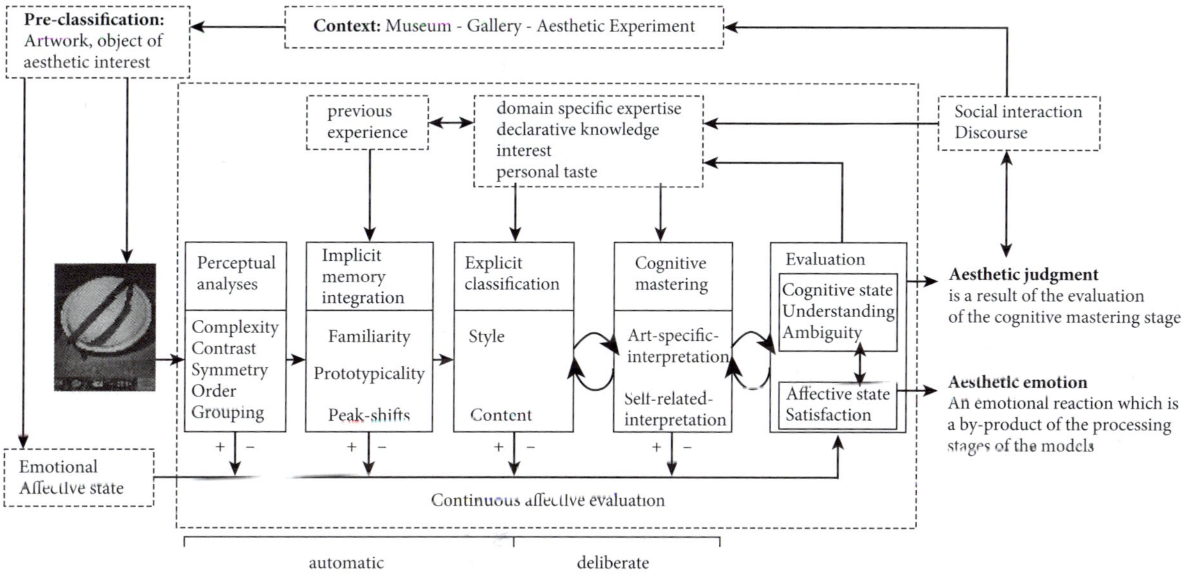

Figure 21.2 Leder and colleagues' model of aesthetic appreciation and aesthetic judgment. Reproduced from Helmut Leder, Benno Belke, Andries Oeberst, and Dorothee Augustin, A model of aesthetic appreciation and aesthetic judgments, *British Journal of Psychology*, 95 (4), pp. 489–508 DOI: 10.1348/0007126042369811 © 2004, The British Psychological Society, with permission.

Table 21.2 Regions active under aesthetic perception, corresponding Brodmann areas and Talairach coordinates. After (A) Jacobsen et al. (2006), and (B) Vessel et al. (2012). Neither of these articles includes analyses of functional connectivity.

		BA area	Talairach coordinates			A	B
Frontomedial cortex	FMC	10	1	54	26	x	
Anterior medial prefrontal cortex	aMPFC	10	–6	38	4		x
Precuneus	PCUN	7	–4	–47	32	x	x
Posterior cingulate cortex	PCC	23/31	1	–18	41	x	
Left posterior cingulate cortex	PCC	23/31	–9	–49	18		x
Superior frontal gyrus	SFG	10	22	45	26	x	
Frontomedial/anterior cingulate	FMC/AC	9/32	1	23	32	x	
Left inferior frontal gyrus	lIFG	44/45/47	–46	17	0	x	
Right inferior frontal gyrus	rIFG		46	24	0	x	
Left temporal pole	lTP	38	–43	2	–29	x	
Right tempoparietal junction	rTPJ	39/40/42	46	–56	32	x	
Left tempoparietal junction	lTPJ		–41	–59	35	x	
Superior frontal gyrus	SFG	6	–5	19	62		x
Left substantia nigra	SN		18	–12	–6		x
Left hippocampus	HC		–30	–21	–10		x

Data from Thomas Jacobsen, Ricarda I. Schubotz, Lea Höfel, and D. Yves v. Cramon, Brain correlates of aesthetic judgment of beauty, *Neuroimage*, 29 (1), pp. 276–85, doi:10.1016/j.neuroimage.2005.07.010, 2006 and Edward A. Vessel, G. Gabrielle Starr, and Nava Rubin, The brain on art: Intense aesthetic experience activates the default mode network, *Frontiers in Human Neuroscience*, 6, (66), doi: 10.3389/fnhum.2012.00066, 2012.

advanced by Jacobsen and collaborators (2006) and Vessel, Star, and Rubin (2012) of connections between medial parts of the frontal cortex (FMC), the precuneus (PCUN) and the posterior cingulate cortex (PCC) among other regions (see list in Table 21.2). These interconnected areas coincide in part with what is known as the default mode network (DMN).

Activation of the DMN during aesthetic perception seems surprising. This network was identified by Raichle and collaborators as a baseline state of the brain present under resting conditions and being curtailed when some concrete action is performed (Fox et al. 2005; Raichle et al. 2001). Aesthetic judgment is just such a task. Jacobsen and colleagues (2006) instructed participants to rate visual stimuli in order to answer the question, "is this pattern beautiful?" comparing the obtained results with those of a second question, "is this pattern symmetric?" In turn, Vessel, Starr, and Rubin's (2012) asked participants to rate visual stimuli on a 1 to 4 scale answering the question "how strongly does this painting move you?" Thus, in both studies, concrete cognitive processes demanding attention were implied, those of (1) seeing stimuli, (2) appreciating their aesthetic content; (3) rating it, and (4) formulating the result as aesthetic judgment.

Why does the DNM remain activated in neuroaesthetics experiments? The answer to this question requires exploring the characteristics of neural connectivity. Claiming that one neural network is formed when performing a cognitive task leads to the assumption that some kind of connection exists between the implied areas. The notion of connectivity has been largely considered by anatomists, neurologists, and psychologists to be at the core of explanations of consciousness. Following von der Malsburg (von der Malsburg and Schneider 1986), connectivity is associated with the presumed synchronization of neuronal assemblies—synchronous "firing." Eventually, distributed local networks of neurons would be transiently linked by reciprocal dynamic connections (Varela et al. 2001).

21.3 Functional connectivity

Statistical studies on the synchrony of firing can identify the existence of a "functional connectivity" of brain areas (Friston et al. 1993; Friston et al. 1995), defined as the statistically temporal dependency of neuronal activation patterns of anatomically separated brain regions (van den Heuvel and Hulshoff Pol 2010). Using functional connectivity to posit the existence of neural networks is a common procedure. In order to obtain such connectivity, temporal series of activation/deactivation of presumably synchronized neurons must be compared. As is well known, such temporal series can be obtained by means of functional magnetic resonance imaging (fMRI). This technique detects changes in blood oxygen content in brain areas—what, following Ogawa and colleagues (1990), is called blood oxygen-level-dependent (BOLD) signals—due to activity along a determinate span of time.

Combining temporal series of BOLD signals and advanced statistical means of analysis such as graph methods, brain architecture consisting of modules of neural functional connectivity can be determined. A review of the literature on the resting state functional architecture carried out by Lee, Smyser, and Shimony (2012) reported the identification of seven networks. Apart from the DMN, these networks correspond to different particular processes ranging from relatively simple perception to higher cognitive achievements. Figure 21.3 shows the correspondence between He and colleagues' (2009) modules and cognitive systems.

If the cognitive process, whose neural correlate is sought, maintains its activity during a considerable length of time, then fMRI can be used as an accurate technique to obtain connectivity patterns. However, compared with the relatively stable condition of subjects' brain activity during resting state, aesthetic perception refers to cognitive processes taking place over a much shorter period (Cela-Conde et al. 2004). Variations of BOLD signals along several seconds are not accurate enough to detect such changes, since networks might be modified in a relatively narrow time during aesthetic processing tasks. Some studies have focused on dynamic changes in brain networks using fMRI techniques (Bassett et al. 2011), but their time-windows cover several minutes. Therefore, to reach brain activation related to the perception of beauty it is mandatory to use high temporal-resolution techniques, like electroencephalography (EEG) or magnetoencephalography (MEG).

EEG and MEG have high temporal resolution but less spatial resolution; for fMRI the reverse is true. We are thus facing a methodological dilemma: either (1) we use a high spatial resolution to determine the distribution of active modules in the brain, but only averaging

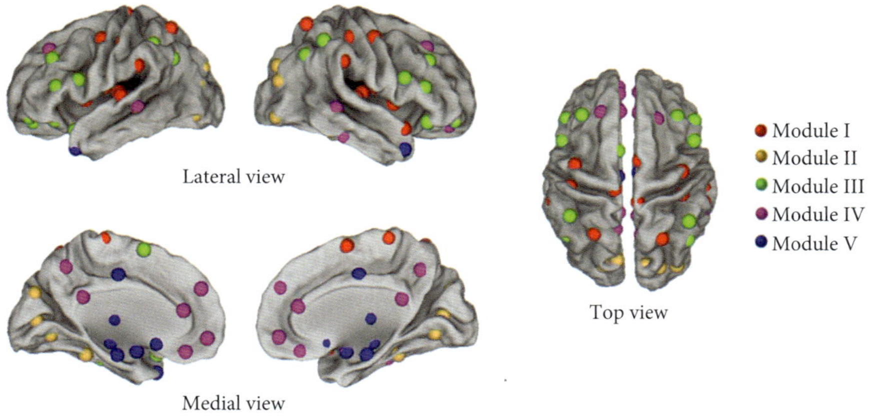

Figure 21.3 Association between modules and cognitive systems. Reprinted from *PLoS One*, 4 (4), He, Y., Wang, J., Wang, L., et al., e5226, doi: 10.1371/journal.pone.0005226. User rights governed by an Open Access license.
Module 1: somatosensory, motor, and auditory functions
Module 2: visual processing
Module 3: default network
Module 4: attention processing
Module 5: limbic/paralimbic and subcortical systems

a large period of time, or (2) we choose to detect dynamic changes across short time-spans, but thus can only obtain the modular structure of the cortical surface.

A study of the dynamics of brain networks in aesthetic appreciation carried out by means of MEG offers some clues for explaining the persistence of DMN in aesthetic tasks (Cela-Conde et al. 2013). The authors obtained temporal series of brain activity for 24 participants during resting state and judgment of beauty of 400 diverse visual stimuli. They then estimated the synchronization in the beta band of the temporal series by means of Pearson correlation coefficient and phase locking value (PLV) (Mormann 2000; Pereda et al. 2006). The MEG signals were split into three temporal windows (Figure 21.4, top):

- TW0, 500 ms (milliseconds) prior to stimuli projection
- TW1, 250–750 ms after stimuli projection
- TW2, 1000–1500 ms after stimuli projection

Inter-window comparisons evaluated the differences in connectivity between temporal windows along each condition. Inter-condition comparisons evaluated differences in connectivity between beautiful and not-beautiful stimuli in each temporal window (Figure 21.4, bottom).

Cela-Conde and colleagues' (2013) results show that the neural connectivity present in the resting state is curtailed in the first temporal window (TW1; Figure 21.5, top), being

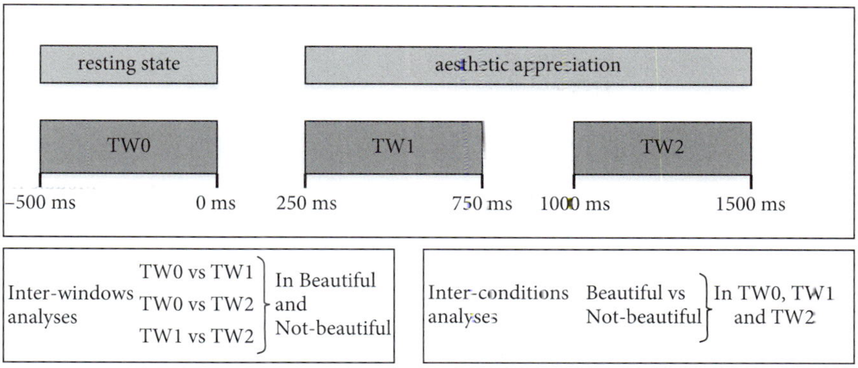

Figure 21.4 Temporal windows and conditions compared in the analysis of brain dynamics during aesthetic appreciation. Reproduced from Camilo J. Cela-Conde, Juan García-Prieto, José J. Ramasco, Claudio R. Mirasso, Ricardo Bajo, Enric Munar, Albert Flexas, Francisco del-Pozo, and Fernando Maestú, Dynamics of brain networks in the aesthetic appreciation, *Proceeding of the National Academy of Science of the USA*, 110 (Supplement 2), pp. 10454–61, Figure 21.1, doi: <http://www.pnas.org/cgi/doi/10.1073/pnas.1302855110> © 2013, National Academy of Sciences.

replaced by what the authors call the "initial aesthetic network." This network mainly connects occipital regions (Figure 21.5, middle).

The initial aesthetic network is nearly the same while viewing beautiful and not-beautiful stimuli. However, during the second temporal window (TW2) significant differences depending on the judgment of beauty appear. The "delayed aesthetic network" identified in Cela-Conde and colleagues' (2013) experiment consists of synchronized activity mainly present along medial occipital, lateral occipital, lateral posterior parietal, medial parietal, medial frontal, and dorsolateral prefrontal in the left hemisphere, as well as in the right lateral parietal (Figure 21.5, bottom).

The dynamic scenario during aesthetic appreciation seems to be, thus, as follows:

1 a starting point of high synchronization during the resting state (TW0);
2 this connectivity is curtailed during the first temporal window (TW1), and is replaced by a different network;
3 part of the resting state networks is later recovered during the second temporal window (TW2).

Both beautiful and not-beautiful conditions share, during the second temporal window, a bilateral higher synchronization along frontal-parietal-temporal-occipital lateral regions. This pattern matches the similar bilateral synchronization during the resting state. Due to its lateral position, this connectivity would have little relationship with the medially placed DMN. Hypothetically speaking, it is sound to claim that attentional tasks, obviously engaged in the aesthetic appreciation, would be responsible for this network.

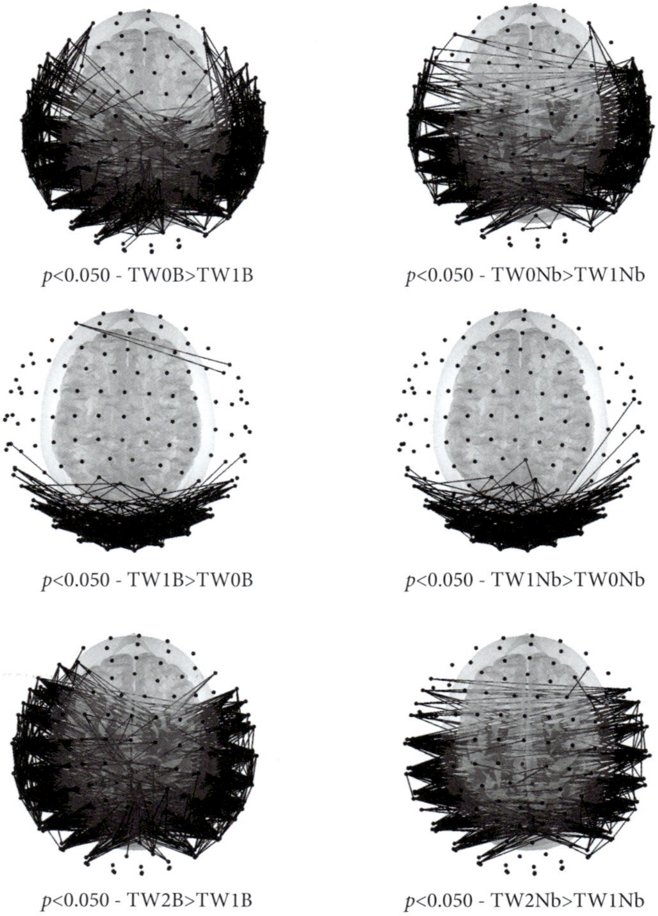

Figure 21.5 Differences in synchronization while viewing beautiful (left) and not-beautiful (right) stimuli. Top: TW1>TW0. Bottom: TW2>TW1. Reproduced from Camilo J. Cela-Conde, Juan García-Prieto, José J. Ramasco, Claudio R. Mirasso, Ricardo Bajo, Enric Munar, Albert Flexas,Francisco del-Pozo, and Fernando Maestú, Dynamics of brain networks in the aesthetic appreciation, *Proceedings of the National Academy of Science of the USA*, 110 (Supplement 2), pp. 10454–61, Figures 21.2 and 21.3, doi: <http://www.pnas.org/cgi/doi/10.1073/pnas.1302855110> © 2013, National Academy of Sciences.

In turn, differences of synchronization in favor of beautiful stimuli mainly affect medial parts of the brain. These differences are better shown in the TW2 inter-condition analysis, which manifests that a higher connectivity exists in favor of the beautiful stimuli (Table 21.3 and Figure 21.6, left). Not-beautiful stimuli do not show any higher synchronized link (Table 21.3 and Figure 21.6, right).

We are looking at the delayed aesthetic network and Figure 21.7 shows it from different perspectives.

Table 21.3 Number of MEG sensors and links more synchronized in the inter-condition comparisons at $p < 0.050$.

	Beautiful > Not-beautiful		Not-beautiful > Beautiful	
	Sensors	Links	Sensors	Links
TW0	6	3	0	0
TW1	2	1	0	0
TW2	19	10	0	0

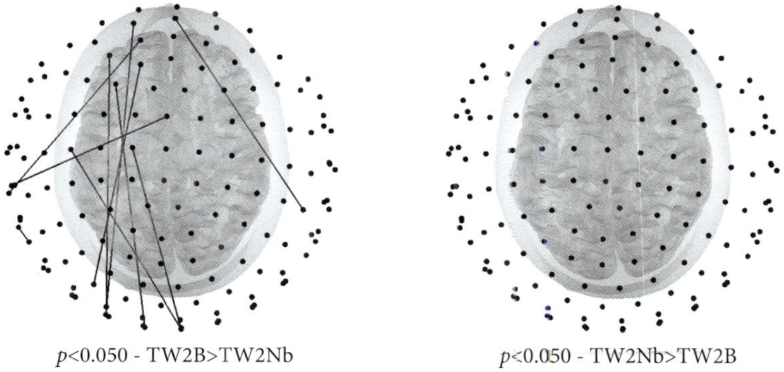

Figure 21.6 Inter-condition comparison in the second temporal window (TW2). Left: higher synchronization for beautiful stimuli. Right: higher synchronization for not-beautiful stimuli.

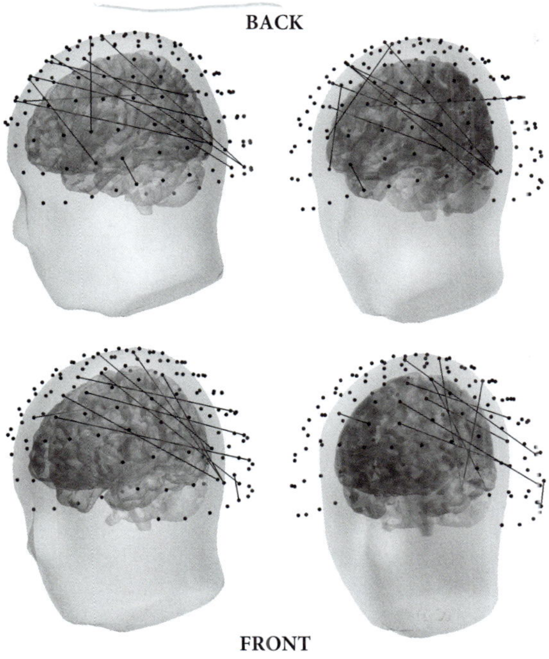

Figure 21.7 The delayed aesthetic network from different perspectives. The delayed aesthetic network matches, in part at least, the default mode network of the resting state. As we will see, this coincidence may throw some light on the elusive question of the evolution of the human aesthetic capacity.

21.4 Perceiving beauty as a human trait

Neuroaesthetics experiments have been normally carried out with very specific groups of subjects: college students from developed countries in most cases. However, aesthetics (i.e. artworks) are found everywhere, impinging on people the ability to recognize products of other cultures as artworks. This universality is, in the opinion of Carroll (2004), an argument in favor of considering the proposition that art may serve universal adaptive purposes.

Theoretically speaking, a trait can be universal because it has been inherited from an ancestor who fixed it—plesiomorphy—thus being shared with other species belonging to our human lineage. Alternatively, we may consider that the trait was developed as a new feature—apomorphy—quite early in our species. In this second case, all modern humans would have the capacity for appreciating beauty, but no other species would share this trait. However, this universality poses a problem. As it is well known, brain tissues are metabolically expensive. Resting state activity most likely accounts for the major cost, in terms of energy, of running the brain (Raichle 2011). In the awake resting state, the brain accounts for 20% of the total oxygen consumption of the body, despite the fact that it represents only 2% of body weight (Gusnard and Raichle 2001). It is difficult to justify how any brain-related capacity would have become fixed during human evolution without explaining its benefits (Aiello and Wheeler 1995). The adaptive advantages of aesthetic appreciation remain to be accounted for—although a different possibility, considering it as an exaptation, exists. We will examine it later.

The power of artworks to build communities of sentiments, promoting the cohesion of groups, seems to offer a conspicuous advantage (Dissanayake 1992, 2007). Aesthetics would be a cohesive factor, thus adaptively convenient. Also, the emotional (i.e. hedonic) reward of art offers an explanation for the eventual selection of the capacity for appreciating beauty (see, for instance, Dutton 2003). Both reward and cohesion factors are good hypotheses to account for the current relationships existing between beauty and art (Dissanayake 2009). But what can we say about their evolutionary journey?

Although the phylogenetic appearance of the brain correlates for appreciating beauty cannot be ascertained, producing beauty seems less elusive. Works of art and decorative objects persist after the disappearance of their creators. Since painters' or sculptors' works are durable, but dancing movements and music sounds are not, we will henceforth refer only to objects as artworks when speaking of the evolution of the human aesthetic capacity.

When did this capacity appear? Although "beauty," "art," and "aesthetics" are different concepts that might be even in opposition—consider, for example, the movement known as "Ugly Art"—we will take these three concepts as being equivalent. A cautionary note must be stated. Since natural objects can be considered as beautiful by spectators too, the capacity for appreciating beauty might have evolved earlier and across more locations than did the production of actual artistic objects, created with the purpose of being aesthetically moving. In a similar manner, some spectators could consider as "beautiful" an object—a lithic tool, for instance—manufactured for practical purposes.

The lack of correspondence between artists' intentions and spectators' responses is well known, with many examples present in the recent history of art. Some examples include the popular reaction to the first Impressionists' exhibitions, or the initial reception of the Eiffel Tower. Therefore, the current consideration of lithic tools, or of any other object, as "beautiful" is not proof of either the intention of the artisans that produced them, or the eventual aesthetic appreciation of them by spectators at that time. Bearing this caution in mind, authors dealing with the phylogeny of aesthetics distinguish between objects that can be used as tools, like handaxes, and objects that do not have any "practical" benefit—thus "symbolic"—such as decorative objects. In spite of the obvious utility of decorative pieces to express, for instance, social hierarchy, the symbolic criterion is useful. Non-practical objects belonging to the archaeological record are items of proof pointing to the appearance of aesthetic appreciation.

A long-lived controversy between archaeologists, paleontologists, and anthropologists refers to the question of whether Neanderthals had the capacity to manufacture artworks. Some Neanderthal sites contain decorative objects; for instance, the Grotte du Renne (Arcy-sur-Cure, France) has yielded a series of up to 36 objects, such as carved ivory pieces and perforated bones, the sole purpose of which must have been decorative (Hublin et al. 1996). Hublin and colleagues (1996) interpreted the Arcy-sur-Cure artifacts as the result of cultural exchange. D'Errico and colleagues (1998) arrived at a different conclusion: those objects were the result of an independent and characteristically Neanderthal cultural development, which had managed to cross the threshold of the symbolism inherent in decorative objects. White (2001) has offered an alternative interpretation: "It seems implausible that . . . Neandertals and Cro-Magnons independently and simultaneously invented personal ornaments manufactured from the same raw materials and using precisely the same techniques." Consequently, White defended the proposition that the Châtelperronian ornaments from the Grotte du Renne are Aurignacian and were produced by modern humans.

In addition to Arcy-sur-Cure, two other Neanderthal sites have provided perforated shells: Cueva de los Aviones (genus *Acanthocardia*, *Glycimeris*, and *Spondylus*, with traces of colored pigments into the shell in *Spondylus*) and Cueva Antón (genus *Pecten*), both in Murcia in Spain (Zilhão et al. 2010).

Had Neanderthals reached this cognitive level? Dating of paintings found in Altamira, El Castillo, and Tito Bustillo caves in Spain suggested provenances for some symbols and marks found of between 35 600 and 40 800 years (Pike et al. 2012), and some suggest they may be even older (Appenzeller 2013). If this is the case, arguments that suggest that the Neanderthals' decorative items are mere copies of similar objects manufactured by modern humans cannot be true; our species would not have reached southern Europe by that time.

Whether Neanderthals discovered decorative objects by themselves, or they imitated the manufacture and use from modern humans is not something that it easy to determine. Some proof of cultural sharing during the early Upper Paleolithic do, however, exist. Karavanic and Smith (1998) documented the presence of two contemporary sites at Hrvatsko

Zagorje in Croatia which are close to each other. The Vindija Cave has yielded Neanderthal remains, while Velika Pécina has only produced remains of anatomically modern humans. The authors believe that the similarities exhibited by the tools from both sites are due to imitation or even commercial exchange. These Croatian sites do not include ornaments but they provide remarkable indications of cultural exchange. Nevertheless, the cognitive capacity linked to perception of beauty does not depend on skills for manufacturing decorative objects. If Neanderthals appreciated pendants enough to imitate, steal, or exchange them, it would be beyond doubt that they considered perforated shells and bones to be "beautiful objects."

21.5 **Patterns of the evolution of the brain**

If Neanderthals and modern humans shared a similar capacity for appreciating aesthetic objects, the question would be whether we can link this capacity to Neanderthal brain characteristics.

Emilano Bruner, Giorgio Manzi, and Juan Luis Arsuaga (2003) analysed the transition from a generalized archaic pattern of the brain within the genus *Homo* to the modern and Neandertal morphologies. Comparing metric variables and landmark data on physical endocasts and from virtual reconstructions based on computed tomography and three-dimensional image analysis, Bruner, Manzi, and Arsuaga (2003, p. 15338) concluded that:

> archaic and Neandertal specimens share a common endocranial model, in which a large amount of variation is based on a single allometric trend. In this case, encephalization (viewed as cerebral volumetric expansion) structurally influences the variation in endocranial shape. This trajectory represents therefore a continuous gradation, ranging from archaic small specimens . . . to archaic larger ones . . . up to the extremely encephalized Würmian Neandertals.

The "archaic" brain pattern shared by Neanderthals involves a relative reduction of the length and width of the occipital lobes, a vertical development, an enlargement of the frontal breadth, and the shortening of the parietal chord. The main difference with the modern human pattern of the brain refers to the parietal development of the modern morphotype, something that Bruner, Manzi, and Arsuaga (2003) consider may have represented "a key to surpass the encephalization constraints imposed by the archaic structural model."

Expansion of the frontal lobes is an evolutionary trait shared by Neanderthals and modern humans (Bruner and Holloway 2010). On the other hand, modern humans developed a neomorphic hypertrophy of the parietal volumes, leading to a dorsal growth and ventral flexion (convolution) characteristic of the cranial globularity of our species (Bruner 2004).

Roseman, Weaver, and Stringer (2011) analysed cranial morphological integration—covariance of traits—in a sample of 2524 modern humans and 20 Neanderthals and concluded that the overall patterns are significantly different between the species. Nevertheless, Neanderthals' morphology is consistent with a modern human pattern of integration for more than three-quarters of the traits. To what extent these similarities and differences could affect neurological configuration cannot be assessed yet.

21.6 **Comparative studies**

Human-derived features may also be distinguished by means of comparative approaches with non-human primates.

Rilling and collaborators (2007) offered a comparison between the brain activation in humans and chimpanzees during the resting state. These authors analysed, by means of positron emission tomography (PET), anesthetized chimpanzees that, during the awake resting state, would have completely fixed radioactive glucose in the brain. According to their results, both humans and chimpanzees seem to coincide in the activation of medial and dorsolateral OFC, as well as medial parietal cortex, with the highest level of activity placed more dorsally in humans (BA 9, BA 32) and more ventrally (BA 10) in chimpanzees. As Rilling and colleagues (2007) found, during the resting state "the strongly left-lateralized activity related to language and conceptual processing in humans was absent in chimpanzees."

Watanabe (2011) obtained, also by means of PET, the default network curtailed in awake rhesus monkeys when attention-demanding cognitive tasks were performed. Watanabe (2011) shows that "[s]imilar to the human default system, all monkeys showed higher rest-related activity in the medial prefrontal and medial parietal areas. . . ." Moreover, considering that the human default mode network is related to internal thought processes, Watanabe stated that default activity in the medial brain areas suggests that "there might be internal thought processes in the monkey."

Mentioning the chimpanzees case, Northoff and Panksepp (2008) assumed that high degrees of self-relatedness correspond to high resting-state neuronal activity. If Watanabe and Northoff and Panksepp are right, the similarities between monkeys, apes, and humans might indicate a self-relatedness capacity, shared to some degree. However, aesthetic capacity does not equal self-relatedness. The more interesting aspect of the comparative approaches is probably the strong left-laterality identified by Rilling and colleagues (2007) as a human trait not present in apes during the resting state.

Let us return to the identification between the DMN and the aesthetic networks.

21.7 **Aesthetic appreciation as an exaptation**

It has been posited that aesthetic perception does not need to imply any adaptive advantage, since it could profit from other previously evolved cognitive characteristics with their own adaptive benefits. In other words, aesthetics might be just an exaptation. For instance, Kaplan (1987) states that "[i]t would be adaptive for animals to like the sort of settings in which they thrive." Thus, preference for landscapes would have led to preference for ornaments like gardens. Focusing on positive-valence aesthetic appraisal, Brown and collaborators (2011) hold that "such a system evolved first for the appraisal of objects of survival advantage, such as food sources, and was later co-opted in humans for the experience of artworks for the satisfaction of social needs."

Obviously, any hypothesis about the subject would be difficult to test. However, a complementary justification for the evolution of capacities for appreciating beauty may be

DMN - default mode network - resting brain state

given by pointing out the coincidence between the delayed aesthetic network and the default mode network.

A fundamental function of the DMN is facilitating responses to stimuli. As Raichle and Snyder (2007) held, "intrinsic activity instantiates the maintenance of information for interpreting, responding to and even predicting environmental demands." This functional capacity seems adaptive enough to justify by itself its metabolic costs, and it is performed by means of "mind-wandering" processes, in which DMN seems to play a fundamental role.

A DMN phylogenetically fixed and linked to aesthetic perception is sufficient to justify such human capacity for appreciating beauty in objects. A different question is that of explaining how this link between DMN and aesthetic perception appeared or, in other words, what characteristic of the default mode network might lead to the sudden experiences of the beauty of a picture or a landscape.

An added function of the DMN relates to "mind-wandering" processes. Mind wandering refers to images, thoughts, voices, and feelings that the brain spontaneously produces in the absence of external stimuli (stimulus-independent thoughts, SIT) (Mason et al. 2007). The SIT are what we might call "the mind talking to itself."

Three possible explanations of the functional significance of mind wandering were offered by Mason and colleagues (2007). SIT enables subjects to maintain an optimal level of arousal. Alternatively, SIT adds coherence to one's experience—past and present. A non-adaptive meaning was also posited. SIT could be a by-product of a general ability to manage concurrent mental tasks obtained during human evolution. Although mind wandering can be useful, "the mind may wander simply because it can" (Mason et al. 2007).

Aesthetic perception is not stimulus-independent. Except in the case of recalling past experiences, detecting beauty depends on external stimuli. However, the aesthetic perception might be a by-product of that general capacity for mind wandering. Mind wandering is a general process of perception neither guided by any goal, nor directed to any particular aspect. It obviously applies to the aesthetic appreciation of the environment. According to Kaplan (1987), the move from appreciating landscapes to recreating them as artworks is supported by the coincidence between DMN and the aesthetic delayed network.

A close-to-mind-wandering capacity for assigning beauty or ugliness to visual stimuli, from landscapes to artworks, could thus lead to continuous and very quick processes of aesthetic perception. Sudden comprehension that solves a problem or a perceptual ambiguity has been detected, combining EEG and fMRI (Kounios and Beeman 2009; Kounios et al. 2008), as the culmination of a series of neural processes at different time scales leading to "Aha!" moments. Regarding aesthetic perception, our current study suggests that the appreciation of beauty might be an "Aha!" moment too, which appears at early temporal stages of the perceptive process, and is not guided by goal-directed tasks but works instead in an almost-holistic way. In turn, the obvious advantages of the capacity for the appreciation of beauty, moving from the inner hedonic complacence to the satisfaction of social needs (Brown et al. 2011), would add adaptive advantages to further uses of such a trait. Not surprisingly, aesthetic perception also activates regions, like the dorsolateral prefrontal cortex (DLPFC), linked to executive functions (Jacobsen 2006; Vessel et al. 2012).

21.8 The question of quale

The internal, but stimulus-dependent visual appreciation of beauty is an individual instance of subjective and conscious experience—what philosophers call "quale" (Lewis 1929). The intriguing problem of qualia (i.e. the mind–brain relationship) can be stated as follows: how does the brain produce qualitative subjectivity? (Searle 2011).

In their study of consciousness, Crick and Koch put aside the "hard problem" of quale, the subjective content of the mental states: "no one has produced any plausible explanation on how the experience of redness or red could arise from the actions of the brain" (Crick and Koch 2003). Instead, they focused on the "soft problem," the neural correlates of consciousness. Regarding aesthetic appreciation, this "soft problem," consisting of the localization of brain areas active when subjects gauge the beauty of a visual object, has been already solved, in part at least, by means of neuroaesthetics.

Moreover, some aspects of the dynamics of aesthetic appreciation could help scratch the surface of the "hard problem" too. By means of a combination of fMRI, MEG, and behavioral studies of impaired subjects, it seems that the way in which the experience of beauty could stem from the actions of the brain may begin to move within our reach. However, current approaches to the hard problem yield only partial solutions. On the one hand, it seems that the *structure* of the qualia, consisting of a description of mental processes leading to the appreciation of the beauty of an object, can be accessed by means of scientific procedures. On the other hand, many personal circumstances, from previous experiences to character traits, health, age, perhaps gender, as well as the cultural and historical particularities of each subject and epoch, surely contribute to the experience of appreciating beauty. These aspects would modify, in a still not detailed fashion, subjective feelings. For the time being, the *content* of the qualia—the eventual result of beauty, or its absence, as an inner sensation—exceeds our grasp.

References

Aiello, L.C. and Wheeler, P. (1995). The expensive tissue hypothesis: The brain and the digestive system in human and primate evolution. *Current Anthropology* **36**, 199–221.

Appenzeller, T. (2013). Old masters. *Nature* **497**, 302–4.

Bassett, D.S., Wymbs, N.F., Porter, M.A., et al. (2011). Dynamic reconfiguration of human brain networks during learning. *Proceedings of the National Academy of Sciences of the USA* **108**(18), 7641–6.

Brown, S., Gao, X., Tisdelle, L., et al. (2011). Naturalizing aesthetics: Brain areas for aesthetic appraisal across sensory modalities. *Neuroimage* **58**, 250–8.

Bruner, E. (2004). Geometric morphometrics and paleoneurology: brain shape evolution in the genus Homo. *Journal of Human Evolution* **47**, 279–303.

Bruner, E. and Holloway, R.L. (2010). A bivariate approach to the widening of the frontal lobes in the genus Homo. *Journal of Human Evolution* **58**(2), 138–46.

Bruner, E., Manzi, G., and Arsuaga, J.L. (2003). Encephalization and allometric trajectories in the genus *Homo*: Evidence from the Neandertal and modern lineages. *Proceedings of the National Academy of Sciences of the USA* **100**, 15335–40.

Carroll, N. (2004). Art and human nature. *The Journal of Aesthetics and Art Criticism* **62**(2), 95–107.

Cela-Conde, C. and Ayala, F.J. (2007). *Human Evolution. Trails from the Past*. New York, NY: Oxford University Press.

Cela-Conde, C.J., García-Prieto, J., Ramasco, J.J., et al. (2013). Dynamics of brain networks in the aesthetic appreciation. *Proceedings of the National Academy of Sciences of the USA* 10(Supplement 2), 10454–61. doi: www.pnas.org/cgi/doi/10.1073/pnas.1302855110.

Cela-Conde, C.J., Marty, G., Maestú, F., et al. (2004). Activation of the prefrontal cortex in the human visual aesthetic perception. *Proceedings of the National Academy of Sciences of the USA* **101**, 6321–5.

Chatterjee, A. (2003). Prospects for a cognitive neuroscience of visual aesthetics. *Bulletin of Psychology and the Arts* **4**, 55–60.

Crick, F. and Koch, C. (2003). A framework for consciousness. *Nature Neuroscience* **6**, 119–26.

d'Errico, F., Zilhão, J., Julien, M., et al. (1998). Neanderthal acculturation in western Europe? A critical review of the evidence and its interpretation. *Current Anthropology* **39**, supplement, 1–44.

Dissanayake, E. (1992). *Homo Aestheticus: Where art comes from and why*. New York, NY: Free Press.

Dissanayake, E. (2007). What art is and what art does: An overview of contemporary evolutionary hypotheses. In C. Martindale, P. Locher, and V.M. Petrov (eds), *Evolutionary and neurocognitive approaches to aesthetics, creativity and the arts*. Amityville, NY: Baywood, pp. 1–14.

Dissanayake, E. (2009). The Artification Hypothesis and its relevance to cognitive science, evolutionary aesthetics, and neuroaesthetics. *Cognitive Semiotics* Fall 2009(5), 148–73.

Dutton, D. (2003). Aesthetics and evolutionary psychology. In J. Levinson (ed.), *The Oxford Handbook for Aesthetics*. New York, NY: Oxford University Press, pp. 693–705.

Fox, M.D., Snyder, A.Z., Vincent, J.L., et al. (2005). The human brain is intrinsically organized into dynamic, anticorrelated functional networks. *Proceedings of the National Academy of Sciences of the USA* **102**(27), 9673–8.

Friston, K., Frith, C.D., Liddle, P.F., et al. (1993). Functional connectivity: the principal-component analysis of large (PET) data sets. *Journal of Cerebral Blood Flow Metabolism* **13**, 5–14.

Friston, K.J., Holmes, A.P., Worsley, K.J., et al. (1995). Statistical parametric maps in functional imaging: A general linear approach. *Human Brain Mapping* **2**, 189–210.

Gusnard, D.A. and Raichle, M.E. (2001). Searching for a baseline: Functional imaging and the resting human brain. *Nature Reviews Neuroscience* **2**(10), 685–94. doi: 10.1038/35094500.

He, Y., Wang, J., Wang, L., et al. (2009). Uncovering intrinsic modular organization of spontaneous brain activity in humans. *PLoS One* **4**(4), e5226. doi: 10.1371/journal.pone.0005226.

Hublin, J.J., Spoor, F., Braun, M., et al.(1996). A late Neanderthal associated with Upper Palaeolithic artefacts. *Nature* **381**, 224–6.

Jacobsen, T. (2006). Bridging the arts and sciences: A framework for the psychology of aesthetics. *Leonardo* **39**, 155–62.

Jacobsen, T., Schubotz, R.I., Höfel, L., et al. (2006). Brain correlates of aesthetic judgment of beauty. *Neuroimage* **29**, 276–85.

Kaplan, S. (1987). Aesthetics, affect, and cognition. *Environment and Behavior* **19**, 3–32.

Karavanic, I. and Smith, F.H. (1998). The Middle/Upper Paleolithic interface and the relationship of Neanderthals and early modern humans in Hrvatsko Zagorje, Croatia. *Journal of Human Evolution* **34**, 223–48.

Kawabata, H. and Zeki, S. (2004). Neural correlates of beauty. *Journal of Neurophysiology* **91**, 1699–705.

Kounios, J. and Beeman, M. (2009). The *Aha!* moment. The Cognitive Neuroscience of insight. *Current Directions in Psychological Science* **18**, 210–6.

Kounios, J., Fleck, J.I., Green, D.L., et al. (2008). The origins of insight in resting-state brain activity. *Neuropsychologia* **46**, 281–91.

Leder, H., Belke, B., Oeberst, A., et al. (2004). A model of aesthetic appreciation and aesthetic judgments. *British Journal of Psychology* **95**, 489–508.

Lee, M.H., Smyser, C.D., and Shimony, J.S. (2012). Resting-state fMRI: A review of methods and clinical applications. *American Journal of Neuroradiology* 34, 1866–72.

Lewis, C.I. (1929). *Mind and the World Order*. New York, NY: Charles Scribner's Sons.

Mason, M.F., Norton, M.I., Van Horn, J.D., et al. (2007). Wandering minds: The default network and stimulus-independent thought. *Science* **315**(5810), 393–5.

Mormann, F. (2000). Mean phase coherence as a measure for phase synchronization and its application to the EEG of epilepsy patients. *Physica D* **144**, 358–69.

Northoff, G. and Panksepp, J. (2008). The trans-species concept of self and the subcortical–cortical midline system. *Trends in Cognitive Sciences* **12**(7), 259–64.

Ogawa, S., Lee, T.M., Kay, A.R., et al. (1990). Brain magnetic resonance imaging with contrast dependent on blood oxygenation. *Proceedings of the National Academy of Sciences of the USA* **87**(24), 9868–72.

Pereda, E., Quiroga, R.Q., and Bhattacharya, J. (2006). Nonlinear multivariate analysis of neurophysiological signals. *Progress in Neurobiology* **77**, 1–37.

Pike, A.W.G., Hoffmann, D.L., García-Diez, M., et al. (2012). U-series dating of Paleolithic art in 11 caves in Spain. *Science* **336**(6087), 1409–13.

Raichle, M.E. (2011). The restless brain. *Brain Connect* **1**(1), 3–12. doi: 10.1089/brain.2011.0019.

Raichle, M.E., MacLeod, A.M., Snyder, A.Z., et al. (2001). A default mode of brain function. *Proceedings of the National Academy of Sciences of the USA* **98**(2), 676–82.

Raichle, M.E. and Snyder, A.Z. (2007). A default mode of brain function: A brief history of an evolving idea. *NeuroImage* **37**, 1083–90.

Rilling, J.K., Barks, S.K., Parr, L.A., et al. (2007). A comparison of resting-state brain activity in humans and chimpanzees. *Proceedings of the National Academy of Sciences of the USA* **104**(43), 17146–51.

Roseman, C.C., Weaver, T.D., and Stringer, C.B. (2011). Do modern humans and Neandertals have different patterns of cranial integration? *Journal of Human Evolution* **60**(6), 684–93. doi: 10.1016/j.jhevol.2010.04.010

Searle, J.R. (2011). The Mystery of Consciousness Continues. *The New York Review of Books* 9 June.

van den Heuvel, M.P. and Hulshoff Pol, H.E. (2010). Exploring the brain network: A review on resting-state fMRI functional connectivity. *European Neuropsychopharmacology* **20**(8), 519–34. doi: http://dx.doi.org/10.1016/j.euroneuro.2010.03.008.

Varela, F., Lachaux, J.-P., Rodriguez, E., and Martinerie, J. (2001). The brainweb: Phase synchronization and large-scale integration. [10.1038/35067550]. *Nature Reviews Neuroscience* **2**(4), 229–39.

Vartanian, O. and Goel, V. (2004). Neuroanatomical correlates of aesthetic preference for paintings. *Neuroreport* **15**, 893–7.

Vessel, E.A., Starr, G.G., and Rubin, N. (2012). The brain on art: Intense aesthetic experience activates the default mode network. *Frontiers in Human Neuroscience* 6.

von der Malsburg, C. and Schneider, W. (1986). A neural cocktail-party processor. *Biological Cybernetics* **54**, 29–40.

Watanabe, M. (2011). Are there internal thought processes in the monkey?—Default brain activity in humans and nonhuman primates. *Behavioural Brain Research* **221**(1), 295–303. doi: http://dx.doi.org/10.1016/j.bbr.2011.02.032.

White, R. (2001). Personal ornaments from the Grotte du Renne at Arcy-sur-Cure. *Athena Review* **2**(4), 41–6.

Zilhão, J., Angelucci, D.E., Badal-García, E., et al. (2010). Symbolic use of marine shells and mineral pigments by Iberian Neandertals. *Proceedings of the National Academy of Sciences of the USA* **107**, 1023–8. doi: 10.1073/pnas.0914088107.

Sousa

Chapter 22

Art as a human "instinct-like" behavior emerging from the exaptation of the communication processes

Luigi F. Agnati, Diego Guidolin, and Kjell Fuxe

22.1 Introduction

One extraordinary capability of the human brain is the production and appreciation of artworks (Young 1974). This capacity has usually been associated with the concept of "beauty," a term notoriously difficult to define[1] and possibly related, in a broad sense, with the selection of palatable food and attractive mates that is also present in animals.

This may be one of the reasons why appreciation of artworks and beauty itself are not coincident, why neuroaesthetics is not yet a well understood field, and why important contributions from artists themselves are of a fundamental importance. Thus, we agree with Zeki's (2009) suggestion that "the artist is in a sense, a neuroscientist, exploring the potentials and capacities of the brain but with different tools. How such creations can produce aesthetic experiences can only be fully understood in neural terms. Such an understanding is now well within our reach" (see also Maffei and Fiorentini 2008). Painters, sculptors, architects, and also novelists may be regarded as a special type of neuroscientist (see also Agnati et al. 2012a; Lehrer 2007). In this chapter, Fernando Pessoa's masterpiece, *The Book of Disquiet* (*Livro de desassossego por Bernardo Soares*), will be cited as it gives an extraordinary insight into the thinking of one of the greatest poets of the past century.

Consistent with Zeki's quotation our chapter will analyse if there are specific brain areas that are used in art appreciation, as indicated by neuroimaging investigations. A well-known and specific type of neuron that lies in these areas offers a special contribution in this regard. In view of the fact that the human brain is the result of a long evolutionary process, some general principles deduced from evolutionary studies may shed some light on the peculiarity of the human brain, in spite of similarities with other mammalian brains, especially in the primate order of mammals.

The two concepts of exaptation (Gould and Vrba 1982) and of tinkering (Jacob 1970) will be introduced together with a brief discussion on theoretical models of brain organization as well as on intercellular communication in the brain, something our group proposed in the last few decades (Agnati and Fuxe 1984; Agnati et al. 2005, 2012b).

Against this background, we propose a new interpretation of the morphological and neurochemical investigations on the human brain areas that according to neuroimaging studies are involved in the appreciation of artworks.

22.2 General premise

22.2.1 Two basic concepts deduced from studies on evolution

The exaptation concept and the Russian doll model of living beings' organization have both been applied to the modeling of the central nervous system (CNS) and offer a heuristic hypothesis on human brain-specific capabilities such as self-awareness (Agnati et al. 2012b). In this chapter these concepts will be used to offer insights into unique human products, namely artworks.

22.2.1.1 Exaptation

A subtle but important distinction should be made between "adaptation" and "exaptation" with regard to their evolutionary roles. While adaptation refers to a feature produced by natural selection for its current function (such as echolocation in bats), exaptation has been defined as a feature that performs a function that was not produced by natural selection for its current use (e.g. feathers that might have originally arisen in the context of selection for insulation and not for flying; see e.g. Xu et al. 2012). However, natural selection may subsequently operate on such a new function to improve its adaptation to possible new environmental demands. Related to this concept, and focused on brain function, is Anderson's (2007, 2010) concept of redeployment (or reuse) of a neural structure for a new function. There are obvious evolutionary advantages to redeployment of brain areas and Anderson analyses this phenomenon in relation to cognition.

We propose that the concept of exaptation can be used to shed some light on the neuropsychological aspects of artwork production by human beings from the Palaeolithic to the present day. In addition, the redeployment of some brain areas and special types of neurons (namely, von Economo's neurons, VENs) may be exaptated to contribute to this peculiar human capability.

22.2.1.2 The Russian doll model of the CNS and the concept of functional module

The concept of the functional module (FM) to some extent derives from the proposal of the modular organization of the cortex cerebri proposed already in the pioneering work of Lorente de No (1938) and Mountcastle (Mountcastle et al. 1957), who introduced the concept of the "cortical column." The existence of cortical integrative structures is not completely proven (Jones 2000; Tanaka 2003) as anatomical columns are not solid structures and modular structures occur also in non-cortical structures like the striatum (Olson et al. 1972; Rockland and Ichinohe 2004; Rockland 2010).

The relevance of the concept of a modular organization of brain circuits can be deduced by the theoretical consideration that such an organization can perform basic operations providing both compartmentalization and redundancy, which reduces the interdependence of components, enhances robustness, and facilitates behavioral adaptation. Furthermore, a modular organization also confers evolvability to a system by reducing constraints on change.

Our proposal has combined the modular concept with Jacob's proposal that from a structural standpoint the living organism is organized as a nest of "Russian Matryoshka Dolls" (Jacob 1970), implying that smaller structures are buried within larger ones. Thus, our group has proposed that FMs are processing units organized according to a "nested" hierarchic criterion from cellular down to molecular level (Agnati and Fuxe 1984; Agnati et al. 2012b). It should be noted that the hierarchic organization as well as the elements at each level are dynamically arranged and interconnected in order to carry out a task (i.e. the "building of the Russian doll") and thus change with the functional state. Hence FMs are structural and functionally plastic modules without well delimited anatomical borders, as seems to be the case for the cortical columns. Each FM is formed mainly, but not solely, by neurons and astrocytes, and at each miniature level the involved elements (e.g. cells, dendritic spines, or proteins) form a "mosaic."

The term "mosaic" has been introduced in order to convey the concept of topology. This may have a potentially large impact on the integrative functions of the assembly since it highlights the possibility that with the same set of "tesserae" (i.e. elements) markedly different mosaics can be assembled at a given level of miniaturization, which then become capable of a different way of handling information. The elaboration of the information can occur simultaneously at different miniaturization levels (Guidolin et al. 2011) but also sometimes in different temporal domains. The highest level of a "horizontal elaboration" is formed by mosaics of FMs, while the lowest level occurs at the molecular level (e.g. within multimeric protein complexes).

In line with Hebb's classical hypothesis about the possible existence of cell assemblies interconnected via reverberating circuits (Hebb 1949; Wang 2001), it has been suggested that different FMs can be transiently interconnected to form a higher-order mosaic and, as indicated above, a "3-dimensional elaboration of the information" is continuously in operation in the CNS (Agnati et al. 2012b).

22.2.2 Volume and wiring transmission

In 1986 we published our original proposal on the existence of two main modes of intercellular communication potentially connecting all types of cells in the CNS: the wiring transmission (WT) and the volume transmission (VT) (Agnati et al. 1986). Our proposal was influenced by important contributions on communication in the CNS (Golgi 1914; Guillemin 1978; Nicholson 1979; Schmitt 1984), and based on a number of observations, especially on the central monoamine neurons (for reviews see Fuxe and Agnati 1991; Agnati and Fuxe 2000; Descarries et al. 2008; Nieuwenhuys 2000; Bach-Y-Rita 2005; Agnati et al. 2010). The concept of VT introduced the extracellular space and the ventricular system as important channels for chemical transmission in the CNS complementary to synaptic transmission and VT is basically due to diffusion and flow of transmitters, ions, and trophic factors in the extracellular fluid (ECF) and cerebrospinal fluid (CSF). The basic dichotomous classification of intercellular communication in the brain proposed more than two decades ago is still valid. However, evidence of the existence of new specialized structures for intercellular communication, such as microvesicles (see Cocucci et al. 2009) and tunneling nanotubes (Rustom et al. 2004; Baluška et al. 2006; Goncharova and Tarakanov 2008), called for an updating of our original conceptual model (Agnati et al. 2010).

The criteria which can be applied to characterize WT and VT and their sub-classes have been deduced not only from structural and neurochemical findings but also from concepts and from the lexicon offered by informatics (Hopcroft and Ullman 1979). The sole and main criterion which allows differentiating WT from VT is the characteristics of the communication channel and more precisely the physical boundaries of the channel, which are well delimited for WT but not for VT.

The classification can be further detailed by taking into account the other signal features. Some of these can be summarized as follows:

> Signal privacy—a signal is characterized by high privacy (i.e. a "reserved signal") if only cells endowed with a specific recognition/decoding apparatus (as, for instance, specific receptors) respond to the signal. On the contrary, we deal with a low privacy signal (i.e. an "open signal") when any cell reached by the signal can have access to it (e.g. CO_2, H^+).
>
> Signal safety—as far as safety is concerned, we deal with a "safe signal" if it is not altered during its diffusion from the source to the target cell and with an "unsafe signal" if it can be altered during its pathway. This occurs, for instance, to some VT signals that can be broken down or modified (e.g. by enzymes) or trapped in a cul-de-sac in the ECS pathways (Nicholson 1979).
>
> Connectivity—if the connections between cells can be rapidly formed or removed they provide a "dynamic network." On the contrary, the structure of a communication network is "static" when the pattern of connections is almost stable in time.

Thus, WT may be better suited for a phasic, promptly acting, safe signal transmission to few precisely spatially defined targets. VT may instead be better suited for a tonic, slow acting, potentially unsafe signal transmission to many spatially disseminated targets. Accordingly, the synaptic type of WT in particular may mainly be present in the sensory input and motor output pathways where a prompt acquisition of environmental information and a prompt motor response is needed.

On the contrary, VT is mainly active in the resetting of the mood tone and, in general, in delayed and long-lasting emotional responses. However, it is most likely that all integrative actions in the brain involve strict cooperation between the WT and VT modes for intercellular communication.

22.3 **Why human beings are artwork makers?**

This is central to the current chapter and will be afforded attention on the basis of the conceptually far-reaching words of a great British biologist, J.Z. Young. In an extraordinary book, Young (1974) pointed out the fundamental relevance for the human beings of communication and especially of communication by symbols:

> Whatever status we may decide to ascribe to the mind there is no question that the individual human being with his capacity for collection, storage, and transmission of information is a unit of

> exceptional biological importance. After all it is by individuals that the decisions are made that regulate human life. It is true that each can (in practice) survive only as a member of a society, but this is the corollary of the fact that he lives by acquiring and transmitting information. (Young 1974, p. 10)

Furthermore, he points out the special relevance of human artworks:

> But man's particular genius is for the transfer of information between individuals and many subtle features are involved in his homeostasis. Communication by symbols is improved by all the agencies that we include as art, literature, and aesthetics, and these, far from being 'impractical,' are major contributors to human homeostasis. (Young 1974, p. 12)

> It is clear that the creation of beautiful and symbolic objects is a characteristic feature of the human way of life. They are as necessary to us as food or sex. (Young 1974, p. 360)

These inspiring sentences may help in introducing the concepts of "instinct-like behavior" and of "psychic homeostasis."

22.3.1 The special human "instinct-like" for intra-species communication by symbols

The essential drive for intra-species communication by symbols that are vehicles of information not immediately related to basic needs seems to be a unique feature of the human evolutionary line, stretching back to ancestors of *Homo sapiens* living 164 000 years ago in South Africa (Marean et al. 2007) and forward until the present times. This can be thought of as an evolutionary exaptation of the communication processes that are also present in other species for more basic aims such as survival of the single living being and reproduction of that species.

In general terms, instinct or innate behavior is the inherent inclination of a living organism toward a particular behavior. The simplest example of an instinctive behavior is a fixed action pattern, in which a very short- to medium-length sequence of actions, without variation, is carried out in response to clearly defined stimuli.

It is proposed that the exaptation of the intra-species communication processes in humans has led to the emergence of "all the agencies that we include as art, literature, and aesthetic ... as necessary to us as food or sex." These agencies are an essential human drive that can be thought of, in a broad sense, as a "new instinct." As a matter of fact, the role of instincts in determining the behavior of animals varies from species to species. It is true that the higher up the evolutionary ladder the greater is the role of the cerebral cortex. Social learning and basic instincts play a lesser role but the highly evolved human brain could also allow, according to our hypothesis, the emergence of new uniquely human instinct-like behaviors.

The fact that humans are bipedal, have hands to make tools, and possess complex brains are all biological features that have favored the emergence of the new instinct-like behavior of communication by symbols. However, these are necessary but not sufficient conditions to completely account for this emergence. A more probable and important aspect to consider concerns the process used by the brain to acquire knowledge about the world. As pointed out by Zeki (2004), such a process is basically a diffuse process of abstraction in which each FM possesses an abstractive machinery which transforms an incoming fragment of sensation into a particular fragment of perception (Agnati et al. 2012b); that is, it extracts a specific attribute of a sensory modality. The primary visual cortex represents a well-studied example of this strategy (see Carreira-Perpiñan and Goodhill 2002). Mechanisms of binding (see Agnati et al. 2012b) then allow these different components to be gathered into one picture, generating general models of the external world (Blakemore 1976). This process of generalization is a very efficient way in which a finite mind grasps the infinity of particulars and also frees the brain from total dependence upon a memory system (Zeki 2004). However, there is also a price to be paid for this, which most likely became significant when the complexity of the brain permitted the realization of sophisticated abstraction processes. In fact, abstraction leads to models, ideas, and concepts, but our experience remains in the particular which may not always satisfy the ideal model formed in and by our brain. One way to obtain that satisfaction is to "download" (Zeki 2004) the ideas formed in the brain into a work of art. Paul Klee, a great artist as well as a commentator on art, wrote that art does not reproduce the visible; rather, it makes the invisible visible (Klee 1920; see Figure 22.1).

The peculiar importance of this communication of the human internal theater can also be understood with reference to the concept of "psychic homeostasis." As pointed out in previous papers (see below, and also Agnati et al. 2011, 2012a), humans' special status is the result of a continuous interplay between two types of homeostasis, the somatic and the psychic, which are strictly interconnected.

22.3.2 Psychic homeostasis versus somatic homeostasis

The classical term "homeostasis" originates from Claude Bernard (1813–1878) and Walter Cannon (1871–1945). Both proposed that the functions of apparatuses of a multicellular organism have as their final goal the maintenance of the constancy of composition of the internal environment. This is mainly composed of the liquid which surrounds the cells of the organism and by means of which cells exchange solutes and metabolites. This condition of dynamic constancy is called homeostasis.

The maintenance of homeostasis implies an energy cost. Thus, McEwen (2003, Ganzel et al. 2010), in his consideration of homeostasis, introduced three important additional concepts:

- allostasis: the process by which the organism actively adjusts the activity of its apparatuses in order to maintain the stability of the internal environment, notwithstanding the challenges due to changes of the external environment;
- allostatic load: which represents the cumulative energy cost to the body for the allostatic process;
- allostatic overload: a state in which the compensatory actions of the organism are overwhelmed and as a consequence serious pathologies can develop.

The anthropocentric belief in the uniqueness of the human being is philosophically stated by the contraposition between *res extensa* and *res cogitans* (Descartes 1960) that has led to a distinction between two elements of the human holistic complexity, soma and psyche. Accordingly, it may be proposed that not only is there a somatic homeostasis (which

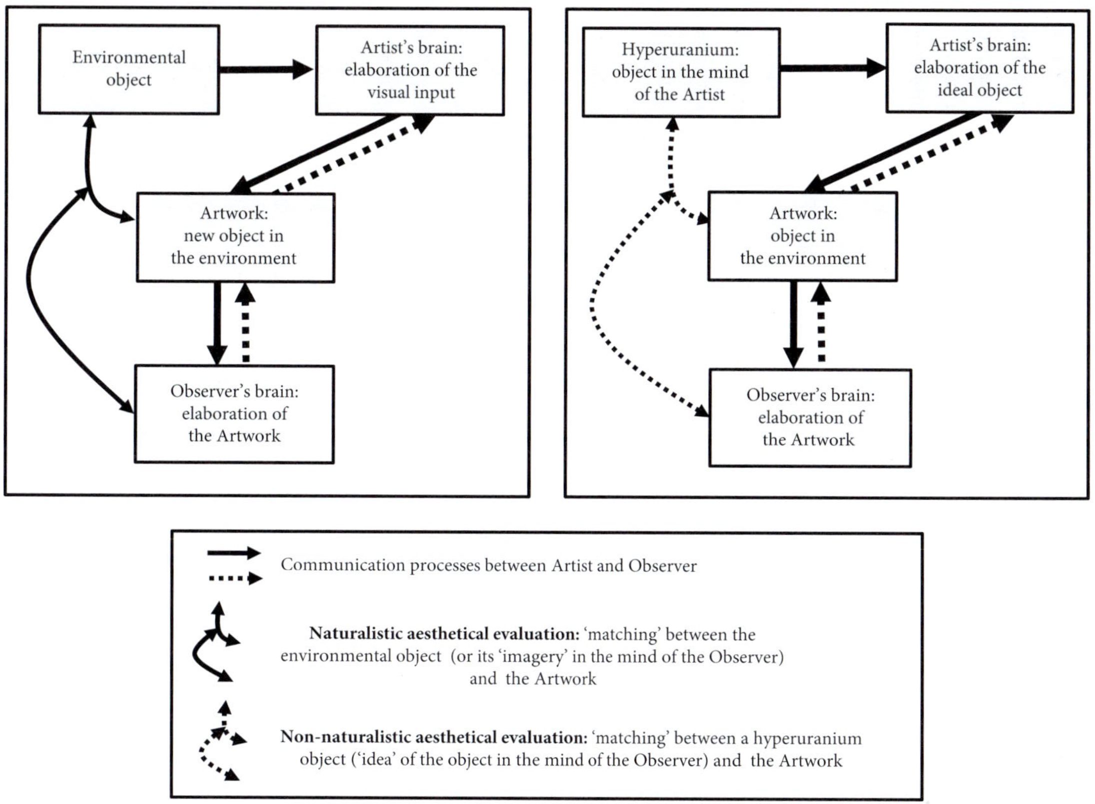

Figure 22.1 Schematic representation of the communication channels between artist, artwork, and observer. The two panels give the possible basic steps involved in the "imagery style" and in the "sensitive style" (see Herbert Kuhn in Jaffé 1983). The former communicates the author's fantasy or her personal experience expressed according to modules not immediately derived from reality and sometimes derived from her ideal world or even from her dreams. The latter one usually is more an attempt to give an faithful interpretation of the nature. In other words, the left panel represents the possible basic steps involved in an aesthetical evaluation of naturalistic artworks (sensitive style). The right panel represents the possible basic steps in an aesthetical evaluation of non-naturalistic artworks (imagery style).

is directed towards optimal somatic function), but there is also a psychic homeostasis (directed towards an equilibrated mental function; Agnati et al. 2011, 2012a).[2] Thus, the human organism in toto has as its main goal not simply the homeostasis of the internal environment but also a "psychic homeostasis" which represents a state of emotional balance (eudemonia) that can be markedly altered during psychic disturbances, such as depression, paranoia, and post-traumatic anxiety.

Recently, the concept of psychic homeostasis has been used to explain a possible functional role of dreams (Agnati et al. 2011). In their relationship with sleep and dreaming, the homeostatic components operate via adjustments, using multiple dynamic feedbacks, of the various neuro-endocrine systems. Such a resetting of the control and effector neural (vegetative) and endocrine systems also prepares the subject for future challenges (Horowski et al. 2004). This aspect was expanded (Agnati et al. 2011) with the concept of "predictive psychic homeostasis" whereby the various feedback mechanisms bring the subject to a state of preparedness in anticipation for potential threats that may precipitate an allostatic psychic overload. Predictive psychic homeostasis is based on communicative social contacts, and it plays a fundamental role not only in the regulation of the internal somatic and psychic milieu, but also in relation to behavioral adaptations and social interactions (Agnati et al. 2011; Schulkin 2011). As a matter of fact, psychic homeostasis depends greatly on social contacts where communication processes by means of symbols play a fundamental role. In this context, the relevance of production and appreciation of artworks that are instrumental for harmonious social contacts, and hence for the psychic homeostasis of the individual, becomes immediately clear.[3]

22.3.3 Brain–body holistic system

The special two-faced "homeostasis" of the individual that can be deduced from the previous discussion leads to the view of the human being as a unique body–brain holistic system which operates owing to a specialized interface based on basal hypothalamus (BH, especially paraventricular and supraoptic nucleus; PVN, SON) and solitary tract nucleus (STN). As a matter of fact, STN receives direct neural visceral afferences from most of the peripheral organs while many blood-borne signals reach BH directly and STN indirectly. Hence, these brain regions are continuously informed about the body state as far as chemical parameters (e.g. glycaemia, osmolarity, pH) and hormones (e.g. glucagon, insulin, adiponectin, cholecystokinin) are concerned. Thus, by affecting these interface regions, peripheral signals (neural, chemical, and hormonal) can orchestrate inter alia the central monoamine neuron systems and the release of oxytocin and vasopressin from neuroendocrine neurons, with oxytocin and vasopressin acting both as neurotransmitters and endocrine signals. These are major connections between the CNS and the body through which many psychosomatic processes occur. The importance of bi-directional (afferent, efferent) communication should be emphasized: both directions are important. Thus, the hypothesis has been introduced of a fundamental role of BH and STN and their reciprocal interconnections to maintain homeostasis via their combined actions on limbic structures (e.g. amygdala) and cortical structures (e.g. prefrontal cortex) (Agnati 2012c).

These basic mechanisms involved in the somatic homeostasis in mammals have been exaptated in humans to play a role in maintaining not only somatic homeostasis but also psychic homeostasis. Recently, a very important aspect of this complex control system has been discovered by Insel and colleagues (see Insel 2010) demonstrating that vasopressin and oxytocin released from central vasopressin and oxytocin pathways (see Romero-Fernandez et al. 2012) play a major role in social interactions especially pair bonding.

It has also been demonstrated by the Insel group (Insel 2010) that pair bonding may be brought about by increases in vasopressin (male) and oxytocin (female) transmission. Fuxe and co-authors (Fuxe at al. 2012) have pointed out the role of vasopressin and oxytocin VT in critical limbic regions of the mammalian brain: ventral pallidum (arginin vasopressin receptor binding; male) and nucleus accumbens shell (oxytocin receptor binding; female). Species differences in oxytocin- and vasopressin-receptor binding densities have been described (Gimpl and Fahrenholz 2001). These results are consistent with the hypothesis that changes in neuroendocrine systems may evolve via changes in receptor distribution rather than by restructuring the presynaptic pathway (Wang et al. 1996). Thus, the increase of oxytocin and vasopressin VT could be brought about by a rise of oxytocin and vasopressin receptor densities without a change in the density of oxytocin and vasopressin terminal networks (Fuxe et al. 2012). This means that behavioral differences such as pair bonding found in different species of the same genus may be caused by increasing the extra-synaptic receptor densities in critical limbic circuits for the diffusing VT signals, in this case the diffusing oxytocin and vasopressin signals from their nerve terminals in these limbic networks. As pointed out earlier, by 1986 Agnati and Fuxe proposed that monoamine and peptides (such as oxytocin and vasopressin) can also use the CSF as a pervasive diffusion pathway to reach target areas.

It can therefore be stated that the reciprocal interaction between body and brain (mind) is a key concept in the study of integrative mechanisms of the human CNS not only for social interactions but also for the capacity to communicate by symbols. Thus, it is suggested that the holistic brain–body system is likely involved in the production/appreciation of artworks.

22.4 Artwork production/appreciation: a hub role for insula and anterior cingulate cortex?

Neurology investigations in art (i.e. "experimental neuroaesthetics") have focused on artwork appreciation despite the fact that the field should investigate both production/appreciation. Neuroimagining techniques have provided important findings (Nadal et al. 2008). However, further investigations on the neurobiological substrate should be carried out to clarify the following aspects at the very least:

- brain areas activated or inhibited in the appreciation of artworks;
- extension and boundaries of these brain areas;
- sequence according to which the brain areas are involved;
- time course of the electrical phenomenon within each brain area involved.

As pointed out by Cela-Conde and colleagues (2011), until now investigations have been focused only on some selected aspects of aesthetic appreciation and have analysed only some of the main neuroimaging features of the brain areas involved. Thus, current findings cannot do justice to the evaluation of the whole panorama of what constitutes the appreciation of beauty by the mind/brain. A further problem to be considered concerns the modern art forms, such as conceptual art (Barilli 2005), where the prime role of beauty has diminished, and is often actively challenged (Cela-Conde et al. 2011). Thus, researchers interested in the biological underpinnings of aesthetic experiences are faced with the problem of dealing with the several integrated mechanisms that are likely to underlie aesthetic appreciation versus beauty appreciation, and the underlying processes of perception, memory, attention, decision-making, reward, and emotion involved (Chatterjee 2004; Leder et al. 2004; Nadal and Pearce 2011).

Methodological reasons can justify some contradictory results of the neuroimaging investigations. However, as Chatterjee (2011) has rightly pointed out, a further challenge common to all neuroimaging studies should be considered, namely the basic problem of the reverse inference. In other words, it is not clear if such techniques can adequately deal with the interplay of such varied cognitive, affective, personal, social, and cultural factors, and whether or not it is sufficient to consider activity in specific brain regions as proof of their involvement in a given cognitive process.[4]

It should also be considered that somewhat different neuroimaging findings could be the result of a partially different (temporal and spatial) assembly of mosaics of FMs in different subjects, since this plastic process can be affected by the memories, emotional conditions, and cognitive capabilities of each subject.

However, a revision of the available data has been proposed suggesting that aesthetic appreciation of painting, music, and dance involves a common pattern of at least three different kinds of measurable brain activity (Nadal and Pearce 2011): (1) An enhancement of low-level cortical sensory processing; (2) high-level top-down processing and activation of cortical areas involved in evaluative judgment; (3) an engagement of the reward circuit, including cortical (anterior cingulate, orbitofrontal, and ventromedial prefrontal) and subcortical (ventral tegmental area and nucleus accumbens) regions, as well as some of the regulators of this circuit (amygdala, thalamus, hippocampus).

Let us focus on the orbitofrontal cortex (OFC), anterior insula (frontal insula, FI) and anterior cingulated cortex (ACC), as these have been found by several researchers to be the most commonly involved brain areas in aesthetics appreciation (Brown et al. 2011).

The OFC occupies the ventral surface of the frontal part of the brain plays a prominent role in general sensory processing, being a secondary sensory cortex for both olfaction and taste (Kringelbach 2004; Rolls 2004). The orbitofrontal cortex projects to several brain areas such as cingulate cortex (Carmichael and Price 1996), the ventral striatum (Ferry et al. 2000), head of the caudate nucleus (Kemp and Powell 1970; Haber et al. 2006), ventral tegmental area (Johnson et al. 1968), and the entorhinal cortex (Barbas 2007) providing a route for reward information to reach the hippocampus (Rolls and Xiang 2005) and preoptic region (Rolls et al. 1976).

The insular cortex is a portion of the cerebral cortex folded deep within the lateral sulcus between the temporal lobe and the frontal lobe and it is divided into two parts, the larger anterior insula and the smaller posterior insula. It is approximately 30 per cent enlarged in the human relative to the macaque monkey (Semendeferi and Damasio 2000).

The most strongly activated region by artworks corresponds with the anterior insula proper, a region associated with self-awareness, cognitive functioning, and interpersonal experience. According to several studies, this brain area is deeply involved in interoceptive awareness/insight (Critchley et al. 2004; Critchley 2005), a sense of the "feeling self" (Craig 2009), core affect (Barrett et al. 2007), the subjective experience of emotions such as pain, sadness, anxiety, and disgust (Liotti et al. 2000), and the capacity to anticipate the impact of emotional events on the body's responses (Singer et al. 2009). In particular, Craig (2005) argued that the right anterior insula is associated with sympathetic autonomic activity whereas the left anterior insula is associated with parasympathetic activity. By this reasoning, aesthetic processing should most likely be associated with the right insula due to its arousing nature.

The anterior cingulate cortex can be divided based on cognitive (dorsal), and emotional (ventral) components. The dorsal part of the ACC is connected with the prefrontal cortex and parietal cortex as well as the motor system and the frontal eye fields, making it a central station for processing top-down and bottom-up stimuli. By contrast, the ventral part of the ACC is connected with amygdala, nucleus accumbens, hypothalamus, and anterior insula, and is involved in assessing the salience of emotion and motivational information.

Devue and collaborators (Devue et al. 2007) have found foci of activity in FI and ACC in subjects when viewing their own faces as compared with viewing familiar faces of colleagues, suggesting that these foci may be involved in discriminating self from other. It could also be proposed that these brain areas are involved in judging human figures in paintings, above all portraits, and perhaps particularly one's own portrait.

The presence of FI and ACC activations in the analysis of artworks raises an important question: are there special types of neurons and/or a possible special organization of FMs in the insula and ACC that can be hypothesized as playing a role in the production/appreciation of works of art? In fact, a special class of neurons, the so-called von Economo's neurons (VENs), are interesting candidates for a special role in the functional organization of insula and ACC.

22.4.1 On the functional role of the von Economo's neurons (VENs)

22.4.1.1 Architecture of VENs

The VENs are large, bipolar cells that can be found in the cerebral cortex of only a group of species, including *Cetacea* (Butti et al. 2009), *Elephantidaes* (Hakeem et al. 2009), great apes, and humans (Nimchinsky et al. 1999), characterized by relatively large brains. This finding led to the suggestion that VENs are a possible obligatory neuronal adaptation in very large brains, permitting fast information processing and transfer along highly specific projections and that evolved in relation to emerging social behaviors (Butti et al. 2009). In the human brain they are located in layer 5 of only two cortical regions, the ACC and FI cortex (Fajardo et al. 2008). More interestingly for their possible functional correlates are the observations on the qualitative and quantitative aspects of the VENs since they may possibly serve to identify differences between the human brain and the brain of other species (Allman et al. 2005; Premack 2007). The VENs have a narrow dendritic arborization that spans the layers of the cortex and may be able to sample and rapidly relay the output from a columnar array of neurons. The apical and basal dendrites of the VENs are remarkably symmetrical; this architecture suggests that the VENs may be comparing inputs to these two symmetrical dendrites.

The cell bodies of VENs in ACC are, on average, 4.6 times larger than that of layer 5 pyramidal cells in this area (Nimchinsky et al. 1999). The VENs' large size suggests that they bear large, rapidly conducting axons, which is a characteristic feature of large neurons in layer 5 elsewhere in the cerebral cortex (Sherwood et al. 2003). Watson (Watson et al. 2006) has described in Nissl-stained VENs the origin of the axon from the side of the soma.

The dendritic architecture of neurons reflects the way in which they integrate information (Vetter et al. 2001). Both spines (Sabatini et al. 2001) and branches (Polsky et al. 2004) can operate as processing compartments and, compared with their layer 5 pyramidal counterparts, VENs have fewer of both, which suggests that the von Economo cells receive, and therefore integrate, fewer inputs than pyramidal neurons.

The radially polarized structure of the VENs could be a specialization that facilitates the rapid radial signal transmission, providing an output response that reflects the integration of few, but likely, highly relevant inputs from the neurons of the neighboring FMs.

22.4.1.2 Localization and possible functions of VENs

As mentioned earlier, the FI is involved in self-awareness and decision making under uncertain conditions and it has been found to be consistently activated by peripheral autonomic changes. The activity of anterior insula also varied as a function of uncertainty during the anticipatory period in a gambling task. It has, therefore, been proposed that the VENs and related circuitry are involved in rapid intuition which, like perceptual recognition, involves immediate effortless awareness rather than the engagement of deliberative processes (Allman et al. 2005, 2011).

It is interesting to note that in all great apes and in postnatal human brains, VENs are around 30 per cent more numerous in the fronto-insular cortex of the right hemisphere compared with their number in the corresponding cortex of the left side.

The larger number of VENs in the right hemisphere may be related to asymmetry in the central control of the autonomic nervous system in which the right hemisphere is preferentially involved in sympathetic activation, resulting from negative feedback and subsequent error-correcting behavior; the left hemisphere is preferentially involved in parasympathetic activity associated with reduced tension or calming responses. Thus, it seems that a preferential leftward activation in FI and ACC involve positive and affiliative emotions. This autonomic asymmetry is consistent with the proposal that the right hemisphere responds to the unexpected and the left hemisphere to more routine stimuli.

It may therefore be speculated that VEN predominance in the right hemisphere in the postnatal period is related to the right hemisphere specialization in humans, at least to the social emotions that are of basic importance for self-consciousness (Allman et al. 2005, 2011).

An extension of this line of thinking suggests that the central control of the sympathetic system involves VENs in FI, especially in the right hemisphere that is associated with sympathetic autonomic activity (Craig 2005). Thus, the VEN predominance in the right hemisphere supports a possible role of these neurons in the aesthetic processing. Actually, aesthetic appreciation always has the kind of arousing nature found associated with the right insula activation. Hence, this brain region may both allow a first integration of the visual inputs from the artwork inspection with the "feeling of self" (see above and Craig 2009). Thereafter, especially via their large, rapidly conducting axons, VENs may rapidly relay this information to other parts of the brain and hence play an important role in the assembly of the mosaics of FMs involved in the artworks production/appreciation.

22.4.1.3 VENs as target neurons of VT signals

As already mentioned, important contributions by Insel, Young, and colleagues indicate that the oxytocin and vasopressin V1a receptors mediate social bonding and the rewarding aspects of social bonding (Young et al. 2001; Lim et al. 2004; Insel 2010). Allman and collaborators tested adult human ACC and FI tissue for reactivity to antibodies raised against the vasopressin and oxytocin receptors. These authors have demonstrated that antibodies specific for the V1a receptor label a subpopulation of VENs, as well as pyramidal neurons in layers 2/3 and 5 of ACC and FI. The pattern of labeling was interesting since apical dendrites labeled with the V1b receptor antibody formed columns that spanned layer 5 to layer 1 (Allman et al. 2005, 2011). It has also been shown that VENs are strongly labeled with antibodies to 5-HT2B receptor (see below) and to the dopamine D3 receptor, which is a high affinity dopamine receptor that has been proposed to signal the expectation of reward under uncertainty (Sokoloff and Schwartz 2002). It should be noted that the 5-HT2B receptor, strongly expressed on the VENs, is rarely expressed elsewhere in the CNS (see Baumgarten and Göthert 1997), but it is also strongly expressed in the human gut. It has been proposed that serotonin might serve as an antagonistic signal to dopamine, with serotonin signaling punishment and dopamine signaling reward (Daw et al. 2002). The activation of the 5-HT2B receptor on VENs might, therefore, be an opponent to the dopamine D3 signal of reward expectation (Allman 2005).

In agreement with these neurochemical findings activation of FI and ACC increases with the degree of situations in which the subject sustains a gambled loss (punishment) and then switches to a different behavioral strategy, implying that in normal subjects these areas are involved in adaptive decision making and cognitive flexibility (Carter et al. 2012). As shown in Figure 22.2, the several receptors described by Allman and co-workers in VENs can be targets for monoamines and neuropeptides working as WT and/or VT signals and contributing to the control of forementioned behaviors.

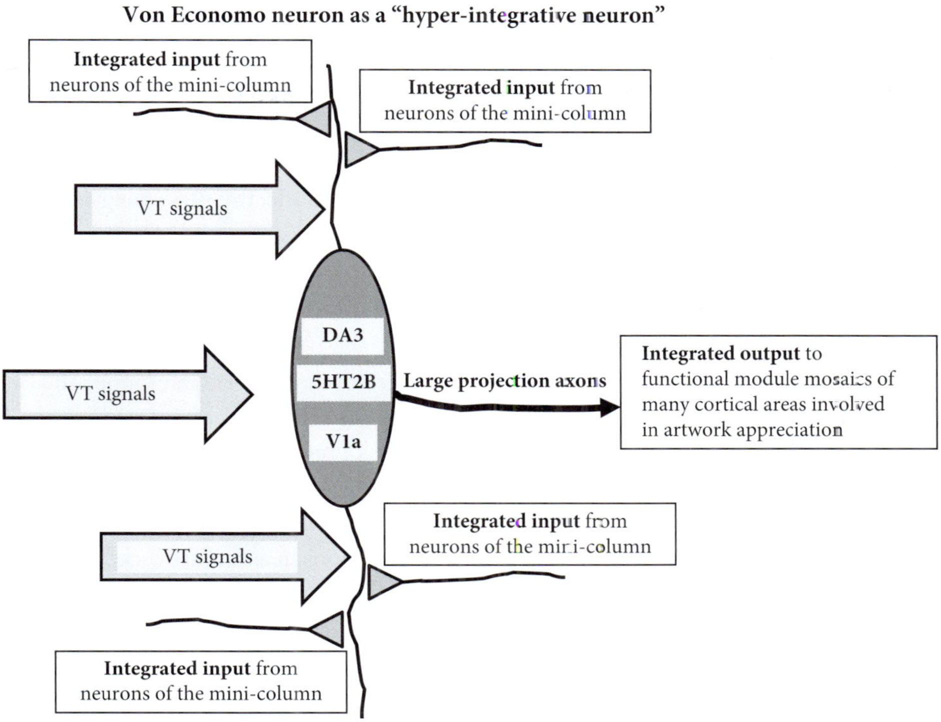

Figure 22.2 Schematic representation of a von Economo neuron (VEN). The main wiring and volume transmission inputs are indicated as well as the main output via the large rapidly conducting, axon.

It should be noticed that peripheral inputs either via the AP-NTS and/or the AV3V-BH can reach ACC and FI that (see also Allman et al. 2005) play an important role in interoception and hence in conscious awareness of visceral activity (Allman 2005). This could be the neurobiological substrate that gives a "somatic" component to the aesthetics appreciation.[5] This proposal is in agreement with Damasio's theory of "somatic states," since he proposed that such monitoring of sensations arising from the body is crucial to adaptive decision making (Damasio 1995; Rainville et al. 2006). Thus, according to our view it may be considered that the complexity of aesthetics appreciation can be traced to the depths of human homeostasis, somatic and psychic homeostasis, and both WT and VT (brain-borne and peripherally borne) play important and complementary roles.

As a matter of fact:

- vasopressin can be, inter alia, also a CSF VT-mediated signal since, as pointed out recently for oxytocin (Veening et al. 2010), it can reach many brain areas (Sewards and Sewards 2003). Thus, as suggested by our group in 1986 (Agnati et al. 1986, 2010; Agnati and Fuxe 2000; see also the General Premise), CSF is used as a "broadcasting channel" to send coordinated messages (monoamines and peptides) to a wide variety of nearby and distant brain areas contributing considerably to the VT communication processes affecting mainly but not only peri-paraventricular brain areas and deeply affecting behavioral states.

- as discussed by Nadal and collaborators (Nadal et al. 2008) and pointed out earlier, the seemingly varied results of the neuroimaging studies are compatible with a general model linking aesthetics to the neuroscience of visual and affective processing as well as reward systems and decision making (Chatterjee 2011). Thus, engaging visual properties of paintings increases activity within ventral visual cortices. Aesthetic judgments activate parts of the dorsolateral prefrontal and medial prefrontal cortices. In addition, emotional responses to these stimuli activate orbitofrontal as well as anterior cingulate cortices.

Here we have proposed a more complex frame where both the peripheral inputs by the soma, and not only the visual inputs, affect the eudemonia of the subject and are integrated with the brain-borne signals which are also the results of previous experiences and of the psychic state and susceptibility of the subject.

An "ugly" or even a "terrifying" painting can activate the peculiar, complex neurochemical machinery of the human brain which, by judging an object as an "artwork," can often trigger extraordinary effects on the psychic homeostasis of a sensitive subject.[6]

22.4.2 Possible role of VENs in the functional organization of insula and ACC

Understanding the VENs' distinctive relationship to other cortical neuron types as well as comparing VENs in humans versus other mammals could shed light on their neurobiological role in the ACC and in the FI cortex and hence on their contribution to some specific cognitive functions. Among the morphological and functional human differences, Premack (2007) cites evidence showing that our species has not only many more and larger VENs than apes, but also far wider perimeters of dendrites (an average of 51 μm in humans compared with 36 μm in chimpanzees and monkeys). Watson and colleagues (2006) suggest that VENs are designed to perform relatively simple integrative functions within a cortical column before conveying output to other brain regions for further processing. These findings are highly interesting but we would like to propose a different view. As already outlined in section 22.4.1.3, we propose that VENs are a special type of "hub neurons" of FMs in the FI and ACC capable of integrating few but selected WT inputs (mainly from neighboring neurons) and several VT signals (also broadcasted via the CSF channel). Thus, they may carry out an extensive integration that can be rapidly communicated via large VEN axons to several brain areas (Stimpson et al. 2011). This special role of VENs can also be deduced by the cited observation that human VENs are located in only two parts of the human brain, the ACC and the FI cortex, which are involved in socially crucial tasks, such as empathy, feelings of guilt, and embarrassment. In agreement with such a view, neuropathological data demonstrate that the selective destruction of VENs, as in the early stages of fronto-temporal dementia, is associated to profound deficits in empathy, social awareness, and self-control (Seeley et al. 2006).

22.4.2.1 Possible role of von Economo's neurons in the redeployment of insula and anterior cingulate cortex

Anderson has put forward the interesting proposal of the creative reuse of existing neural components, which may have played a significant role in the evolutionary development of cognition. Thus, Anderson (2010) analyses the problem of regional brain specialization under an evolutionary perspective, taking into account the "tinkering principle" proposed more than three decades ago by Jacob (1977). Jacob claimed that evolution is a process that tinkers, and novelties come from previously unseen associations of already available material.

As Anderson points out, over the past several years it has been assessed that, for example, although Broca's area is associated with language processing, it is also involved in many different action- and imagery-related tasks (Anderson 2010; Poldrack 2006). There are obvious evolutionary advantages of redeployment of brain areas and Anderson (2010) analyses this phenomenon in relation with cognition by a careful recent meta-analyses of imaging results that support the view that most regions of the brain, even fairly small regions, appear to be activated across diverse task categories. The meta-results also suggest that the brain achieves its variety of functions by putting the same regions together in different patterns of functional cooperation. This view chimes with our proposal of the mosaic assembly of FMs. The tinkering principle suggests that special types of neurons, such as VENs, could have been exaptated with a "plastic" rebuilding of the circuits allowing peculiar functional "reuse" of insula and ACC. Taken together, it is proposed that the exapted functions of ACC and FI in artworks production/appreciation may be the result of a more complex dynamic and plastic reassemblage of FM mosaics, in particular of those having VENs as a hyper-integrative output. It should, however, be noted that such an innovative assemblage of the available material may occur at various miniaturization levels of these FMs. In other words, the "tinkering process" of VENs controlled FMs can allow the "emergence" of novelties that, via VENs' projections, rapidly and deeply affect several and also distant brain areas which could be involved in FM mosaics carrying out emergent functions, such as the appreciation of art.

22.5 Conclusion

The main questions posed by this chapter have been:

(1) Why artworks?

(2) Artworks and eudemonia: psychic vs somatic homeostasis.

22.5.1 Why artworks?

An interesting comment is made by Nadal and Pearce (2011): "The first and most basic aspect is to propose theories about how artists might use an implicit understanding of the brain's perceptual systems to engage their audience in their work" (see also Maffei and

Fiorentini 2008). According to our hypothesis the transfer of a complex message via artworks depends on the unique instinct-like communication of human beings. Nadal and Pearce (2011) continue: "The second approach goes beyond such observations to neuropsychological and clinical examinations of how disrupting brain function affects artistic creation and appreciation".

Nadal and Pearce suggest that beauty is of little relevance to many manifestations of art, both around the world and throughout history, which may have been created to intimidate, induce sorrow, to broadcast grandeur, to make us reflect on our own existence. Indeed, contemporary Western art includes many instances of works that were explicitly conceived as a reaction against the concept of beauty (Barilli 2005). Further, a given artwork often serves a multitude of purposes for different people, in different contexts, at different times. It is, therefore, interesting that Brown and collaborators proposed "naturalizing aesthetics" (Brown 2011) by observing that human neuroimaging studies have shown convincingly that the brain areas that mediate aesthetic responses to artworks overlap those that mediate the appraisal of objects of evolutionary importance, such as the desirability of food items or the relative attractiveness of potential mates. Hence, it is likely that artworks are appreciated by exaptation of neural systems that evolved to evaluate objects of more primary survival importance.

Our proposal, while not discarding this view, takes also into account the human communication processes that, thanks to the abstraction capacity of the human brain and the bipedal conditions with the possibility of using hands to make tools, have been exaptated, leading to the emergence of an instinct-like need to communicate by symbols and to produce artworks, which as Young points out "are as necessary to us as food or sex." Furthermore, we suggest that artworks are an artist's attempt to communicate a message present in her inner world that may or may not be derived from an environmental input (Agnati et al. 2007).[7] Thus, artworks can convey messages that extend beyond the physical confines of humans in time and space,[8] and the appreciation of art by other human beings is due to the fact that they gain access to these messages.

According to our view the complexity of the human brain has, therefore, allowed the emergence of an instinct-like need to communicate not simply the features of the outer world but also aspects of the subject's inner world. A description of the detachment from the outer world to communicate a message of the inner world is described in a famous novel by Bernhard.[9]

The current hypothesis may also suggest a possible original point of view on recent artworks that break with the previous art tradition. These Pop artworks, indeed, do not appear to look for an ideal beauty or want to communicate a message about some complex intellectual aspect of the artist's inner world; rather, they seem to represent a crude denouncement of some "invasion" by the mass media of the inner world of the human beings. This view could be grasped by comparing, for example, an artwork conveying mainly a message of the inner-world of an artist with an artwork (in particular Pop-art products) conveying mainly a message anchored to the modern outer world[10] (see Figures 22.3a and 22.3b). The comparison may suggest a colonization of the inner world by the outer world occurs, likely thanks to mass media.

(a)

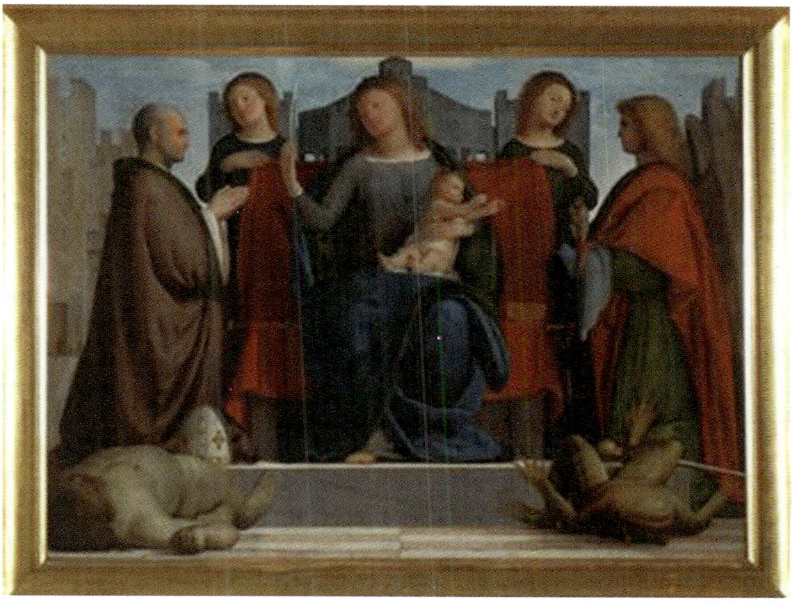

Figure 22.3 Panel a. Bartolomeo Suardi's (called Bramantino 1465 c.–1530) Trittico di San Michele (1505) is an example of a non-naturalistic artwork which makes visible an artist's inner thought process. Two dead figures appear in the foreground: the heretic Ario defeated by Sant'Ambrogio and the Devil (as a toad) defeated by San Michele. These two figures inspire a feeling of pity both for both "toads" and heretics. One of Emily Dickinson's poems suggests that a toad can die if flooded by a supernal light, but the same can be true for a man. It should be noted that a supernal light had the power of reducing to ashes also in the Semele mythos. Some Fathers of the Church believe in the final spiritual salvation of all the creatures as well as sinners and Devil.[11]

Madonna and Child Enthroned with Saint Ambrose and Saint Michael, by Bartolomeo Suardi, known as Il Bramantino, circa 1520, tempera and oil on panel, Bramantino, Bartolommeo Suardi (1450/55–c.1536) / Pinacoteca Ambrosiana, Milan, Italy / De Agostini Picture Library / © Veneranda Biblioteca Ambrosiana—Milano / The Bridgeman Art Library.

(b)

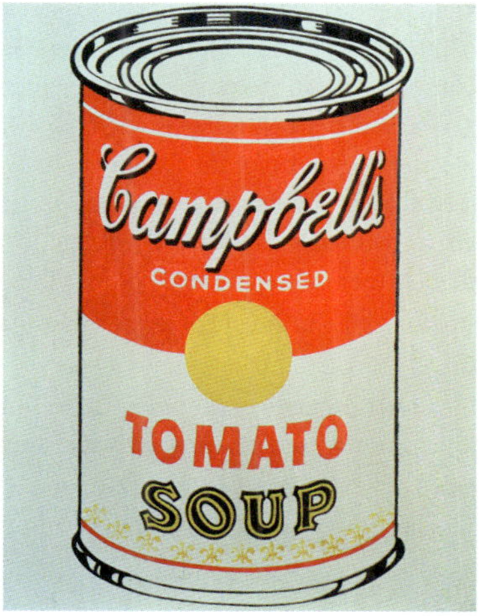

Panel b. Warhol's *Campbell's Soup Can* (1962) exemplifies the "invasion" by the mass media of the artist's inner world. Some modern artworks, in particular Pop-art, may be the result of such an invasion and use it as a subject for denunciation.

Campbell's Soup Can, 1962 (screen print), Warhol, Andy (1928–87) / Saatchi Collection, London, UK / The Bridgeman Art Library © DACS 2015.

22.5.2 Artworks and eudemonia: psychic versus somatic homeostasis

The concept of psychic homeostasis is in agreement with Levine's noting (2009) that several psychologists, starting with Kurt Lewin and Abraham Maslow in the 1930s and 1940s, expanded the idea of "needs" beyond the purely physiological to include needs for social connections, aesthetic and intellectual stimulation, esteem and self-fulfilment.

Thus, psychic homeostasis can be affected by production and appreciation of artworks as is well documented by autobiographies of artists (see e.g. Agnati et al. 2012a), and by the very existence of art therapy (Johnson and Sullivan-Marx 2006; Karlik and Baurek 2011; Caddy et al. 2012).

It is, therefore, possible to conclude that artworks are artists' attempts to communicate a message which is present in their inner worlds that may or may not be derived from environmental inputs. The communication of this message causes a profound "resonance" between artists' and observers' inner worlds. Thus, sensitive observers may not simply find their inner world enriched by experiencing an artwork, but may also experience a positive action on their psychic homeostasis, favoring eudemonia.

Communicating one's own inner world by means of artworks can offer a significant refuge for the human being confronted with life and death.[12] As Zeki (2004) points out:

> A refuge lies in recreating the brain's ideal in art and through art . . . The translation of concepts in the artist's mind onto canvas, or into music or literature constitutes art. Great art is that which corresponds to as many different concepts in as many different brains over as long a period of time as possible. Ambiguity is such a prized characteristic of all great art because it can correspond to many different concepts.

Dedication

This chapter is dedicated to Professor Eugenio Riva Sanseverino with gratitude and warm friendship.

Notes

1 Ceronetti comments in his inspiring book *L'Occhiale malinconico* (1988) on a letter that he received from China after an interesting article that he published in a newspaper criticizing the embarrasing behavior of the President of China in front of Goya's painting *La Maja Desnuda*. Ceronetti discusses how deeply ideologies and social conventions can restrain also powerful men in their judgments on beauty. The letter that Ceronetti received deals with the problem of "beauty" citing the Chinese philosopher Chuang-tzu: "The male monkey looks for the female monkey; the male deer fights for the female deer; the eel looks for other eels; Mao-quiang and Li-ji are beautiful women beloved by men, but the fish that comes close to them immediately sinks into the deepest water, the bird flies rapidly away and the deer runs far from them. Is there someone that really knows what ideal beauty is?"

2 It should be noticed that Freud recognized the principle of psychic homeostasis as a basic concept in *The Interpretation of Dreams* (1913).

3 Nikolai Gogol (*The Portrait*, 1835): "Illustrious artworks appear in the human society to alleviate and pacify human beings. Art calms down the soul and as a prayer in aeternum strive to ascend to God."

4 Fernando Pessoa: "I cannot establish a clear link between a visible mass of gray matter and this 'thing' that behind my gaze see skies and can even think and conceive not existing skies" (*Livro de desassossego por Bernardo Soares*).

5 The presence of 5-HT2B receptors on the VEN that is otherwise rare in the brain, but common in the viscera, suggests an interesting extension of the concept that these areas are monitoring activity in the gut and likely contribute to "gut level" or intuitive decision making in a given behavioral context. In agreement with such a view, ACC and FI are known to have an important role in interoception or the conscious awareness of visceral activity (Craig 2004).

6 See, e.g. the novel by Nikolai Gogol *The Portrait* (1835).

7 This assumption is partly in agreement with Herbert Kuhn's distinction between "imagery style" and "sensitive style." The former communicates an author's fantasy or his personal experience expressed according to modules not immediately derived from reality and sometimes derived from dreams. The latter one usually is more an attempt to give a faithful reproduction of the nature (citation from Jaffé 1983).

8 Fernando Pessoa: "I fail in my attempt to communicate messages out from my inner world since I am always here, enclosed by the high walls of my self-consciousness" (*Livro de desassossego por Bernardo Soares*).

9 Thomas Bernhard: "by means of thin lines he was painting a world that annihilates itself . . . all of what he was painting was his terrifying reality" (*Verstörung*, 1967).

10 Emerging in the late 1950s in England and America, pop art sought to depict the images of mass media advertising, comic books, and consumer products. Pop art, much like pop music, aimed to appeal to a broad audience and rebelled against the idea of high or elitist culture. The Andy Warhol (1928–1987) pop-art phenomenon may be best characterized by his *Campbell's Soup Cans* silkscreen work. In fact, Warhol's widespread recognition came with the 1962 Los Angeles exhibition of his now-famous depictions of the soup cans. Drawn from mass-market culture, *Campbell's Soup Cans* appeared larger than life and made a definite statement about the consumer-driven society that had become so prevalent in America.

11 Such a "pietas" for the Devil, who shares the same fate of sin and pain as the human being, has been discussed by Giovanni Papini in his book *Il Diavolo* (1953). Papini suggests that even the Devil is not condemned in "aeternum." Papini's book was brusquely rejected by *L'Osservatore Romano*, even if Papini's view was in agreement with the great theologian Origene (184/185–253/254), who believed in the universal reconciliation at the end of the cosmos story thanks to God's infinite clemency.

12 Art gives us an illusory freedom from the sordidness of life. When we feel the anguish and experience the invectives of Hamlet, prince of Denmark, we forget our own despair (Pessoa).

References

Agnati, L.F., Agnati, A., Mora, F., et al. (2007). Does the human brain have unique genetically determined networks coding logical and ethical principles and aesthetics? From Plato to novel mirror networks. *Brain Research Reviews* **55**, 68–77.

Agnati, L.F., Barlow, P.W., Baluška, F., et al. (2011). A new theoretical approach to the functional meaning of sleep and dreaming in humans based on the maintenance of 'predictive psychic homeostasis'. *Communicative & Integrative Biology* **4**, 640–54.

Agnati, L.F., Barlow, P., Ghidoni, R., et al. (2012a). Possible genetic and epigenetic links between human inner speech, schizophrenia and altruism. *Brain Research* 1476, 38–57. doi:10.1016/j.brainres.2012.02.074.

Agnati, L.F. and Fuxe, K. (1984). New concepts on the structure of the neuronal networks: the miniaturization and hierarchical organization of the central nervous system. *Bioscience Reports* 4, 93–8.

Agnati, L.F. and Fuxe, K. (2000). Volume transmission as a key feature of information handling in the central nervous system possible new interpretative value of the Turing's B-type machine. *Progress in Brain Research* **125**, 3–19.

Agnati, L.F., Fuxe, K., and Ferré, S. (2005). How receptor mosaics decode transmitter signals. Possible relevance of cooperativity. *Trends in Biochemical Sciences* **30**, 88–193.

Agnati, L.F., Fuxe, K., Zoli, M., et al. (1986). A correlation analysis of the regional distribution of central enkephalin and beta endorphin immunoreactive terminals and of opiate receptors in adult and old male rats. Evidence for the existence of two main types of communication in the central nervous system: the volume transmission and the wiring transmission. *Acta Physiologica Scandinavica* **128**, 201–7.

Agnati, L.F., Guidolin, D., Cortelli, P., et al. (2012b). Neuronal correlates to consciousness. The "Hall of mirrors" metaphor describing consciousness as an epiphenomenon of multiple dynamic mosaics of cortical functional modules. *Brain Research* **1476**, 3–21.

Agnati, L.F., Guidolin, D., Guescini, M., et al. (2012c). Aspects on the integrative actions of the brain from neural networks to "brain-body medicine." *Journal of Receptors and Signal Transduction Research* **32**, 163–80.

Agnati, L.F., Guidolin, D., Guescini, M., et al. (2010). Understanding wiring and volume transmission. *Brain Research Reviews* **64**, 137–59.

Allman, J.M., Tetreault, N.A., Hakeem, A.Y., et al. (2011). The von Economo neurons in the frontoinsular and anterior cingulate cortex. *Annals of the New York Academy of Sciences* **1225**, 59–71.

Allman, J.M., Watson, K.K., Tetreault, N.A., et al. (2005). Intuition and autism: a possible role for Von Economo neurons. *Trends in Cognitive Sciences* **9**, 367–73.

Anderson, M.L. (2007). Massive redeployment, exaptation, and the functional integration of cognitive operations. *Synthese* **159**, 329–45.

Anderson, M.L. (2010). Neural reuse: a fundamental organizational principle of the brain. *Behavioral and Brain Sciences* **33**, 245–313.

Bach-Y-Rita, P. (2005). Emerging concepts of brain function. *Journal of Integrative Neuroscience* **4**, 183–205.

Baluška, F., Volkmann, D., and Barlow, P.W. (2006). Cell–cell channels and their implications for Cell Theory. In F. Baluška, D. Volkmann, and P.W. Barlow (eds), *Cell–Cell Channels*. New York: Landes Bioscience, Georgetown TX/Springer Science, pp. 1–18.

Barbas, H. (2007). Flow of information for emotions through temporal and orbitofrontal pathways. *Journal of Anatomy* **211**, 237–49.

Barilli, R. (2005). *L'arte contemporanea*. Milano: Feltrinelli.

Barrett, L.F., Mesquita, B., Ochsner, K.N., et al. (2007). The experience of emotion. *Annual Review of Psychology* **58**, 373–403.

Baumgarten, H.G. and Göthert, M. (eds) (1997). *Serotoninergic Neurons and 5-HT Receptors in the CNS*. Berlin, Heidelberg, New York, NY: Springer-Verlag.

Blakemore, C. (1976). *Mechanics of the Mind*. Cambridge: Cambridge University Press.

Brown, S., Gao, X., Tisdelle, L., et al. (2011). Naturalizing aesthetics: brain areas for aesthetic appraisal across sensory modalities. *Neuroimage* **58**, 250–8.

Butti, C., Sherwood, C.C., Hakeem, A.Y., et al. (2009). Total number and volume of Von Economo neurons in the cerebral cortex of cetaceans. *Journal of Comparative Neurology* **515**, 243–59.

Caddy, L., Crawford, F., and Page, A.C. (2012). "Painting a path to wellness": correlations between participating in a creative activity group and improved measured mental health outcome. *Journal of Psychiatric and Mental Health Nursing* **19**, 327–33.

Carmichael, S.T. and Price, J.L. (1996). Connectional networks within the orbital and medial prefrontal cortex of macaque monkeys. *Journal of Comparative Neurology* **371**, 179–207.

Carreira-Perpiñan, M.A. and Goodhill, G.J. (2002). Are visual cortex maps optimized for coverage? *Neural Computation* **14**, 1545–60.

Carter, R.M., Bowling, D.L., Reeck, C., et al. (2012). A distinct role of the temporal-parietal junction in predicting socially guided decisions. *Science* **337**, 109–11.

Cela-Conde, C.J., Agnati, L., Huston, J.P., et al. (2011). The neural foundations of aesthetic appreciation. *Progress in Neurobiology* **94**, 39–48.

Ceronetti, G. (1988). *L'occhiale malinconico*. Adelphi, Milan.

Chatterjee, A. (2004). The neuropsychology of visual artistic production. *Neuropsychologia* **42**, 1568–83.

Chatterjee, A. (2011). Neuroaesthetics: a coming of age story. *Journal of Cognitive Neuroscience* **23**, 53–62.

Cocucci, E., Racchetti, G., and Meldolesi, J. (2009). Shedding microvesicles: artefacts no more. *Trends in Cell Biology* **19**, 43–51.

Craig, A.D. (2004). Human feelings: why are some more aware than others? *Trends in Cognitive Sciences* **8**, 239–41.

Craig, A.D. (2005). Forebrain emotional asymmetry: a neuroanatomical basis? *Trends in Cognitive Sciences* **9**, 566–71.

Craig, A.D. (2009). How do you feel—now? The anterior insula and human awareness. *Nature Reviews Neuroscience* **10**, 59–70.

Critchley, H.D. (2005). Neural mechanisms of autonomic, affective, and cognitive integration. *Journal of Comparative Neurology* **493**, 154–66.

Critchley, H.D., Wiens, S., Rotshtein, P., et al. (2004). Neural systems supporting interoceptive awareness. *Nature Neuroscience* **7**, 189–95.

Damasio, A.R. (1995). *Descartes' Error*. New York, NY: Harper Perennial.

Daw, N.D., Kakade, S., and Dayan, P. (2002). Opponent interactions between serotonin and dopamine. *Neural Networks* **15**, 603–16.

Descartes, R. (1960). *Discourse on Method and Meditations*. L.J. Lafleur (trans.). New York, NY: The Liberal Arts Press.

Descarries, L., Bérubé-Carrière, N., Riad, M., et al. (2008). Glutamate in dopamine neurons: synaptic versus diffuse transmission. *Brain Research Reviews* **58**, 290–302.

Devue, C., Collette, F., Balteau, E., et al. (2007). Here I am: the cortical correlates of visual self-recognition. *Brain Research* **143**, 169–82.

Dickinson, E. (1997). *Colloqui con le ombre*. Tavernola: Libreria Giunti al Punto Spa.

Fajardo, C., Escobar, M.I., Buriticá, E., et al. (2008). Von Economo neurons are present in the dorsolateral (dysgranular) prefrontal cortex of humans. *Neuroscience Letters* **435**, 215–8.

Ferry, A.T., Ongür, D., An, X., et al. (2000). Prefrontal cortical projections to the striatum in macaque monkeys: evidence for an organization related to prefrontal networks. *Journal of Comparative Neurology* **425**, 447–70.

Freud, S. (1913). *The Interpretation of Dreams*. New York, NY: Macmillan.

Fuxe, K. and Agnati, L.F. (1991). Two principal modes of electrochemical communication in the brain: volume vs. wiring transmission. In K. Fuxe and L.F. Agnati (eds), *Volume Transmission in the Brain, Novel Mechanisms for Neural Transmission, Vol. 1*. New York, NY: Raven Press, pp. 1–9.

Fuxe, K., Borroto-Escuela, D., Romero-Fernandez, W., et al. (2012). On the role of volume transmission and receptor–receptor interactions in social behavior: focus on central catecholamine and oxytocin neurons. *Brain Research* **1476**, 119–31. doi:10.1016/j brainres.2012.01.062.

Ganzel, B.L., Morris, P.A., and Wethington, E. (2010). Allostasis and the human brain: Integrating models of stress from the social and life sciences. *Psychological Review* **117**, 134–74.

Gimpl, G. and Fahrenholz, F. (2001). The oxytocin receptor system: structure, function, and regulation. *Physiological Review* **81**, 629–83.

Gogol, N. (1835). *I racconti di Pietroburgo*. Turin: Einaudi.

Golgi, C. (1914). La moderna evoluzione delle dottrine e delle conoscenze sulla vita. Milano: *Rendiconti Regio Istituto Lombardo* **XLVII**, 1.

Goncharova, L.B. and Tarakanov, A.O. (2008). Nanotubes at neural and immune synapses. *Current Medicinal Chemistry* **15**, 210–8.

Gould, S.J. and Vrba, E.S. (1982). Exaptation: a missing term in the science of form. *Paleobiology* **8**, 4–15.

Guidolin, D., Albertin, G., Guescini, M., et al. (2011). Central Nervous System and computation. *Quarterly Review of Biology* **86**, 265–85.

Guillemin, R. (1978). Peptides in the brain: the new endocrinology of the neuron. *Science* **202**, 390–402.

Haber, S.N., Kim, K.S., Mailly, P., et al. (2006). Reward-related cortical inputs define a large striatal region in primates that interface with associative cortical connections, providing a substrate for incentive-based learning. *Journal of Neuroscience* **26**, 8368–76.

Hakeem, A.Y., Sherwood, C.C., Bonar, C.J., et al. (2009). Von Economo neurons in the elephant brain. *Anatomical Record (Hoboken)* **292**, 242–8.

Hebb, D.O. (1949). *The Organization of Behavior*. New York, NY: Wiley.

Hopcroft, J.E. and Ullman, J.D. (1979). *Introduction to Automata Theory, Languages and Computation*. Boston, MA: Addison-Wesley.

Horowski, R., Benes, H., and Fuxe, K. (2004). Striatal plasticity and motor learning importance of circadian rhythms, sleep stages and dreaming. *Parkinsonism and Related Disorders* **10**, 315–7.

Insel, T.R. (2010). The challenge of translation in social neuroscience: a review of oxytocin, vasopressin, and affiliative behavior. *Neuron* **65**, 768–79.

Jacob, F. (1970). *La logique du vivant. Une histoire de l'hérédité*. Paris: Editions Gallimard.

Jacob, F. (1977). Evolution and tinkering. *Science* **196**, 1161–6.

Jaffé, A. (1983). Il simbolismo nelle arti figurative. In C.G. Jung (ed.), *L'Uomo e i suoi Simboli*. Milano: Raffaello Cortina Editore, p. 246.

Johnson, T.N., Rosvold, H.E., and Mishkin, M. (1968). Projections from behaviourally-defined sectors of the prefrontal cortex to the basal ganglia, septum and diencephalon of the monkey. *Experimental Neurology* **21**, 20–34.

Johnson, C.M. and Sullivan-Marx, E.M. (2006). Art therapy: Using the creative process for healing and hope among African American older adults. *Geriatric Nursing (New York)* **27**, 309–16.

Jones, E.G. (2000). Micro-columns in the cerebral cortex. *Proceedings of the National Academy of Sciences of the USA* **97**, 5019–21.

Karlic, H. and Baurek, P. (2011). Epigenetics and the power of art. *Clinical Epigenetics* **2**, 279–82.

Kemp, J.M. and Powell, T.P. (1970). The cortico-striate projection in the monkey. *Brain* **93**, 525–46.

Klee, P. (1920) Schöpferische konfession. Berlin: Erich Reiss.

Kringelbach, M.L. (2004). Food for thought: hedonic experience beyond homeostasis in the human brain. *Neuroscience* **126**, 807–19.

Leder, H., Belke, B., Oeberst, A., et al. (2004). A model of aesthetic appreciation and aesthetic judgements. *British Journal of Psychology* **95**, 489–508.

Lehrer, J. (2007). *Proust was a Neuroscientist*. New York, NY: Houghton Mifflin Harcourt.

Levine, D.S. (2009). Brain pathways for cognitive-emotional decision making in the human animal. *Neural Networks* **22**, 286–93.

Lim, M.M., Murphy, A.Z., and Young, L.J. (2004). Ventral striatopallidal oxytocin and vasopressin V1a receptors in the monogamous prairie vole (Microtus ochrogaster). *Journal of Comparative Neurology* **468**, 555–70.

Liotti, M., Mayberg, H.S., Brannan, S.K., et al. (2000). Differential limbic-cortical correlates of sadness and anxiety in healthy subjects: implications for affective disorders. *Biological Psychiatry* **48**, 30–42.

Lorente de Nò, R. (1938). Architectonics and structure of the cerebral cortex. In J.F. Fulton (ed.), *Physiology of the Nervous System*. New York, NY: Oxford University Press, pp. 291–330.

Maffei, L. and Fiorentini, A. (2008). *Arte e Cervello*. Bologna: Zanichelli.

Marean, C.W., Bar-Matthews, M., Fisher, E., et al. (2007). Early human use of marine resources and pigment in South Africa during the middle Pleistocene. *Nature* **449**, 905–9.

McEwen, B.S. (2003). Interacting mediators of allostasis and allostatic load: towards an understanding of resilience in aging. *Metabolism* **52**, 10–6.

Mountcastle, V.B., Davies, P.W., and Berman, A.L. (1957). Response properties of neurons of cats somatic sensory cortex to peripheral stimuli. *Journal of Neurophysiology* **20**, 374–407.

Nadal, M., Munar, E., Capó, M.A., et al. (2008). Towards a framework for the study of the neural correlates of aesthetic preference. *Spatial Vision* **21**, 379–96.

Nadal, M. and Pearce, M.T. (2011). The Copenhagen neuroaesthetics conference: prospects and pitfalls for an emerging field. *Brain and Cognition* **76**, 172–83.

Nicholson, C. (1979). Brain cell microenvironment as a communication channel. In F.O. Schmitt and F.G. Worden (eds), *The Neurosciences: Fourth Study Program*. Cambridge MA: MIT Press, pp. 457–76.

Nieuwenhuys, R. (2000). Comparative aspects of volume transmission, with sidelight on other forms of intercellular communication. *Progress in Brain Research* **125**, 49–126.

Nimchinsky, E.A., Glissen, E., Allman, J.M., et al. (1999). A neuronal morphologic type unique to humans and great apes. *Proceedings of the National Academy of Sciences of the USA* **96**, 5268–73.

Olson, L., Seiger, A., and Fuxe, K. (1972). Heterogeneity of striatal and limbic dopamine innervation: highly fluorescent islands in developing and adult rats. *Brain Research* **44**, 283–8.

Papini, G. (1953). *Il Diavolo. Appunti per una futura diabologia*. Florence: Vallecchi.

Poldrack, R.A. (2006). Can cognitive processes be inferred from neuroimaging data? *Trends in Cognitive Sciences* **10**, 59–63.

Polsky, A., Mel, B.W., and Schiller, J. (2004). Computational subunits in thin dendrites of pyramidal cells. *Nature Neuroscience* **7**, 621–7.

Premack, D. (2007). Human and animal cognition: continuity and discontinuity. *Proceedings of the National Academy of Sciences of the USA* **104**, 13861–7.

Rainville, P., Bechara, A., Naqvi, N., et al. (2006). Basic emotions are associated with distinct patterns of cardiorespiratory activity. *International Journal of Psychophysiology* **61**, 5–18.

Rockland, K.S. (2010). Five points on columns. *Frontiers in Neuroanatomy* **4**, 22.

Rockland, K.S. and Ichinohe, N. (2004). Some thoughts on cortical minicolumns. *Experimental Brain Research* **158**, 265–77.

Rolls, E.T. (2004). Convergence of sensory systems in the orbitofrontal cortex in primates and brain design for emotion. *The Anatomical Record. Part A, Discoveries in Molecular, Cellular, and Evolutionary Biology* **281**, 1212–25.

Rolls, E.T., Burton, M.J., and Mora, F. (1976). Hypothalamic neuronal responses associated with the sight of food. *Brain Research* **111**, 53–66.

Rolls, E.T. and Xiang, J.Z. (2005). Reward-spatial view representations and learning in the primate hippocampus. *Journal of Neuroscience* **25**, 6167–74.

Romero-Fernandez, W., Borroto-Escuela, D., Agnati, L.F., et al. (2013). Evidence for the existence of dopamine D2-oxytocin receptor heteromers in the ventral and dorsal striatum with facilitatory receptor-receptor interactions. *Molecular Psychiatry* 18, 849–50. doi: 10.1038/mp.2012.103.

Rustom, A., Saffrich, R., Markovic, I., et al. (2004). Nanotubular highways for intercellular organelle transport. *Science* **303**, 1007–10.

Sabatini, B.L., Maravall, M., and Svoboda, K. (2001). Ca(2+) signaling in dendritic spines. *Current Opinions in Neurobiology* **11**, 349–56.

Schmitt, F.O. (1984). Molecular regulators of brain function. A new view. *Neuroscience* **13**, 991–1001.

Schulkin, J. (2011). Social allostasis: anticipatory regulation of the internal milieu. *Frontiers in Evolutionary Neuroscience* **2**, 1–15.

Seeley, W.W., Carlin, D.A., Allman, J.M., et al. (2006). Early frontotemporal dementia targets neurons unique to apes and humans. *Annals of Neurology* **60**, 660–7.

Semendeferi, K. and Damasio, H. (2000). The brain and its main anatomical subdivisions in living hominoids using magnetic resonance imaging. *Journal of Human Evolution* **38**, 317–32.

Sewards, T.V. and Sewards, M.A. (2003). Fear and power-dominance motivation: proposed contributions of peptide hormones present in cerebrospinal fluid and plasma. *Neuroscience and Biobehavioral Reviews* **27**, 247–67.

Sherwood, C.C., Lee, P.W., Rivara, C.B., et al. (2003). Evolution of specialized pyramidal neurons in primate visual and motor cortex. *Brain, Behavior and Evolution* **61**, 28–44.

Singer, T., Critchley, H.D., and Preuschoff, K. (2009). A common role of insula in feelings, empathy and uncertainty. *Trends in Cognitive Sciences* **13**, 334–40.

Sokoloff, P. and Schwartz, J. (2002). The dopamine D3 receptor and its implications in neuropsychiatric disorders and their treatments. In G. Di Chiara (ed.), *Dopamine in the CNS* (***Vol. 1***). Berlin: Springer, pp. 185–222.

Stimpson, C.D., Tetreault, N.A., Allman, J.M., et al. (2011). Biochemical specificity of von Economo neurons in hominoids. *American Journal of Human Biology* **23**, 22–8.

Tanaka, K. (2003). Columns for complex visual object features in the inferotemporal cortex: clustering of cells with similar but slightly different stimulus selectivities. *Cerebral Cortex* **13**, 90–9.

Veening, J.G., de Jong, T., and Barendregt, H.P. (2010). Oxytocin messages via the cerebro-spinal fluid: behavioral effects; a review. *Physiology and Behaviour* **101**, 193–210.

Vetter, P., Roth, A., and Häusser, M. (2001). Propagation of action potentials in dendrites depends on dendritic morphology. *Journal of Neurophysiology* **85**, 926–37.

Wang, X-J. (2001). Synaptic reverberation underlying mnemonic persistent activity. *Trends in Neuroscience* **24**, 455–63.

Wang, Z., Zhou, L., Hulihan, T.J., et al. (1996). Immunoreactivity of central vasopressin and oxytocin pathways in microtine rodents: a quantitative comparative study. *Journal of Comparative Neurology* **366**, 726–37.

Watson, K.K., Jones, T.K., and Allman, J.M. (2006). Dendritic architecture of the von Economo neurons. *Neuroscience* **141**, 1107–12.

Xu, X., Wang, K., Zhang, K., et al. (2012). A gigantic feathered dinosaur from the lower cretaceous of China. *Nature* **484**, 92–5.

Young, J.Z. (1974). *An Introduction to the Study of Man*. Oxford: Oxford University Press.

Young, L.J., Lim, M.M., Gingrich, B., et al. (2001). Cellular mechanisms of social attachment. *Hormones and Behavior* **40**, 133–8.

Zeki, S. (2004). Neural concept formation and Art: Dante, Michelangelo, Wagner. In F.C. Rose (ed.), *Neurology of the Arts: Painting, Music, Literature*. London: Imperial College Press, pp. 13–42.

Zeki, S. (2009). *Statement on neuroesthetics* [Online]. Available at: <http://www.neuroesthetics.org/statement-on-neuroesthetics.php>.

Section 7

Integrative approaches

Chapter 23

Neurobiological foundations of art and aesthetics

Edmund T. Rolls

23.1 Introduction

A theory of the neurobiological foundations of aesthetics and art is described. This has its roots in emotion, in which what is pleasant or unpleasant, a reward or punisher, is the result of an evolutionary process in which genes define the (pleasant or unpleasant) goals for action (Rolls 2005, 2014). It is argued that combinations of multiple such factors provide part of the basis for aesthetics. To this is added the operation of the reasoning syntactic brain system which evolved to help solve difficult multi-step problems, and the use of which is encouraged by pleasant feelings when elegant, simple, and hence aesthetic solutions are found that are advantageous because they are parsimonious, and follow Occam's Razor. The combination of these two systems, and the interactions between them provide an approach to understanding aesthetics that is rooted in evolution and its effects on brain design and function (Rolls 2011a, 2012).

I start by considering how affective value is generated in the brain as a solution to the problem of how genes can specify useful goals for actions. This is more efficient and produces more flexible behavior than by specifying the actions themselves. Then, in sections 23.5 and 23.6, I develop this theory further into a theory of the foundations of aesthetics and art.

23.2 Emotions as states elicited by rewards and punishers

Emotions can usefully be defined (operationally) as states elicited by rewards and punishers that have particular functions (Rolls 1999, 2005, 2014). The functions are defined below and include working to obtain or avoid the rewards and punishers. A reward is anything for which an animal (which includes humans) will work. A punisher is anything that an animal will escape from or avoid. A diagram summarizing some of the emotions associated with the delivery of a particular reward or punisher or a stimulus associated with them, or with the omission of a reward or punishment, is shown in Figure 23.1. It is emphasized that this shows states elicited by any one reward or punisher, and that there are many different rewards and punishers. This helps to account for many different emotions (Rolls 1999, 2005, 2014).

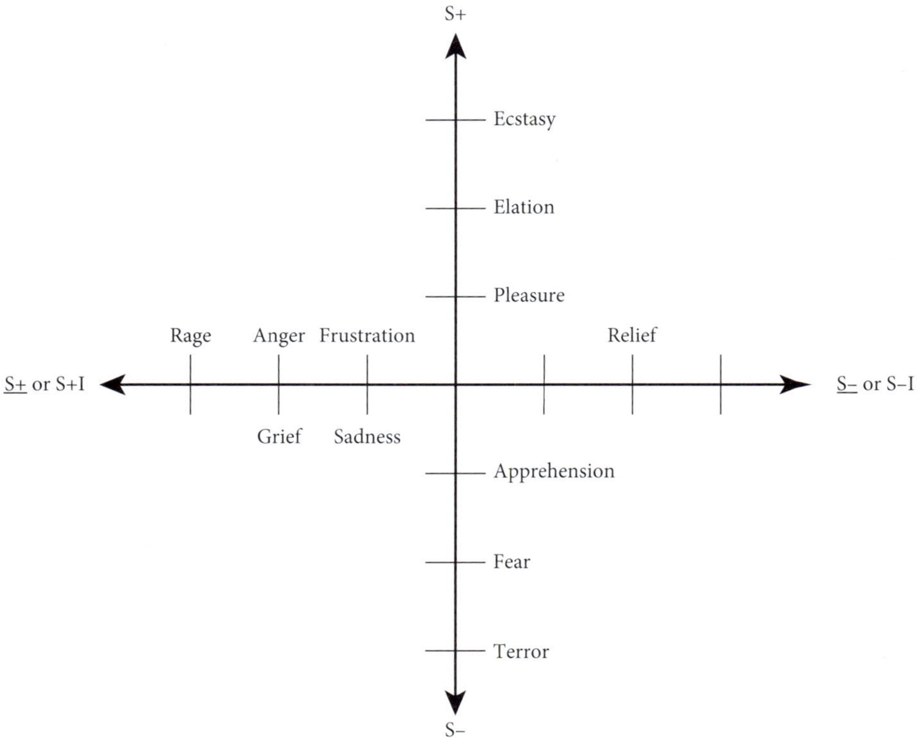

Figure 23.1 Some of the emotions associated with different reinforcement contingencies are indicated. Intensity increases away from the center of the diagram, on a continuous scale. The classification scheme created by the different reinforcement contingencies consists of (1) the presentation of a positive reinforcer (S+), (2) the presentation of a negative reinforcer (S–), (3) the omission of a positive reinforcer (S+) or the termination of a positive reinforcer (S+!), and (4) the omission of a negative reinforcer (S–) or the termination of a negative reinforcer (S-!).

The proposal that emotions can be usefully seen as states produced by instrumental reinforcing stimuli follows earlier work by Millenson (1967), Weiskrantz (1968), Gray (1975, 1987), and Rolls (1986a, 1986b, 1990, 1999, 2000, 2005). (Instrumental reinforcers are stimuli which, if their occurrence, termination, or omission is made contingent upon the making of a response (action), alter the probability of the future emission of that response.) Some stimuli are unlearned reinforcers (e.g. the taste of food if the animal is hungry, or pain), while others may become reinforced by learning because of their association with such primary reinforcers, thereby becoming "secondary reinforcers."

This foundation has been developed (see Rolls 1986a,1986b, 1990, 1999, 2000, 2005, 2014) to show how a very wide range of emotions can be accounted for, as a result of the operation of a number of factors, including the following:

(1) The *reinforcement contingency* (e.g. whether reward or punishment is given, or withheld) (see Figure 23.1).

(2) The *intensity* of the reinforcer (see Figure 23.1).

(3) Any environmental stimulus might have a *number of different reinforcement associations* (e.g. a stimulus might be associated both with the presentation of a reward and of a punisher, allowing states such as conflict and guilt to arise).

(4) Emotions elicited by stimuli associated with *different primary reinforcers* will be different. A list of some primary reinforcers to illustrate some of the different affective states is provided in *Emotion and Decision-Making Explained* (Rolls 2014) and in *Neuroculture* (Rolls 2012).

(5) Emotions elicited by *different secondary reinforcing stimuli* will be different from each other (even if the primary reinforcer is similar).

(6) The emotion elicited can depend on whether an *active or passive behavioral response* is possible. (For example, if an active behavioral response can occur to the omission of a positive reinforcer, then anger might be produced, but if only passive behavior is possible, then sadness, depression, or grief might occur.)

By combining these six factors, it is possible to account for a very wide range of emotions (for elaboration see Rolls 2014).

23.3 The functions of emotion

The most important functions can be summarized as follows (Rolls 1990, 1999, 2005, 2014):

(1) The *elicitation of autonomic responses* (e.g. a change in heart rate) and *endocrine responses* (e.g. the release of adrenaline). These prepare the body for action.

(2) *Flexibility of behavioral responses to reinforcing stimuli.* Emotional (and motivational) states allow a simple interface between sensory inputs and action systems. The essence of this idea is that goals for behavior are specified by reward and punishment evaluation. When an environmental stimulus has been decoded as a primary reward or punishment, or (after previous stimulus-reinforcer association learning) a secondary rewarding or punishing stimulus, then it becomes a goal for action. The person can then perform any action (instrumental response) to obtain the reward, or to avoid the punisher. Thus there is flexibility of action.

The emotional state intervenes between delivery of the stimulus and its decoding as rewarding or punishing, which produces the emotional state, and the learning and performance of the action, which may only be possible with some delay. In this sense, for goal-directed action an intervening state is required. For over-learned stimulus-response habit-based responses, no intervening state is necessary, and emotional states need not be present. This is one of the reasons why I propose that emotions are part of a brain/behavior system in which arbitrary actions must be learned to reinforcing stimuli in order to obtain goals. This is an important reason why I relate emotions to the evolution of instrumental actions to rewarding and punishing stimuli, as intervening states are needed in this process (Rolls 2014). The motivation that is part of the intervening state is to obtain the reward or avoid the punisher, and animals must be built to obtain certain rewards and avoid certain punishers. Further, and very importantly, for this shows why emotions have evolved, primary or unlearned rewards and punishers are specified by genes which effectively specify the goals for action. This is the solution which natural selection has found for how genes can influence behavior to promote their fitness (as measured by reproductive success), and for how the brain could interface with sensory systems to action systems, and it is an important part of Rolls' theory of emotion (1990, 1999, 2005, 2014).

Selecting between available rewards with their associated costs, and avoiding punishers with their associated costs, is a process that can take place both implicitly (unconsciously) and explicitly using a language system to enable long-term plans to be made (Rolls 2005, 2008). These many different brain systems, some involving implicit evaluation of rewards, and others explicit, verbal conscious evaluation of rewards and planned long-term goals, must all enter into the selector of behavior (see Figure 23.2). This selector is poorly understood, but it might include a process of competition between all the competing calls

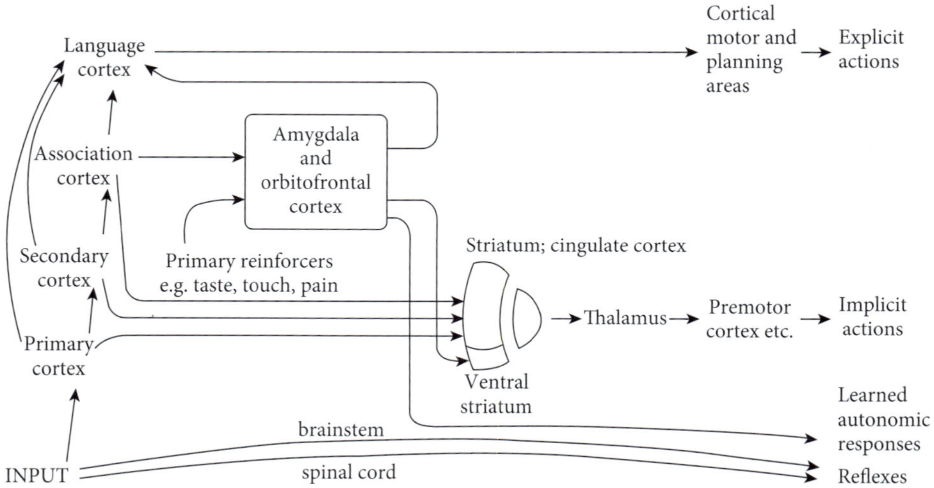

Figure 23.2 Dual routes to the initiation of action in response to rewarding and punishing stimuli. The inputs from different sensory systems to brain structures such as the orbitofrontal cortex and amygdala allow these brain structures to evaluate the reward- or punishment-related value of incoming stimuli, or of remembered stimuli. The different sensory inputs enable evaluations within the orbitofrontal cortex and amygdala based mainly on the primary (unlearned) reinforcement value for taste, touch, and olfactory stimuli, and on the secondary (learned) reinforcement value for visual and auditory stimuli. In the case of vision, the "association cortex" which outputs representations of objects to the amygdala and orbitofrontal cortex is the inferior temporal visual cortex. One route for the outputs from these evaluative brain structures is via projections directly to structures such as the basal ganglia (including the striatum and ventral striatum) to enable implicit, direct behavioral responses based on the reward- or punishment-related evaluation of the stimuli to be made. The second route is via the language systems of the brain, which allow explicit decisions involving multi-step syntactic planning to be implemented.

on output, and might involve the anterior cingulate cortex and basal ganglia in the brain (Rolls 2005, 2008, 2014) (see Figure 23.2).

23.4 Dual routes to action: gene-defined goals, and syntactic reasoning

The first route is via the brain systems that have been present in non-human primates such as monkeys, and to some extent in other mammals, for millions of years, and have built in the brain a system for defining these goals. Achieving these goals feels pleasant or unpleasant. The goals may be primary reinforcers, or stimuli associated with them by learning.

The second route in humans and also perhaps closely related animals involves a computation with many "if . . . then" statements to implement a plan to obtain a reward. In this case, the reward may actually be *deferred* as part of the plan, which might involve working first to obtain one reward, and only then to work for a second more highly valued reward, if this was thought to be overall an optimal strategy in terms of resource usage (e.g. time). In this case, syntax is required, because the many symbols (e.g. names of people) that are part of the plan must be correctly linked or bound. Such linking might be of the form: "if A does this, then B is likely to do this, and this will cause C to do this. . . ." The requirement of syntax for this type of planning implies that an output to language systems in the brain is required for this type of planning (see Figure 23.2). Thus the explicit language system in humans may allow working for deferred rewards by enabling use of a one-off, individual, plan appropriate for each situation.

The question then arises of how decisions are made in animals such as humans that have both the implicit, direct, reward-based instrumental action, and the explicit, rational, planning systems (see Figure 23.2) (Rolls 2008, 2014). One particular situation in which the first, implicit, system may be especially important is when the interests of the genes are being maintained. In contrast, when the implicit system continually makes errors, it would then be beneficial for the organism to switch from automatic, direct action based on obtaining what the orbitofrontal cortex system decodes as being the most positively reinforcing choice currently available, to the explicit conscious control system that can evaluate with its long-term planning algorithms what action should be performed next.

The second route to action allows, by reasoning, decisions to be taken that might not be in the interests of the genes, might be longer-term decisions, and might be in the interests of the individual. Thus we may speak of the choice as sometimes being between the "Selfish Genes" (Dawkins 1989) and the "Selfish Phenes" (Rolls 2011b, 2012, 2013a).

23.5 A theory of the neurobiological foundations of aesthetics and art

23.5.1 Introduction to and outline of the theory

Now that we have a fundamental, Darwinian approach to the value of people, objects, relationships, etc., I propose that this provides a fundamental neurobiological approach to understanding aesthetics and art. I propose that while the gene-specified rewards and punishers define many things that have aesthetic value, the value that we place on items is enhanced by the reasoning, rational system which enables what produces aesthetic value to become highly intellectualized, as in music.

I emphasize at the outset that this does not at all reduce aesthetics to a common denominator. Genetic variation is essential to evolution by natural selection, and this is one reason why we should expect different people to assign aesthetic value differently. But rational thought, which will lead in different directions in different people, partly because of noise caused by random neuronal firing times in the brain (Rolls and Deco 2010; Rolls 2014), and because of what they have learned from the environment, and because different brain areas will be emphasized in different people, will also be different between individuals, so that the rational system will also contribute to differences between individuals in what is considered aesthetic.

Indeed, although the theory presented here on the origin of aesthetics is a reductive explanation in that it treats the underlying bases and causes, it should not be seen to "reduce" aesthetics. Far from it. When we understand the underlying origins and bases of aesthetics, we see that the processes involved are elegant and beautiful, as part of a Darwinian theory. But the approach also provides important pointers about how to enhance aesthetics. For example, by understanding that verbal-level cognitive factors that can be produced by reasoning have a top-down modulatory influence on the first cortical area where value (reward) is made explicit in the representation, the orbitofrontal cortex (de Araujo et al. 2005; Grabenhorst et al. 2008; McCabe et al. 2008; Rolls 2013b, 2014), we can see ways in which we can enhance our aesthetic feelings. (For example, if love be the thing, then it can be heightened by explicitly choosing the musical treatment of it in Wagner's *Tristan and Isolde*.)

I should also emphasize that aesthetic value judgments will usually be influenced by a number of different value factors, so that while accounting for an aesthetic judgment by just one of the value factors I describe is and will often seem too simple, it does seem that aesthetic value judgments can be understood by combinations of some of the factors I describe.

I emphasize that rewards contribute to what makes stimuli or brain processing positively aesthetic, beautiful; and that the punishers contribute to what makes stimuli or processing in the brain aesthetically negative, lacking beauty, ugly, or distasteful. Both rewards and punishers are needed for the theory of aesthetics.

The overall theory of the origin of aesthetics I propose is that natural selection, whether operating by "survival or adaptation selection," or by sexual selection, operates by specifying goals for action, and these goals are aesthetically and subjectively attractive or beautiful (Rolls 2005, 2014), or the opposite, and provide what I argue here is the origin of many judgments of what is aesthetic.

23.5.2 "Survival" or "adaptation" selection (natural selection in a narrow sense)

"Natural selection" encompasses in its broad sense both "survival or adaptation selection," and sexual selection. Both are processes now understood to be driven by the selection of genes, and it is gene competition and replication into the next generation that is the driving force of biological evolution (Dawkins 1986, 1989). The distinction can be made that with "survival or adaptation selection," the genes are selected to make the individual stronger, healthier, and more likely to survive and reproduce, whereas sexual selection operates by sexual choice, selecting for genes that may or may not have survival value to the individual but enable that individual to be selected as a mate or to compete for a mate in intra-sexual selection and thus pass on the genes selected by intra-sexual or inter-sexual selection to the offspring. More generally, we might have other types of selection as further types of natural selection, including selection for good parental care, and kin selection.

Many of the reward and punishment systems described by Rolls (2014) deal with reward and punishment decoding that has evolved to enable genes to influence behavior in directions in a high-dimensional space of rewards and punishments that are adaptive for survival and health of the individual, and thus promote reproductive success or fitness of the genes that build such adaptive functionality. We can include kin-related altruistic behaviors because the behavior is adaptive in promoting the survival of kin, and thus promoting the likelihood that the kin (who contain one's genes, and are likely to share the genes for kin altruism) survive and reproduce. We can also include reciprocal altruism as an example of "survival or adaptation" selection. Tribalism can be treated similarly, for it probably has its origins in altruism. Resources and wealth are also understood at least in part as being selected by "survival" natural selection, in that resources and wealth may enable the individual to survive better. As we will see next, resources and wealth can also be attractive as a result of sexual selection.

23.5.3 Sexual selection

Darwin (1871) also recognized that evolution can occur by sexual selection, when what is being selected for is attractive to potential mates (*inter-sexual selection*, for example the peacock's "tail," and a slim, young-looking physique in females as a signal of fertility), or helps in competing with others of the same sex (*intra-sexual selection*, e.g. the deer's large antlers, and a strong male physique).

Overall, Darwinian natural or survival selection increases health, strength, and potentially resources, and survival of the individual, and thus ability to mate and reproduce, and to look handsome or beautiful. Inter-sexual sexual selection does not make the individual healthier, but does make the individual more attractive as a mate, as in female choice, an example of intersexual selection. Intrasexual sexual selection does not necessarily help survival of the individual, but does help in competition for a mate, for example in intimidation of one male by another (Darwin 1871; Kappeler and van Schaik 2004). The differences between "survival" and sexual selection are elaborated elsewhere (Rolls 2011a, 2012, 2014).

23.5.4 Beauty in men and women

Given this background in the processes that drive evolution to make certain stimuli and types of brain processing rewarding or punishing, in this section I examine how they contribute to what factors make men and women aesthetically beautiful, and that influence their depiction in art.

23.5.4.1 Female preferences: factors that make men attractive

Factors that across a range of species influence female selection of male mates include the following.

23.5.4.1.1 **Athleticism** This is the ability to compete well in mate selection (including being healthy and strong), as this will be useful for her genes when present in her male offspring. Consistently, women show a strong preference for tall, strong, athletic men (Buss and Schmitt 1993).

23.5.4.1.2 **Resources, power, and wealth** In species with shared parental investment (which include many birds and humans), having power and wealth may be attractive to the female, because they are indicators of resources that may be provided for her young. Consistently, women place a greater premium on income or financial prospects than men (Buss 1989, 2008).

23.5.4.1.3 **Status** Status correlates with the control of resources (e.g. alpha-male chimpanzees take precedence in feeding), and therefore acts as a good cue for women. Women should therefore find men of high status attractive (e.g. rock stars, politicians, and tribal rulers), and these men should be able to attract the most attractive partners. Consistent with this, cross-culturally, women regard high social status as more valuable than do men, and attractive women often marry men of high status (Buss 1989, 2008). Status may be attractive because of direct effects (e.g. as an indicator of resources for children), or because of indirect effects (because high status implies good genes for offspring).

23.5.4.1.4 **Age** Status and higher income are generally only achieved with age, and therefore women should generally find older men attractive. Cross-culturally, women prefer older men (3.42 years older on average, and marriage records from 27 countries show that the average age difference was 2.99 years) (Buss 1989).

23.5.4.1.5 **Ambition and industriousness** These may be good predictors of future occupational status and income and are attractive to women (Buss 1989).

23.5.4.1.6 **Testosterone-dependent features** These may also be attractive. These features include a strong (longer and broader) jaw, a broad chin, strong cheekbones, defined eyebrow ridges, a forward central face, and a lengthened lower face (secondary sexual characteristics that are a result of pubertal hormone levels).

23.5.4.1.7 **Symmetry** This may be attractive (in both males and females), in that it may reflect good development *in utero*, a non-harmful birth, adequate nutrition, and lack of disease and parasitic infections (Thornhill and Gangestad 1999).

23.5.4.1.8 **Dependability and faithfulness** These may be attractive, particularly where there is paternal investment in bringing up the young, as these characteristics may indicate stability of resources (Buss et al. 1989).

23.5.4.1.9 **Risk-taking by men** This may be attractive to women, perhaps because it is a form of competitive advertising: surviving the risk may be an honest indicator of high-quality genes (Barrett et al. 2002).

23.5.4.1.10 **Other** Characteristics that may not be adaptive in terms of the survival of the male, but that may be attractive because of inter-sexual sexual selection, are common in birds, perhaps less common in most mammals, though present in some primates (Kappeler and van Schaik 2004), and may be present in humans (see section 23.5.3).

23.5.4.1.11 **Odor** The preference by women for the odor of symmetrical men is correlated with the probability of fertility of women as influenced by their cycle (Gangestad and Simpson 2000). Another way in which odor can influence preference is by pheromones that are related to major histocompatibility complex (MHC) genes, which may provide a molecular mechanism for producing genetic diversity by influencing those who are considered attractive as mates (Rolls 2011a, 2012, 2014).

23.5.4.2 Male preferences: what makes women attractive and beautiful to men

Males are not always indiscriminate. When a male chooses to invest (e.g. to produce offspring), there are preferences for the partner with whom he will make the investment. Accurate evaluation of female quality (reproductive value) is therefore important, and males look out for cues to this, and find these cues attractive, beautiful, and rewarding. The factors that influence attractiveness include the following (Barrett et al. 2002; Rolls 2011a, 2012, 2014).

23.5.4.2.1 **Youth** As fertility and reproductive value in females is linked to age (reproductive value is higher when younger, and actual fertility in humans peaks in the 20s), males (unlike females) place a special premium on youth. It is not youth per se that men find attractive, but indicators of youth; for example, neotenous traits such as blonde hair and wide eyes. An example of this preference is that male college students preferred an age difference on average of 2.5 years younger (Buss 1989). Another indicator of youth might be a small body frame, and it is interesting that this might contribute to the small body frame of some women in this example of sexual dimorphism.

23.5.4.2.2 **Beautiful features** Features that are most commonly described as the most attractive tend to be those that are oestrogen-dependent; for example, full lips and cheeks, and short lower facial features (oestrogen caps the growth of certain facial bones).

Why do women apparently compete for men by paying attention to their own beauty and fashion? Perhaps the answer that males who are willing to make major investments of time and resources in raising the children of a partner are a somewhat limiting resource (as other factors may make it advantageous genetically for men not to invest all their resources in one partner), and because women are competing to obtain and maintain this scarce resource. Faithful men may be a limited resource because there are alternative strategies that may have a low cost, whereas women are essentially committed to a considerable investment in their offspring.

Given that men are a scare resource, and that women make such a major investment in their offspring that they must be sure of a man's commitment to invest before they commit in any way, we have a scientific basis for understanding why women are reserved and more cautious and shy in their interactions with men, which has been noticed to be prevalent in visual art, in which men look at women, but less vice versa (Berger 1972).

23.5.4.2.3 **Body fat** The face is not the only cue to a woman's reproductive capacity, her attractiveness, and beauty. Although the ideal body weight varies significantly with culture (in cultures with scarcity, obesity is attractive and relates to status, a trend evident in beautiful painting throughout its history), the ideal distribution of body fat seems to be a universal standard, as measured by the waist-to-hip ratio (which cancels out effects of actual body weight). Consistently, across cultures, men preferred an average ratio of 0.7 (small waist/bigger hips) when rating female figures (line drawings and photographic images) for attractiveness (Singh and Luis 1995). At a simpler level, a low waist-to-hip ratio is an indication that a woman is not already pregnant, and is thus a contributor to attractiveness and beauty.

23.5.4.2.4 **Fidelity** The desire for fidelity in females is most obviously related to concealed ovulation—see the following and *Emotion and Decision-Making Explained* (Rolls 2014)—and therefore the degree of paternity uncertainty that males may suffer.

23.5.4.2.5 **Attractiveness and the time of ovulation** Although ovulation in some primates and in humans is concealed, it would be at a premium for men to pick up cues to ovulation and find women highly desirable (and beautiful) at this time. Possible cues include an increased body temperature reflected in the warm glow of vascularized skin (Vandenberghe and Frost 1986), and pheromonal cues. Another possibly unconscious influence might be on the use of cosmetics and the types of clothes worn, which may be different close to the time of ovulation.

In humans, male investment in caring for the offspring means that male choice has a strong effect on intra-sexual selection in women. Female cosmetic use and designer clothing could be seen as weapons in this competition, and perhaps are reflected in extreme female self-grooming behavior such as cosmetic surgery, or pathological disorders such as anorexia, bulimia, and body dysmorphic disorder. By bombarding people with images of beautiful women, the media may heighten intra-sexual selection even further, pushing women's competitive mating mechanisms to a major scale.

23.5.5 Pair-bonding, love, and a beautiful partner

Attachment to a particular partner by pair bonding in a monogamous relationship, which in humans becomes manifest in love between pair-bonded parents, and which occurs in humans in relation to the advantage to the man of investing in his offspring, may have special mechanisms to facilitate it. One is oxytocin, a hormone released from the posterior pituitary, whose other actions include the milk let-down response, which is released during mating and which promotes attachment, making a partner attractive (Lee et al. 2009).

Are similar mechanisms at work in humans to promote pair bonding and love (and what is found to be aesthetically attractive, and to influence depictions in art)? There is as yet no definitive evidence for this, but in humans, oxytocin is released by intercourse, and especially at the time of orgasm, in both women and men (Meston and Frohlich 2000; Kruger et al. 2003).

23.5.6 Parental attachment: beautiful children

Many mammal females make strong attachments to their own offspring and this is also facilitated in many species by oxytocin. In humans, oxytocin is released during natural childbirth, and rapid placing of the baby to breast feed and release more oxytocin (Uvnas-Moberg 1998) might further facilitate maternal attachment to her baby. Prolactin, the female hormone that promotes milk production, may also influence maternal attachment and how beautiful a mother thinks her child is. It is certainly a major factor in humans that bonding can change quite suddenly at the time that a child is born, with women having a strong tendency to shift their interests markedly towards the baby as soon as it is born (probably in part due to hormonal influences), and this can result in relatively less attachment behavior to the husband.

The tendency to find babies beautiful is not of course restricted to parents of their own children. Part of the reason for this is that in the societies in which our genes evolved with relatively small groups, babies encountered might often be genetically related, and the tendency to find babies beautiful is probably a way to increase the success of selfish genes. One may still make these aesthetic judgments of babies in distant countries with no close genetic relationship, but this does not of course mean that such judgments do not have their evolutionary origin in kin-related advantageous behavior.

23.5.7 Synthesis on beauty in humans

We see that many factors are involved in making humans attractive, and beautiful. All may contribute to different extents, and differently in different individuals, and moreover we may not be conscious of some of the origins of our aesthetic judgments but may confabulate reasons for what we judge to be aesthetic.

When there is a biological foundation for art, for example when it is figurative, and especially when it is about human figures, there may be a basis for consensus about what is good art—art that stimulates our rational system and at the same time speaks to what we find beautiful due to our evolutionary history. However, if art becomes totally abstract, we lack the biological foundation for judging whether it is aesthetically beautiful, and judgments may be much more arbitrary and driven by short-term fashion. Some abstraction away from very realistic and figurative in art can of course have advantages because it allows the viewer to create in his or her own experience of a work of art by adding his or her own interpretation of it.

There is an important point here about the separation between art and the world. Objects of art can idealize beauty, and enhance it. An example is the emphasis on thin bodies, long limbs, and athletic poses found in some Art Deco sculpture, for example in the works of Lorenzl. Here what is beautiful can be made super-normal, one might say in the literal sense super-natural. Another example is in the emotion in the music of Wagner's *Tristan and Isolde*. We see that art can emphasize and thus idealize some of the properties of the real world, and lose other details that do not enhance, or distract from it. This abstraction of what we find beautiful due to evolution can be seen in some semi-figurative/semi-abstract art, as in some of the line drawings of humans by Matisse and Picasso. It is also found in the sculptures of human forms of Brancusi. What I argue is that if art becomes too abstract then it loses the aesthetic value that can be contributed by tapping into these evolutionary origins. Interesting cases are found in the sculptures of Barbara Hepworth and Henry Moore. In the case of Barbara Hepworth, I find it interesting that she often retains sufficient figurative allusions in her sculpture to tap into evolutionary origins, in that some of her sculptures do seem to have some relation to male and female forms and relations. Much of the sculpture of Henry Moore is clearly figurative, and where it becomes apparently very abstract it may lose what is gained by tapping into evolutionary origins, but may gain by association and interpretation in relation to his more figurative work. Where art becomes very abstract, as in some of the work of Mark Rothko, perhaps those especially interested are those who have expertise themselves in what is being achieved technically, such as the painting of colors by Rothko.

23.5.8 Sexual selection of mental ability, survival or adaptation selection of mental ability, and the origin of aesthetics

Miller (2000, 2001) has developed the hypothesis that courtship provides an opportunity for sexual selection to select non-sexual mental characteristics such as kindness, humor, the ability to tell stories, creativity, art, and even language. He postulates that these are "courtship tools, evolved to attract and entertain sexual partners."

Miller (2000, 2001) also suggests that art, language, and creativity can be explained by sexual selection, and that they are difficult to account for by survival selection. He suggests that art develops from courtship ornamentation, and uses bowerbirds as an evolutionary example. Male bowerbirds ornament their often enormous and structurally elaborate nests or bowers with mosses, ferns, shells, berries, and bark to attract female bowerbirds. The nests are used just to attract females, and after insemination the females go off and build their own cup-shaped nests, lay their eggs, and raise their offspring by themselves with no male support. In this sense, the bowers are useless ornamentations that do not have survival value. Darwin (1871) himself viewed human ornamentation and clothing as outcomes of sexual selection. Sexual selection for artistic ability does not mean of course that the art itself needs to be about sex. This example helps to show that sexual selection can lead to changes in what is valued and found attractive in areas that might be precursors to art in humans. In Miller's (2001) view, the fine arts are just the most recent and pretentious manifestations of a universal human instinct for visual self-ornamentation, which in turn is a manifestation of sexual selection's universal tendency to ornament individuals with visual advertisements of their fitness. Thus, the human capacity for visual artistry is viewed as a "fitness indicator," evolved like the peacock's tail and the bowerbird's bower for a courtship function. So although inherently useless, the bower or work of art is seen as attractive because it is difficult to produce, and might only be made by a brain that is very competent in general, and thus the bower or work of art may act as a fitness indicator.

A useful point (Miller 2001) is that although artworks are now commodified and distributed widely so that we may not necessarily know the artist producing the ornament, when we seek the evolutionary origins of art we should remember that any artwork our prehistoric ancestors would have been able to see would have probably been made by a living individual with whom they could have interacted socially or sexually. The artist was never far from his or her work, or else the work could not have functioned as the artist's extended phenotype.

Miller (2000) further suggests that language and creativity may be related to systems that can explore random new ideas and they also may be courtship devices in males to attract females. My view, elaborated here and elsewhere (Rolls 2008; Rolls and Deco 2010; Rolls 2012, 2014), is that language and creativity have functions that have survival value and thus are not just sexually selected.

Indeed, a criticism of the approach of Miller (2000) is that many of these characteristics (e.g. language, creative solutions, originality, problem solving) may have survival value, and are not purely or primarily sexually selected. For example, syntax and language have many uses in problem solving, planning ahead, and correcting multi-step plans that are likely to be very important to enable immediate rewards to be deferred, and longer-term goals to be achieved (Pinker and Bloom 1992; Rolls 2008, 2012, 2014). In relation to aesthetics, I argue that when syntax is used successfully to solve a difficult problem we feel aesthetic pleasure, and I argue that the generation of pleasure generated by the survival value of good ideas contributes to the appeal of those ideas, and that sexual selection of the ideas as mental ornaments is not the only process at work in aesthetics.

Moreover, the notion (Miller 2000, 2001) that art has to do with useless ornaments (useless in the sense that sexual selection is for characteristics that may not have "survival" value, but may be attractive because they are "indicators of fitness") does not have much to say about the utilitarian arts such as simplicity of design in architecture. Perhaps the structure of a piece of music can appeal and be pleasing because it taps into our syntactic system that finds that elegant and simple solutions to problem solving produce pleasure. Further, interest in social relations and knowledge about them is adaptive as it may help to understand who is doing what to whom, and more generally help to understand what can happen to people. A great deal of literary fiction addresses these issues and is not primarily ornamental and without inherent value. Thus, although Miller (2000, 2001) may well be right that there are aspects of art that may be primarily ornamental and useless and these are simply indicators of general mental fitness, though attractive to members of the opposite sex in courtship, I suggest that much art has its roots in goals that have been specified as pleasurable or unpleasurable because of their adaptive or survival value, whether as primary reinforcers, other stimuli associated by learning with these, or rewards of a more cognitive origin that accrue when difficult cognitive, syntactic problems are solved (Rolls 2011a, 2012, 2014).

Miller might predict that men should be specialized to have artistic creativity, to provide an ornament that women might find attractive because it is a fitness indicator. Evidence on this is difficult to evaluate, because there have been fewer opportunities available for women in the past, as argued for so beautifully by Virginia Woolf in *A Room of One's Own* (1929), and I come to no conclusions, but have the following thoughts. Whereas Virginia Woolf argues about circumstances, one can consider in addition the possibility that women's and men's brains have been subject to different selective pressure in evolution, and that this might contribute to differences in the ways in which they are creative. In terms of artists, composers of music, poets, writers of drama and non-fiction, there appears to be on average a preponderance of men relative to women. This is on average, and there are individual women who given the distribution around the average are undoubtedly highly creative in these areas, and have made enormous contributions. If this is the case (and it might take a long time into the future to know, given the imbalance of opportunity in the past), does this mean that sexual selection is the underlying process? I suggest that this would not necessarily be the case. Such a "sexual dimorphism" could occur by natural (adaptation) selection, not by sexual selection, in that women might have specialized for an environmental niche to emphasize child rearing, cultivation including food gathering and preparation, fashioning of clothing, and creating peaceful order among siblings and parents. On the other hand, men might have specialized for an environmental niche to emphasize spatial problem solving, useful for producing and using tools, building shelters, creating structures, etc., and navigational problem-solving useful for hunting, all of which would be good for survival. Interestingly, the same (narrow) natural selection pressure might have provided a survival advantage for men to have a stronger physique which is likely to be advantageous when manufacturing items useful for survival such as shelters. Thus interestingly, one of the predictions of sexual selection, sexual dimorphism, including human mental problem-solving as well as physique, could in this case have its origin at least partly in adaptation and survival.

There is however a possible exception to the generalization that at least in the past men have been more likely to be creative in "art" than women, and this is the area of literary fiction, where there are many women with high reputations as novelists (e.g. Jane Austen, George Eliot, Virginia Woolf). If women take more to this area of creative art, might this be because of the adaptive value of gossip to women, so knowing about who is doing what to whom, and having an interest and expertise in this, could be adaptive, perhaps helping a woman, and her children, to survive better (Dunbar 1996)? If this were the case, there might even be a prediction that women might be relatively better, on average, in areas of fiction, such as novels, where this interest and expertise in mind-reading and gossip, might be especially engaged. More generally, the evolutionary survival value approach might argue that women have adapted to relational, social, caring, problem solving, and the novel, particularly the novel of manners, is ideally suited to displaying this specialization. Indeed, the specialization for a caring role is consonant with Carol Gilligan's argument in *In a Different Voice* (1982) that women's sense of morality concerns itself with the activity of "care . . . responsibility, and relationships."

The overall point I make is that natural selection, sometimes operating by "survival or adaptation selection," and sometimes by sexual selection (and sometimes both), operates by specifying goals for action, and these goals are aesthetically and subjectively attractive or beautiful (Rolls 2012, 2014), or the opposite, and provide what I argue here is the origin of many judgments of what is aesthetic. Many examples of these rewards and punishers, many of which operate for "survival or adaptation selection," and many of which contribute to aesthetic experience and judgments, are described elsewhere (Rolls 2011a, 2012, 2014).

23.5.9 Fashion and memes

We have seen that sexual selection can provide runaway selective pressure for what is not something that is produced by "survival or adaptation" selection. In a sense, a fashion or useless ornament (which may indicate fitness) can be selected for genetically.

However, fashion is a strong characteristic of many human aesthetic judgments, and we may ask if there are further reasons for this that are not to do with genetic variation (which necessarily takes place over generations), but that instead operate over timescales of months to years. Such fashions (in, for example, clothing) may occur because they fit adaptations of the human mind, themselves the result of adaptive pressure in evolutionary history. For example, the human mind will be attracted towards new ideas (of clear adaptive value, for it is only by exploring new ideas that advantage may be gained partly as a result of finding a match with one's own genetically influenced capacities) (Rolls 2014). In this way, there may be runaway changes that do not necessarily make the individual better adapted to the environment. Of course, many factors, again frequently of evolutionary origin, influence fashion, including its cost (of which the brand is an indicator) which helps to make it attractive as it indicates wealth, plentiful resources, and status, and the elegance and simplicity of the idea, which as argued below, the human mind finds attractive because simplicity often is a good indicator of a correct and useful solution to a problem. It is argued that memes (Blackmore 1999), ideas that follow some of the rules of fashion, fit these properties of the human mind.

23.5.10 The elegance and beauty of ideas, and solving problems in the reasoning system

Solving difficult problems feels good, and we often speak about elegant (and beautiful) solutions. What is the origin of the pleasure we obtain from elegant ideas, what makes them aesthetically pleasing? It is suggested that solving problems should feel good to us, to make us keep trying, because being able to solve difficult problems that require syntactic operations may have survival value (Rolls 2014). But what is it that makes simple ideas and solutions (those with fewest premises, fewest steps to the solution, and fewest exceptions for a given level of complexity of a problem) particularly aesthetically pleasing, so much so that physicists may use this as a guide to their thinking? It is suggested that the human brain has become adapted to find simple solutions aesthetically pleasing because they are more likely to be correct (Rolls 2012, 2014), and this is exactly the thrust of parsimony and Occam's Razor—the principle or heuristic that entities and hypotheses should not be multiplied needlessly; the simplest of two competing and otherwise equally effective theories is to be preferred. The principle states that the explanation of any phenomenon should make as few assumptions as possible, eliminating those that make no difference in the observable predictions of the explanation or theory.

This finds expression in art: in, for example, the structure of a piece of music, in the solution of how to incorporate perspective into painting (which developed over hundreds of years and was helped by the camera obscura), and in the interest by Vitruvius and Leonardo in the proportions of the human body (tapping into our gene-based appreciation of that) to provide rules for proportions in architecture. Of course, focus on intellectual aspects of art can lead to art that we may find fascinating and revealing, if not conventionally physically beautiful, as in some of the work of Francis Bacon's. Factors such as cultural heritage and familiarity with the rules of a system can also make a style of architecture more appealing than something very unfamiliar. Some of the history of ecclesiastical architecture in England from the eleventh to the fifteenth century (from Norman through Early English and Decorated to Perpendicular styles) can also be seen as solutions to difficult architectural problems, of how to increase the light and feeling of space in a building, and how to create an impression of grand and daring height.

23.5.11 Cognition and aesthetics

Not only can operation of our reasoning, syntactic, explicit, system lead to pleasure and aesthetic value, as just described, but also this cognitive system can modulate activity in the emotional, implicit, gene-identified goal system. This cognitive modulation, working from the level of word descriptions, can have modulatory effects right down into the first cortical area, the orbitofrontal cortex, where affective value, including aesthetic value such as the beauty found in a face, is first made explicit in the representation (O'Doherty et al. 2003; de Araujo et al. 2005; Grabenhorst et al. 2008; McCabe et al. 2008; Rolls and Grabenhorst 2008; Grabenhorst and Rolls 2011; Rolls 2012, 2014). Indeed, cognition and attention can be used to enhance the emotional aspect of aesthetic experience, as described in section 23.5.1.

The human mind may create objects such as sculpture and painting in ways that depend to different extents on the explicit reasoning system and the more implicit emotional system. I know at least one sculptor who intentionally reduces cognitive processing by turning off attention to cognitive processing when creating works of art, and then follows this with an explicit, conscious, reasoning stage in which selections and further changes may be made, with the whole creation involving very many such cycles.

Because cognition can by top-down cortico-cortical back projections influence representations at lower levels, it is possible that training, including cognitive guidance, can help to make more separate the representations of the representations of stimuli and their reward value at early levels of cortical processing (Rolls and Treves 1998; Rolls and Deco 2002; Rolls 2008). This top-down effect may add to the bottom-up effects of self-organization in competitive networks that also through repeated training help representations of stimuli to be separated from and made more different to each other (Rolls and Treves 1998; Rolls and Deco 2002; Rolls 2008). These effects may be important in many aesthetic judgments that are affected by training, including the appreciation of fine art, architecture, and wine.

23.5.12 Wealth, power, resources, and reputation

As already described, wealth, power, resources, and status are attractive qualities, aesthetically attractive, because resources are likely to be beneficial to the survival of genes. Reputation is similar, in that guarding one's reputation can be important in reproductive success: trust is important in a mate, or in reciprocal altruism, and hormones such as oxytocin may contribute to trust (Lee et al. 2009). This provides some insight into the history of Western art, in which individual and family portraits frequently have as one of their aims the display of wealth, power, and resources. The clothes and background depicted in these paintings often show how wealthy and powerful their owners are. Commissioned portraits thus frequently emphasize beauty, status, wealth, and resources. Interestingly, because self-portraits are rarely commissioned they are less likely to emphasize these characteristics (Cumming 2009); they can reflect subjective knowledge of the person portrayed. An additional property that can add aesthetic value to a portrait is that an image of someone held dear is associated with the owner and the attraction of photographic images is an example of this. Religion and states of spiritual ecstasy often aimed at achieving everlasting happiness should also be recognized as drivers of art.

23.5.13 The beauty of scenery and places

Many topological features of landscapes may be aesthetically attractive because they tap into brain systems that evolved to provide signals of safety, food, etc. Open space may be attractive because potential predators can be seen; cover may be attractive as a place to hide (Appleton 1975); a verdant landscape may be attractive because it indicates abundant food; flowers may be attractive as predictors of fruit later in the season. The color blue is preferred by monkeys, and this may be because blue sky, seen from the canopy, is an indicator of a safe place away from predators on the ground (Humphrey 1971). A clear red/orange sunset may be attractive as a predictor of good weather, and of safety overnight without bad weather. These factors do not operate alone to produce beauty, but may contribute to aesthetic beauty which I argue is influenced by many of the factors described in this theory of the origin of aesthetics.

23.5.14 The beauty of music

Vocalization is used for emotional communication between humans, with an origin evident in other primates (Rolls et al. 2006). Examples include warning calls, war-like encouragement to action, and a soothing lullaby or song to an infant. It is suggested that this emotional communication channel is tapped into by music, and indeed consonant versus dissonant sounds differentially activate the orbitofrontal cortex (Blood et al. 1999; Blood and Zatorre 2001) involved in emotion (Rolls 2014). Of course, the reasoning system then provides its own input to the development, pleasure, and aesthetic value of music in ways described in sections 23.5.10 and 23.5.11.

What underlies the greater pleasure and aesthetic value that many people accord to consonant as opposed to dissonant music? I suggest that consonance is generally pleasant because it is associated with natural sounds, including vocal ones, with a single source that is naturally harmonic. A good example is a calm female voice. Dissonance may often occur when there are multiple unrelated sources, such as those that might be produced by a catastrophe such as an earthquake, or boulders grinding against each other (or strings on a violin that are not tuned to be harmonics of each other). Further, a human voice when angry, shouting, etc. (and therefore by evolutionary adaptation affectively unpleasant) might have non-linearities, in, for example, vocal cords due to over-exertion, and these may be harmonically much less pure than when the voice is calm and softer.

23.5.15 Beauty, pleasure, and pain

If a mildly unpleasant stimulus is added to a pleasant stimulus, sometimes the overall pleasantness of the stimulus, its attractive value, and perhaps its beauty, can be enhanced. A striking example is the sweet, floral scent of jasmine, which as it occurs naturally in *Jasminum grandiflorum* contains typically 2–3 per cent of indole, a pure chemical which on its own at the same concentration is usually rated as unpleasant. The mixture can, at least in some people (and this may depend on their olfactory sensitivity to the different components), increase the pleasantness of the jasmine compared to the same odor without the indole. Why might this occur? One investigation has shown that parts of the brain such as the medial orbitofrontal cortex that represent the pleasantness of odors (Anderson et al. 2003; Rolls et al. 2003; Grabenhorst and Rolls 2009) can respond even more strongly to jasmine when it contains the unpleasant component indole, compared to when it only contains individually pleasant components (Grabenhorst et al. 2007). Thus one brain mechanism that may underlie the enhancement effect is a principle that brain areas that represent the pleasantness of stimuli can do so in a way that is at least partly independent of unpleasant components, thereby emphasizing the pleasant component of a hedonically complex mixture.

A second factor that may contribute to the enhanced pleasantness of the mixture of jasmine and indole is that the indole may produce a contrast effect in the brain areas that represent the pleasant components of the mixture. An indication of this was found in increased activations in the medial orbitofrontal cortex (which represents the pleasantness of many stimuli) when the jasmine–indole mixture was being applied, compared to just the jasmine alone (Grabenhorst et al. 2007). To the extent that the representation of pleasantness may drive hedonic experience separately from unpleasantness representations (Grabenhorst et al. 2007), and this might be facilitated by paying attention selectively to the pleasantness of a stimulus versus its unpleasantness (Rolls et al. 2008), then a factor might be the increased activation of pleasantness representations if there is a component to the stimulus that is unpleasant, and can enhance the pleasantness representation by a contrast effect. Another example of enhancement of pleasant by unpleasant stimuli occurs when an odor become more pleasant if it is preceded by an unpleasant (compared to a pleasant) odor, an effect represented in the human orbitofrontal cortex (Grabenhorst and Rolls 2009).

A third factor is that the interaction between the pleasant (jasmine) and unpleasant (indole) components makes the complex hedonic mixture (jasmine + indole) capture attention (which in turn may enhance and prolong the activation of the brain by the complex hedonic mixture), and evidence for the capture of attentional mechanisms in the brain by the pleasant–unpleasant mixture has been found (Grabenhorst et al. 2011).

These principles may of course operate in most areas where pleasant and unpleasant stimuli combine. Examples might include the pleasure we get from demanding terrain (high cliffs, high mountains, high seas), from spicy food that activates capsaicin (hot somatosensory) as well as gustatory and olfactory receptors (Rolls 2007), from tragedy in literature, although empathy makes a large contribution here, and from difficult feats, such as those performed by Odysseus (Rolls and Deco 2010).

Let us consider the paradox of tragedy. For Aristotle, tragedy purged one of anxieties (Herwitz 2008). Somehow the depiction of tragedy in drama, which raises unpleasant emotions such as sadness at the tragedy, can also as drama afford pleasure. Hume's explanation was that the beauty of the language and the eloquence of the artist's depictive talents are the source of pleasure (Hume 1757; Yanal 1991). Is there more to say about this? Schadenfreude, gloating, pleasure at the distress of an envied person, is associated with activation of brain areas that respond to pleasant stimuli (Shamay-Tsoory et al. 2007; Takahashi et al. 2009), and I suggest is related to the evolutionary origin of competition between individuals, and winning the competition. It is probably not an important factor in the appreciation of tragedy in drama. What may be more important is first that we (and this is especially strong in women) always want to know what is happening to whom, and gossip has evolutionary value (Dunbar 1996) in that this can provide information about how others are likely to treat one, and more generally, about the things that can happen to people in life, and from which lessons can potentially be derived. Second, the ability to empathize with another's emotions, and indeed to be good at empathy and find it rewarding, may also be important in communities in order to facilitate kin or reciprocal altruism (Ridley 1996). Third, the ability to consider a theory of other people's minds is adaptive in facilitating prediction of their behavior (Frith and Singer 2008), and fascination with this subject should again in an evolutionary context be rewarding and be associated with pleasure. It is suggested that these three factors are at least important contributors to the pleasure that people find in tragedy in drama. The same factors, I suggest, are also important contributors to the popularity of novels. In the cases of both drama and novels, we know that they are fiction, or at least that the events described are not happening to the spectator or reader, and this helps to make them particularly rewarding ways to learn about social relations and life events because there is no risk to the spectator or reader.

Knowing that the work of art (music, literature, painting, sculpture) is a fiction may also account for why the "aesthetic" emotions are not as long-lasting, and are not as motivating, as the goals in real life.

23.5.16 Absolute value in aesthetics and art

The approach described here proposes that what we find aesthetic has its roots and origins in two main processes, gene-specified goals, rewards, and punishers, and the value that is felt when our reasoning system produces, and understands, elegant and simple solutions to problems (Rolls 2012). What implications does this have for absolute aesthetic value? The implication is that while there is no absolute aesthetic value that is independent of these processes, we will nevertheless find considerable agreement between individuals, especially when the aesthetic value being judged has its roots in the two main processes described. However, as illustrated here, there will be variation for good evolutionary reasons between what different individuals find of value, and there will be variation in individuals' thought processes caused by their cultural heritage, and by noise in the brain which is an important component to creativity (Rolls and Deco 2010; Rolls 2012). For these reasons, and because aesthetic value is multifactorial (i.e. is influenced by multiple conscious and unconscious processes), we must expect variation in aesthetic value across people, time, and place, with no absolute aesthetic value.

23.6 Is what is attractive also beautiful and aesthetic?

I wish to counter a possible objection to the theory of the origin of aesthetics described here. This is that some of the goals specified by our genes, such as the reward value and pleasantness of a high-energy high-fat diet, might seem rather unsavory, and not quite aesthetic. The point I make is that it is not just the gene-specified rewards and punishers that cause stimuli to have aesthetic value. My proposal is that the reasoning (rational) system also contributes to aesthetic value and in a number of ways. It makes rather longer-term goals attractive. It introduces the further goal that innovation is attractive, as this is likely to help solve difficult problems and move the person into a new part of state space where the person may have an advantage. It introduces the use of syntactic relational structure to provide another way of computation. Problem solving using this reasoning system is encouraged by simple elegant solutions which are rewarding and which have aesthetic value, as described above. These factors would help the sophisticated structure in a Bach partita and fugue to contribute to what we judge as aesthetically pleasing, because such music taps not only into our emotional systems but also into the systems that provide intellectual pleasure because difficult and complex structural problems are posed, and solutions to these difficult structural problems are provided which provide aesthetic pleasure.

In this sense, aesthetic value may have its roots partly in gene-specified rewards (and punishers), but also in the pleasure that the rational system can provide when it is posed, and finds, elegant and simple solutions (which by parsimony are likely to be correct) to complex problems. For this reason, emotions may not be perfectly aligned with aesthetic value. Although both have their origin in gene-specified rewards, emotions may be produced by any one of a large number of reinforcers, whereas aesthetic value usually includes contributions of the reasoning (rational) system, as just described.

Art as a whole is a larger issue than aesthetics and beauty. The content of "Art" might, I suggest, be seen as the result of multiple separate trajectories through a state space in which each trajectory is guided by the origins of aesthetics (products of adaptations for survival and of sexual selection for useless sometimes handicapping ornament, and rational thought to develop structure in which an elegant and simple solution is pleasing), and depends on each previous trajectory, the history of art in each culture. Each trajectory is not itself deterministic because it is influenced by noise (Rolls and Deco 2010; Rolls 2012) (as is Darwinian evolution). Thus, future trajectories cannot be predicted. In each trajectory, though, a number of factors dominate, including newness (which is biologically attractive as argued earlier), a quality of wildness (as in Beethoven's late string quartets), as well as what we rationally find aesthetic (as described earlier), and what survival and sexual selection have also provided in us as some of the origins of aesthetics.

23.7 Comparison with other theories of aesthetics

Much research I have performed shows that there is a perceptual representation of objects formed in cortical areas that is kept separate from the representation of the affective value of objects, which happens further on in processing, in brain regions such as the orbitofrontal cortex (and in an area to which it projects, the anterior cingulate cortex) and the amygdala (see Figure 23.2) (Rolls 2008, 2014).

There are good functional and adaptive reasons for separate representations of objects and of their affective value. We can still see and recognize objects (including tastes, smell, the sight of objects, etc.) even when they are not rewarding to us; for example, if they are foods and we are not hungry. We are not blind to objects when they are not rewarding or punishing. Moreover, it is adaptive to be able to learn about where we have seen objects, people, etc., even if they are not currently rewarding, so that we can find them later when they are needed. Thus there is strong neuroscientific evidence, and sound biological arguments, for separate representations of perceptual objects and of their affective value. Baumgarten (1750) expressed this thought in his *Aesthetica* when he suggested that sensation, the use of the five senses, is separate from sensibility, which is something more, a "kind of intuition/cognition/formulation of the thing which judges it beautiful," and in doing so gave rise to the term aesthetics (Herwitz 2008). Before this, abstract questions such as "what is beauty," "what is art," had not been discussed generically in philosophy (although Aristotle had discussed the social role of drama as purging us of ever present anxiety, and Plato had dismissed poetry as obfuscating by sending the mind reeling into hypnotic trances instead of focusing on rational deductions and argument) (Herwitz 2008).

David Hume (1777) takes a broad view of taste (which engages beauty), and argues for five standards of ("delicacy of") taste that might be shown by experts: "Strong sense, united to delicate sentiment, improved by practice, perfected by comparison, and cleared of all prejudice, can alone entitle critics to this valuable character; and the joint verdict of such, wherever they are to be found, is the true standard of taste and beauty." Hume's difficulty is that he believes taste is objective, because delicacy is the probing instrument for truth; but instead, taste is a circular and constructivist enterprise (Herwitz 2008). My approach has in contrast a clear foundation for aesthetics in brain function and its evolutionary design, with clear views about how it includes rational thought which provides its own pleasures, and about how art can idealize beyond the normal world by building on these foundations and origins.

Immanuel Kant (1724–1804) distinguishes between liking something and finding it beautiful. According to Kant, when I find a painting beautiful this is not conditioned by any causal relation between its properties and my pleasures. For Kant, a judgment of beauty carries the weight of "ought," that others should judge it beautiful too, so his theory has moral implications. His judgment of beauty is a "disinterested" judgment, one that is not peculiar to him. He wants the beauty to be in the person, but not causally dependent on the properties of the object in the world such as the pleasure it produces (Kant 1790). He thus appears to be committed to an objective and universal view of art, with exactly how this view is arrived at not at all clear. The biological and neuroscientific view that I propose indicates that, in contrast, art is not universal or objective, but instead can be judged to be good art if it taps into many of the human rational and gene-based reward systems (see further section 23.6), with therefore individual differences expected, as described in section 23.5.16.

Darwin (1871) recognized that evolution can occur by sexual selection, when what is being selected for has no inherent adaptive or survival value, but is attractive to potential mates (inter-sexual selection), or helps in competing with others of the same sex (intra-sexual selection). His view was that natural beauty arose through competition to attract a sexual partner. His process of sexual selection through mate choice—the struggle to reproduce, not to survive—drove the evolution of visual ornamentation and artistry, from flowers through bird plumage to human self-adornment. Many have developed or ascribed to this idea (including Thorstein Veblen (1899), Ernst Gombrich (1977), Amotz Zahavi (1978), and Denis Dutton (2009)), and Miller (2000, 2001) has proposed a sexual selection theory of art. The implication of Miller's theory is that art has to do with what are frequently useless ornaments (useless in the sense that sexual selection is for characteristics that do not have "survival" value, but are usually just attractive because they are handicaps and are indicators of fitness). I agree that useless, handicapping ornamentation produced by sexual selection does play a role in aesthetics. However, the sexual selection theory does not therefore have much to say about the utilitarian arts such as simple design in architecture. Perhaps the structure of a piece of music can appeal, and be pleasing, because it taps into our syntactic system that finds that adaptive, survival value-related, elegant and simple solutions to problem-solving produce pleasure. As I argued above, interest in social relations and knowledge about them is adaptive and this has survival value as it may help to understand who is doing what to whom, and more generally to understand what can happen to people, and much fictional literature addresses these issues, and is not purely ornamental and without inherent value. Thus although Miller may well be right that there are aspects of art that may be primarily ornamental and useless, though attractive to members of the opposite sex in courtship, I suggest that a great deal of art has its roots in goals that have been specified as pleasurable or unpleasurable because of their "survival or adaptive" value, whether as primary reinforcers, other stimuli associated by learning with these, or rewards of a more cognitive origin that accrue when difficult cognitive, syntactic, problems are solved. I also emphasize that some of the characteristics emphasized by sexual selection may have some inherent survival value (sections 23.5.3.1 and 23.5.3.2).

My theory (the Rolls' theory) of the foundations of aesthetics and art thus specifies the roles of Darwinian "survival or adaptive" selection and sexual selection in aesthetics (Rolls 2011a, 2012). It is thus thoroughly Darwinian. A key idea is that many of the things that provide pleasure, or its opposite, do so because they are, or are related to, the gene-specified goals for action. Motivational states arise when trying to obtain these goals, and emotional or affective states when these goals are obtained, or are not obtained. These states are associated with affect and value, and with subjective pleasantness or unpleasantness, because it is an efficient way in which genes can influence their own (reproductive) success ("fitness"), and much more efficient and effective as a Darwinian process than prescribing that the animal should make particular responses to particular stimuli (Rolls 2014). The theory is that aesthetic value has its roots partly in these gene-specified rewards that have survival or adaptive value, but also in the pleasure that the rational system can provide when it is posed, and finds, elegant and simple solutions (which by parsimony are likely to be correct and hence adaptive) to complex problems, and to some extent in sexual selection. What makes good art can be influenced by many factors, as described here, so it is complex and multifaceted, and these factors must include whether the effect of the art is for good or for ill. It also follows that attempts in aesthetics to produce a systematic account based on consistent explicit beliefs will not succeed because many factors that are not necessarily consistent with each other are involved in aesthetic values, and because some of these factors operate at least partly unconsciously and non-propositionally/non-syntactically; that is, using computational systems in the brain that do not involve reasoning (Rolls 2012, 2014).

Acknowledgments

Discussions with M.S. Dawkins, B.K. Scott, and P. Wheatley, were very helpful.

References

Anderson, A.K., Christoff, K., Stappen, I., et al. (2003). Dissociated neural representations of intensity and valence in human olfaction. *Nature Neuroscience* **6**, 196–202.

Appleton, J. (1975). *The Experience of Landscape*. New York, NY: Wiley.

Barrett, L., Dunbar, R., and Lycett, J. (2002). *Human Evolutionary Psychology*. Basingstoke: Palgrave.

Baumgarten, A.G. (1750). *Aesthetica*.

Berger, J. (1972). *Ways of Seeing*. Harmondsworth: Penguin.

Blackmore, S.J. (1999). *The Meme Machine*. Oxford: Oxford University Press.

Blood, A.J. and Zatorre, R.J. (2001). Intensely pleasurable responses to music correlate with activity in brain regions implicated in reward and emotion. *Proceedings of the National Academy of Sciences of the USA* **98**, 11818–23.

Blood, A.J., Zatorre, R.J., Bermudez, P., et al. (1999). Emotional responses to pleasant and unpleasant music correlate with activity in paralimbic brain regions. *Nature Neuroscience* **2**, 382–7.

Buss, D.M. (1989). Sex differences in human mate preferences: evolutionary hypotheses tested in 37 cultures. *Behavioural and Brain Sciences* **12**, 1–14.

Buss, D.M. (2008). *Evolutionary Psychology: The New Science of the Mind* (3rd edn). Boston, MA: Allyn and Bacon.

Buss, D.M. and Schmitt, D.P. (1993). Sexual strategies theory: an evolutionary perspective on human mating. *Psychological Review* **100**, 204–32.

Cumming, L. (2009). *A Face to the World: on Self-portraits*. London: Harper.

Darwin, C. (1871). *The Descent of Man, and Selection in Relation to Sex*. London: John Murray (reprinted in 1981 by Princeton University Press).

Dawkins, R. (1986). *The Blind Watchmaker*. Harlow: Longman.

Dawkins, R. (1989). *The Selfish Gene* (2nd edn). Oxford: Oxford University Press.

de Araujo, I.E.T., Rolls, E.T., Velazco, M.I., et al. (2005). Cognitive modulation of olfactory processing. *Neuron* **46**, 671–9.

Dunbar, R. (1996). *Grooming, Gossip, and the Evolution of Language*. London: Faber and Faber.

Dutton, D. (2009). *The Art Instinct*. Oxford: Oxford University Press.

Frith, C.D. and Singer, T. (2008). The role of social cognition in decision making. *Philosophical Transactions of the Royal Society of London* **363**, 3875–86.

Gangestad, S.W. and Simpson, J.A. (2000). The evolution of human mating: trade-offs and strategic pluralism. *The Behavioral and Brain Sciences* **23**, 573–87; discussion pp. 587–644.

Gilligan, C. (1982). *In a Different Voice*. Cambridge, MA: Harvard University Press.

Gombrich, E. (1977). *Art and Illusion: a study in the psychology of pictorial representation* (5th edn). London: Phaidon Press.

Grabenhorst, F. and Rolls, E.T. (2009). Different representations of relative and absolute value in the human brain. *NeuroImage* **48**, 258–68.

Grabenhorst, F. and Rolls, E.T. (2011). Value, pleasure, and choice in the ventral prefrontal cortex. *Trends in Cognitive Sciences* **15**, 56–67.

Grabenhorst, F., Rolls, E.T., and Bilderbeck, A. (2008). How cognition modulates affective responses to taste and flavor: top down influences on the orbitofrontal and pregenual cingulate cortices. *Cerebral Cortex* **18**, 1549–59.

Grabenhorst, F., Rolls, E.T., and Margot, C. (2011). A hedonically complex odor mixture captures the brain's attention. *NeuroImage* **55**, 832–43.

Grabenhorst, F., Rolls, E.T., Margot, C., et al. (2007). How pleasant and unpleasant stimuli combine in different brain regions: odor mixtures. *Journal of Neuroscience* **27**, 13532–40.

Gray, J.A. (1975). *Elements of a Two-Process Theory of Learning*. London: Academic Press.

Gray, J.A. (1987). *The Psychology of Fear and Stress* (2nd edn). Cambridge: Cambridge University Press.

Herwitz, D. (2008). *Aesthetics*. London: Continuum.

Hume, D. (1757). *Four Dissertations: Of Tragedy*.

Hume, D. (1777). *Selected Essays: Of the Standard of Taste*.

Humphrey, N. (1971). Colour and brightness preferences in monkeys. *Nature* **229**, 615–7.

Kant, I. (1790). Critique of Judgement.

Kappeler, P.M. and van Schaik, C.P. (2004). Sexual selection in primates: review and selective preview. In P.M. Kappeler and C.P. van Schaik (eds), *Sexual Selection in Primates*. Cambridge: Cambridge University Press, pp. 3–23.

Kruger, T.H.C., Haake, P., Chereath, D., et al. (2003). Specificity of the neuroendocrine response to orgasm during sexual arousal in men. *Journal of Endocrinology* **177**, 57–64.

Lee, H.J., Macbeth, A.H., Pagani, J.H., et al. (2009). Oxytocin: the great facilitator of life. *Progress in Neurobiology* **88**, 127–51.

McCabe, C., Rolls, E.T., Bilderbeck, A., et al. (2008). Cognitive influences on the affective representation of touch and the sight of touch in the human brain. *Social, Cognitive and Affective Neuroscience* **3**, 97–108.

Meston, C.M. and Frohlich, P.F. (2000). The neurobiology of sexual function. *Archives of General Psychiatry* **57**, 1012–30.

Millenson, J.R. (1967). *Principles of Behavioral Analysis*. New York, NY: Macmillan.

Miller, G.F. (2000). *The Mating Mind*. London: Heinemann.

Miller, G.F. (2001). Aesthetic fitness: How sexual selection shaped artistic virtuosity as a fitness indicator and aesthetic preferences as mate choice criteria. *Bulletin of Psychology and the Arts* **2**, 20–5.

O'Doherty, J., Winston, J., Critchley, H., et al. (2003). Beauty in a smile: the role of medial orbitofrontal cortex in facial attractiveness. *Neuropsychologia* **41**, 147–55.

Pinker, S. and Bloom, P. (1992). Natural language and natural selection. In J.H. Barkow, L. Cosmides, and J. Tooby (eds), *The Adapted Mind*. New York, NY: Oxford University Press, pp. 451–93.

Ridley, M. (1996). *The Origins of Virtue*. London: Viking.

Rolls, E.T. (1986a). Neural systems involved in emotion in primates. In R. Plutchik and H. Kellerman (eds), *Emotion: Theory, Research, and Experience. Vol. 3. Biological Foundations of Emotion*. New York, NY: Academic Press, pp. 125–43.

Rolls, E.T. (1986b). A theory of emotion, and its application to understanding the neural basis of emotion. In Y. Oomura (ed.), *Emotions. Neural and Chemical Control*. Basel: Karger, pp. 325–44.

Rolls, E.T. (1990). A theory of emotion, and its application to understanding the neural basis of emotion. *Cognition and Emotion* **4**, 161–90.

Rolls, E.T. (1999). *The Brain and Emotion*. Oxford: Oxford University Press.

Rolls, E.T. (2000). Précis of The Brain and Emotion. *Behavioral and Brain Science* **23**, 177–233.

Rolls, E.T. (2005). *Emotion Explained*. Oxford: Oxford University Press.

Rolls, E.T. (2007). Sensory processing in the brain related to the control of food intake. *Proceedings of the Nutrition Society* **66**, 96–112.

Rolls, E.T. (2008). *Memory, Attention, and Decision-Making: A Unifying Computational Neuroscience Approach*. Oxford: Oxford University Press.

Rolls, E.T. (2011a). A neurobiological basis for affective feelings and aesthetics. In E. Schellekens and P. Goldie (eds), *The Aesthetic Mind: Philosophy and Psychology*. Oxford: Oxford University Press, pp. 116–165.

Rolls, E.T. (2011b). Consciousness, decision-making, and neural computation. In V. Cutsuridis, A. Hussain, and J.G. Taylor (eds), *Perception-Action Cycle: Models, Algorithms and Systems*. Berlin: Springer, pp. 287–333.

Rolls, E.T. (2012). *Neuroculture. On the Implications of Brain Science*. Oxford: Oxford University Press.

Rolls, E.T. (2013a). What are emotional states, and why do we have them? *Emotion Review* **5**, 241–247.

Rolls, E.T. (2013b). A biased activation theory of the cognitive and attentional modulation of emotion. *Frontiers in Human Neuroscience* **7**, 74. doi: 10.3389/fnhum.2013.00074.

Rolls, E.T. (2014). *Emotion and Decision-Making Explained*. Oxford: Oxford University Press.

Rolls, E.T., Critchley, H.D., Browning, A.S., et al. (2006). Face-selective and auditory neurons in the primate orbitofrontal cortex. *Experimental Brain Research* **170**, 74–87.

Rolls, E.T. and Deco, G. (2002). *Computational Neuroscience of Vision*. Oxford: Oxford University Press.

Rolls, E.T. and Deco, G. (2010). *The Noisy Brain: Stochastic Dynamics as a Principle of Brain Function*. Oxford: Oxford University Press.

Rolls, E.T. and Grabenhorst, F. (2008). The orbitofrontal cortex and beyond: from affect to decision-making. *Progress in Neurobiology* **86**, 216–44.

Rolls, E.T., Grabenhorst, F., Margot, C., et al. (2008). Selective attention to affective value alters how the brain processes olfactory stimuli. *Journal of Cognitive Neuroscience* **20**, 1815–26.

Rolls, E.T., Kringelbach, M.L., and de Araujo, I.E.T. (2003). Different representations of pleasant and unpleasant odors in the human brain. *European Journal of Neuroscience* **18**, 695–703.

Rolls, E.T. and Treves, A. (1998). *Neural Networks and Brain Function*. Oxford: Oxford University Press.

Shamay-Tsoory, S.G., Tibi-Elhanany, Y., and Aharon-Peretz, J. (2007). The green-eyed monster and malicious joy: the neuroanatomical bases of envy and gloating (schadenfreude). *Brain* **130**, 1663–78.

Singh, D. and Luis, S. (1995). Ethnic and gender consensus for the effect of waist-to-hip ratio on judgment of women's attractiveness. *Human Nature—an Interdisciplinary Biosocial Perspective* **6**, 51–65.

Takahashi, H., Kato, M., Matsuura, M., et al. (2009). When your gain is my pain and your pain is my gain: neural correlates of envy and schadenfreude. *Science* **323**, 937–9.

Thornhill, R. and Gangestad, S.W. (1999). Facial attractiveness. *Trends in Cognitive Sciences* **3**, 452–60.

Uvnas-Moberg, K. (1998). Oxytocin may mediate the benefits of positive social interaction and emotions. *Psychoneuroendocrinology* **23**, 819–35.

Vandenberghe, P.L. and Frost, P. (1986). Skin Color Preference, Sexual Dimorphism and Sexual Selection—a Case of Gene Culture Coevolution. *Ethnic and Racial Studies* **9**, 87–113.

Veblen, T. (1899). *The Theory of the Leisure Class*. New York, NY: Macmillan.

Weiskrantz, L. (1968). Emotion. In L. Weiskrantz (ed.), *Analysis of Behavioural Change*. London: Harper and Row, pp. 50–90.

Woolf, V. (1929). *A Room of One's Own*. London: Hogarth Press.

Yanal, R.J. (1991). Hume and others on the paradox of tragedy. *Journal of Aesthetics and Art Criticism* **49**, 75–6.

Zahavi, A. (1978). Decorative patterns and the evolution of art. *New Scientist* **19**, 182–4.

Chapter 24

Aesthetic evaluation of art: A formal approach

Alexander J. Huston and Joseph P. Huston

24.1 Introduction

Every compound stimulus that impinges upon the organism can be perceived in multiple ways. For example, a visual scene will be unlikely to have an invariant impact on a population of viewers, as each will "perceive," "sense," "associate," "respond to" the display in a different way, depending on the history of experience, on momentary state variables, and probably genetic determinants. Even the repetitive presentation of the same display to one observer will be likely to have different effects, each presentation being influenced by the effects of prior presentations (as we will discuss below, perceptual systems, attention mechanisms, and cognitive processes habituate or adapt to repeated input and, thus, change their response to the input). An important set of determinants of variation to visual (or auditory, tactile, olfactory, taste) input between viewers and within viewers themselves is the notion of multiple attributes of complex stimuli. For example, any photograph or painting can at any given time be viewed on the basis of many different parameters depending on the momentary state (e.g. mood, expectation) and history (education, cultural background, experience) of the viewer. This holds for the viewing of any complex stimulus, including any work of art. A work of art can be considered as a compound of attributes or qualities on the basis of which it will be perceived as well as evaluated. These attributes can be relatively simple variables such as color, size, and grain. They can be emotionally charged variables designated by terms such as "novelty," "beauty," "spiritual," "sublime," "monumental," or formal aspects of a work that can be defined by basic compositional terms such as form, space, balance, and harmony. They can also be information with associative value based on representational and figurative content loaded with "meaning," as in expressionistic and representational art. Then there are complex qualities related to experimentation, playfulness, innovation, and conceptual art (see section 24.4 below). The number of possible attributes of a compound stimulus is endless and a list of the attributes of a particular work can be many or few, depending on the viewer's state and experience. A sophisticated viewer will be able to perceive as well as verbalize more attributes than a novice.

The viewer's evaluation of a work of art is related to the experienced attributes of the work. Here we can already ask whether or not these functional attributes must be "consciously perceived" to affect the evaluative process. Since we have no notion of an answer to this, we will simply assume that any experience of an object—be it "conscious," with the viewer being aware of it, or "unconscious"—has a corresponding representation in the brain which can lead to an action (i.e. a response to the experience) and thereby influence the evaluative process.

We proceed on the basis of the postulate that the multiple attributes of a work determine the viewer's evaluative response to the presentation. It follows that if one can define the functional attributes of a work and their relative importance and integrity, we should be able to predict how the work is evaluated and potentially have a measure of its worth for the particular viewer. Accordingly, we will here present an attempt to formalize the relationship between the multiple attributes of an image in determining the viewer's overall evaluation of the work.

Our approach to modeling the aesthetic judgment can be seen in the context of earlier attempts at computational aesthetics as, for example, expressed by the mathematician G. D. Birkhoff's formula or "aesthetic measure" which was based on the assumption that aesthetic objects derive their quality from their numerical or geometrical characteristics; that is, that the pleasure derived from an image equals the ratio between the order (0) and the complexity (C) in the object: M = O/C (Birkhoff 1933). Most of the many attempts at formalizing aesthetic experience dealt with the analysis of individual attributes of a work or subsets of such, as for example the debate on the role of the *golden ratio* (phi φ) as a source of architectural aesthetics, whereby two quantities are said to be in the golden ratio if the ratio of the sum of the quantities to the larger quantity is equal to the ratio of the larger quantity to the smaller one:

$$\frac{a+b}{a}=\frac{a}{b}\equiv\varphi \text{ with the following solution: } \varphi=\frac{1+\sqrt{5}}{2}=1.6180339887$$

Baumgarten (1739) already declared the laws of beauty, like those of nature, to be amenable to systematic, empirical investigations via a new science, which he named *aesthetic* (the judgment or estimation of taste or of the beautiful), the object of which was to assess perception whereby particular representations are combined into a whole; that is, to translate the unity of perception into measurable variables. Subsequent formal mathematical approaches (Lappin et al. 1995; Wenger and Townsend 2001) can be said to derive from the Gestalt psychologists' attempts to account for particular aspects of our perception of motion, faces, texture, etc., on the basis of our ability to generate and recognize whole forms from simple elements, with the emphasis on *the whole being greater than the sum of the parts*.

Unlike the many attempts to model particular attributes of a work and their role in aesthetics, we will present a formal theory of their individual and compound roles in the resultant total evaluation (appreciation) of a display or work. We believe that such a formal approach to dissecting the elements of displays and objects can serve as a *conceptual guide* to our active viewing and understanding of the complexities of evaluation of works of art.

24.2 Qualia (Qs) and aesthetic evaluation

From here on we will arbitrarily focus on visual presentations and designate their attributes as "qualia." The term "*qualia*" (singular "quale") is a loose designation for a concept that encompasses how we experience the world (e.g. Lewis 1929; Chalmers 1995; Horgan 1984). Synonyms for qualia might include features, attributes, percepts, qualities, sensation of, or subjective experience of an image. There is no consensus in the use of this term, which has engaged many philosophers in their attempts to account for issues of phenomenology related to perception, subjective experience, and the mind–brain relationship. We use the term "quale" to refer to a strictly subjective experience of the observer in response to a stimulus. A quale is a virtual property—solely an inner experience that has no physical reality in the sense of being a property of the stimulus, display or object itself in the physical world that is being viewed (e.g. Schrödinger 2001; Perkins 1983).

The major variables we postulate as determinants of the subjective evaluation (E) of an image or any work of art are its attributes or qualia (Q). Each quale has a value (g), which is a measure of its *integrity* or *goodness*, as well as a value *weight* (w), which is its relative *importance* for the particular work. The weight and importance of the quale determine its level of *dominance* or primacy over other qualia and also the nature of the *interactions* between the constituent qualia in their influence on the momentary subjective experience of the work.

How do these variables relate to each other in deciding the subjective evaluation (E) of the work or image?

We can arbitrarily posit the following relationship (equation 1), which states that the subjective evaluation of an image (E = subjective overall quality) is a function of the number (n) of inherent qualia (Q) that are considered to compose the work, whereby each quale has a value (g), which is a measure of its *integrity* or *goodness* and a value *weight* (w), which determines its relative *importance* for the work (values for w and g can be conceptualized to be on a scale, e.g. from 1–10, 1–100, see below).

$$E = \sum_{i=1}^{n} g_i w_i Q_i \tag{24.1}$$

We have arbitrarily presented a simple additive effect of the multiple qualia on the overall evaluation. Such an additive relationship could hold, for example, for the purpose of comparing different works with each containing the same number of qualia, such as multiple works by the same producer. Different goodness values (g) and weights (w) for identical qualia would then decide different degrees of the overall evaluation of the compared works. However, a simple additive relationship cannot hold for comparisons between works with different overall numbers of qualia, since it would imply that the greater the number of qualia experienced, the higher the overall evaluation would be. We must consider this to be unlikely as a general law, since it would imply that simply adding qualia to a work would improve/increase the E of a work, which cannot be true.

Therefore, we must assume some kind of limit to the maximum level possible for E. Consequently, for purposes of comparing images or works with divergent numbers of qualia, we postulate a limit to the maximal possible *E* for any particular work. For this purpose, as a first approximation, we posit that not the sum, but the average or *mean* of the number of qualia and their properties determines the overall *E*, by dividing sum by the total number of qualia (*n*), as in equation 24.2:

$$E = \frac{\sum_{i=1}^{n} g_i w_i Q_i}{n} = E = \frac{1}{n}\sum_{i=1}^{n} g_i w_i Q_i \tag{24.2}$$

We have also arbitrarily stated a multiplicative relationship between *w* and *g of* a quale, whereby the value of each in goodness (*g*) is multiplied by the value of its weight or importance (*w*). Another perhaps more reasonable relationship could be, for example, an additive one, whereby the rating of goodness (*g*) of a quale is added to the value of its weight (w) as in equation 24.3:

$$E = \frac{1}{n}\sum_{i=1}^{n} (g_i Q_i + w_i Q_i) \tag{24.3}$$

We could hypothesize other more complex relationships between the individual *Q* in their influence on *E*. For example, one could argue that *the whole is more than the sum or mean of its separate parts*, a reasonable assumption which will have to wait for an adequate formula. However, as we discuss below, the qualia that are high in goodness and importance interact, harmonize, and compete to create the "whole" *E* (see section 24.9). Thus, the result of the interaction between qualia in terms of synchrony, harmony, etc., can become by itself an important quale.

In the meantime, in order to illustrate the usefulness of this type of model, we will continue to employ the multiplicative relationship between *w* and *g* as expressed in equation 24.2 as a basis for our argumentation.

Models as we have presented them can be useful for conceptualizing relationships between variables. However, a model will have significant utility only if it can be tested and disproven and eventually be adapted to empirical reality. As we will argue later, this model may have such heuristic value in making empirically testable predictions regarding the evaluation and composition of works of art.

In an experimental test of the relationship between qualia in deciding *E*, posited above, one could query a viewer as to, for example, which qualia are inherent in the work, and to rank individual discernible *Q*s in terms of their importance (*w*) and goodness (*g*) via a rating scale of, say, between 0–10. Works of art such as a painting or photographic image can surely be analysed or discussed in terms of the weight and goodness of perceived *Q*s. One can thus, systematically *fragment* and *decompose* the image or work—dismantle it by identification of its functional (perceived) qualia.

Such a systematic dissection of a work could also be performed in the service of testing the validity of the relationships postulated here. For example, it would be possible to test the utility or validity of the formulae empirically by comparing the evaluation of raters on a scale based simply on their intuitive appeal and one based on explicitly defined parameters of qualia inherent in the formulae. The degree of concordance between these evaluations could be a basis for hypotheses as to how to improve the formulae to improve the approximation of "reality" in aesthetic judgment.

24.3 Violation and extraction of qualia

We can now ask how the evaluation E of an image would be influenced by the elimination of or damage to a particular quale—Q.

From the formulae above it follows that *the smaller the number of Qs that comprise an image, the more dependent is the overall appraisal (E) of the work on the integrity (g) and weight (w) of its individual components.* Conversely, as the number of Qs in a work increases, the less impact will w and g of any individual quale have on the overall subjective evaluation. (This prediction is independent of whether we take the overall number of Qs or the mean number of qualia to constitute E, and independent of whether the relationship between w and g is additive or multiplicative.)

The relationship between the relative weight of an individual Q to the total number of Qs in a work: *we hypothesize that the number of Qs in a work will also determine the relative weight of each in deciding the overall E.* Thus, we should adapt the equation to account for this relationship, which is likely to be an inverse correlation between number of Qs and their individual weights.

We can now ask as to the influence on E of eliminating or damaging a particular quale. We postulate that the subjective evaluation E of an image that has been modified by the extraction of a quale Q (x, with $x = 0, \ldots, n$), will be a function of the number, integrity, and weight of the remaining Q relative to the original total number of defined Qs and their integrity and weight. This relationship can be expressed as follows in equation 24.4:

$$E=\frac{\frac{1}{n-x}\sum_{i=1}^{n-x}g_iw_iQ_i}{\frac{1}{n}\sum_{i=1}^{n}g_iw_iQ_i}=\frac{\left(\sum_{i=1}^{n-x}g_iw_iQ_i\right)}{(n-x)}\times\frac{n}{\sum_{i=1}^{n}g_iw_iQ_i}=\frac{n\sum_{i=1}^{n-x}g_iw_iQ_i}{(n-x)\sum_{i=1}^{n}g_iw_iQ_i} \tag{24.4}$$

It follows from this relationship that the more qualia that are inherent in a work, the more resistant will be the value of E to an extraction or violation of any particular Q, since more Qs are left over to "carry" the image (i.e. its level of evaluation E). Conversely, the smaller the number of Qs in the image, the more vulnerable will be the residual overall evaluation of the work E to elimination or violation of any individual quale. The same holds for the violation of a particular Q in a work due to a low level or failure of goodness g or impact w: that is, *the fewer the definable* Qs *in a work, the more impact will any particular* Q *have on* E. Therefore, a low value of g and/or w of any Q will have a more critical influence on E than in a work that is characterized by a higher number of qualia.

We can perform virtual as well as real experiments to test the predicted relationship between number of Qs and the impact of modifying or eliminating a particular quale on the evaluation of the work. For example, we could change the characteristics of an obvious source of a quale in works composed of relatively large numbers of associative Qs (Hieronymus Bosch) and compare the relative effects on the evaluation *E* with the influence of a violation of a *Q* (e.g. color) in minimal artworks (e.g. by Barnett Newman or Ad Reinhardt) that contain fewer obvious discernible Qs. Equation 24.4 predicts that the evaluation of a minimalist work will be influenced more by such a violation or extraction of a *Q*. A greater impact of a violation of a *Q* in art which is characterized by the use of few but powerful Qs would actually imply a higher degree of difficulty in creating such work in comparison to work embellished with more Qs.

In fact, Greenberg has espoused the deliberate refusal of particular qualia and the systematic delimitation of such in the development of Modernism (Greenberg 1961). His analysis of the gradual development of Modernist art can be seen in terms of the dispensing with of traditional qualia, such as three-dimensionality. Painting, he states, has made itself abstract and achieved autonomy by divesting itself of everything it might share with sculpture; painting should be self-critical and address only its inherent properties or Qs, namely, flatness and color.

Looking at the content of art from the standpoint of its Qs can contribute to an understanding of the work independent of the language used in descriptive analysis. For example, minimal art has been defined as being the maximal expression of formalism by its elimination of emotional elements (Qs) characteristic of Informal art and abstract expressionism. Such works can be characterized as (1) containing few discernible Qs beyond the simple formal ones (e.g. proportion), (2) devoid of Qs with common associative properties and the accompanying emotional effects, and yet, (3) containing abstract Qs that may be difficult to label, other than with descriptors such as "purity," "spirituality," "peacefulness," "simplicity," "elegance," etc., which may have emotional components, but not of the kind associated with everyday meaningful and identifiable concepts, ideas, objects, or events. The achievement of this category of Qs may define the teleology of minimalism. A similar analysis applied to other art forms will most likely arrive at different categories of dominant Qs that define it as a style or direction of art.

24.4 **Taxonomy of qualia**

Such an individual analysis of an artwork can encompass (1) "simple" basic qualia, such as color, size, brightness, texture, and grain, brushstroke, transitions—technique; (2) emotionally loaded Qs related to novelty, conflict, beauty, and the standard emotions such as love, hate, disgust, fear, desire, and laughter; (3) the formal compositional Qs, such as form, space, perspective, balance, harmony (or disharmony) that are taught as fundamental tools of the craft; (4) Qs with associative value of items or ideas, based on representational and figurative content; (5) Qs derived from the flow, interaction, and balance between concurrent Qs (see section 24.9), and (6) the "higher order" poetic, aesthetic, philosophical, ideological, and spiritual Qs. It should be possible to establish a classification scheme of the types of Qs to be found in images or works of art and on that basis describe artworks "objectively," independent of epochs, "schools of art," and content descriptors.

24.5 Viewing art on the basis of qualia

We will from here on use the term qualia rather loosely to refer to particular attributes or qualities of a work of art or architecture. The notion of qualia can also guide us in understanding the development of a work of art. To illustrate such an approach, we can take an actual example of a work of art and trace its development in stages to finality, as the artist defines it. The first two images in Figure 24.1 present actual steps in the course of the production of a work of photographic art (there were many other steps in between these). In the top left, the picture is still structured in layers (the plastic ground, the red surface, blue surface, and back wall). In step 2 (top right), the focus is on the red-plus-blue surface and the scaffold in between giving the structure more monumentality. In the final version below, the structure is lower and appears heavier, whereas the space is clearer and takes on a function in the picture, and the numbers become important in filling the space and defining the background. We can see the progressive flattening of the image approaching that

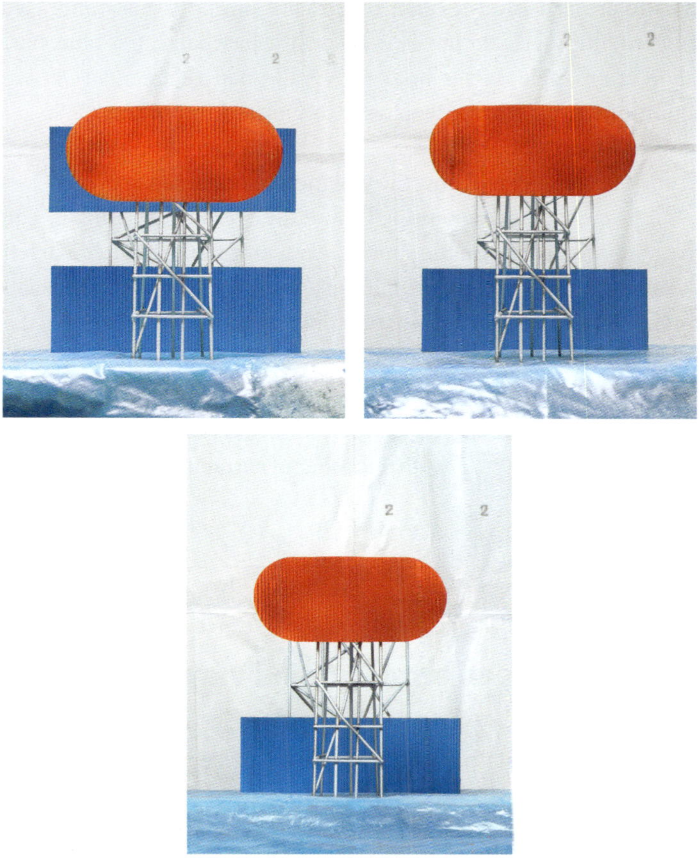

Figure 24.1 Steps on the way to a final photograph by Hannes Norberg, 2012 (175 × 140 cm), bottom image is the final. © VG Bildkunst, Bonn / Hannes Norberg, 2012.

of a painting. Whereas the steel scaffold and the colors remain invariant, major changes lie in the surface of the base perhaps to approximate an impression of water, the adjustment in hue of the colors and the proportions of the elements to each other and the frame. We can, on the basis of this progression infer some of the invariant Qs in the artist's *a priori* intentions.

24.6 **Eliminating a Q**

We can also perform *post hoc* experiments and violate qualia in works of art and assess the influence on our evaluation. For example, in Figure 24.2 we extracted the postcard from the center of the photograph. We can see that its removal has a profound effect on the image in terms of balance, proportion, depth, and color—fundamental Qs in artworks. However, the extraction also influences other more subtle Qs in this work: The postcard takes up various central themes in the picture, for example, art-historical Qs—the postcard is of a painting by Jan Vermeer and is dominantly positioned in an arrangement where there are various references to art, music, and literature. Aging as a Q—the postcard is yellowed and creased and the other objects are similarly dusty and used. The hat—the lady wears a red, fluffy, elegant hat at an angle; the baseball cap is blue, hard-rimmed, and common and at an angle on the books. Overall poetic elements (Qs) are influenced by the extraction of the postcard. We have also violated important Qs in two architectural masterpieces in Figures 24.3 and 24.4.

We can now ask whether and how the violation or elimination of one Q will influence the poignancy and weight of the remaining Qs and the overall evaluation *E*. Equation 24.4, as a first approximation, is based on the simplified assumption that after a deletion of any individual Q, the values *g* and *w* of the remaining Qs stay the same. An alternate and more valid assumption might be that after elimination or subtraction of a quale (*x*), the *g* and *w* values of the remaining Qs will change, with the consequence that the overall *E* might even increase, instead of always decreasing as in equation 24.4. Such a relationship can be expressed in the manner shown in equation 24.5. Here $y = n-x$, the number of remaining qualia Q; $y0$ = the original state; $x = 0, \ldots, n$, the number of extractions: gi and wi are a *function of y*.

$$E(y)=\frac{\frac{1}{n-x}\sum_{i=1}^{n-x} g_i(y)w_i(y)Q_i}{\frac{1}{n}\sum_{i=1}^{n} g_i(y_0)w_i(y_0)Q_i}=\frac{n\sum_{i=1}^{n-x} g_i(y)w_i(y)Q_i}{(n-x)\sum_{i=1}^{n} g_i(y)w_i(y)Q_i} \qquad (24.5)$$

24.7 **Rapport between producer and viewer**

Interestingly, we can employ our basic model to also account for the "failure" or "success" of a work of art. For example, a lower evaluation *E* given by the viewer than that expected by the producer can be a result of a discrepancy between the qualia that are intended/perceived by the artist and perceived by the viewer. Accordingly, the bigger the discrepancy between the evaluations of artist and viewer, the less likely it is that the work will be "understood." We can measure the degree of dissonance between viewer and producer by

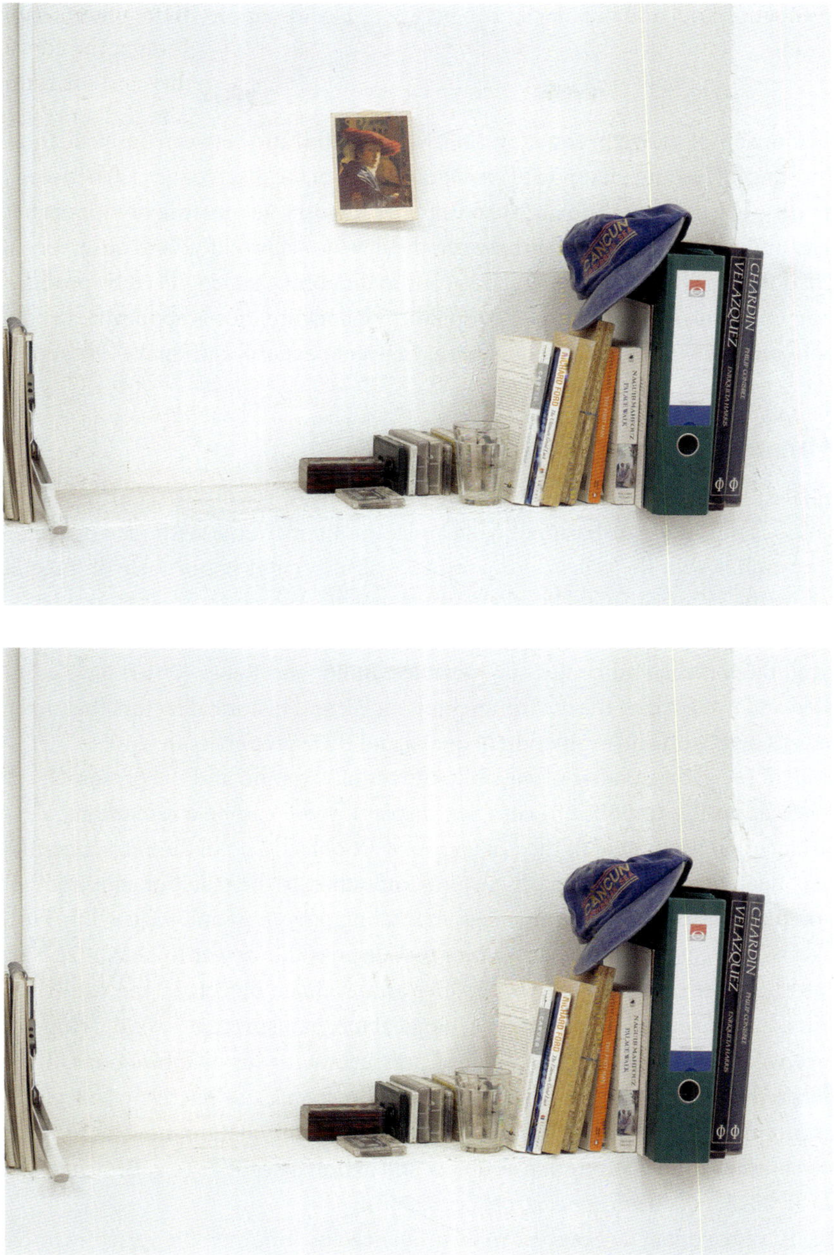

Figure 24.2 Extraction of part of a photograph by Christopher Muller, *Studio*, 2012 (69 × 95 cm), top: original, bottom: defaced. © Christopher Muller.

taking equation 24.4 and making a ratio between the evaluations of the producer (p) and viewer (v):

$$dEs = Ep / Ev \tag{24.6}$$

This relationship is shown in equation 24.6, whereby the ratio between the evaluations (Es) of the producer (Ep) and viewer (Ev) provide a metric of the discrepancy (d) between these, namely dEs. The degree of proximity to value 1 provides the measure of rapport between producer and viewer. A divergence between the E values of producer and observer can be a result of different w and/or g of the same Q, or to different Qs recognized by producer and observer. Of course, this measure of dissonance or concordance is applicable to comparisons between any other populations, such as between "expert" and "naïve" viewers.

24.8 Invariance of qualia

As we have already stated, we assume that when we view a work we consciously or subconsciously assess it on the basis of qualia and their interactions. With experience we may become familiar with the criteria for goodness (quality) and importance of the qualia as dictated by the experts (e.g. the market, critics, artists). Which Qs are discerned in a work and how they are weighted and evaluated are, of course, influenced by cultural, political, historical, theoretical, and contextual variables. Bullot and Reber (2013) have admirably compiled and categorized the environmental, social, and historical factors that determine the viewer's detection and evaluation of qualia and their properties in works of art.

In spite of the great number of possible sources of variation and the presumed subjectivity involved in the response to complex displays, there can be a remarkable degree of concordance and agreement in the evaluation of art objects and in the extraction of qualia and their relative impact. Such consensus is undoubtedly the result of acquired notions, common history, and cultural influences, marketing pressures, and suggestion. However, there is also evidence for genetic variables that dispose the viewer to recognize, value, or avoid attributes of images. How we perceive and evaluate objects in the world is partly determined and limited by built-in physiological mechanisms that have developed in accordance with evolutionary and ecological pressures with survival functions. These can be built predispositions in the perception of and responses to relatively "simple" qualia such as proportions, balance between elements, color combination, novelty, etc., but also to more complex information such as object and action categories (e.g. Brown et al. 2011; Huth et al. 2013; Ishizu and Zeki 2011; Zeki 2000).

Since, as we have stated above, E can be influenced by innumerable variables, including context, history of exposure, and current state variables of the viewer (e.g. time of day, transient mood, concentration level, health), we can ask to what degree a measure of E can be reliable; that is, what is the variability in the measure of E with repeated viewing of the same display by the same observer? We can also ask about the concordance of E within a particular population of viewers; to what extent does a sub-population concur in the inherent qualia and eventual E of a work and how consistent is such a population E? We could hypothetically as well as empirically also compare the mean Es of different subpopulations of viewers (e.g. based on levels of education, demographics, etc.). It should also be possible to detect invariance in the properties of works by the same artist, to identify Qs that define or characterize the oeuvre of a particular artist or class of artists. An example of how an individual artist's work can be discussed on the basis of idiosyncratic Qs has been carried out elsewhere (Huston 2002a, 2002b).

24.9 The problem of concurrent qualia: poignancy, fading, and interplay of Q

If the perception, registration, and judgment of a work are a function of the actions of its multiple qualia on the observer, we must ask whether and to what degree multiple parameters, attributes, or characteristics of a display can be simultaneously processed to result in the *E*. We cannot rule out the possibility that the nervous system can simultaneously process different qualia in parallel. Even if we accept this possibility, we can ask whether a viewer can simultaneously also attend to multiple Qs in an image or work (whether consciously or unconsciously).

24.10 Poignancy

To address the issue of the processing and impact of multiple concurrent Qs on the observer, we can make the following assumptions or hypotheses: (1) qualia with high values of weight (w) will predominate over Qs with relatively low weight in *poignancy*. Accordingly, the probability of a quale having an impact on the observer will correlate with its weight. (2) Similarly, qualia with high quality or goodness (g) will predominate over Qs with low values of g and, as with weight, will have a hierarchical probabilistic advantage correlated with goodness. (3) As we have posited an interaction between w and g in determining the value of judgment *E*, it follows that this interaction (in our case stated arbitrarily as multiplicative) provides a compound measure with a relative probability in terms of relative dominance over other Qs in the impact on the observer, which we can designate as *poignancy*, a *Q* in itself which will determine processes related to detection of and attention to the Qs.

As we have emphasized, the fewer the number of Qs in a work or display, the more important each individual *Q* will be in deciding overall *E*, and most likely, the more poignant it is, the higher will be its weight; that is, the number of Qs in a work will also determine the relative weight of each in deciding the overall *E*. Also, in reference to the earlier section on violation of *Q* and equation 24.5, we emphasize that violation or manipulation of one *Q* is likely to influence the weight and poignancy of the remaining ones, with a probable effect on total *E*.

24.11 Fading or decay of Q

We hypothesize that an attribute of any stimulation will not remain invariant over the duration of viewing. Thus, although the poignancy of a *Q* may increase as a function of enhanced recognition and attention to it, it is also likely to fade or decay with time of observation. Such fading of Qs can be a consequence of habituation, satiation, fatigue, and boredom, or due to shift in attention to competing Qs. Such processes are well known to influence perception and can occur at various levels of the nervous system, including the basic receptor level, such as the retina. The issue of fading qualia has also been treated by Chalmers (1995).

We posit that *each Q has a specific rate of decay or fading*. The rate of decay determines its dominance over competing Qs over the time of viewing (as well as its *interaction* with Qs that replace it—see section 24.12. The rate of decay of a Q is likely to be related to its combined weight and goodness. Furthermore, the rate of fading is also likely to be a function of the competing Qs in the work; that is, the fewer the competing Qs, the slower should be the rate of decay. Such fading of Qs is related to processes of "attention": the viewer's attention may shift to competing Qs. It follows that the fewer competing Qs there are, the less likely that attention will shift and the fewer Qs to which can attention be shifted.

Such fading allows a second-level Q to assume a higher level of poignancy. As new dominant Qs also fade out, there will be an emergence of new and previously faded Qs. This process of substitution of Qs enhances the novelty of Qs and in itself provides an overall dynamic process, which may be considered to influence the overall impact of a work of art (or of any complex display) on the viewer. Thus, the subjective impact of a display is not a static all-or-none effect on the observer and on his or her nervous system, but a continuous process involving the flow of information of the Qs that define or constitute the experience.

24.12 **Interplay of Qs**

The concept or process of fading of qualia underlies an important form of interaction between Qs. Qualia can interact in several ways; for example (1) they may influence each other in powerful ways to create new compound Qs. Their virtual simultaneity can have additive properties to define a "whole." (2) They also serve as each other's context. (3) Due to the fading out we hypothesized above, we can postulate that interactions between Qs are not merely parallel, but also sequential and dynamic.

For example, in the case of an art object that is presented to the observer statically (painting, photography), an important process pertains to the interplay between Qs during the process of evaluating the image. Such an interaction between competing dominant Q can be conceptualized as a dynamic reciprocal flow between them that can perhaps determine *harmonic* properties of the work. When we confront an image our perception thereof is not invariant, but on the contrary, is determined by such a flow between the images' Qs, whereby one or a set of such can dominate and recede in rapid succession. For example, when viewing Figure 24.1, the associative elements or Qs, such as the impression of water or the steel scaffold, can predominate at one point in time, whereas compositional Qs (color, proportion) may dominate at another moment. Such a switching between component Qs can be "willed," since we can consciously decide to focus on one or another, or it can be uncontrolled due to fading of Qs, as one quale displaces another in predominance. Uncontrolled switching between perceptual elements or Qs is illustrated by the well-known reversible figures of Gestalt psychology. Of course, the effect or memory of one Q will hereby influence the interpretation or effect of the subsequent Q. There are, deliberate goals in the visual arts to present the unitary "whole" in a work as one Q instead of as a composition of parts, particularly in so-called minimalist works, that attempt to reduce a concept to its simplest and purest form via a minimal use of geometric shapes, colors, lines, and progressions.

24.13 Qualia in architecture

For art forms that are dependent on temporal or sequential properties of *Q*s (such as music, architecture. and literature) or dependent on different perspectives for viewing (such as architecture and sculpture), the exact nature of the interaction between elements in their sequence and order can be critical. A work of architecture can potentially be seen from an infinite number of perspectives as it is four-dimensional (three dimensions in space plus time); one can enter the structure, traverse, touch, and use it. It cannot be viewed primarily from the standpoint of visual art. Architectural design can involve a string of logical and causally related sequential processes. Thus, the *Q*s in architecture can display a mutual dependency in terms of purpose and idea. The preservation of ideas and conceptual intentions throughout the process of design and building stages to the finished product is a major complex *Q*. The degree of understanding of these cardinal design ideas will strongly influence the evaluation of the product. For example, the process of employing as little as possible to achieve as much as possible is a major *Q* of minimalism in architecture, with the notion that "less can be more," as fewer *Q*s gain more weight in the impact of the whole. From this perspective architecture is more complex than, say, a painting as it cannot be evaluated in toto from one point of view. Its *Q*s include functional aspects, material, and tectonic elements, aesthetic as well as contextual properties. Parts of a building have specific functions which can be viewed in isolation but also in relation with the other sections of a building, its environment, and history. Consequently the possible bases for evaluation can be complex or simple, depending on the viewer's focus (on the primary poignant *Q*s).

We will here examine some architectural *Q*s with examples taken from different architectural traditions.

24.14 Antique

To consider just one *Q*, namely proportion, as emphasized around 25 BC by Marcus Vitruvius Pollio in *De architectura:*

> Proportion is a correspondence among the measures of the members of an entire work, and of the whole to a certain part selected as standard. From this result the principles of *symmetry*. Without symmetry and proportion there can be no principles in the design of any temple; that is, if there is no precise relation between its members as in the case of those of a well shaped man. (Vitruvius 1914)

Proportion in architecture refers to the ratio of the length, width, and height of a building, a façade, or a component. The Parthenon in Figure 24.3 illustrates the importance of proportion in classical architecture and the effects of violating such in width and height (see also section 24.1 on the *golden ratio* (φ), as a basis for architectural proportion and aesthetics). We would predict that the mean *E* of a set of observers will be higher in value when presented with the proportions of the original (Figure 24.3, top) than those of the modified Parthenon (bottom). In the middle image we changed the number of columns from

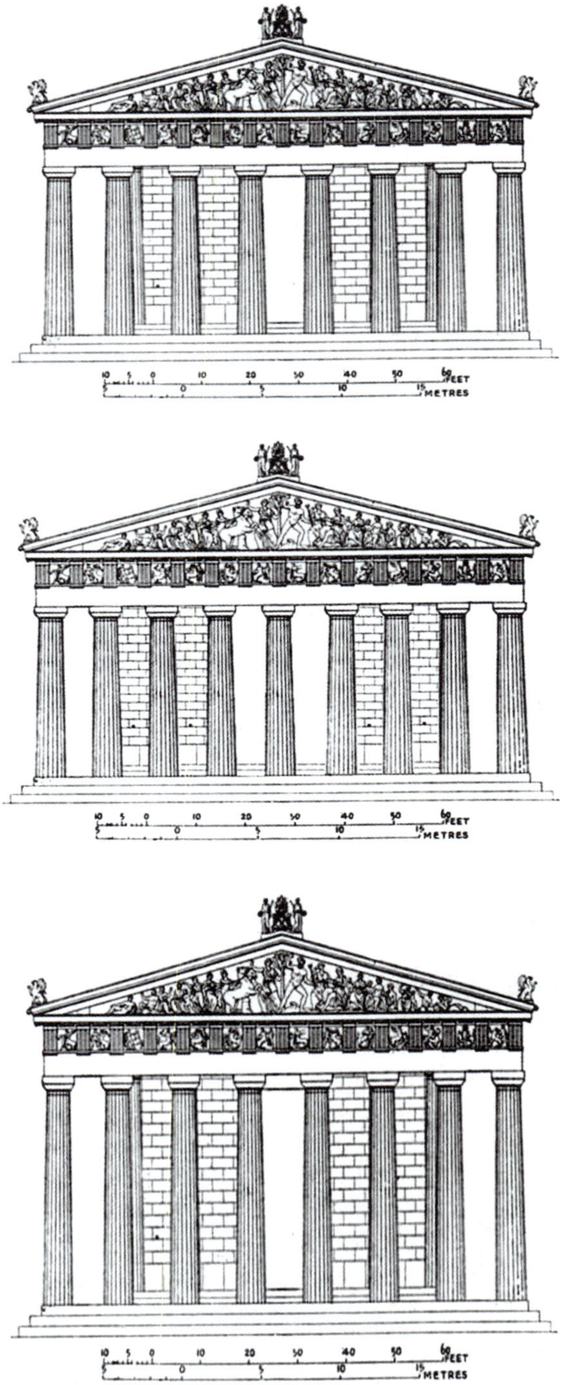

Figure 24.3 The *Parthenon* on the Athenian Acropolis, Greece. Top view represents the original; in the middle a column has been added; bottom view shows the result of vertical stretching.

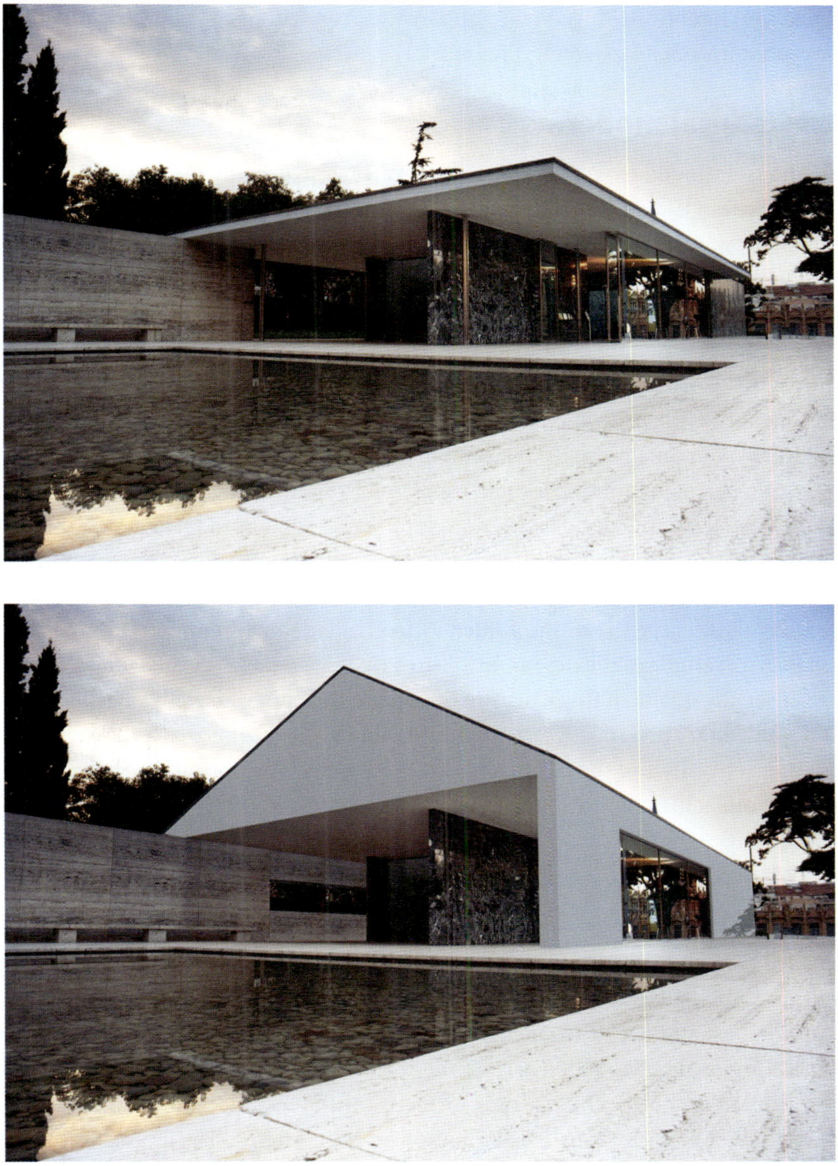

Figure 24.4 The *Pavilion* by Mies van der Rohe built for the 1929 International Exposition in Barcelona. Top: original; bottom: violated. (Photo by Ashley Pomeroy, Creative Commons Attribution Non-Commercial License.)

eight to the uneven nine, which also influences the overall proportions of the building and eliminates the central space, which is an important *Q* in ancient Greek architecture. In architecture the proportions are closely related to structural variables such as the loads or forces applied to a structure or its components. These variables, which influence the necessary thickness of walls, columns, and floors, depend on building materials, regional, historical, and intentional variables (compare the proportions of the Parthenon with those of Mies van der Rohe's *Pavilion* in Figure 24.4).

24.15 Classical modern

The Bauhaus, founded by Walter Gropius in Weimar, Germany (1919–1933) propagated the avant-garde of Classical Modernism in all of the free and applied arts. One of its guiding principles (*Q*) was the creation of a "total" work of art, a "Gesamtkunstwerk" in which all of the arts, including architecture, would eventually be brought together. Modular elements came to be employed in industrial structures and for creating affordable housing (major *Q*s being industrial prefabrication, simple, cost-efficient manufacturing). In his *Vers une architecture*, Le Corbusier proclaimed the principles of the "New Architecture" to be based on functionality, with an aesthetic based on pure form (Le Corbusier 1927). His buildings, starting with *Villa Savoye* in 1931, exemplify his five principles (*Q*s) of architecture; for example, *Q*1, the main building is often separated from the ground, supported by concrete stilts, allowing *Q*2, a free façade with non-supporting walls, and *Q*3, an open floor-space. In the upper floors he employed large windows providing a wide-open view (*Q*4). Roof gardens compensated for the greenery that was displaced by the building (*Q*5). Le Corbusier also followed the use of the golden ratio *Q* blended with human proportions in his buildings. An example of the new architecture is Mies van der Rohe's *Pavilion* in Barcelona (Figure 24.4): the qualia in this work are not only visual, but especially of a conceptual nature. Free-flowing, open spaces were made possible by the removal of load-carrying exterior and interior walls and the use of slim, steel columns. The bottom of Figure 24.4 shows the effect of replacing conventional outer walls and roofs to this building.

24.16 Minimalism

Minimalism has been practised in the past inter alia in Japanese architecture, following ideas of Zen philosophy or the Japanese "Ma" (negative space). Modern minimalism evolved from the Bauhaus school which adopted Zen ideas of a radical stripping down—ridding of ballast and excess information not germane to the main naked structure, message, and composition. As stated earlier, the common denominator of minimalism in the fine arts, literature, and architecture is to focus on the "essence" of a work—its abstract quality via the elimination of superfluous concepts and items. The emphasis in architecture was on *Q*s of simplicity, basic geometric forms, cold and hard lighting, open living spaces, minimal use of materials, objects, and decoration, with the intention to reduce to the point where no further reduction could add to the concept. Note the aphorisms "less is more" (Mies van der Rohe) or "less but better" (Dieter Rams), by which the simplest and fewest elements are to be applied to produce a maximum effect. A great deal of attention came to be focused on clean spaces, natural light and air in interaction with the inhabitant to achieve abstract qualities (*Q*s) related to essential harmony, order, and simplicity, independent of tradition and historical burdens. Contrasting Wang Shu's minimal *Ceramic House* with Frank Gehry's *Art Museum* in Figure 24.5 highlights the obvious important

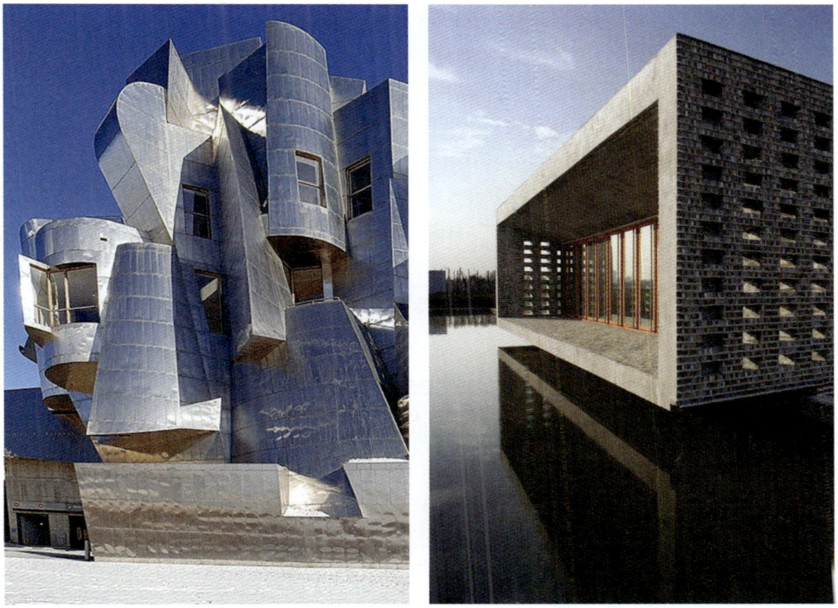

Figure 24.5 Left: detail of Weisman Art Museum, Minnesota, USA. Architect: Frank Gehry. (Photo by William J. Falk, at <http://www.flickr.com/photos/bill_in_stl/>). Right: detail of Ceramic House, Jinhua, China. Architect: Wang Shu, 2012 Pritzker Architecture Prize (Photo by Lv Hengshong, courtesy of Wang Shu and Amateur Architecture Studio, Hangzhou, The People's Republic of China).

Qs in minimal architecture. Gehry's building is highly compositional (like a Cubist painting) with Qs of balance and the questioning of balance in form, light, shadow, and statics. It exemplifies modernist tendencies to Qs of individualism, freedom of expression, and technological experimentation. Its parts could be arranged differently without sacrifice; its design process is formal—from outside to inside—in contrast to Shu's design, which is causal and conceptual. The former design might invoke more sensual emotional Qs in the viewer than Shu's with its emphasis, on the one hand, on conceptual intellectual mathematical Qs and, on the other hand, on Qs related to serenity, purity, the spiritual, the sublime, the poetic. Which brings us back to equation 24.4 and the prediction that the smaller the number of Qs in a work, the more vulnerable will be the residual overall evaluation of the work E to elimination or violation of any individual quale, and we can speculate as to which of these two works could more likely withstand an extraction or defacement of a Q.

As emphasized earlier, such considerations and goals hold also for expressions of minimal art in painting, music, film, literature, and theater, where minimalism can be contrasted with past art forms often cluttered with decoration, symbols (of religion and power, etc.), information, and facade. "It isn't necessary for a work to have a lot of things to look at, to compare, to analyze one by one, to contemplate. The thing as a whole, its quality as a whole, is what is interesting. The main things are alone and are more intense, clear and powerful" (Judd 1965).

We can now refer back to equation 24.4, which predicts an inverse relationship between the impacts of *Q*s in relation to their total number in the evaluation of a work, and ask whether this formulation does not provide a mathematically based "explanation" for the appeal of minimalism in art and architecture independent of the prevalent ideological/psychological rationales presented above. It follows that the equation would also allow comparisons between different art forms, such as schools of art, based on relative number of inherent qualia. Such congruence between predictions from this model with the reality of human evaluation argues, both, for the heuristic value of such modeling and for the utility of the concept of qualia in accounting for art evaluation.

24.17 How does the brain process qualia?

As part of a collection of essays on aesthetics and the brain, this chapter should attempt to relate our approach to aesthetic evaluation to the prospect of eventually being able to understand the neurophysiological mechanisms that must underlie the processes of perception, recognition, and evaluation of qualia as espoused in our model. How can the nervous system cope with such complex processes and how can we hope to discover the physiological bases for these processes? Each *Q* extracted from an image constitutes a separate complex representation in the brain at the moment of its recognition. If our viewing and evaluation of a complex image depends on the interplay between individual attributes or *Q*s of an image, our nervous system must possess the mechanisms necessary to attend to, encode, recognize, store, and respond to such representations. Some of the required mechanisms involved include attention to the image, its registration and recognition, plus the processing of associative, emotional, reflective, and evaluative responses to the image. A vast research effort is engaged in delineating the anatomical, physiological, molecular, and genetic mechanisms involved in each of these categories of central nervous system information processing, which, in toto, encompass the totality of neuroscience research, which subsumes many highly specialized disciplines. It can be expected that as these individual mechanisms become known, it will be possible to extrapolate to what happens in the brain during confrontation with and response to an image.

In section 24.4 we presented a rough classification of *Q*s based on (1) "simple" basic stimulus properties; (2) emotional *Q*s; (3) formal compositional *Q*s; (4) *Q*s with associative value of items or ideas; (5) *Q*s, related to interaction between concurrent *Q*s; and(6) "higher-order" *Q*s. For our purposes it may be useful to attempt to relate these categories of *Q*s to known major separable central neurophysiological subsystems involved in information processing. Some progress in understanding processes involved in our first three categories can be attributed to imaging experiments. The main contribution so far has been in analyzing aesthetic preference, that is, in discerning between "liked" versus "less liked" presentations on the basis of activation of corresponding structures in the brain (summarized in Cela-Conde et al. 2011). For example, neural activity can be registered in various brain areas while observers rate aspects of the presentation (such as "beauty" or symmetry) on some scale and such studies have identified structures activated during evaluation of preferred images, such as the orbitofrontal cortex (Kawabata and Zeki

2004), left dorsolateral prefrontal cortex (Cela-Conde et al. 2004), left anterior cingulate gyrus and bilateral occipital gyri (Vartanian and Goel 2004), and fronto-median and anterior cingulate cortex (Jacobsen et al. 2006). Given the realm of Qs that can comprise the evaluation of images, reducing aesthetics to the experience "beauty" is, of course, a gross oversimplification of what will be required to account for the experiences derived from, for example, visual images or works of art. "Beauty" represents an overall evaluative response to an image and probably subsumes emotional properties related to bliss, joy, and reward, but the concept masks the constituent elements (Qs) and processes that lead to such a response, and which must be the objects of an ultimate neuronal accounting for aesthetics.

Several attempts have been made to delineate the sequence of brain areas activated during the presentation of an image with some success in identifying different areas engaged within milliseconds of exposure in correlation with processes of registration of, attention to it, aesthetic evaluation of, emotional response to, and engagement with the presentation (Chatterjee 2004; Jacobsen and Höfel 2003). Other brain imaging studies have focused on neural systems involved in recognition processes, the role of the brain's reward system, emotional responses, and the role of experience in viewing artworks (summary in Cela-Conde et al. 2011). A meta-analysis of neuroimaging studies of positive-valence aesthetic appraisal led to the hypothesis that a system for aesthetic processing initially evolved for appraising objects with survival value (e.g. food) and was later co-opted in humans for aesthetic experience to satisfy social needs (Brown et al. 2011).

A number of studies have also dealt with the question of how object and action categories are represented in the brain. For example, functional magnetic resonance imaging (fMRI) was used to measure brain activity elicited by presentation of movies in order to investigate the cortical representation of 1705 different object and action categories (Huth et al. 2013). The authors reasoned that, given the vast number of such categories humans can recognize, the limited size of the human brain would preclude the possibility that every single category is represented in a specific brain area. Accordingly, they found that object and action categories are represented in a semantic space consisting of at least four dimensions, and that this space is organized into smooth gradients covering most of the visual and nonvisual cortex. The cortical organization of this visualized semantic space was invariant across different individuals. Thus, the brain represents object and action categories into a continuous space onto the cortical sheet so that adjoining points in the cortex represent semantically similar categories. However, other studies have shown that some categories of objects and actions are represented in specific cortical areas, such as for faces (Rajimehr et al. 2009), body parts (Peelen and Downing 2005), and outdoor scenes (Epstein and Kanwisher 1998). A large effort is dedicated also to understanding how conscious perception evolves following presentation of a stimulus (Van Rullen and Koch 2003) and the mapping of brain regions responsible for the encoding and retrieval of semantic and perceptual associations (Prince et al. 2005).

Although some progress has been made in identifying brain systems involved in viewing and evaluating images, a major challenge for the area of neuroaesthetics research relates to the competition between different qualia that comprise the perception and evaluation of the image. We postulated earlier that qualia fade in and out as the viewer habituates to one or tires of it. Potentially, the kind of semantic organization of categories in the brain as demonstrated by Huth and colleagues (2013) can provide a basis for investigating how the brain handles such processes—the rapid activation and alteration of representations of different classes of qualia. However, the brain's response to presentation of a display involves many separate neural systems with a continuous interaction between the perceptual, mnemonic, associative, attention, and emotional systems. An ultimate understanding of the brain's role in perceiving and responding to a complex display with multiple *Q* will have to invoke all of the systems involved in conscious and unconscious perception, information processing, emotional response, and internal mentation.

Acknowledgments

This essay is dedicated to Eleonore Huston née Mauerer. Our thanks to Bianca Topic and M. Angelica de Souza Silva for their help with this project.

References

Baumgarten, A. (1739). *Metaphysics*. In: E. Watkins (ed.) (2009). *Kant's Critique of Pure Reason: Background Source Materials*. Cambridge: Cambridge University Press, pp. 87–131.

Birkhoff, G.D. (1933). *Aesthetic Measure*. Cambridge, MA: Harvard University Press.

Brown, S., Gao, X., Tisdelle, L., et al. (2011). Naturalizing aesthetics: brain areas for aesthetic appraisal across sensory modalities. *Neuroimage* **58**(1), 250–8.

Bullot, N.J. and Reber, R. (2013). The artful mind meets art history: toward a psycho-historical framework for the science of art appreciation. *Behavioral and Brain Sciences* **36**(2), 123–37.

Cela-Conde, C.J., Agnati, L., Huston, J.P., et al. (2011). The cognitive and neural foundations of aesthetic appreciation. *Progress in Neurobiology* **94**, 39–48.

Cela-Conde, C.J., Marty, G., Maestu, F., et al. (2004). Activation of the prefrontal cortex in the human visual aesthetic perception. *Proceedings of the National Academy of Sciences of the USA* **101**, 6321–5.

Chalmers, D.J. (1995). Absent Qualia, Fading Qualia, Dancing Qualia. In T. Metzinger (ed.), *Conscious Experience*. Exeter: Imprint Academic.

Chatterjee, A. (2004). Prospects for a cognitive neuroscience of visual aesthetics. *Bulletin of Psychology and the Arts* **4**, 55–60.

Epstein, R. and Kanwisher, N. (1998). A cortical representation of the local visual environment. *Nature* **392**(6676), 598–601.

Greenberg, C. (1961). Modernist painting. *Arts Yearbook* **4**, 1–8.

Horgan, T. (1984). Functionalism, qualia, and the inverted spectrum. *Philosophy and Phenomenological Research* **44**, 453–69.

Huston, J.P. (2002a). Decomposing stimulus complexes and dissecting art: a theory of evaluation. *Physiological Research* **51**, 105–8.

Huston, J.P. (2002b). Quality flow. In S. Gronert, R. Pfab, and J.P. Huston (eds), *Christopher Muller. Seeing Things*. Nürnberg: Verlag für Moderne Kunst, pp. 110–122.

Huth, A.G., Nishimoto, S., Vu, A.T., et al. (2013). A continuous semantic space describes the representation of thousands of object and action categories across the human brain. *Neuron* **76**(6), 1210–24.

Ishizu, T. and Zeki, S. (2011). Toward a brain-based theory of beauty. *PLoS One* **6**, e21852.

Jacobsen, T. and Höfel, L. (2003). Descriptive and evaluative judgment processes: behavioral and electrophysiological indices of processing symmetry and aesthetics. *Cognitive Affective Behavioral Neuroscience* **3**, 289–99.

Jacobsen, T., Schubotz, R.I., Höfel, L., et al. (2006). Brain correlates of aesthetic judgment of beauty. *Neuroimage* **29**, 276–85.

Judd, D. (1965). Specific objects. *Arts Yearbook* **8**, 74–82.

Kawabata, H. and Zeki, S. (2004). Neural correlates of beauty. *Journal of Neurophysiology* **91**, 1699–705.

Lappin, J.S., Ahlström, U.B., Craft, W.D., et al. (1995). Spatial primitives for seeing three-dimensional shape from motion. In T.V. Papathomas, C. Chubb, and A. Gorea (eds), *Early Vision and Beyond*. Cambridge, MA: The MIT Press, pp. 145–53.

Le Corbusier, C-E.J. (1927). *Towards a New Architecture*. London: The Architectural Press.

Lewis, C.I. (1929). *Mind and the World Order: Outline of a Theory of Knowledge*. London: Courier Dover Publications.

Peelen, M.V. and Downing, P.E. (2005). Within-subject reproducibility of category-specific visual activation with functional MRI. *Human Brain Mapping* **25**(4), 402–8.

Perkins, M. (1983). *Sensing the World*. Indianapolis, IN: Hackett Pub. Co.

Prince, S.E., Daselaar, S.M., and Cabeza, R. (2005). Neural correlates of relational memory: successful encoding and retrieval of semantic and perceptual associations. *Journal of Neuroscience* **25**, 1203–10.

Rajimehr, R., Young, J.C., and Tootell, R.B. (2009). An anterior temporal face patch in human cortex, predicted by macaque maps. *Proceedings of the National Academy of Sciences of the USA* **106**(6), 1995–2000.

Schrödinger, E. (2001). *What is Life?: the Physical Aspects of the Living Cell* (Repr. edn). Cambridge: Cambridge University Press.

VanRullen, R. and Koch, C. (2003). Is perception discrete or continuous? *Trends in Cognitive Sciences* **7**, 207–13.

Vartanian, O. and Goel, V. (2004). Neuroanatomical correlates of aesthetic preference for paintings. *Neuroreport* **15**, 893–7.

Vitruvius (1914). *The Ten Books on Architecture*. Morris Hicky Morgan (ed.). Cambridge, MA: Harvard University Press.

Wenger, M.J. and Townsend, J.T. (eds) (2001). *Computational, Geometric and Process Perspectives on Facial Cognition*. Mahwah, NJ: Lawrence Erlbaum Associates Inc.

Zeki, S. (2000). *Inner Vision: an Exploration of Art and the Brain*. Oxford: Oxford University Press.

Chapter 25

Tempos of eternity: Music, volition, and playing with time

Barbara G. Goodrich

25.1 Introduction

How is it that some works of music, the most temporal of all art forms, can elicit an experience of timelessness? In this chapter I explore three pieces known to have such an effect: the haunting fifth movement of Olivier Messiaen's *Quartet for the End of Time*; Morten Lauridsen's quietly ecstatic choral masterwork *O Nata Lux*; and *I'm Daffy over You*, a tune finely crafted to frustrate and annoy the audience, written by Chico Marx and Sol Violinsky. The effects of these works on listeners can be explained only if we acknowledge that our experience of subjective time is more complex than usually thought; in particular, it may include not only sensation but also a purposive stance, particularly involving the anticipation that is crucial to music. Fortunately, one of the traditions in musical aesthetics can help us begin to articulate this experience, and it is complemented by recent neurophysiological insights. Such an approach may also clarify the remarkably enduring gifts of Clive Wearing as described by Oliver Sacks.

25.2 Preliminary considerations

In traditional aesthetics, Formalist theories hold that the pleasure we get from music is a *cognitive* pleasure; we enjoy recognizing the structure of music. There is some neurophysiological support for something similar to this: Chenier and Winkielman (2009) offer a hypothesis of aesthetic pleasure as arising from processing fluency; that is, sensory perception and its related information processing occurring with increasing ease. This fluency increases with repetition, perhaps by lowering the threshold for stimulation of the relevant neural pathways, perhaps by successively pruning away responses to auditory elements regarded as irrelevant. We are also more fluent at processing groupings that are structurally related rather than arbitrarily clustered (e.g. in language we more easily recognize "grammatical" strings of letters rather than arbitrary ones). Symmetrical or otherwise simple sensory objects—which thus have less information—lend themselves to processing as well, as do prototypes or "averages" of categories; the first measures of a song that is recognizably a ballad or a patter song or some other genre render it more predictable in a pleasant way.

Competing with the Formalist theories, there are Expressionist theories of musical aesthetics which explain aesthetic pleasure as pleasure in somehow sharing the *emotions* that the composer expresses through music. This approach too has some neurophysiological evidence. In one recent study, listeners' reports of subjective pleasure in listening to their favorite music and to neutrally valued music was strongly correlated to various objective measures of autonomic nervous system arousal (such as electrodermal activity, heart rate, respiratory rate, temperature, and blood-volume pulse amplitude), and thus, presumably, to emotional arousal (Salimpoor et al. 2009). Interestingly, intense pleasure could be taken in either happy or sad music; the pleasure's intensity was related to the how emotionally arousing the music was, regardless of which emotion was involved. Another study found that openness to sharing the emotion represented also increased the emotional response to music. Opera listeners in a cognitive "state of empathy" with the characters singing responded with greater reported feelings of emotion and greater physiological parameters of emotional arousal (Miu and Baltes 2012).

These two kinds of theories now appear to be more complementary than competing. But even taken together, they may not be sufficient as a way of explaining this phenomenon.

A third major tradition in musical aesthetics, influential within the classical music world, has received surprisingly short shrift outside that community. This tradition emphasizes the *driving force over time* that we experience in music, the anticipations, conflicts, and resolutions we live vicariously through music. The nineteenth-century philosopher Arthur Schopenhauer built his metaphysics to accommodate music as one direct expression of a universal *will*, just as he regarded life itself as another direct expression of this will (Schopenhauer 1966). Not surprisingly, many prominent musicians, including Wagner, Brahms, Dvořák, Mahler, Rimsky-Korsakov, Schönberg, and Prokofiev, have admired Schopenhauer's philosophy and the importance it places on music, whether they accepted it as metaphysically true or merely as an astute phenomenological description.[1] American Pragmatist philosophers avoided such metaphysical speculation, but their interests in psychology led to similar themes and influenced musicologists such as Leonard Meyer and David Huron to develop analyses of music based on listeners' anticipation of and desire for impending notes and harmonies. These approaches could perhaps be loosely categorized together as emphasizing the *purposiveness* or future-directedness of music, whether we think of it in terms of listeners' anticipation, or the teleology implicit in a work's cohesion over time, or the music as guiding our attention, our volition, or even our movement. These approaches would incorporate both the intellectual musical structure of Formalist theories and the emotions of Expressionist theories as important aspects of any work of music, as contributing to its overall momentum.

These last aesthetic theories are particularly valuable for current neurophysiological explorations since they allow us, more than the other theories, to articulate the temporal aspects of music and the structure of experienced time, and time is a dimension increasingly vital to neurophysiology.

In this chapter, I shall briefly summarize how these approaches to music could be combined with recent neurophysiological data to clarify how our rhythm, anticipation, and motility interact with music. These approaches solve one apparent paradox about how certain musical works elicit a sense of *timelessness* or eternity in the listener. In exploring two such works (and one song designed to elicit the opposite sense, frustration, in the listener), I shall suggest a hypothesis about the structure of subjective time as involving an interaction between perceptual processes and volitional or movement-initiation processes.

25.3 Rhythm, anticipation, and motility/volition

Listen to Tchaikovsky's *1812 Overture*. At some points, your heart rate will probably speed up considerably. The *Adagietto* from Mahler's *Fifth Symphony*, on the contrary, seems to slow a listener's perception down to the precision of time-lapse photography. Rimsky-Korsakov's *Flight of the Bumblebee* may make you feel as if you have had too much coffee, but need even more, *and like it*. On the other hand Pachelbel's *Canon in D Major*, if aimed at telephone customers trapped on hold, can make 30 seconds feel like eternal damnation as imagined by a James Joycean priest. For good or ill, music can alter our sense of time's passing: its duration, its rate, its continuity, its quality.

25.3.1 Rhythm

Recent advances in physiology present plausible lines of inquiry about how this might occur. Thaut (2005) suggests that when we perceive musical rhythm, we aren't passively counting durations of external timekeepers, or comparing external durations to some absolute internal counter, as some traditional theories of time perception hold. Instead we are *interpreting* the rhythm through our own complex and often changing biological rhythms, the ongoing to-and-fro oscillations of countless ion gradients in neurons and muscle cells, of respiratory cycles, and so on.[2] Listening to music is not a disembodied "god's point of view" affair. When we listen to music we aren't counting the beat; we are interacting with it, even living it.

For example, pulse rates close to the physiological heart rate, between 60 and 150 beats per minute (bpm), are easily interpreted as musical tempos. Pulses faster than 300 bpm tend to be perceived as subdivisions of a tempo; pulses slower than 30 bpm tend to be heard as multiples of a tempo (Thaut 2005, p. 9). Accelerandos and ritardandos are more complicated. With them, we must recognize the pulses as a continuation of the rhythm despite their change in duration, and we must simultaneously recognize that they *are* slowing down or speeding up, which requires some implicit background of a steady beat (possibly from neural oscillations) contrasting with the rubato.

It may help to remind ourselves of the gestalt structure of perception. In normal perception there is no single "datum," no single isolated pixel or indivisible moment of sound. Instead, we perceive patterns, comparisons, and contrasts.[3] We focus our attention on some sensory input as a foreground against the rest of our perceptual field, which remains in the background, preconscious but nonetheless framing the foreground, influencing how we perceive it. The rhythm of music is usually part of that background, framing and structuring the melody, which is typically the foreground on which we are concentrating.[4] But as twentieth-century philosopher and psychologist Maurice Merleau-Ponty points out, our bodies themselves provide a *second* perceptual background, composed of our heartbeats, our respiration, our habits of movement and perspective, language, and—now we can add—neuronal oscillations and language.[5] These frames of any perception, implicitly comparing and contrasting with it, are a large part of what gives a sensation its timing, its role in the perceptual field, and its meaning.

In the case of the Tchaikovsky, the tempo of some sections may even directly entrain the heart rates of listeners, slowing and then speeding them up to a rate associated with excitement. The dazzling speed of the piece by Rimsky-Korsakov makes it a challenge even to follow, but the insistent, amusing melody seizes our attention again and again, in a playfully disorienting way similar to the effect of watching such an insect—without the risk of being stung. We want to follow that sound, even if it evades us from time to time.

25.3.2 Anticipation

What of the melody and harmony? Many musicologists regard *anticipation* of impending notes and harmonies as central to music. For two theorists, Meyer (1961) and Huron (2006), the core of music is its setting up our expectations of the next notes and chords, and either fulfilling or thwarting those expectations. The biological mechanisms underlying these effects are now beginning to be clarified.

For example, our response to a surprise—even merely an unexpected musical harmony or rhythm—involves a single afferent pathway which then diverges at the thalamus into two tracks. The faster track goes directly to the amygdala, causing reactions ranging from fear triggering the autonomic release of epinephrine to the mild disorientation of perceptual field that can accompany a failed prediction. The slower track takes a detour through the sensory cortex before arriving at the amygdala, and incorporates conscious appraisals of the situation. These more considered responses can modify the fast track's automatic reactions, for example, changing a negative valence to a positive one. However, these two overlapping tracks are also interacting, in that a greater surprise can *increase* the pleasure of a pleasant outcome by its contrast. (Huron's example is of someone startled by a surprise birthday party, with momentary fear changing quickly to pleasure.)

Musical tension, the delay of an anticipated resolution, is in many ways the opposite of surprise. In Western music, much of what drives a melody is our preference for ending on the tonic note, the "home base" note of the key. We don't feel closure or satisfaction until this is reached. Certain chords (e.g. the dominant seventh of any particular key) make us expect an immediate return to the tonic, and immediate resolution. Other chords and phrases suspend the resolution for beats, bars, even minutes, and manipulate our expectations in subtle and sophisticated ways. Just as when listening to someone speak, we constantly reinterpret a spoken syllable according to the most likely word it may be introducing, we also reinterpret individual notes and chords in the light of possible phrases and resolutions to come. Delaying the expected resolution draws our attention and prompts us to imagine that resolution, to want it more.[6] The *Adagietto* from Mahler's *Fifth Symphony* is built from lengthy suspensions in which our anticipations of harmonic resolution are frustrated by unexpected dissonances and eventually rewarded only when we are minutely engaged with the melody. Perhaps this is why every second seems so full while we are listening to it.

Why do we feel this pull towards musical resolution? That is still unknown, but researchers have recently discovered preferences for certain pitches and intervals, and have correlated consonant intervals with greater neural synchronization, and dissonant intervals with less neural synchronization (Bidelman and Krishnan 2009; Lots and Stone 2008). There is a measurable neural substrate for the preference for harmonious resolutions. And insofar as musical conflicts, surprises, tensions, and resolutions mirror our nervous and endocrine systems' tendency towards healthy homeostasis, towards pleasure and happiness, they may be experienced as mirroring our own volition, allowing us a vicarious sense of fulfillment.[7]

25.3.3 Motility/volition

For humans and some bird species, at least, music can also recruit our volition in the sense of encouraging us to initiate movement. Our motor areas, including the cerebellum and basal ganglia, are activated not only when we are already moving in response to music, but even when we are merely *perceiving* it (Grahn 2009). Because of these close connections of auditory pathways and motor pathways, we can synchronize our movements to anticipated beats by a "feed-forward" mechanism (anticipation, again), entraining us into those famously contagious motions of hand-clapping, nodding, marching, or dancing to music (Thaut 2005, p. 45).[8] After all, music works by the same rhythms that serve as a scaffold for our own movement and sequential skills (Conway et al. 2009). We spontaneously want to move to it, to synchronize with it. Because of this predisposition, Thaut (2005) and others (e.g. Sacks 2007) have been able to use music with striking success in helping Parkinson's disease patients recover initiative in movement.

The effect of music on our tendency to move may reflect an even deeper influence by music on us than usually thought. Two leading neurophysiologists, Buszáki (2006) and Llinás (2001), have hypothesized that consciousness itself evolved as an aid to motility. If this is correct, it would account for how crucial future-oriented purposiveness or volition is to our experience, as well as this drive's elusiveness to reflective self-awareness; it would be *too* primordial to be easily represented or verbalized (Goodrich 2010). Because of the primacy Buszáki attributes to motility, he suggests that we reconsider how we think of sentient animals in their environments. Rather than a circuit of perception triggering a behavioral reaction whose results are then perceived and so on, it may be more accurate to think of the ongoing circuit as beginning with subjects acting in their environment and then perceiving the effects of their action, which prompt more action from them. In any case, there is a finely tuned ongoing circuit of action and perception in any sentient animal's interaction with its environment, from a bird's balancing its weight on a branch in a strong breeze to two humans waltzing with each other in time to music.

Overall, then, humans are influenced profoundly by music in many ways. Some music is able somehow to cause direct emotional arousal in many people, complete with chills and dopamine release at the listener's favorite moments. It can alter our biological rhythms, and our perception of time. But more than that, by engaging with and altering our expectations, many musical works are able somehow to recruit our volition itself, seducing our attention, entraining even our movements. If it were not so widespread as to be taken for granted, this effect would appear to be magical.

25.3.4 Note on volition

Some researchers have avoided the notion of *volition* or *will* (terms used synonymously here), partly because it and its sources aren't terribly clear even to the individual willing, partly because it resists measurement, and partly because the term is used for a variety of meanings: agency, a sense of agency, initiating movement, drives, *telos*, forming long-term plans, directing attention, choosing among options, inhibiting possible actions, and so on. These are genuine problems. However, in some topics the notion is unavoidable, and it appears to be gradually undergoing a rehabilitation as a needed and legitimate focus of research.

One *misguided* reason for avoiding it is the problems that arise when we imagine that one's will must be fully *conscious and free* (and disembodied).[9] Many Westerners still believe, with Kant, that such a radically free will is required for ethical and legal responsibility. But almost invariably the existence of "free/conscious will" is undermined by empirical data and even by our own embodied experience. Further, Westerners tend to devalue semi-conscious, preconscious, emergent, and habituated actions and intentions as mere mechanical processes, overlooking the subtly but profoundly different gestalt possibilities inherent in any complex sentient organism. Thus Libet (2004), after finding data inconsistent with the notion of a fully conscious free will, bit the bullet and ingeniously salvaged what he thought he must to retain human responsibility: the conscious mind's ability to veto subconscious urges as "free won't."

I suggest that we abandon this problematic notion of radically free will and instead explore human volition using the criterion of a sense of agency; that is, whether we experience volition as *ours*, as expressing ourselves (as opposed to us being constrained or undergoing simple reflexes). One's volition would include subconscious or preconscious forces such as predetermined instincts (parental instincts are "mere" instincts, but are also part of the deepest core of oneself), both socialized habits and self-made habits (the latter allowing a measure of "freedom" since one's future self is in part a creation of one's present self), and autobiographical learnings, as well as reflective decisions about long-term goals.

With regard to physiology, perhaps the main determinant of our sense of agency, of whether our actions are due to our own volition, is the *fluency* with which we select actions, monitored as we select them by the inferior parietal cortex (Chambon et al. 2012; Haggard 2008). This seems roughly to indicate how *congruent* we feel about the action, and, significantly, that fluency is not limited to conscious and non-stimulus-prompted actions. Yet after all, unconscious information processing can be flexible, adaptive, sophisticated (Wokke et al. 2011). Admittedly, this sense of agency is not infallible since it can be manipulated (Chambon et al. 2012), and inevitably there are gray areas. Even these weaknesses, however, contribute to its fitting the ordinary use of the terms "volition" and "will" surprisingly well.[10]

Incidentally, commercial law is ahead of many philosophers with this notion; for example, in the law of economic duress what matters is not the absolute freedom of the contracting party, but rather the nature of the pressure to which they are subjected (*Atlas Express v Kafco (Importers & Distributors) Ltd* [1989] QB 833). In other words, the legal arbiters of human responsibility do not require the kind of absolute freedom of will for which some psychologists and philosophers are still searching.

25.4 Music and the temporality of consciousness

One of the important insights from these physiological accounts is that there is *no present moment isolated from past and future*. There is not even an instantaneous present moment for the listening brain; as Kraus (2011) demonstrates, even a single click lasting 0.1 millisecond is processed in the auditory brainstem over about 7 milliseconds. The experienced present is saturated with memory and anticipation as well. How we perceive a current chord depends on previous chords and phrases, anticipated ones, genre, etc., analogous to how we perceive a single syllable.[11] We are not timeless beings observing a world unfolding before us; we are unfolding through time ourselves.

There is a major puzzle, though. Certain works of music famously elicit a paradoxical sense of *eternity or timelessness* in many listeners. How can this occur? Let us examine two of those works. I shall present a hypothesis as to how they could have such a baffling effect. If it is correct, it would have important implications for time perception as a whole. I recommend that readers listen to recordings of each piece before reading the descriptions following.

25.4.1 Messiaen's *Quatuor pour la fin du temps*: escaping time into eternity?

Olivier Messiaen's moving, dissonant 1941 *Quartet for the End of Time* is one of the key works of twentieth-century music. The quartet was written for an odd combination of instruments—violin, clarinet, cello, and piano—because those were the instruments to which Messiaen had access in a concentration camp during the Second World War. It is an assertion of a pacifist Catholic's faith from the depths of war, although it has transcended its specific original religious context to establish new norms in Western classical music in general. Nonetheless, Messiaen conceived of mundane time as emerging from a timeless eternity, and sought to change his listeners' focus from particular events in time, such as the War, to the larger timelessness of mysticism. This is especially true of the fifth movement, the only one with "eternity" in its title: *Louange à l'Éternité de Jésus*.

This movement is a duet; soft piano chords play a background pulse at a tempo marking of "infinitely slow, ecstatic," and the cello draws out a poignant, slow, sustained melody line. The effect on the listener of escaping time is intense enough that one eloquent philosopher of aesthetics does not attempt to explain it naturalistically, but regards it as gesturing towards the possibility of something timeless beyond nature (Hoekema 2009).

Music scholars have analysed the *Quartet for the End of Time* for its extraordinary use of unusual rhythms, tempos, and tonalities to disrupt the normal sense of time continuity.[12] The rhythms in the *Quartet* are frequently disconnected from the meter, and in the fifth movement, there *is* no explicit meter. The beats per measure vary, as do the length of individual beats. The piano's groups of four 16th notes expand to five, shrink to three, and so on. Nonetheless, the piano part consists solely of 16th notes, played at an extremely slow $x = 44$, limiting any sense of obvious rhythmic phrases but keeping a slow pulse. Below is a representative line (see Figure 25.1). The cello part is one long melodic arc, incorporating isorhythmic phrases, accents both on and off beats, slow dynamic changes, and a slow return to the tonic that gives a sense of peaceful resolution but not finality.

The different analyses of Messiaen's techniques explain the sense of temporal disorientation we experience at different points in the *Quartet*, the mismatch between expected rhythms and actual ones, the loss of a temporal anchor, but to my knowledge there is no account for the fifth movement's paradoxical sense of *escaping* time.

Merleau-Ponty's description of the perceptual gestalt, in particular with a second background of one's body, may be able to do so. Normally, a background beat serves to mark time, and against it the melody stands out, but the very slow background beat of the soft piano chords, played as slow as 44 beats per minute, is close to the lower limits of the human ability to recognize a beat as a beat (Thaut 2005, p. 9), and it is slower than most normal human heart rates. If a listener's heart rate could be entrained by such a slow tempo, then the beat, synchronized to the body's own "second background" pulses, might conceivably fade into that second background enough that it no longer functions to interact with the cello melody in the conventional way. As Thaut (2005) writes, notes occurring with a piece's underlying pulses are regarded as low-level information and tend to be overlooked. If the music's almost subliminal pulses have converged with the listener's own pulse, they could be difficult to notice, despite continuing to function as pulse markers. Instead, one's conscious auditory field can be filled entirely by what would normally be only the perceptual foreground; that is, the slow, sustained, deeply emotional cello line. We come almost to identify with the cello's sustained note. Though it moves and changes, our attention is carried along with it *without the usual rhythmic point of reference*.

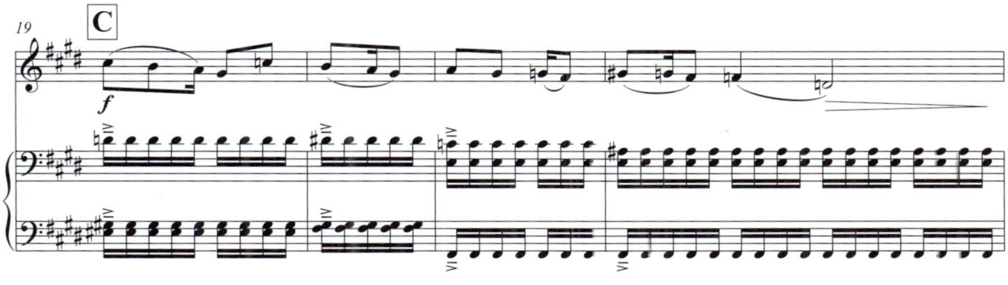

Figure 25.1 Excerpt from Messiaen's *Louange à l'Éternité de Jésus*.

Here is my suggestion: experienced time is not caused by any single pacemaker or any single oscillation. It is caused by the *interaction* between at least two different processes, causing a sense of "friction," as it were, between them. It is premature to speculate about what the neural mechanisms might be (though one suspects dopamine may be involved at some point). Phenomenologically, though, as Merleau-Ponty (1962) describes, we tend to sense a kind of friction or intersection between a more receptive aspect of ourselves, formed by perceptions and memories, and a more active, future-oriented aspect of ourselves, formed by anticipations, intended goals, initiative, and motility.

Normally in each moment of consciousness there simultaneously exist memories and habits from the past, present perceptions, *and* aspects of future-oriented volition, be they motions, plans, anticipations, decisions, intentions, attention.[13] However, these seem to be organized into a kind of gestalt. The foreground of this temporal gestalt is the future-directed aspect of ourselves against the background of memories and perceptions. We are, Merleau-Ponty (1962, p. 428) writes, simultaneously passive and active, recipient and agent. Our temporality "is the basis both of our activity or individuality, and our passivity or generality—that inner weakness which prevents us from ever achieving the density of an absolute individual."

This description accommodates our subjective time speeding up or slowing down. Our usual sense of time might result from internal interactions, the friction, so to speak, between some of these different processes. When our sense of time is altered, this might be caused by a change in how those two kinds of process are interacting, disrupting each other, generating greater or less internal "friction." When we are waiting in line, there is more friction, so to speak, and time seems to go more slowly. When we are sipping a good whisky with friends, there is less metaphorical friction and time seems to fly by.[14] More rarely, if one of these two aspects—roughly speaking, we could perhaps refer to them as perception and volition—temporarily overwhelms the other, the whole structure of the interaction can be changed or even temporarily collapse. This is a useful model for thinking of pathological distortions of experienced time, for example, in psychological shock, in schizophrenia, in Parkinson's disease, and in clinical depression, and it is telling that in each of these pathological cases there are also disturbances in initiative.

In the case of the Messiaen, the "friction" is removed by the background rhythm fading into our own biological rhythms. An actual solo for cello might invite the listener's imagination to fill in the pulses of the piece, so that the listener might have a stronger than usual sense of its rhythms, and its progression through time.[15] But a slow, barely perceptible background pulse such as in this work, by discreetly blending in with the cardiac oscillations, seems to remove entirely the experienced "friction" between intention and perception, temporarily dismantling the structure of subjective time itself.

By this account, we've indicated how the effect of escaping normal time into a seemingly supernatural eternity could be caused by natural means—*precisely because it is based in the rhythms of the changing organism*, adapting to the, well, almost heart-stoppingly beautiful cello line.

25.4.2 Lauridsen's *O nata lux*: glimpses beyond time?

Morten Lauridsen is a contemporary US composer, not yet fully celebrated. *O Nata Lux* is an *a cappella* motet, the most meditative of the songs from his 1997 *Lux Aeterna*. Like his more well-known 1994 *O Magnum Mysterium*, it deftly combines understated features to stunning effect: the long elastic phrases and subtle hesitations of Gregorian chant, simple Renaissance harmonies in counterpoint, and modern touches. Upon their debuts, these works were instantly acknowledged as making choral history, and have since broken out of ecclesiastical circles to achieve critical renown and growing popularity in the larger secular classical music world. Lauridsen's style might be compared very roughly to Arvo Pärt's in its surface simplicity and reverence, but I find it far more nuanced and unself-conscious, despite, or perhaps because of, its roots in the oldest surviving traditions of Western music. The quiet mastery of Lauridsen's compositional techniques avoids the sentimentality risked by any work aiming unabashedly at pure beauty.

This section incorporates insights into this piece's remarkable effects from distinguished choral director Allison Olsson.[16] Some of the descriptions are inevitably technical, but non-musicians will be able to understand the main points.

The motet is set for a full *a cappella* choir of sopranos, altos, tenors, and basses, and each of these four voices splits into two parts at some point or other, frequently having five or six notes sung simultaneously. The gentle, weightless melody line of isochronous eighth notes in 4/4 meter is repeated by different voices at different times, often overlapping each other in a wave-like pattern. In his advice to musicians, Lauridsen emphasizes that this motet is based on chant. He insists that the singers should not be thinking vertically, in chords, but linearly, with long melodies in a flowing give and take, a push forward and fall back to make the moment at the end of the phrase last a bit longer (Lauridsen et al. 2008). This hesitation is part of traditional Gregorian style. It also, significantly, matches the slow human respiratory rate. When the overlapping voices end a phrase simultaneously, this occurs in an isolated 3/4 measure, drawing back very briefly and holding the final note to create a moment of something like suspended animation. The effect is almost of another 4/4 measure, but not quite, and this difference is slightly disorienting even as it aligns itself with our breath *more* precisely and familiarly than a straight 4/4 beat.

The tonality at first seems traditional and simple, but there are ambiguities and sophisticated changes within it. The overall form of the song is a modified AABA form: twelve measures in the key of D Major, then a middle section of ten measures of ambiguous tonality functioning mainly as A Major, then a three-measure transition back to a restatement of the opening passage again in D Major, finally resolving and ending with a seven-measure coda.

Olsson points out that Lauridsen only twice uses the tonic D chord in the standard root position (for non-musicians: think of the usual bottom-most note, the "home-base" note): at measure 23 when returning to the key of D Major, and at the piece's final resolution and coda beginning at measure 35, in other words, at the only two times when the melody is hinting at a sense of completion. Elsewhere, the D chords nearly always have the third,

F#, on the bottom, leaving the tonic root absent. This is a device Lauridsen borrows from Renaissance music, and it gives a floating, unanchored feeling. This is not the feeling of a question that needs to be answered, or a puzzle that must be resolved, but it is the sense of not-quite-completeness, of something waiting for finality.

Lauridsen's signature chord, for this piece and for others, includes this third (here, F#) on the bottom, then the tonic an octave higher than expected (here a D), then the key's ninth (E), and the dominant fifth (A) at the top. Including the ninth in a tonic chord is a twentieth-century innovation, found frequently in the playful, nearly disruptive final chords of jazz songs. However, when hidden in the middle voices as it is in this particular chord, this added note is more gentle. It enriches the sound while also hinting at an ambiguity of tonal centers: The unexpected E note may be the unconventional ninth to our tonic D, but it is also the important dominant fifth overtone of D's *own* dominant fifth overtone, A, which is emphasized in this chord at the top. In this way, even when stopping on the resting tone, D, the listener still hears hints of another tonality, A (see Figure 25.2). It is almost analogous to what are known as ambiguous *Gestalten*, visual images such as a "duck-rabbit" or cube that can be interpreted in either of two ways.

What we would need to complete the transition from the key of D to the key of A is to change the G natural to G#. The occasional accidental of G# appears briefly several times in the bass before it finally becomes the norm in the soprano line to mark the start of the middle section at measure 13. After the return to the key of D at measure 23, it recurs several more times in the bass. Lauridsen often uses the added G# to build an E Major chord which, when juxtaposed with the D Major chord, gives hints of a whole-tone scale, adding to the feeling of indeterminacy as in measures 9 and 10. (Whole-tone scales are symmetrical throughout, so they lack any hierarchy. Because of that, they lack a clear "home-base" note, lending a feeling of dreamy indeterminacy to the sound.)

Olsson speculates that another harmonic factor fostering ambiguity is Lauridsen's almost complete avoidance of the explicit A Major chord in the D Major sections of the piece (A Major is the dominant, V, chord for the key of D Major). Lauridsen builds many of the chords in these sections with notes from the tonic chord *and* from the IV, subdominant, chord, G major (G natural, here). These two chords share the note D, yet have two other different notes each. The tonic sound is reinforced, but with extra notes blurring the sense of harmonic progression usually found in Western music.

Lauridsen also places more emphasis than usual on the alto voice, so that the highest soprano voice is no longer the obvious melody but functions as an overtone to the alto line. As a result, the often-overlooked middle voices assume a greater importance, and the

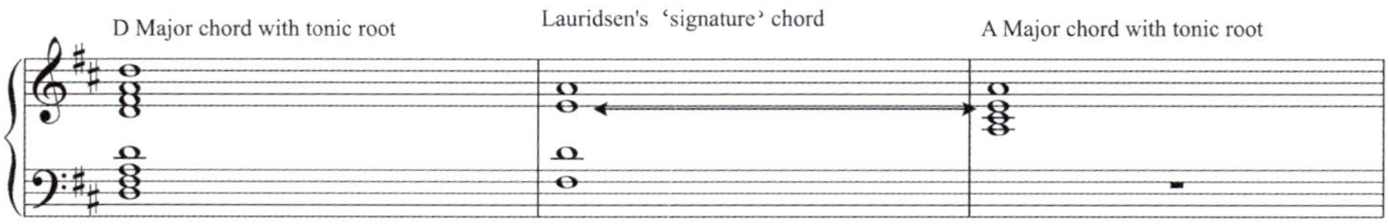

Figure 25.2 Lauridsen's signature chord.

simultaneous melodies of the polyphony prevent any single voice from being highlighted over the others for more than a measure or two. There is no single lead voice and there are no back-up singers; all the voices are integral.

Throughout the piece, much of the momentum occurs half-hidden in the middle voices; for example, measure 31 has a series of chords incorporating second intervals in them, leading to another iteration of the signature chord with an extra A in the bass at measure 32. The listener is engaged in the experience of movement, but it is not in reaction to obvious change; instead, the soft-focus tonalities merely shift among themselves, in the background, so to speak, rather than in any foreground melody.

These many tonal ambiguities do not demand resolution, however. The consonant harmonies, the repetition in overlapping waves, the sense of floating throughout, allow us to relax into them. Even when there are hints of a whole-tone scale, the overall order of the simple, intuitive Renaissance harmonies introduces us to them with enough familiarity that we feel not presented with a problem to be solved, but rather gently invited to peek into the new possibilities. They function almost as a shimmering of possible gestalt shifts, with, for example, the key of A Major partly hidden behind that of D Major, and each of the two keys assigning a different meaning to each note. The result is subtle, almost subliminal, but profound. Psychologists speak of ambiguous visual *Gestalten* as "unstable," but the experience here is too orderly and peaceful to be described in that way. Instead, the hinted D Major key is presented almost as an additional new dimension that one could enter, depending on the next few notes. And it is this chord that is frequently formed when all four voices end a phrase together, and we feel a glimpse into possibilities usually beyond our perception.

Lauridsen has masterfully balanced numerous different tendencies in the listener so that no single instance of melody springs univocally into the foreground. The melody is effortlessly recognizable, but it also functions as overtones to other phrases, and as echoes of other phrases. It is part of the larger flow, and that whole flow is indicating possibilities of tones beyond itself. We find ourselves *releasing* our usual constraining focus of attention, our agenda-driven volition, our control over our attention in order to absorb the nuances that we cannot consciously trace out. As we do, the everyday gestalt structure of our perception is subtly but radically altered: The *musical foreground and background blend together*, and as they do, the distinction between listener and music as well seems to dissipate. The perceptual field temporarily suffuses what Merleau-Ponty called the "second background" of one's body. Then, when the flow of music holds back, we are brought with it into a momentary stillness of rare tranquility and openness to our surroundings, with an attitude of wonder—at least temporarily. This experience is precisely *not* escaping the changing natural world into an eternity beyond it. It is engaging fully *with* the world, within time.

25.4.3 Chico Marx's theme: from the sublime to the ridiculous

This third musical example is Chico Marx and Sol Violinsky's *I'm Daffy over You*, as played by Chico in *Animal Crackers*. This is a deliberately, brilliantly execrable little ditty designed to elicit something like road-rage in the helpless listener, and worth purchasing the film to hear. At one point in Chico's performance, he is audibly counting the beats. Groucho, in the audience, impatiently tells Chico if he gets near a song, to play it. Chico explains that he can't think of the finish. Groucho: "That's strange, I can't think of anything else." Chico: "You know whadda I think, I think I went past it." Groucho: "Well, when you come around again, jump off." We do not get a sense of eternity with this song; we are merely afraid that it may take forever.

Chico achieves almost the maximum annoyance possible. He bangs out the repetitive melody slowly in high octaves, and with the tonic firmly, intractably in place in the left hand. Not only does he avoid playing with any kind of phrasing, he actually emphasizes the last note of the first phrase with an ad hoc tremolo, played with muscular gusto. The first grace note in the next phrase gets our hopes up for a reprieve from the monotony, but it is quickly revealed as the first of several measures of an odious little filler. When he reaches the end of those four bars, we are desperate for another change. And with the reappearance of the horrible octaves, we realize—with a burst of laughter at Chico's slyness—that we should have predicted *exactly their return*.

Note that the song isn't just unpleasant to hear. It's frustrating, *thwarting*. It is an affront not merely to our perceptions, but also to our sense of agency, to our volition. It's the same sense as one's blood pressure rising in response to a traffic jam delaying an important appointment. And each additional obstacle, especially a predictable one which one can anticipate *but not prevent*, grinds time more and more slowly, as the status quo exerts more aggravating friction on one's plans. So it's not surprising that a musician with more schadenfreude than the amiable Chico can increase the effect of listeners' frustration by slowing down the tempo even more.[17]

This is different from the much more common "earworm" phenomenon, in which a fragment of an annoying tune is repeated intrusively in one's aural imagination against one's will.[18] In the case of earworms, the culpable tune is typically just some recent and overplayed popular song. (The granddaddy earworm, now thankfully obscure, is King and Harold's 1930 *Goofus*.[19]) In some cases, the vividness and intractability of similar unwanted "musical imaging" can approach hallucination status.[20] Yet these cases are more mundane inner conflicts; the *structure* of volition or attention itself remains intact in each of its different foci. ("Terrible music, must be dispelled!" versus "Catchy tune, must repeat!") Chico's ditty can function as an earworm, but it also has the more sophisticated power of *recruiting some aspect of our volition against it*, dragging the flow of subjective time.[21]

25.5 **Clive Wearing's enduring musical gifts**

The importance of volition for the experience of time and for musical performance is seen in the case of British composer and conductor Clive Wearing, a case made famous by Sacks (2007, pp. 201–31). In 1985, encephalitis destroyed about 20 years of Wearing's personal memories, and, most devastatingly, almost all of his capacity for even short-term episodic memory. While his semantic memory of, for example, language and facts, was partly unaffected (up to his life in the 1960s), he lost the ability to remember where he was, whom he was with, and so on beyond just a few seconds. However, Wearing's partly functioning semantic memory and sociability have enabled him to maintain some social connection, helping to ward off his agonizing disorientation in time. Even with a terribly constricted memory, he converses cheerfully.

Wearing's procedural memory for many activities has fortunately remained intact; he shaves, dresses, retrieves cups from the kitchen, makes coffee, dances. Astonishingly, his procedural memory for playing music is also intact. While he can't remember having played or learned any piece if shown his old sheet music, once someone else initiates his playing it, he can continue uninterrupted until the end, many minutes later, far longer than his episodic memory could track. He can conduct lengthy pieces. Even more astonishing, his formidable capacity for playful musical improvisation remains.

The creativity in Wearing's music is unexpected, since procedural memory is normally thought to be somewhat mechanical, a set sequence of movements, rendered automatic by concentration and repetition. Thus Sacks regards this intact creativity as puzzling:

> Each time Clive sings or plays the piano or conducts a choir, automatism comes to his aid. But what comes out in an artistic or creative performance, though it depends on automatisms, is anything but automatic. The actual performance reanimates him, engages him as a creative person; it becomes fresh and alive, and perhaps contains new improvisations or innovations. Once Clive starts playing, his 'momentum,' as Deborah writes, will keep him, and keep the piece, going. (Sacks 2007, p. 224)

Deborah Wearing describes an instance of this:

> The momentum of the music carried Clive from bar to bar Within the structure of the piece, he was held, as if the staves were tramlines and there was only one way to go. He knew exactly where he was because in every phrase there is context implied, by rhythm, key, melody. It was marvelous to be free. When the music stopped Clive fell through to the lost place. But for those moments he was playing he seemed normal. (Sacks 2007, pp. 224–5)

How can "automatic" procedural memory enable this expansion beyond his usual limits, this return to his exceptional creativity? Sacks's answer is that a piece of music is an organic whole, and each present note strongly implies the past and future ones, more than our usual thoughts and anticipations do. Playing one chord of a piece will bring to mind, as it were, the rest of the piece, in a way that is not captured by the usual notion of memory.

> When we 'remember' a melody, it plays in our mind; it becomes newly alive. There is not a process of recalling, imagining, assembling, re categorizing, re-creating, as when one attempts to reconstruct or remember an event or a scene from the past. We recall one tone at a time and each tone entirely fills our consciousness, yet simultaneously it relates to the whole. It is similar when we walk or run or swim—we do so one step, one stroke at a time, yet each step or stroke is an integral part of the whole, the kinetic melody of running or swimming. Indeed, if we think of each note or step too consciously, we may lose the thread, the motor melody. / It may be that Clive, incapable of remembering or anticipating events because of his amnesia, is able to sing and play and conduct music because remembering music is not, in the usual sense, remembering at all. Remembering music, listening to it, or playing it, is entirely in the present. (Sacks 2007, pp. 227–8)

It is entirely in the present in the sense that we are fully reliving it, bringing the past into the present moment and letting it direct our attention towards the future.

If we apply Merleau-Ponty's description of consciousness in time, with its dual aspects of receptiveness and volition, Wearing's continued musical gifts make even more sense. Playing memorized music is not merely executing a program contained within memory. It is instead goal-directed behavior, within the composer's given but not overly strict rules, in which there is a clear temporal gestalt. Within those rules, personal creativity and expression are permitted, even facilitated, similar to acting or dancing a dramatic role after one has memorized the character's script or choreography. Because a piece of music is more integrated than, say, an impromptu conversation or even monologue, the goal of continuing the gestalt exerts a stronger "pull" and can coax more continuing action via more momentum. By the same token, beginning a piece might be more difficult, hence Wearing's need for others to initiate his playing. Wearing's case also may illuminate how highly integrated the volition–perception circuit is. It is crucial to this circuit that it be ongoing, and in his case it is precisely the ongoing *long-term*, non-musical feedback from his environment that is interrupted. But as his memory gradually improves, and as his musical activities continue, Wearing is reclaiming his life.

25.6 **Conclusion**

By drawing from Schopenhauer's and Huron's musical aesthetics as well as Merleau-Ponty's psychology and much recent neurophysiological data, we can begin to build an account of how the most temporal of art forms can, in certain cases, elicit a paradoxical experience of timelessness. This account relies upon the notions of anticipation and purposiveness or volition, which are involved in listening to music as well as in performing it. This account also suggests that our perception of time is derived not from a single "counter" or oscillation, but from the *interaction* of at least two kinds of rhythmic biological processes, perhaps one associated with motility or volition, and one associated with perception.

Acknowledgments

I am grateful to a number of colleagues and friends for their help with this paper. In particular I want to mention two people. Allison Olsson, an extraordinary choral director and musician, generously elucidated Lauridsen's music more than I had thought possible. Aristotle scholar and spouse William A. Simpson was willing to debate the nature of teleology and time over endless meals in an eloquent and enlightening way. With editing, historical comparisons, and encouragement, he helped immensely even while not agreeing fully—the mark of a true philosopher.

Notes

1 Richard Wagner's seminal opera *Tristan und Isolde* was written explicitly as a musical expression of Schopenhauer's philosophy.

2 Thaut's "oscillator" approach to rhythm perception is similar to other oscillator-based approaches, including that of Buszáki (2006). For a review of some competing theories, see Grondin, 2010.

3 Even the most primitive form of vision, light-sensitive cells responding to undifferentiated sunlight, involves a contrast to a different direction, location, or time in which there is no light. After all, the evolutionary use of perception is to make these distinctions.

4 In the last section of Ottorino Respighi's *The Pines of Rome*, background and foreground become reversed to great effect; the steady percussion begins as a mere rhythm in the background of the horns' melodies, but gradually the drumbeats get louder, at some point usurping the status of foreground. It is at that point that we are struck by the thrilling image of the Roman troops marching towards us on the Appian Way.

5 Lidji and colleagues (2011) found that monolingual and bilingual English speakers, attuned to a more stress-timed language, were able to tap along to perceived speech rhythms more frequently and at higher hierarchical levels of rhythms than monolingual French speakers, attuned to a more syllable-timed language.

6 For readers who have not formally studied music: imagine your favorite piece of music, cut short just before the last note, or just "Shave and a haircut, two . . ." The resulting feeling of dissatisfaction, of incompleteness, of lack of closure, is an example of what might be called the driving force we feel with music.

7 Non-human animals have varying responses to music per se. Tempo seems to affect many species, often dramatically. A variety of mammals relax in the presence of "soothing" music, to the extent that animal shelters are beginning to play classical music radio stations or even specially chosen and simplified classical piano music. The Larimer Humane Society's shelter in northern Colorado reports that classical music dramatically reduces barking in their kennels (13 September 2012, personal communication).

However, more detailed responses to music seem species-specific. For example, Tamarins don't respond much to most music, perhaps because their voices are several octaves higher than humans' and their heartbeats are much faster. But a curious composer in one new study developed Tamarin-specific music based on patterns and rhythms of different Tamarin vocalizations, and the pieces based on Tamarin affiliation vocalizations did indeed trigger greater relaxation, social behavior, and foraging, and those based on Tamarin threat cries did increase anxious behavior and huddling. In contrast, these Tamarins responded with typical indifference to most analogous human music, that is, based on the inflections of human speech (Snowdon and Teie 2009). Some samples of the Tamarin music can be found at <http://www.news.wisc.edu/17030>. The composer, David Teie, has also written some amusing pieces geared to the ranges of interest (e.g. mouse vocalizations) and biological tempos of cats. Anecdotally, they appear to be met with approval.

White-throated sparrows respond with exquisite discrimination to their conspecifics' birdsongs: Male birdsong played to females in breeding season activated the analogous structures in their mesolimbic reward system to those in a human enjoying music. Male birdsong played to other males, though, activated only part of this system, similar to a human listening to unpleasant music (Earp and Maney 2012).

8 Dissanayake (2006) even argues that music was originally developed as part of ceremonies involving group participation, including dancing, singing, and the visual arts.

9 A widely accepted philosophical alternative is "compatibilism," according to which determinism would in fact be compatible with our common-sense, non-absolute notions of freedom.

10 Indeed, it fits our ordinary sense of volition better than Haggard himself realizes.

11 Presumably other vertebrates also interpret and anticipate sensory data in an ongoing way. It is worth emphasizing that the experienced present seems *already* future-oriented in normal mammals, since it follows from this that the philosophical questions of continuity of consciousness/personhood and diachronic personal identity don't arise except with unwieldy levels of abstraction or cases of severe brain damage.

12 Composer David Doty's blog entries on the *Quartet* at <http://daviddotymusicblog.blogspot.com/> are brief but outstanding, and explore the tension between static and dynamic elements in Messiaen. Steve Hicken's "Out of Time" is another composer's very insightful essay, at <http://www.thehighhat.com/PopsClicks/006/Messiaen_Hicken.html>.

13 The classical Greeks did not have had a term for will, but, as Thaut (2005) mentions, they weren't satisfied with *chronos*, *χρονος*, mere clock-time. They also developed the term *kairos*, *καιρος*, i.e. opportunity, tim*ing*. So even they acknowledged something like an active aspect of experienced time.

14 This notion of metaphorical friction or smoothness might be supported by the study of Chambon and colleagues (2012) correlating action-selection fluency with the experience of agency.

15 Grahn (2009) states that the beat-perception activity in the basal ganglia, e.g. putamen, is strongest during *un*accented beats. Without external accents, the brain must generate internal, subjective, accents.

16 I first heard Lauridsen's *O Nata Lux* at a local concert a few years ago, conducted by Olsson. She has performed and directed this particular piece for more than ten years, and I have asked her to publish a detailed analysis of this work as a choral director. In the meanwhile, she agreed to share some of her insights for this paper (personal communication, 4 June 2012).

17 Don't ask me how I know this.

18 Sacks (2007) eloquently describes earworms, also known as "brainworms."

19 Wayne King and William Harold, 1930. For those readers with more curiosity than prudence, a particularly skillful and thus caustic performance is by the Cumberland Ridge Runners.

20 I would have been skeptical of this, but when I recently began a challenging veterinary medicine program after many years in philosophy and music, for the first year of veterinary classes I was woefully afflicted with near-constant earworms of loud music scales as if played with both hands by a beginning music student; I was indeed going back to the basics of this new field.

21 Philip Glass's *Akhnaten* also strives for a fascinating shift in time experience, aiming for a cohesive unity of minimalist style and content: the unchanging traditions of ancient Egypt. I am not covering that work at greater length here because, due to my own limitations no doubt, I experience this work as committing the aesthetic "imitative fallacy," viz. creating a sense of unrelenting sempiternity in the audience rather *too* accurately, and without Chico Marx's iconoclastic flair.

References

***Animal Crackers*, Victor Heerman, dir.** (Paramount Pictures, 1930) (Film).

Bidelman, G.M. and Krishnan, A. (2009). Neural correlates of consonance, dissonance, and the hierarchy of musical pitch in the human brainstem. *The Journal of Neuroscience* **29**(42), 13165–71.

Buszáki, G. (2006). *Rhythms of the Brain*. Oxford: Oxford University Press.

Chambon, V., Wenke, D., Fleming, S.M., et al. (2012). An online neural substrate for sense of agency. *Cerebral Cortex* April, 2012, doi: 10.1093/cercor/bhs059. Available at: <http://www.ncbi.nlm.nih.gov/pubmed/22510529>.

Chenier, T. and Winkielman, P. (2009). The origins of aesthetic pleasure: processing fluency and affect in judgment, body, and the brain. In M. Skov and O. Vartanian (eds), *Neuroaesthetics*, Amityville, NY: Baywood Publishing Company, pp. 279–81.

Conway, C.M., Pisoni, D.B., and Kronenberger, W.G. (2009). The importance of sound for cognitive sequencing abilities: the auditory scaffolding hypothesis. *Current Directions in Psychological Science* **18**(5), 275–9.

Dissanayake, E. (2006). Ritual and ritualization: musical means of conveying and shaping emotion in humans and other animals. In S. Brown and U. Voglsten (eds), *Music and Manipulation: On the Social Uses and Social Control of Music*. Oxford: Berghahn Books, pp. 31–56.

Earp, S.E. and Maney, D.L. (2012). Birdsong: is it music to their ears? *Frontiers in Evolutionary Neuroscience* **4**, Article 14.

Goodrich, B.G. (2010). We do, therefore we think: time, motility, and consciousness. *Reviews in the Neurosciences* **21**, 331–61.

Grahn, J.A. (2009). The role of the basal ganglia in beat perception: neuroimaging and neuropsychological investigations. *The Neurosciences and Music III—Disorders and Plasticity, Annals of the New York Academy of Sciences* **1169**, 35–45.

Grondin, S. (2010). Timing and time perception: a review of recent behavioral and neuroscience findings and theoretical directions. *Attention, Perception & Psychophysics* **72**(3), 561–82.

Haggard, P. (2008). Human volition: towards a neuroscience of will. *Nature Reviews Neuroscience* **9**, 934–46.

Hoekema, D. (2009). Music as natural and supernatural, The 67th annual meeting of the American Society of Aesthetics in Denver, CO.

Huron, D. (2006). *Sweet Anticipation: Music and the Psychology of Expectation*. Cambridge (MA): M.I.T. Press.

Kraus, N. (2011). Listening in on the listening brain. *Physics Today* **40**, 40–5.

Lauridsen, M., Salamunovich, P., and Jordan, J., *Evoking Sound Dialogues: Musical Conversations between Composers and Conductors* Vol. 1 (GIA Publications CD-773, 2008).

Libet, B. (2004). *Mind Time: The Temporal Factor in Consciousness*. Cambridge, MA: Harvard University Press.

Lidji, P., Palmer, C., and Peretz, I. (2011). Listeners feel the beat: entrainment to English and French speech rhythms. *Psychonomic Bulletin and Review* **18**, 1035–41.

Llinás, R. (2001). *I of the Vortex*. Cambridge, MA: M.I.T. Press.

Lots, I.S. and Stone, L. (2008). Perception of musical consonance and dissonance: an outcome of neural synchronization. *Journal of the Royal Society Interface* **5**, 1429–34.

Merleau-Ponty, M. *Phénoménologie de la perception* (Paris, 1945); tr. Colin Smith as *Phenomenology of Perception*, trans. Colin Smith (1962), London: Routledge & Kegan Paul Ltd.

Meyer, L. (1961). *Emotion and Meaning in Music*. Chicago, IL: University of Chicago Press.

Miu, A.C. and Baltes, F.R. (2012). Empathy manipulation impacts music-induced emotions: a psychophysiological study on opera. *PLoS One* **7**(1), e30618.

Sacks, O. (2007). *Musicophilia: Tales of Music and the Brain*. New York, NY: Vintage Books.

Salimpoor, V.N., Benovoy, M., Longo, G., et al. (2009). The rewarding aspects of music listening are related to degree of emotional arousal. *PLoS One* **4**(10), e7487.

Schopenhauer, A. *Die Welt als Wille und Vorstellung* (2nd edn). Leipzig, 1844; tr. Payne E.F.J as *The World as Will and Representation*, vols. 1 and 2 (1966). New York, NY: Dover Publications.

Snowdon, C.T. and Teie, D. (2009). Affective responses in tamarins elicited by species-specific music. *Biology Letters* **6**, 30–2.

Thaut, M.H. (2005). *Rhythm, Music, and the Brain: Scientific Foundations and Clinical Applications*. New York, NY: Routledge.

Wokke, M.E., van Gaal, S., Scholte, H.S., et al. (2011). The flexible nature of unconscious cognition. *PLoS One* **6**(9), e25729.

Index

Please note: "n." indicates "endnote" and is followed by the endnote number.

F

M

O

S

T